T0221450

REINTRODUCTION *of* FISH *and* WILDLIFE POPULATIONS

REINTRODUCTION *of* FISH *and* WILDLIFE POPULATIONS

Edited by

David S. Jachowski, Joshua J. Millspaugh,
Paul L. Angermeier, and Rob Slotow

UNIVERSITY OF CALIFORNIA PRESS

University of California Press, one of the most distinguished university presses in the United States, enriches lives around the world by advancing scholarship in the humanities, social sciences, and natural sciences. Its activities are supported by the UC Press Foundation and by philanthropic contributions from individuals and institutions. For more information, visit www.ucpress.edu.

University of California Press
Oakland, California

Library of Congress Cataloging-in-Publication Data

Names: Jachowski, David, 1977–editor. | Millspaugh, Joshua J., editor. | Angermeier, Paul L., editor. | Slotow, Robert H., editor.
Title: Reintroduction of fish and wildlife populations / edited by David Jachowski, Joshua J. Millspaugh, Paul L. Angermeier, and Rob Slotow.
Description: Oakland, California : University of California Press, [2016] | Includes bibliographical references and index.
Identifiers: LCCN 2016032948 (print) | LCCN 2016034750 (ebook) | ISBN 9780520284616 (cloth : alk. paper) | ISBN 9780520960381 (epub)
Subjects: LCSH: Fishes—Reintroduction. | Wildlife reintroduction.
Classification: LCC QL83.4 R44 2016 (print) | LCC QL83.4 (ebook) | DDC 639.97/7—dc23
LC record available at https://lccn.loc.gov/2016032948

Manufactured in the United States of America

25 24 23 22 21 20 19 18 17 16
10 9 8 7 6 5 4 3 2 1

CONTENTS

CONTRIBUTORS

KIM AARESTRUP
National Institute of Aquatic Resources
Technical University of Denmark
Silkeborg, Denmark

CHRIS S. ALLEN
US Fish and Wildlife Service
Portland, OR, USA

PAUL L. ANGERMEIER
Virginia Cooperative Fish and Wildlife Research
    Unit
US Geological Survey
Blacksburg, VA, USA

DOUG P. ARMSTRONG
Institute of Agriculture and Environment
Massey University
Palmerton North, New Zealand

JOEL BERGER
FWCB - Colorado State University
Fort Collins, CO, USA
and
Wildlife Conservation Society
Bronx, NY, USA

IRIS BIEBACH
Department of Evolutionary Biology and
    Environmental Studies
University Zurich
Zürich, Switzerland

SAMANTHA BREMNER-HARRISON
School of Animal, Rural and Environmental
    Sciences
Nottingham Trent University
Southwell, Nottinghamshire, UK

STEFANO CANESSA
Institute of Zoology
Zoological Society of London
London, UK

ALIENOR L.M. CHAUVENET
Centre for Biodiversity and Conservation
    Science
University of Queensland
St. Lucia, QLD, Australia

SARAH J. CONVERSE
US Geological Survey
Patuxent Wildlife Research Center
Laurel, MD, USA

JASON B. DUNHAM
US Geological Survey
Corvallis, OR, USA

JOHN G. EWEN
Institute of Zoology
Zoological Society of London
London, UK

ANNA L. GEORGE
Tennessee Aquarium Conservation Institute
Chattanooga, TN, USA

ROBERT A. GITZEN
School of Forestry and Wildlife Sciences
Auburn University
Auburn, AL, USA

JAMES C. GODWIN
Auburn University Museum of Natural History
Auburn, AL, USA

SCOTT M. GOETZ
Department of Biological Sciences
Auburn University
Auburn, AL, USA

MATT W. HAYWARD
Schools of Biological Sciences and Environment,
    Natural Resources and Geography
Bangor University
Bangor, UK

DAVID S. JACHOWSKI
Department of Forestry and Environmental
    Conservation
Clemson University
Clemson, SC, USA
and
School of Life Sciences
University of KwaZulu-Natal
Scottsville, South Africa

BARBARA J. KELLER
Missouri Department of Conservation
Columbia, MO, USA

LUKAS F. KELLER
Department of Evolutionary Biology and
    Environmental Studies
University Zurich
Zürich, Switzerland

DEBORAH M. LEIGH
Department of Evolutionary Biology and
    Environmental Studies
University Zurich
Zürich, Switzerland
and
Swiss Institute of Bioinformatics
Quartier Sorge - Batiment Genopode
Lausanne, Switzerland

NATASHA A. LLOYD
Centre for Conservation Research
Calgary Zoological Society
Calgary, AB, Canada

R. LEE LYMAN
Department of Anthropology
University of Missouri
Columbia, MO, USA

BRUCE G. MARCOT
US Forest Service
Pacific Northwest Research Station
Portland, OR, USA

HAMISH MCCALLUM
Griffith School of Environment and
Environmental Futures Research Institute
Nathan, QLD, Australia

MELISSA A. MILLER
Department of Biological Sciences
Auburn University
Auburn, AL, USA

JOSHUA J. MILLSPAUGH
Wildlife Biology Program
University of Montana
Missoula, MT, USA

AXEL MOEHRENSCHLAGER
Centre for Conservation Research
Calgary Zoological Society
Calgary, AB, Canada

ERIN MUTHS
US Geological Survey
Fort Collins Science Center
Fort Collins, CO, USA

SHAUN J. RILEY
Department of Fisheries and Wildlife
Michigan State University
East Lansing, MI, USA

ESTELLE A. SANDHAUS
Santa Barbara Zoo
Santa Barbara, CA, USA

CAMILLA SANDSTRÖM
Department of Political Science
Umeå University
Umeå, Sweden

PHILIP J. SEDDON
Department of Zoology
University of Otago
Dunedin, New Zealand

DAN SHIVELY
US Forest Service
Washington, DC, USA

ROB SLOTOW
School of Life Sciences
University of KwaZulu-Natal
Scottsville, South Africa

KASIA SLUZEK
Department of Evolutionary Biology and
    Environmental Studies
University Zurich
Zürich, Switzerland
and
Swiss Institute of Bioinformatics
Quartier Sorge - Batiment Genopode
Lausanne, Switzerland

DAVID A. STEEN
Department of Biological Sciences
Auburn University
Auburn, AL, USA

ROLLIE WHITE
Assistant Regional Director - Ecological Services
Pacific Region, US Fish and Wildlife Service
Portland, OR, USA

FOREWORD

Joel Berger

Earth has unassailable beauty, from physical scapes 8,000 m below the ocean's surface and 8,000 m above to species unimaginable. A single bowhead whale attains the size of a bison herd numbering 50. Some mammals lay eggs; some lizards are legless. Bats catch fish. Birds catch bats. Kingfishers fish. Wood frogs in Alaska have two-thirds of their body tissues turn to ice, but survive winter. Yet as sea and land temperatures warm, change is everywhere. Ice will continue to melt at high latitude and high elevation. Snow deepens in some places, and sublimates in others. The adaptations of many species will no longer offer them the protection it once did.

Across all the vastness of the planet, we humans have done a marvelous job to erase what has come before. It's easy to be grim, and to understand why the impoverished turn a blind eye, a behavior that also is too common in wealthy countries. People summit peaks like K-2 or Denali for adventure; ultramarathoners endeavor on different continents. We soar into space and spy on neighbors with drones. Technology offers real visits to the North Pole and virtual ones to our past. This odd mix between individual achievement and modern technology offers opportunity to enrich human lives, yet erodes our curiosity about the biological world. Most of us are detached from nature, but no one is immune to its recoil.

Fortunately, the challenges presented by our human needs coupled with our callousness are not the topic of this prescient volume. Instead, it is the achievable hope accompanied by action that counteracts some of our appalling past treatment of populations of wild vertebrates.

The beauty of our world is that there is hope. People still love nature and love animals, and vast realms remain free of human meddle. Optimism should flourish and, indeed, it does, showcased here in *Reintroduction of Fish and Wildlife Populations* where science is increasingly present to help guide our future. It brings us an improved understanding of ecological baselines. It divulges the lives of species, evinces fascination in the process of adaptation, and teaches us about the relevance of ecosystem function. Yet—and this is critical—*Reintroduction of Fish and Wildlife Populations* fuses science with conservation to net real gains for the species who co-share our planet.

Restoration is dictated by geography, history, and culture. As a consequence, it wears many hats—from reintroduction to removal, and introduction to augmentation to translocation. The tools are many: zoos, museums, parks, media, books, and education. But the bottom lines, which typically are considered ecological, demographic, and genetic, are not where this volume stops. It deals with the human milieu, for without it conservation cannot truly happen.

Less than 30 years ago, few might have imagined large carnivores expanding in a country with 300+ million people where wolves and grizzly bears had been under assault for two centuries. Now Montana, Wyoming, and Idaho have more than 1,500 wolves and 2,000 grizzlies. Black-footed ferrets, once extinct in the wild, are back on the ground spanning terrain from Mexico to Canada. Condors fly over Arizona and California. Bolson tortoises are back in the Chihuahua Desert. Lions are in Rwanda, and leopards into the forests above Sochi in Russia. Fences have been removed from parts of Kruger; alien predators are gone from more than 800 islands around the world, actions which offer petrels and boobies and kiwis

chances to do better. Europe has re-wilded; brown bears and wolves are in parts of Italy and France, Spain, and Germany. Tigers and rhinos move through fields and forests of India's once degraded lands but now with some rebound. The reasons: restoration, education, kindness, and governance.

The faces of success come in many forms. Science is a fundamental one that has bettered our world. Among others is an important reminder, a communique which rings as loudly now as it did nearly half a century ago. During his 1968 speech in New Delhi, Nigerian Ambassador Baba Dioum suggested what is required to keep us on track: "We will conserve only what we love, we will love only what we understand, and we will understand only what we are taught." *Reintroduction of Fish and Wildlife Populations* is a groundbreaking effort that will do much to help us understand, teach, and put more into practice.

ONE

Animal Reintroduction in the Anthropocene

David S. Jachowski, Rob Slotow, Paul L. Angermeier, and Joshua J. Millspaugh

AFTER CENTURIES OF WIDESPREAD persecution and extinction of fish and wildlife species at the hands of humans, biodiversity conservation is entering what E. O. Wilson (1992) has termed the "era of restoration." Among restoration techniques, reintroductions are unique by going beyond the traditional conservation objective of holding the line against adverse anthropogenic impacts, and more radically push the line backward by bringing species back to the landscape. Such tactics are likely to become increasingly mainstream as we fully enter the Anthropocene, which brings the specter of the sixth, and perhaps most precipitous, mass extinction in our planet's history.

There has been a boom, over the past 20 years in particular, in the reintroduction of animal species (Chapter 2). Reintroductions have occurred across the globe, and reintroduced species span the spectrum of vertebrate taxa, ranging from crested toads (*Peltophryne lemur*) in Puerto Rico to Arabian oryx (*Oryx leucoryx*) in Jordan. At the same time, the success of reintroduced populations has increased. Three

decades ago, Griffith et al. (1989) estimated that over half of conservation translocations (a broad grouping that included reintroductions) failed to reestablish self-sustaining populations, leading to wide-scale pessimism about the feasibility of reintroduction in practice. However, the authors of Chapter 2 more optimistically highlight that recent evaluations suggest that only 5% of reintroductions are complete failures, and there are multiple avenues to both define and achieve success in animal reintroduction. Our desire to provide insights into what advances have brought about this rapid improvement, and how to continue this positive trend, is the motivation behind this book.

Reintroduction, the process of releasing a species back to where it historically occurred but had been extirpated by humans, sounds simple enough. However, this progressive side of conservation biology is often expensive and complex. Failed reintroductions are not only wasted conservation resources, but, for threatened species, may further reduce long-term viability through loss of individuals from source populations. Importantly, failed reintroductions can negatively affect public perception of conservation competence, not only in the context of a specific failure, but across conservation practice more broadly, reducing public support and sympathy for biodiversity conservation (Jachowski 2014). We initiated this edited volume to bring together a diverse group of researchers from around the globe to help shed light on which techniques and approaches have worked, which have not, and, more specifically, to provide a synthetic resource for use by practitioners in designing and carrying out reintroductions that promote success.

Most of the techniques and lessons described herein were learned from direct experience and are unreported, or reported in case studies that are widely scattered in published and unpublished sources such as journal articles, other texts, and agency reports. Thus, it is difficult for fish and wildlife biologists and managers considering species reintroduction to be aware of the considerable amount of work that is available in this rapidly advancing field. Furthermore, because experts on terrestrial wildlife may not be aware of findings by experts on fishes, and vice versa, persons considering reintroduction of a specific taxon might not be aware of lessons learned from experience with other species that are likely to be of use in designing strategies to reestablish populations.

These strategies fundamentally begin with selection of the species and determining when and where reintroductions are appropriate (figure 1.1). As discussed in Chapter 3, knowledge of past states of ecosystems is critical when defining the baselines or targets that a reintroduction is meant to restore. Furthermore, while science can reveal environmental baselines, appropriate methodologies, and population trends to inform species restoration, the goals and success criteria of reintroductions are largely derived via social processes. These human dimensions often inherently induce accountability to society, which, given the investments and risks of reintroductions, is imperative for success. Thus, Chapter 4

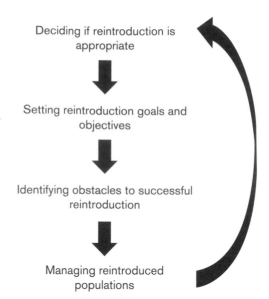

FIGURE 1.1 Conceptual overview of the organizational structure of this volume that reflects four general phases of the reintroduction process. Under the frameworks presented in multiple chapters (particularly Chapters 4, 5, 10, and 12), reintroductions need to be continually reassessed based on information gained on obstacles and opportunities for reintroduction that can enhance probability of success. This information also can lead to the option to abandon or "exit" a reintroduction.

addresses the roles of social attitudes and perceptions in guiding reintroduction programs.

After a species is targeted for reintroduction, as discussed in Part 2 of this book, careful, continued thought needs to be given to the specific goals a reintroduction project is trying to achieve (figure 1.1). For example, will success be measured through attainment of a certain population size or restoring a certain ecological or social function? To accommodate these complex issues, the concept of a "reintroduction landscape" is introduced in Chapter 5, where practitioners are encouraged to borrow from the field of landscape ecology and take a broad view of the inherent social, institutional, and ecological contexts that are necessary parts of successful reintroduction planning. Once it is determined that a reintroduction is to proceed, as discussed in Chapter 6, careful consideration needs to be given to the specific goals and objectives for the reintroduction, as well as a framework and metrics for measuring success. In particular, a common goal is often achieving a certain restored population size, and thus demographic modeling can be a powerful tool to help guide decision-making prior to and following release (Chapter 7).

Evidence from a wide range of taxa illustrates that prerelease planning often is key to achieving success. Accordingly, Chapters 8 through 10 address potential obstacles to reintroduction success that need to be considered prior to releasing animals into the wild. Practitioners need to account for a variety of genetic issues ranging from selection of release stock to inbreeding or outbreeding depression following release (Chapter 8). Similarly, knowledge of likely physiological, behavioral, and community-level responses is essential for maximizing success (Chapter 9). One of the most pressing issues in reintroduction, as in other conservation actions, is mitigating disease risk for both the translocated animals and the receiving community (Chapter 10). Fortunately, there are multiple techniques for facilitating reintroduction success that can be put into action prior to, during, and following reintroduction (Chapter 11).

Of those reintroductions that succeed in establishing populations, management concerns can arise due to impacts of reintroduced species on the receiving ecosystems. Accordingly, there is a need for strategic guidance on reintroductions to ensure successful, cost-effective management of populations. This guidance includes well-developed monitoring protocols, as well as the integration of monitoring with management decisions (Chapter 12). The expectation that reintroduced populations will be self-sustaining is less common than in the past, with most endangered species being instead considered "conservation reliant." This conservation reliance can manifest as many, often controversial, tactics such as fencing, sport hunting and culling, and contraception (Chapter 13). Thus, following release, populations often need to be continually managed in creative ways to meet conservation objectives that also take into account local, regional, and international socioeconomic concerns (figure 1.1).

A major theme that emerged in crafting this book is that the experiences generated through reintroduction biology have direct relevance to a variety of other dimensions of current conservation, including niche modeling, dispersal ecology, population genetics, climate change, captive propagation, and disease ecology. The development of nuanced understanding of animal reintroduction and its linkages to these fields has given rise to, or has direct application to, several other emerging frontiers in conservation biology such as assisted colonization, rewilding, and de-extinction (Chapter 2). In the future, it is likely that the practice of reintroduction biology will continue to evolve in response to new threats, opportunities, and changing social perceptions and values. This includes several new frontiers such as the increased valuation of restoring and conserving ecological processes (Chapter 15). These shifts in emphasis will push the boundaries of conservation biology as we know it, forcing the future of reintroduction biology into novel socio-ecological arenas that require engagement by natural scientists, social scientists, and the public.

A second persistent theme throughout the book is that science can inform decisions and assist in searching for optimal solutions, but it is up to society to dictate what is preserved, brought back, or lost in an increasingly human-dominated world. Accordingly, it is no coincidence that two chapters on the human dimensions of fish and wildlife reintroduction (Chapters 4 and 14) serve as bookends to this book. Only through a concerted effort to broaden long-term stakeholder support, and frame reintroductions within a broader socio-economic as well as ecological construct (Chapters 5 and 15), will we see sustainable establishment and management of reintroduced populations.

Given the major financial and cultural commitments required to successfully restore a species, as highlighted in this volume, we hope that this book indirectly shows the importance and cost-effectiveness of proactive conservation to avoid the need for reintroduction in the first place. It is hard not to be moved by something selfless or noble such as a group of people attempting to right a wrong that was, more often than not, caused by prior generations. In our opinion, this makes animal reintroduction one of the most impressive and ambitious enterprises in conservation biology. However, despite advances illustrating our collective capacity to restore species, the old Benjamin Franklin idiom "an ounce of prevention is worth a pound of cure" surely applies to considering reintroduction within a broader conservation context. The often extreme difficulty in reintroducing species should further emphasize the importance of trying to conserve remaining populations and species in situ.

Finally, as illustrated in the diverse examples provided in this volume, while some reintroductions have taken place in poor countries (largely funded by international donors), a majority of reintroduction have been conducted in developed or rich countries. There are many poor countries that, as development accelerates, are facing local extinction risks, and where the need for future reintroductions may need to become routine. By collecting the knowledge and understanding that has been generated through experience to date, we can broaden agency capacity to deal with reintroductions, as well as provide for effective and efficient implementation in a context of very limited resources.

To tackle such a broad and complex range of topics, we relied heavily on a distinguished group of experts to contribute to this edited volume. We are appreciative of the authors who volunteered their time and effort in crafting chapters. Thanks also to Mark Ryan, who first broached the idea of an edited volume on animal reintroduction at David Jachowski's dissertation defense several years ago. We are grateful to Blake Edgar and Merrik Bush-Pirkle at University of California Press for their support of our idea and seeing it through to fruition. We thank Nigel Adams, Christina Aiello, Chris Baker, Oded Berger-Tal, Dean Biggins, Michael Bruford, Virginia Butler, Jesse Delia, Josh Donlan, David Eads, John Ewen, Andrew Gregory, Brian Irwin, Richard Jachowski, Dylan Kesler, Bob Klaver, Richard Kock, Michelle McClure, Conor McGowan, Anita Morzillo, Mark Stanley Price, Michael Schaub, Phil Seddon, Jeremy Solin, Kelly Swan, Meena Venkataraman, Jack Williams, and Keith Winsten for offering their time and expertise to review chapters in this volume. Finally, we thank Shefali Azad, Robin Eng, Piper Kimprel, Mike Knoerr, Katie Krafte, Glenda Lofink, Nic McMillan, Alec Nelson, Fumika Takashashi, Hillary Thompson, and Wenbo Zhang for reviewing the entire book and offering editorial suggestions that helped improve the cohesiveness of the entire volume.

LITERATURE CITED

Griffith, B., J. M. Scott, J. W. Carpenter, and C. Reed. 1989. Translocation as a species conservation tool: status and strategy. Science 245: 477–480.

Jachowski, D. S. 2014. Wild Again: The Struggle to Save the Black-Footed Ferret. University of California Press, Berkeley.

Wilson, E. O. 1992. The Diversity of Life. Belknap Press, Cambridge, MA.

PART ONE

What Are Reintroductions and When Are They Appropriate?

TWO

Reintroduction and Other Conservation Translocations

HISTORY AND FUTURE DEVELOPMENTS

Philip J. Seddon and Doug P. Armstrong

AUSTRALASIA IS A CURRENT HOT SPOT of conservation translocation, the intentional movement and release of living organisms for conservation purposes (Seddon et al. 2014a). Not surprisingly then, one of the first conservation translocations in the world took place in New Zealand in the 1880s, when large numbers of flightless birds, kakapo (*Strigops habroptilus*) and kiwi (*Apteryx australis*), were moved to an offshore island by Richard Henry, marking the first attempt to protect New Zealand's native species from the impacts of exotic mammalian predators (Seddon et al. 2015, Box 2.1). Henry's attempts ultimately failed because the offshore island release sites were within the swimming range of mainland stoats, also called short-tailed weasels, *Mustela erminea* (Hill and Hill 1987, Miskelly and Powlesland 2013). At about the same time, on the other side of the world, the Tabasco sauce manufacturer Edward (Ned) McIlhenny was conducting ultimately much more successful translocations of captive-bred snowy egrets from eggs sourced from declining populations along the southern Gulf Coast of the United States. Egrets were released into Bird City, a private bird refuge McIlhenny established in 1895 on Avery Island, Louisiana, within the indigenous range of the species. It is possible this very early reintroduction saved the snowy egret from extinction, as birds from Avery Island dispersed and repopulated both the Louisiana and Florida Gulf Coasts (Furmansky 2009).

Reintroduction as more than an individual endeavor, but as an official and organized conservation action, came of age in 1907 when 15 bison were sent by rail and cart from the Bronx Zoo and released into the Wichita Mountains Wildlife Refuge in Oklahoma (reviewed in Beck 2001). This was an initiative of the American Bison Society (ABS), which was founded in 1905 to reintroduce bison into their former range following population declines from over 40 million in 1830 to only around 1,000 animals by 1884. The release in Oklahoma was the first animal reintroduction in North America, and was followed by other releases by the ABS in Montana (1910) and South Dakota (1913). Currently there are over 500,000 plains bison in the United States and Canada, and while most are on private ranches, some 30,000 are in conservation herds. The 1907 reintroduction by the ABS was notable for its comprehensive planning and careful engagement of the public through available media. Since the first use of conservation translocations described above, the use and sophistication of animal reintroductions has increased. Below we review the expansion of reintroduction as a species restoration tool, the creation of the International Union for the Conservation of Nature (IUCN) Reintroduction Specialist Group (RSG), the development of reintroduction guidelines, and the maturation of the emerging discipline of reintroduction biology. The second part of this chapter examines some developments and future issues, including the challenge of assessing reintroduction success, the rise of conservation introductions—the release of organisms outside their indigenous range for conservation benefit—and new or reemerging concepts such as rewilding and de-extinction.

EARLY SUCCESSES

Following the successes of bison reintroductions in the United States, there were several decades with few reintroduction attempts, but several high-profile success stories in the 1960s to 1980s helped raise the profile of reintroduction as a viable population restoration tool (Box 2.1).

South Island Saddleback in New Zealand

Members of the endemic New Zealand wattlebird family (Callaeidae) once filled the forests in both main islands, but were extremely vulnerable to the impacts of exotic mammalian predators such as ship rats (*Rattus rattus*), cats (*Felis catus*), and stoats (*Mustela erminea*). By 1910, one of these species, the South Island saddleback (*Philesturnus carunculatus*), was restricted to only three offshore islands in the far south near Stewart Island. In the early 1960s, ship rats invaded all three islands, causing the extinction of a snipe, a wren, and a bat species. This shocking event was instrumental

in convincing conservation managers of the devastation wrought by invading rats. In 1964, the New Zealand Wildlife Service translocated the 36 last remaining saddlebacks in the world, from Big South Cape, the largest of the three islands, to nearby Big and Kaimohu Islands (Hooson and Jamieson 2003). Well over 30 serial translocations to other islands, and in 2009 to a predator-free protected area on the mainland, have meant the population of South Island saddlebacks probably now exceeds 2,000 birds. From this early crisis-driven start, New Zealand has been one of the world leaders in the application of bird conservation translocations (Miskelly and Powesland 2013, Seddon et al. 2014a).

Peregrine Falcon in North America

Widespread use of organochlorine pesticides, particularly DDT (Dichlorodiphenyltrichloroethane), during the 1950s to 1970s caused eggshell thinning and breeding failure in peregrine falcons (*Falco peregrinus*) in the United States and Canada. The ban of DDT by the early 1970s marked the start of recovery efforts for peregrine falcons in North America, supported by the captive breeding program established by Dr. Tom Cade and The Peregrine Fund. From 1974 to 1999 more than 7,000 captive-bred peregrine falcons were released, and by 1999 the known falcon population in the continental United States had increased from the 1975 low of ~40 pairs to more than 1,600 pairs (Heinrich 2009).

Arabian Oryx in the Middle East

There are not many instances when we are fairly sure of the moment when the death of an animal outside of captivity rendered the species extinct in the wild. In October 1972, D. S. Henderson recorded evidence of the death of a small group of Arabian oryx (*Oryx leucoryx*) at the hands of hunters in the Jiddat al-Harasis in the Omani Central Desert—no free-ranging Arabian oryx were seen subsequently (Hender-

BOX 2.1 · Conservation Translocation Time Line: Key Translocation-Related Events for Wildlife Conservation

This is not an exhaustive list of events or wildlife species translocation projects; it is a time line of important events, including conservation translocation project milestones globally or regionally, policy and guideline developments, and the arrival of influential publications.

Date	Event
1894	Richard Henry started translocation of kiwi and kakapo from the mainland to Resolution Island in Dusky Sound, Fiordland, New Zealand (NZ) (Ormerod 1993).
1895	Edward Avery McIlhenny established Bird City, a private wildfowl refuge on Avery Island in coastal Louisiana, into which he released captive-bred snowy egrets to establish a new population to counter declines elsewhere in the species' range; in 1909 McIlhenny shipped 2,100 snowy egrets from Avery Island, Louisiana, to the Charles Deering estate in Florida where they were held in a flight cage until they started breeding in 1910. Then they were released to the wild and established a nesting colony said to have been largely responsible for the restoration of the species in Florida, following decimation by the millinery trade. [http://en.wikipedia.org/wiki/Bird_City_(wildfowl_refuge)]
1907	Bronx Zoo sent 15 bison by railway for release into the Wichita Mountains Wildlife Preserve in Oklahoma (Beck 2001; Wildlife Conservation Society; American Bison Society). [http://www.wcs.org/saving-wildlife/hoofed-mammals/bison/the-american-bison-society.aspx]
1911	First reintroduction of zoo-born ibexes at the Graue Hörner (St. Gall, Switzerland). A second reintroduction initiated in 1914 at the Piz d'Ela Massif (Grisons) failed, probably due to poaching. [http://www.waza.org/en/site/conservation/waza-conservation-projects/overview/alpine-ibex-reintroduction_1]
1963	Review of Medieval-era animal translocations (Niethammer 1963).
1964	NZ Wildlife Service staff translocated 23 North Island saddleback from Hen Island, Hauraki Gulf, to Whatupuke Island, developing the techniques used the same year to rescue South Island saddleback following invasion of Big South Cape (off Stewart Island) by rats. Thirty-six birds were translocated to nearby Big and Kaimohu Islands. This was the first time that a translocation had prevented the extinction of an island species from anywhere in the world (Miskelly and Powesland 2013).
1967	The Raptor Research Foundation incorporated in South Dakota as an organization devoted to research on birds of prey and, in its early years, specifically to supporting and developing methods of conservation breeding and translocations of peregrine falcons and other raptors for the restoration of extirpated or greatly diminished populations. The foundation is now internationally recognized as a broad-based scientific organization. [http://www.raptorresearchfoundation.org/]
1970	The Peregrine Fund was organized at the Cornell Lab of Ornithology, for the conservation breeding of falcons and the translocation of produced offspring to vacant or depopulated habitats. [http://www.peregrinefund.org/]
1973	Official licensing and academic involvement in goshawk translocations from continental Europe to the United Kingdom, undertaken in the 1960s.
1973	US Endangered Species Act passed.
1974	First US peregrine falcon experimental reintroductions began by hacking in New York State and by fostering to wild parents in Colorado in an attempt to learn techniques to restore extirpated populations in the East and to supplement greatly reduced populations in the West that had resulted from the impacts of pesticides such as DDT (Cade and Burnham 2003).
1976–1977	The entire Chatham Island black robin population (seven birds) was translocated from Little Mangere Island to Mangere Island (Butler and Merton 1992).
1977	Important early conference on translocations and hands-on management (Temple 1978).

1970s	Clandestine release of Eurasian lynx (*Lynx lynx*) in Switzerland (Breitenmoser and Breitenmoser 1999).
1980	Reintroduction first featured in a major work on conservation biology (Campbell and Wilcox 1980).
1982	Extinct in the wild since 1972, Arabian oryx were reintroduced from the San Diego Wild Animal Park back into Jaaluni in the Jiddat al-Harasis, Oman (Stanley Price 2012).
1984	National Zoological Park, Washington, DC, and World Wide Fund for Nature commenced golden lion tamarin reintroductions in Brazil (Kleiman et al. 1986).
1984	First rhino reintroduction in Dudhwa National Park/Tiger Reserve, Uttar Pradesh, India (Sale and Singh 1987).
1984	World Center for Birds of Prey established by The Peregrine Fund in Boise, Idaho, conducted research and management on birds of prey worldwide.
1984	Early review of "Reintroduction as a Method of Conservation" (Cade 1986).
1984	Reintroduction of lake sturgeon (*Acipenser fulvescens*) to the Mississippi and Missouri Rivers began (Drauch and Rhodes 2007).
1985	International conference on peregrine recovery in Sacramento, California, major review of conservation breeding and translocation, celebrating 20th anniversary of the Madison Peregrine Conference (Cade et al. 1988).
1987	IUCN Position Statement on the Translocation of Living Organisms: Introductions, Re-introductions and Restocking (IUCN 1987). [http://www.iucnsscrsg.org/]
1987	First major treatment of reintroduction issues in a leading journal (Lyles and May 1987).
1987	Red wolves released at the Alligator River National Wildlife Refuge, North Carolina, as part of the first attempt in history to restore a carnivore species that had been declared extinct in the wild; first pups born a year later.
1988	Establishment of the IUCN SSC (Species Survival Commission) Reintroduction Specialist Group under the Chair of Mark Stanley Price (Stanley Price and Soorae 2003).
1989	Major review of the outcomes and success factors for 93 species of native birds and mammals (Griffith et al. 1989).
1989	First reintroduction in Mallorcan midwife toad program, which resulted in the species being downgraded from "Critically Endangered" to "Vulnerable" (Griffiths et al. 2008).
1991	Major review of conservation translocations of amphibians and reptiles (Dodd and Seigel 1991).
1991	The introduction of the Pedder galaxias to Lake Oberon in Tasmania. This assisted colonization saved the species but potentially compromised an aquatic ecosystem with no previous exposure to predatory fish (Horwitz 1995, Sanger 2013).
1992	Mission of RSG developed: "To promote re-establishment or reinforcement of self-sustaining populations of plants and animals in the wild" (Stanley Price and Soorae 2003).
1992	Reintroduction of takhi (Przewalski's horse) in Mongolia began in the Gobi Desert around Takhiin Tal in the Great Gobi B Strictly Protected Area (9,000 km²) and in the mountain steppe of Hustai National Park (570 km²) (Walzer et al. 2012).
1993	First use of the term "Reintroduction Biology" in the scientific literature (Viggers et al. 1993).
1994	Reintroduction of Atlantic salmon to tributaries of the Rhine River began (Schneider 2011).
1995	Publication of "Reintroduction biology of Australian and New Zealand fauna" (Serena 1995).
1995	IUCN Guidelines for Reintroductions (IUCN 1998) became official policy. [http://www.iucnsscrsg.org/]
1995	Grey wolves reintroduced to Yellowstone National Park, Montana, Wyoming, and Idaho; first pups born in late spring (Fritts et al. 1997).
1997	Ted Turner and Mike Phillips cofounded the Turner Endangered Species Fund (TESF) and Turner Biodiversity Divisions as the first significant private efforts to emphasize conservation translocations; in 2009 the TESF documented first Bolson tortoise born in the wild in the United States in over 10,000 years due to a captive breeding and reintroduction program; in

2011 the TESF restoration of a desert bighorn sheep population was so successful that for the first time ever the New Mexico Department of Game & Fish was able to use a herd restored to private land as a donor of animals for reintroductions elsewhere in species' historical range. [http://tesf.org/team.html]

1998   Major review of the outcomes of 181 bird and mammal translocation projects (Wolf et al. 1998).

1999   First publication of a framework for ecological replacements (Seddon and Soorae 1999).

1999   The peregrine falcon was officially removed from US list of endangered species, following release and translocation of more than 6,000 captive-produced falcons. A celebration of the announcement by Secretary of the Interior, Bruce Babbit, and a three-day conference in Boise, Idaho, were attended by more than 1,000 people, most of whom had participated in the recovery work (Cade and Burnham 2003).

2000   First translocation to a fenced sanctuary in New Zealand: Zealandia Ecosanctuary, little spotted kiwi (Miskelly and Powesland 2013).

2000   Major review of 20 years of publications on animal translocations (Fischer and Lindenmayer 2000).

2000   Publication of a major update on success or failure of 25 species of raptors involved in translocations (75% success rate), with comments on published criticisms of translocation as a conservation technique (Cade 2000).

2001   Reintroduction of the critically endangered white-winged guan in northwest Peru—the only reintroduction project in this country (Pratolongo 2011).

2004   Asiatic black bear reinforcement was started as the first conservation translocation in Korea (Jeong et al. 2010).

2005   Major conference on hands-on management of the California condor, including conservation breeding and conservation translocations in California, Arizona, and Baja California (Mee and Hall 2007).

2006   Reintroduction of the black-footed ferret in New Mexico; by 2008 the TESF documented the first black-footed ferret born in the wild in New Mexico in over 60 years. [http://tesf.org/team.html]

2006   The pink pigeon (*Nesoenas mayeri*) was down-listed from *critically endangered* to *endangered* on the IUCN Red List following nearly three decades of management that included captive breeding and reintroduction.

2007   First use of the term "assisted migration" in the scientific literature (McLachlan et al. 2007). The related term "assisted colonisation" was suggested by Hunter (2007) to avoid confusion with the concept of annual migration used in animal ecology.

2008   First International Wildlife Reintroduction Conference: Applying Science to Conservation. April 15–16, 2008, Lincoln Park Zoo, Chicago, USA. [http://www.lpzoo.org/reintroworkshop/]

2008   Symposium on Avian Reintroduction Biology: Current Issues for Science and Management, May 8–9, Zoological Society of London, London, United Kingdom [https://static.zsl.org/files/avian-translocations-flyer-502.pdf] (Ewen et al. 2008).

2009   Publication of "Reintroduction of Top-Order Predators" (Hayward and Somers 2009).

2011   Reintroduction of Atlantic salmon to tributaries of Lake Ontario began. [http://cida.usgs.gov/glri/projects/habitat_and_wildlife/restore_aquatic_habitats.html]

2012   Publication of "Reintroduction Biology: Integrating Science and Management" (Ewen et al. 2012).

2013   IUCN Guidelines for Reintroductions and Other Conservation Translocations replaced the 1998 document. [http://www.iucnsscrsg.org/]

2014   Publication of the Scottish Code for Conservation Translocations, based on the revised IUCN Guidelines and designed to be read alongside them (National Species Reintroduction Forum 2014).

2015   Publication of "Advances in Reintroduction Biology of Australian and New Zealand Fauna" (Armstrong et al. 2015).

son 1974). However, the decline of the last population of wild oryx in the southeastern Arabian Peninsula had been noted a few years earlier, and an operation by the Phoenix Zoo and the Fauna and Flora Preservation Society succeeded in capturing four animals in Aden (Yemen) near the Oman border. These animals joined others from captive collections in the Middle East and London to form the nucleus of the World Herd at Phoenix Zoo, numbering 11 animals by mid-1964. Phoenix Zoo sent oryx to other collections, including San Diego Wild Animal Park, from where the first Arabian oryx were released back into the wild in the Jiddat al-Harasis in 1982 (Stanley Price 1989). Captive breeding of oryx started in Saudi Arabia in 1986, with reintroductions into the fenced Mahazat as-Sayd protected area in 1989, and the unfenced 'Uruq Bani Ma'arid protected area in 1995. Despite setbacks in the Omani project (Stanley Price 2012), reintroductions of Arabian oryx have also taken place in Israel, the United Arab Emirates (2007), and Jordan (2009).

UNDOCUMENTED FAILURES

In contrast to the well-planned, well-monitored, and well-documented reintroduction successes described above, there were many poorly planned releases of animals into unsuitable areas where their inevitable failure to survive, breed, and establish a population was largely undocumented. The lack of post-release monitoring or reporting of unfavorable outcomes makes it impossible to summarize these undocumented failures.

The lack of documentation of outcomes could reflect the fact that many reintroductions were viewed as one-off management exercises. A false distinction often exists between management and research, the former involving manipulation to achieve management objectives without necessarily attempting to simultaneously learn about how the systems under management work (McNab 1983). Some management manipulations might lack adequate

monitoring, and without post-release monitoring, nothing can be learned about what variables were important in a successful translocation, no knowledge is gained from undocumented failures, and refinement of release decisions is not possible. (Seddon and Soorae 1999; Moehrenschlager and Lloyd, Chapter 11, this volume). In contrast, conservation managers are rightly critical of researchers who pursue questions with little applied relevance. The greatest gains, however, will come from realization that strategic research can underpin good management, that achieving management objectives often also requires learning to improve outcomes, and that learning can proceed from post-release monitoring that is an integrated part of a reintroduction program (Gitzen et al., Chapter 12, this volume).

THE REINTRODUCTION SPECIALIST GROUP AND THE FIRST GUIDELINES

It was principally a response to rising numbers of ill-conceived reintroduction attempts that led to the IUCN position statement on translocations in 1987 (IUCN 1987) and formation of the IUCN Species Survival Commission (IUCN/SSC) RSG in 1988. The RSG was formed by Mark Stanley Price, the architect of the Arabian oryx reintroduction to Oman (Stanley Price 1989), with the aim of promoting responsible reintroductions (Stanley Price and Soorae 2003). The RSG's first strategic planning workshop was held in 1992, and led to the formulation of a set of reintroduction guidelines (IUCN 1998). By early 2006, the RSG consisted of a volunteer network of over 300 practitioners and maintained a database of nearly 700 reintroduction projects. The 1998 Reintroduction Guidelines were a slim booklet of commonsense suggestions designed to encourage reintroduction practitioners to consider the various aspects of proposed projects, from biological to social, legislative, and economic. They recognized that any reintroduction project is more than just a manipulation of a wildlife population, but that to be successful required the support of stakeholders and a long-term commitment of resources. These guidelines were informed by the first examination of translocation outcomes. In 1989 Brad Griffith and coauthors published a hugely influential review of the factors associated with translocation success, looking at the reintroduction and reinforcement (the addition of individuals to an existing population) of 93 species of native birds and mammals, and identifying habitat quality at the release site, release into the core of a species' range, and total numbers released as determinants of success (Griffith et al. 1989).

THE DISCIPLINE OF REINTRODUCTION BIOLOGY

In no small part due to the work of the IUCN's RSG and the unifying nature of the 1998 guidelines, a discipline of reintroduction biology started to develop from the early 1990s. Reintroduction biology is broadly considered to be the study and associated practice of establishing populations of organisms using conservation translocation tools and maintaining them using ongoing management. Reintroduction projects increasingly were framed as more than just one-off management responses, as practitioners engaged with ecologists, geneticists, population modelers, veterinarians, and social scientists to enhance translocation success. Three areas of development were prominent, as set forth below.

Increasing Post-release Monitoring

Many calls for better post-release monitoring were made during the late 1980s (IUCN 1987, Lyles and May 1987, Griffith et al. 1989, Kleiman 1989), stemming from a frustration with reintroduction attempts from which nothing was being learned, either about the process of successful population establishment or about the timing and causes of the all-too-frequent failures. In many cases even the rationale or the objectives of the project were unclear. Engage-

ment of researchers as reintroduction partners seems to have changed the early management focus and, along with increasing requirements to document outcomes for project funders and other stakeholders, post-release monitoring is now emphasized in many projects.

Developing the Science of Reintroduction Biology

The maturation of a scientific discipline follows three stages: (1) observation guided by intuition and guesswork; (2) organization of observations into categories, and the exploration of observations for patterns; and (3) recognition of underlying causes of patterns and formulation of theories that lead to testing of predictions deduced from these (Williams 1997). Reintroduction biology as a scientific discipline has been moving out of a phase of inductive inference, whereby taxon-specific observations have been used to derive and explore patterns. Much of the early reintroduction-related research has been on components of the reintroduction process that are relatively easy to evaluate, such as release techniques, rather than aspects that are considered critical to establishment and long-term population persistence (Seddon et al. 2007). Recent research involves more general theory drawn from population ecology and other disciplines, and reintroductions can, in turn, provide opportunities for tests of theory (Sarrazin and Barbault 1996).

Strategic Directions

By the 2000s there had been a marked increase in the number of reintroduction-related publications (figure 2.1), facilitated by the generation of data from post-release monitoring. This wealth of information about reintroduction outcomes was used, formally and informally, in reviews seeking general principles and correlates of translocation success (e.g., Wolf et al. 1998, Fischer and Lindenmayer 2000). However, it was apparent that much of the post-release monitoring activity was unfocused,

TABLE 2.1
Key Questions in Reintroduction Biology (after Armstrong and Seddon 2008), and Recent Examples of Published Research Addressing Each Question

---

I. Population level

  A. Establishment

  1. *How is establishment probability affected by size and composition of the release group?*

  Schaub et al. (2009) modeled demographic data for reintroduced bearded vultures (*Gypaetus barbatus*) to assess whether further releases should be conducted.

  2. *How are post-release survival and dispersal affected by prerelease and post-release management?*

  Richardson et al. (2015) found that survival of hihi (*Notiomystis cincta*) held in an aviary for 4 days at the release site was lower than that of birds released immediately.

  B. Persistence

  3. *What habitat conditions are needed for persistence of the reintroduced population?*

  Parlato and Armstrong (2012) obtained probability distributions for growth of reintroduced North Island robin (*Petroica longipes*) populations as a function of rat density and landscape connectivity.

  4. *How will genetic makeup affect persistence of the reintroduced population?*

  Cullingham and Moehrenschlager (2013) measured diversity, effective population size, and genetic connectivity of reintroduced swift fox (*Vulpes velox*) populations over time to assess long-term retention of genetic diversity. Drauch and Rhodes (2007) completed similar analysis for lake sturgeon reintroduction. See also Biebach et al. (Chapter 8, this volume).

II.   Metapopulation level

---

  5. *How heavily should source populations be harvested?*

  Canessa et al. (2014) determined the optimal number of corroboree frog (*Pseudophryne corroboree*) eggs and subadults to release to maximize the sizes of both the reintroduced population and the captive source population.

  6. *What is the optimal allocation of translocated individuals among sites?*

  Rout et al. (2009) showed how active adaptive management could potentially be used to optimize the allocation of bridled nailtail wallabies (*Onychogalea fraenata*) between two reintroduction sites.

  7. *Should translocation be used to compensate for isolation?*

  Gusset et al. (2009) developed an individual-based model based on pack dynamics of African wild dogs (*Lycaon pictus*) to predict the best translocation regime for connecting isolated populations.

III.   Ecosystem level

---

  8. *Are the target taxon and its parasites native to the ecosystem?*

  Walker et al. (2008) traced the source of chytrid fungus (*Batrachochytrium dendrobatidis*) strains affecting the Mallorcan midwife toad (*Alytes muletensis*) to a captive-breeding facility that had been used for reintroduction of the species to its native habitat. See also Muths and McCallum (Chapter 10, this volume).

  9. *How will the ecosystem be affected by the target species and its parasites?*

  Using an enclosure trial, Griffiths et al. (2013) assessed the effectiveness of Aldabra giant (*Aldabrachelys gigantea*) and Madagascan radiated (*Astrochelys radiata*) tortoises for restoring historic plant-animal interactions on Round Island, Mauritius, before introducing these species.

  10. *How does the order of reintroductions affect the ultimate species composition?*

  Work on this issue is not yet well represented in the literature.

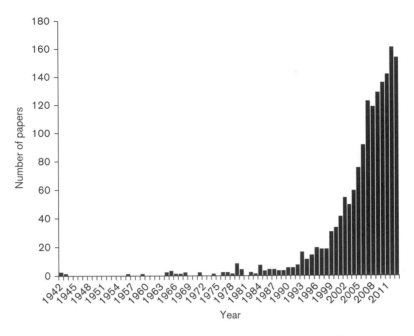

FIGURE 2.1 Numbers of reintroduction-related peer-reviewed outputs published annually between 1942 and 2013 (data after 2005 updates, Seddon et al. 2007). Numbers for the period 2006–2013 were derived following the methods of Seddon et al. (2007), based on a search of the Core Collection of the Web of Science using the term "reintroduction." Only articles, proceedings, and reviews were considered within the topic areas of ecology, biodiversity and conservation, zoology, and environmental science. The search is not exhaustive, but is indicative of publishing trends in this field.

unguided by explicit monitoring objectives, and therefore, while broadly useful, was an inefficient use of resources. Armstrong and Seddon (2008) proposed 10 key questions for reintroduction biologists designed to encourage a more strategic approach to the discipline of reintroduction biology. Examples of research addressing these questions are outlined in Table 2.1, and further examples are covered throughout the remainder of this volume.

Today there is a recognized discipline of reintroduction biology encompassing the science around all forms of conservation translocation (Ewen et al. 2012). Improved translocation procedures, detailed post-release monitoring, and the framing of releases as explicit experimental tests (Kemp et al. 2015) are generating a growing literature that informs reintroduction attempts for a broadening range of species globally (figure 2.1; Seddon et al. 2014a).

The 2013 Reintroduction Guidelines

Although the first Reintroduction Guidelines provided a valuable framework for reintroduction planning, by 2010 it was evident the 1998 booklet was not sufficiently detailed or comprehensive. In particular, it did not fully consider the range of conservation translocation options needed to address the threats of habitat loss and the extinction of keystone species. A task force was formed under the auspices of the IUCN/SSC and the chair of Mark Stanley Price, who had by then passed the leadership of the RSG to Frederic Launay. Because the new guidelines needed to deal with the complexity of translocations outside the indigenous range of species, the task force core membership was drawn from both the RSG and the Invasive Species Specialist Group. The fully revised and much more comprehensive guidelines became official IUCN policy in 2013 (IUCN 2013).

Although this document reflects the tremendous progress made over the previous 20 years, it also emphasizes significant challenges for the future.

FUTURE RESEARCH AND DEVELOPMENT

The remainder of this chapter focuses on future issues that we see as particularly important.

Assessing Success

With the rise of the discipline of reintroduction biology and the closer integration of science and management facilitated by the IUCN RSG, conservation translocation planning and implementation is improving. It is now standard for reintroduction projects to consider feasibility and risks as part of decision-making around whether to proceed, prior to any implementation. A number of tools and approaches are now available to match species to appropriate release areas (Osborne and Seddon 2012). Targeted post-release monitoring is now expected, generating vital information not only to evaluate species project outcomes, but also to inform the wider discipline in general (Ewen et al. 2012). However, there remains an impression that many, or even most, reintroduction attempts fail, although this is based on project summaries that are now over two decades old, when success rates were influenced by poorly planned projects in the period before comprehensive guidelines were available. More recent project evaluations suggest 58% of 250 recent reintroduction projects across all taxa were considered fully successful by all project-specific criteria, and only 5% were classified as complete failures (unpublished data, iucnrsg.org). In addition, reintroduction practitioners are progressively taking on more difficult challenges, so substantial improvement in success rates is not guaranteed.

Any summary of reintroduction success rate should be viewed with caution, since there are no agreed definitive criteria for assessing outcomes as a simple success/failure dichotomy (see also Chauvenet et al., Chapter 6, this vol-

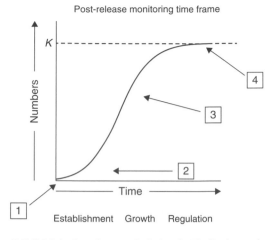

Post-release monitoring time frame

FIGURE 2.2 Growth curve depicting the idealised growth of a reintroduced population. A group of founder animals is released at time zero [1]; initially the population undergoes an establishment phase where growth is limited by the low number of breeding individuals and might be lower than expected due to post-release acclimation, Allee effects and demographic stochasticity [2]; if the population establishes, it undergoes a growth phase facilitated by there being more breeders and little constraint from resource availability [3]; as the population approaches K, the carrying capacity of the habitat, it transitions to the regulation phase [4]. Post-release monitoring should continue until the population reaches the regulation phase, although the intensity of monitoring will typically be reduced. Cessation of monitoring earlier, at [2] or [3], might give an unrealistic impression of population growth potential.

ume). Three challenges remain in this regard: (i) any evaluation of project outcome is time bound since "success" at one period can become "failure" in the future (e.g., Spalton et al. 1999); (2) species-specific post-release monitoring time frames are required to assess project outcomes, dependent on generation times and life-history traits; (3) the typical aspiration of a "self-sustaining" population (IUCN 1998) is vague, partly because it is unclear whether this implies an absence of any ongoing human intervention (usually a fuzzy concept) or the absence of any reinforcements. The most fruitful progress toward criteria for reintroduction success considers three stages: establishment, growth, and regulation (IUCN 2013), with this division reflecting related schemes suggested by Sarrazin (2007) and Armstrong and Seddon (2008) (figure 2.2). To ultimately be successful, a reintroduced population must first establish, overcoming post-release acclimation, demographic stochasticity, and possibly Allee effects; it must then grow and remain sufficiently large to be viable in the long term. Success at these different stages is affected by different process, and requires different monitoring and modeling methods (Armstrong and Reynolds 2012, Nichols and Armstrong 2012, Converse and Armstrong, Chapter 7, this volume). In addition, while it might not be known for a long time whether reintroductions are ultimately successful, useful inferences about establishment and growth can be made over much shorter time frames. Increasing attention is also being given to assessing metrics of success that include wider ecological, social, and even institutional factors (see also Dunham et al., Chapter 5; Gitzen et al., Chapter12, this volume).

Restoration Targets

Reintroduction is a population restoration technique, so reintroduction practitioners are faced with the question: restore to what? What is the target historical state? In Australia and New Zealand, for example, historical restoration targets often relate to some state before European colonization (Jackson and Hobbs 2009). In Europe, restoration targets tend to address more recent species declines. The challenge of setting restoration targets is common to both reintroduction biology and restoration ecology; both disciplines have acknowledged the arbitrary nature of trying to replicate some past condition (Temperton 2007) in the face of a lack of accurate historical records (Hobbs 2007), the dynamic nature of ecological systems (Choi et al. 2008), and the inevitability of environmental change (Hobbs and Harris 2001, Jackson and Hobbs 2009, Hobbs et al. 2013).

Often an implicit assumption is made that because a local extinction has occurred within historic times, reintroduction will focus only on sites within the indigenous range of a species, known or inferred within relatively recent time frames. However, documented records of species presence have some shortcomings (Ponder et al. 2001). A given location might lack records for a given species for a number of reasons other than because it truly is absent, including simply because no one saw it, or that people saw it but never recorded its presence. This will be the case particularly for rare, secretive, or cryptic species. Species distribution maps rely on occurrence records that might be of dubious reliability (Frey 2006), and there might be errors of species identifications or locations, or issues with mislabeled specimens or even falsified records (Boessenkool et al. 2010). Sampling effort is seldom uniform (Maddock and Du Plessis 1999), resulting in distribution maps that depict areas most frequently visited by observers rather than areas of a species' presence. A reliance on historical species distributions to determine reintroduction sites also makes the assumption that environmental conditions have not changed since species extirpation (Seddon 2010).

Habitat is that complex of interacting physical and biotic components that favors persistence of a given species (Hall et al. 1997, Armstrong and Seddon 2008). Species ranges are dynamic (Lyman, Chapter 3, this volume) and environments change over time; thus, there are four critical consequences for reintroductions (after Osborne and Seddon 2012): historical locations of a species' presence might not indicate present-day habitat; present-day locations of a species' presence might not indicate habitat that will allow persistence; present-day locations where a species is absent might not indicate lack of habitat; and present-day locations of a species' presence might not indicate future habitat.

For the last point, an important factor is global climate change, which might significantly alter both the biotic and the abiotic components of a given area. Climate change, in tandem with human-facilitated species invasions and rapid changes in human land use, contributes to the creation of novel ecosystems, systems that differ in composition and function from past systems (Hobbs et al. 2013). Restoration and maintenance of species within their indigenous range will remain a core component of conservation efforts, but it is now widely understood that a return to some arbitrary and supposedly stable, pristine state is not feasible.

Assisted Colonization

Since at least 1985, researchers have recognized that climate change might alter the suitability of habitats for species, and where adaptation or dispersal was not possible, species might become stranded in unsuitable areas (Peters and Darling 1985). There is now global acknowledgment that action needs to be taken to address climate-induced changes in species' habitats, particularly in the situation where individuals are unable naturally to colonize new areas as habitat shifts (McLachlan et al. 2007). In some cases, colonization of new habitat might have to be assisted by translocation of species to sites outside their indigenous range (Hunter 2007, McLachlan et al. 2007, Hoegh-Guldberg et al. 2008). Given the devastating ecological damage caused by exotic species globally, the suggestion of planned introductions for conservation has led to vigorous debate in the literature (Loss et al. 2010, Hewitt et al. 2011, and references therein). The 1998 IUCN Guidelines recognized what was then termed "conservation/benign introduction" as being justified only when no habitat was left in the indigenous range, but did not specifically refer to changes in habitat distribution due to climate change. The 2013 IUCN Guidelines define assisted colonization as "the intentional movement of an organism outside its indigenous range to avoid extinction of populations due to current or future threats" (IUCN 2013). Under this definition, assisted colonization has been applied as a conservation tool for some time (Seddon et al. 2015), for example, through the marooning of kakapo on offshore islands to protect them from exotic mammalian predators (Thomas 2011). More recently, the creation of a disease-free "insurance population" of

Tasmanian devils (*Sarcophilus harrisii*) on an island outside the species' indigenous range (Nogardy 2013) is compatible with the IUCN definition of assisted colonization.

The 2013 IUCN Guidelines place great emphasis on the analysis of feasibility and risk analysis as essential components of any conservation translocation. Given the uncertainties involved in moving species outside their ranges, assisted colonization is considered inherently more risky than "traditional" translocations such as reintroductions. New approaches for understanding and managing risk under uncertainty and with multiple stakeholders are being applied to conservation introduction planning, including quantitative risk analysis (Ewen et al. 2012), structured decision-making (Schwartz and Martin 2013), and active adaptive management (McDonald-Madden et al. 2011). Where protection from threats in the indigenous range is unfeasible, and where appropriate habitat can be identified elsewhere, application of carefully planned assisted colonization might become more acceptable (Larson and Palmer 2013). Although climate change is not the only driving factor, it might be the most compelling justification for assisted colonization in the near future (Richardson et al. 2009).

A key component in planning assisted colonization is the identification of areas that match the biotic and abiotic needs of a focal species under future climate scenarios. Climate-envelope models are being used to determine species' future habitat distributions to guide some of the first assisted colonizations of butterflies in the United Kingdom (Willis et al. 2009). But static bioclimatic envelope models might not adequately account for species' ability to disperse, nor for changing demographic processes as habitat shifts, so more complex integrative climate suitability models will be required (Huntley et al. 2010). New approaches are being developed that explicitly integrate species distribution data with population dynamics or physiology. Stochastic population modeling was combined with habitat suitability models to quantify how climate is predicted to influence the vital rates of the hihi (*Notiomystis cincta*), a New Zealand endemic passerine (Chauvenet et al. 2013). Eco-energetic and hydrological models were integrated to evaluate the long-term suitability of habitat for western swamp tortoise (*Pseudemydura umbrina*), and extended to identify new regions that would meet the tortoise's thermodynamic requirements (Mitchell et al. 2013). Future developments around assisted colonization planning will see the application of fully integrated models, combining climatic and other aspects of habitat suitability, population dynamics, and mechanistic movement models of dispersal, for single species or two or more interacting species (Huntley et al. 2010, Travis et al. 2013).

Ecological Replacements

It is understood that high levels of biodiversity can increase ecosystem stability by buffering the effects of environmental change, resisting species invasions, and preventing secondary extinctions following species losses (MacDougall et al. 2013). Reintroductions are a means to restore biodiversity and ecosystem functions where local extinctions have taken place within the indigenous range (Ripple et al. 2014). With the global extinction of a species, however, restoration of functions might be achieved only through introduction of functionally equivalent exotic species.

The 2013 IUCN Guidelines define ecological replacement as a form of conservation introduction involving the release of an appropriate substitute species "to re-establish an ecological function lost through extinction." This official recognition of ecological replacement, earlier discussed in the literature as "subspecific substitution" (Seddon and Soorae 1999), as a valid conservation tool marks the most significant expansion of the conservation translocation spectrum and attempts restoration of natural processes rather than addressing only extinction risk (Seddon et al. 2014a). An increasing stimulus for ecological replacements could be

the reinstatement of some lost socioeconomic function, for example, restoration of ecosystem services, resurrection of culturally important biological elements, or even replacement of harvestable species.

Ecological replacements have been used to restore herbivory and seed dispersal functions in island ecosystems. Giant tortoises pay an important ecosystem engineering role as specialized herbivores or frugivores and as important dispersers of large seeded plants. Grazing and trampling by tortoises are critical processes in the maintenance of "tortoise turf," the endemic vegetative community once common on Indian Ocean islands (Hansen et al. 2010). Exotic Aldabra giant tortoises *Aldabrachelys gigantea* have been introduced to Ile aux Aigrettes as an ecological replacement for the extinct Mauritian *Cylindraspis* species, in order to restore the seed dispersal of the endangered endemic ebony *Diospyros egrettarum* (Griffiths et al. 2011), and there are plans to use ecologically similar species of giant tortoise to reinstate missing processes in Madagascar and the islands of the Galapagos, Mascarenes, Seychelles, and Caribbean (Hansen et al. 2010).

A future challenge is the identification of suitable replacements to perform desired ecosystem functions within a given system. The longer the time since the extinction of the original form, the greater the uncertainty there will be over selecting a suitable substitute, and because the most suitable replacements could also be at risk of extinction, concurrent ecosystem restoration and species conservation efforts might be necessary. The focus will be more on reinstatement of functions and processes to enhance ecosystem resilience, rather than on restoration to some arbitrary historical state. For any conservation introduction, the risk of unintended effects must be assessed against expected benefits (IUCN 2013). A careful approach will not demand absence of any risks, but will address uncertainty through carefully designed replacements that can be readily monitored (Corlett 2013), and easily removed should there be unwanted ecological, social, or economic effects. Radical replacements, using substitutes that are not closely related taxa, could be warranted providing that risk and uncertainty are adequately evaluated.

Rewilding

Soulé and Noss (1998) introduced the concept of "rewilding," built around the keystone role played by wide-ranging, large animals able to maintain ecosystem structure, resilience, and diversity through top-down trophic interactions (Ripple et al. 2014). Twenty-five years later, the term rewilding is being widely used, or rather, misused, its original meaning having largely been lost. The rise in the popularity of rewilding, and the misapplication of the concept, has been due to controversy around North American "Pleistocene rewilding," the proposed introduction of large wild vertebrates as proxies for megafauna lost 13,000 years ago (Donlan et al. 2005). Pleistocene rewilding is at its core true to the original concept of rewilding, recognizing the important ecosystem-shaping role of large vertebrates. However, it has come to be associated specifically with ecological replacement of long-extinct species rather than restoration of ecosystem function.

A modern interpretation of rewilding involves species translocations to restore ecosystem functioning (Sandom et al. 2013, Seddon et al. 2014a). Translocation for rewilding could entail population restoration through reintroduction, where releases take place in the indigenous range of a species with the primary aim of restoring some ecological function. Alternatively, it could involve a conservation introduction through ecological replacement. The recovery of populations of large carnivores in Europe (Deinet et al. 2013), for example, is viewed as a modern form of rewilding where the aim is not to restore to some arbitrary past state, but rather to reinstate ecological processes such a predator-prey interactions within landscapes shared by humans and wildlife (Boitani and Linnell 2014).

De-extinction

Advances in techniques to manipulate genetic material have, amongst other things, raised the prospect of species de-extinction, the resurrection of once extinct species (Church and Regis 2012). De-extinction hit the headlines only in early 2013 following the TEDx DeExtinction conference co-organized by the Long Now Foundation and the National Geographic Society (http://tedxdeextinction.org). The prospect of resurrecting species such as woolly mammoths (*Mammuthus primigenius*), Tasmanian tigers (*Thylacinus cynocephalus*), and passenger pigeons (*Ectopistes migratorius*) (Brand 2012a) was welcomed by some (Castro 2013). However, it was not universally applauded, with many objections raised, including animal welfare, human health, environmental, political, and moral issues (Gewin 2013, Sherkow and Greely 2013). De-extinction has been called a "quest for redemption" and a "moral imperative" (Heidari 2013) to reverse the species extinctions caused by humans. However, Sandler (2014) argued that it is not possible for a debt of restorative justice to be paid by those who owe it to those who are due it, because both the individual organisms that were harmed during extinction and those who caused the extinction are now absent. A more nuanced view of de-extinction recognizes that due to technological limitations no extinct species could truly be brought back in their genetic, physiological, and behavioral entirety; therefore, a realistic and achievable goal for de-extinction efforts might be the creation of some functional proxy for an extinct taxon (Shapiro 2015).

Despite concerns (Mark 2013, Pimm 2013, Scientific American 2013, Zimmer 2013), de-extinction in some form seems inevitable (Biello 2013). One of the many questions is "which species are the best de-extinction candidates?" The lists of candidates mooted to date tend to be dominated by iconic and charismatic species (www.longnow.org). The stated goal of de-extinction is "deep ecological enrichment" (Brand 2012b), and the enhancement of the resilience of ecosystems in the face of changing environments (Church 2013). Thus the primary ethical argument in favor of de-extinction is the potential promotion of biodiversity (Cohen 2014). This requires restoration of free-ranging populations; thus, before resurrecting individuals that are a functional proxy of an extinct species, it is essential to consider where to release them and the risks or uncertainties of doing so (Jorgensen 2013). De-extinction is a conservation translocation issue and any selection of candidate species must consider the feasibility and risks of release (Jones 2014). Cohen (2014) argued that, in order to reduce the risk of threats to biodiversity, following the existing reintroduction guidelines is a precondition for ethical de-extinction.

Seddon et al. (2014b) translated the relevant sections of the 2013 IUCN Reintroduction Guidelines into a framework of questions to identify any critical uncertainties or unacceptable risks relating to the release of resurrected species, therefore providing a first cut of unsuitable candidates to avoid wasted effort. This pre-selection process assumes that the technical requirements will be met, yielding an acceptable facsimile of the extinct species, and that a sufficient number of genetically diverse individuals can be made available for release. Once a species has been selected for resurrection, the full 2013 Reintroduction Guidelines must be applied to match species to habitat and to consider ecological, ethical, social, economic, legislative, and logistical requirements.

Public Support for Reintroductions

Without public support, particularly local community support, the risk of reintroduction project failure will be increased. To a large extent, what we choose to conserve or restore is the reflection of prevailing social attitudes, environmental awareness, and public support. This has resulted in a marked taxonomic bias in species that are the focus of reintroduction attempts. Unsurprisingly, the larger and more charismatic vertebrate species have been favored, in large part because these are the species for which the necessary public and political support and resources can be obtained (Seddon et al. 2005). But we are starting to see a shift away from restorations of single species chosen for their public appeal, toward multispecies and keystone species restorations that seek to restore ecosystem functions. By necessity this must include keystone species that are ecosystem engineers, species that modify, maintain, create, or destroy structure in their physical environment (Lawton 1994), or predators that regulate the abundance of prey. But some keystone species have the potential to affect human health or livelihood in human-dominated landscapes where people have become accustomed to the absence of endemic elements (Seddon and van Heezik 2013). A future challenge will therefore be to engage an increasingly urbanized public more fully in local restoration projects, to reset public expectations of what the natural world around them should or could look like, and, in so doing, seek to change attitudes and gain public support for the more challenging reintroductions of keystone species (see also George and Sandhaus, Chapter 14, this volume).

SUMMARY

If conservation biology is a discipline born out of the crisis of biodiversity declines, then reintroduction biology is a discipline born out of the management response to that crisis. Reintroduction, the reestablishment of a population of a species within an area of its indigenous range from which it has disappeared, is one of several types of conservation translocation, the intentional movement and release of organisms for conservation benefit. Many early conservation translocations were marked by poor planning, lack of monitoring and failure to establish populations. In 1988, the IUCN created the RSG to promote the use of translocation as a conservation tool, and in 1998 the RSG produced the first guidelines for reintroduction planning. Fifteen years later, in 2013, the RSG published a larger, fully revised version of the guidelines, not only encompassing the increasingly common approaches of population reinforcement and reintroduction, but also considering the more risky and controversial topics of conservation introduction, assisted colonization, and ecological replacement.

The development of the discipline of reintroduction biology paralleled that of the RSG, as reintroduction practitioners sought to involve researchers in conservation translocation planning and assessment. Partly as a result of the efforts of the RSG, and due to some high-profile reintroduction success stories, there has been rapid growth in both the number of reintroduction projects and research outputs. The foundations for reintroduction biology, the study of the establishment, growth, regulation, and ecological role of translocated populations, were laid only in the last two decades, but developed from the first attempts back in the 1800s, to the comprehensively planned and rigorously executed projects that are starting to characterize the discipline today. Phases in the development of translocations as a management approach include early successes that raised the profile of reintroduction as a conservation tool, to a period of ill-planned projects and undocumented failures; to increasing post-release monitoring and the documentation of outcomes enabling retrospective analysis; and to increased strategic direction and the use of experimental approaches and modeling tools.

With official IUCN recognition of a spectrum of conservation translocation possibilities, the emphasis has now shifted to how best to apply these approaches in such a way as to maximize conservation benefit while minimizing the risk of unintended consequences. Particularly for the inherently more uncertain translocations, such as conservation introductions, or those involving resurrected species, a focus will be the development and application of rigorous methods to match species to habitats and to evaluate and manage environmental, social, cultural, and economic risks. From its beginning as isolated projects by committed individuals, to one-off management exercises, through to comprehensive institutionalized programs of species restoration seeking to restore ecosystem functions, the discipline of reintroduction biology continues to develop as conservation managers, biologists, social scientists, and other stakeholders combine forces to improve conservation translocation outcomes, especially in an era of unprecedented anthropologically driven ecological change.

ACKNOWLEDGMENTS

Suggestions for timeline items were provided by Thomas Abeli, D. H. Jeong, Kevin Parker, Markus Gusset, Robert Kenward, Wolfgang Scherzinger, Fernando Angulo Pratolongo, Piet Wit, Satya Sinha, Piero Genovesi, Tom Cade, Mike Phillips, Mark Lintermans, and Richard Griffiths. This chapter was improved by suggestions and comments from Francois Sarrazin, Robert Alexandre, John Ewen, Mark Stanley Price, and Kevin Parker. Ken Miller assisted with figure formatting.

LITERATURE CITED

Armstrong, D. P., M. W. Hayward, D. Moro, and P. J. Seddon, eds. 2015. Advances in Reintroduction Biology of Australian and New Zealand Fauna. CSIRO Publishing, Collingwood, VIC, Australia.

Armstrong, D. P., and M. H. Reynolds. 2012. Modelling reintroduced populations: the state of the art and future directions. Pp.165–222. In J. G. Ewen, D. P. Armstrong, K. A. Parker, and P. J. Seddon, eds., Reintroduction Biology: Integrating Science and Management. Conservation Science and Practice No. 9. Wiley-Blackwell, Chichester, UK.

Armstrong, D. P., and P. J. Seddon. 2008. Directions in reintroduction biology. Trends in Ecology and Evolution 23:20–25.

Beck, B. B. 2001. A vision for reintroduction. Communiqué, September 2001:20–21. American Zoo and Aquarium Association, Silver Spring, MD.

Biello, D. (2013) TEDxDeExtinction 3/15/13 Meeting Report. The Long Now Foundation. Available online at: www.longnow.org (accessed September 18, 2013).

Boessenkool, S., B. Star, P. Scofield, P. J. Seddon, and J. M. Waters. 2010. Genetic analyses suggest fraudulent origins of historic museum penguin specimens. Proceedings of the Royal Society B 277:1057–1064.

Boitani, L., and L. D. C. Linnell. 2014. Bringing large mammals back: large carnivores in Europe. Pp.67–84. In H. Periera and L. Navarro, eds. Rewilding European Landscapes. Springer Science, Dordrecht, Germany.

Brand, S. 2012a. Forward to the Past: De-extinction Projects, Techniques, and Ethics. Meeting Report, Revive & Restore, October 23–24. Available online at: http://longnow.org/revive/1stde-extinction/ (accessed September 17, 2013).

Brand, S. 2012b. Bringing Back the Passenger Pigeon. Meeting convened at Harvard Medical School in Boston on February 8, 2012, Revive & Restore. Available online at: http://longnow.org/revive/passenger-pigeon-workshop/ (accessed September 19, 2013).

Breitenmoser, U., and C. Breitenmoser. 1999. The re-introduction of the Eurasian lynx in the European Alps. Re-introduction News 18:13–14

Butler, D., and D. M. Merton. 1992. The Black Robin: Saving the World's Most Endangered Bird. Oxford University Press, Oxford.

Cade, T. J. 1986. Reintroduction as a method of conservation. Pp. 72–84. In S. E. Senner, C. M. White, and J. R. Parrish, eds., Raptor Research Report No. 5: Raptor Conservation in the Next 50 Years. Raptor Research Foundation, Provo, UT.

Cade, T. J. 2000. Progress in translocation of diurnal raptors. Pp.343–372. In R. D. Chancellor and B.-U. Meyburg, eds., Raptors at Risk. World Working Group on Birds of Prey and Owls, Berlin/Hancock House, Surrey, BC.

Cade, T. J., and W. Burnham. 2003. Return of the Peregrine: A North American Saga of Tenacity and Teamwork. The Peregrine Fund, Boise, ID.

Cade, T. J., J. H. Enderson, C. G. Thelander, and C. M. White, eds. 1988. Peregrine Falcon Populations: Their Management and Recovery. The Peregrine Fund, Inc., Boise, ID.

Campbell, S., and B. A. Wilcox. 1980. Is reintroduction a realistic goal? Pp.263–269. In M. E. Soulé, ed., Conservation Biology: An Evolutionary-Ecological Perspective. Sinauer, Sunderland, MA.

Canessa, S., D. A. Hunter, M. McFadden, G. Marantelli, and M. A. McCarthy. 2014. Optimal release strategies for cost-effective reintroductions. Journal of Applied Ecology 51:1107–1115.

Castro, J. M. 2013. Thoughts on de-extinction. Chemical and Engineering News 91:2.

Chauvenet, A. L. M., J. G. Ewen, D. Armstrong, and N. Pettorelli. 2013. Saving hihi under climate change: as case for assisted colonization. Journal of Applied Ecology 50:1330–1340.

Choi, Y. D., V. M. Temperton, E. B. Allen, A. P. Grootjans, M. Halassy, R. J. Hobbs, M. A. Naeth, and K. Torok. 2008. Ecological restoration for future sustainability in a changing environment. Ecoscience 15:53–64.

Church, G. M. 2013. Please reanimate: reviving mammoths and other extinct creatures is a good idea. Scientific American 309:11–12.

Church, G. M., and E. Regis. 2012. Regenesis: How Synthetic Biology Will Reinvent Nature and Ourselves. Basic Books, New York.

Cohen, S. 2014. The ethics of de-extinction. Nanoethics 8:165–178.

Corlett, R. T. 2013. The shifted baseline: prehistoric defaunation in the tropics and its consequences for biodiversity conservation. Biological Conservation 163:13–21.

Cullingham, C. I., and A. Moehrenschlager. 2013. Temporal analysis of genetic structure to assess population dynamics of reintroduced swift foxes. Conservation Biology 6:1389–1398.

Deinet, S., C. Ieronymidou, L. McRaze, I. L. Burfield, R. P. Poppen, B. Collen, and M. Bohm. 2013. Wildlife Comeback in Europe: The Recovery of Selected Mammal and Bird Species. Final Report to Rewilding Europe by ZSL, BirdLife International and the European Bird Census Council. ZSL, London, UK.

Dodd, C. K., and R. A. Seigel. 1991. Relocation, repatriation, and translocation of amphibians and reptiles: are they conservation strategies that work? Herpetologica 47:336–350.

Donlan, J., H. W. Greene, J. Berger, C. E. Bock, J. H. Bock, D. A. Burney, J. A. Estes et al. 2005. Re-wilding North America. Nature 436: 913–914.

Drauch, A. M., and O. E. Rhodes. 2007. Genetic evaluation of the lake sturgeon reintroduction program in the Mississippi and Missouri Rivers. North American Journal of Fisheries Management 27:434–442.

Ewen, J. G., D. Armstrong, K. Parker, and P. Seddon. 2008. Avian reintroduction biology: current issues for science and management. Avian Biology Research 1:27–50.

Ewen, J. G., D. P. Armstrong, K. A. Parker, and P. J. Seddon, eds. 2012. Reintroduction Biology: Integrating Science and Management. Conservation Science and Practice No. 9. Wiley-Blackwell, Chichester, UK, 73–104.

Fischer, J., and D. B. Lindenmayer. 2000. An assessment of the published results of animal relocations. Biological Conservation 96:1–11.

Frey, J. K. 2006. Inferring species distributions in the absence of occurrence records: an example considering wolverine (*Gulo gulo*) and Canada lynx (*Lynx canadensis*) in New Mexico. Biological Conservation 130:16–24.

Fritts, S. H., E. E. Bangs, J. A. Fontaine, M. R. Johnson, M. K. Phillips, E. D. Koch, and J. R. Gunson. 1997. Planning and implementing a reintroduction of wolves to Yellowstone National Park and Central Idaho. Restoration Ecology 5:7–27.

Furmansky, D. Z. 2009. Rosalie Edge, Hawk of Mercy: The Activist Who Saved Nature from the Conservationists. University of Georgia Press, Athens, GA.

Gewin, V. 2013. Ecologists weigh in on "de-extinction" debate. Frontiers in Ecology and Environment 11:176.

Griffith, B., J. M. Scott, J. W. Carpenter, and C. Reed. 1989. Translocation as a species conservation tool: status and strategy. Science 245:477–480.

Griffiths, C. J., D. M. Hansen, C. G. Jones, N. Zuël, and S. Harris, 2011. Resurrecting extinct interactions with extant substitutes. Current Biology 21:762–765.

Griffiths, C. J., N. Zuël, C. Jones, Z. Ahamud, and S. Harris. 2013. Assessing the potential to restore historic grazing ecosystems with tortoise ecological replacements. Conservation Biology 27:690–700.

Griffiths, R. A., G. García, and J. Oliver. 2008. Re-introduction of the Mallorcan midwife toad, Mallorca, Spain. Pp.54–57. In P. S. Soorae, ed., Global Re-introduction Perspectives: Re-introduction Case Studies from around the Globe. IUCN/SSC Re-introduction Specialist Group, Abu Dhabi, UAE.

Gusset, M., O. Jakoby, M. S. Müller, M. J. Somers, R. Slotow, and V. Grimm. 2009. Dogs on the catwalk: modelling re-introduction and transloca-tion of endangered wild dogs in South Africa. Biological Conservation 142:2774–2781.

Hall, L. S., P. R. Krausman, and M. L. Morrison. 1997. The habitat concept and a plea for standard terminology. Wildlife Society Bulletin 25:173–182.

Hansen, D. M., C. J. Donlan, C. J. Griffiths, and K. J. Campbell. 2010. Ecological history and latent conservation potential: large and giant tortoises as a model or taxon substitutions. Ecography 33:272–284.

Hayward, M. W. and M. Somers, eds. 2009. Reintroduction of Top-Order Predators. Conser-vation Science and Practice No. 5. Wiley-Black-well, Chichester, UK.

Heidari, N. 2013. Reviving the dead. Chemical and Engineering News 91:34.

Heinrich, W. 2009. Peregrine falcon recovery in the continental United States, 1974–1999, with notes on related programs of the Peregrine Fund. Pp.431–444. In J. Sielicki and T. Mizera, eds., Peregrine Falcon Populations: Status and Perspectives in the 21st Century. Turul/Poznan University of Life Sciences Press, Warsaw, Poland.

Henderson, D. S. 1974. Were they the last Arabian oryx? Oryx 12:347–350.

Hewitt, N., N. Klenk, A. L. Smith, D. R. Bazely, N. Yan, S. Wood, J. I. MacLellan, C. Lipsig-Mumme, and I. Henriques. 2011. Taking stock of the assisted migration debate. Biological Conserva-tion 144:2560–2572.

Hill, S., and J. Hill. 1987. Richard Henry of Resolution Island. John McIndoe, Dunedin, New Zealand.

Hobbs, R. J. 2007. Setting effective and realistic restoration goals: key directions for research. Restoration Ecology 15:354–357.

Hobbs, R. J., and J. A. Harris. 2001. Restoration ecology: repairing the Earth's ecosystems in the new millennium. Restoration Ecology 9:239–246.

Hobbs, R. J., E. S. Higgs, and C. Hall, eds. 2013. Novel Ecosystems: Intervening in the New Ecological World Order. Wiley-Blackwell, West Sussex, UK.

Hoegh-Guldberg, O., L. Hughes, S. McIntyre, D. B. Lindenmayer, C. Parmesan, H. P. Possingham, and C. D. Thomas. 2008. Assisted colonization and rapid climate change. Science 321:345–346.

Hooson, S. and I. G. Jamieson. 2003. The distribu-tion and current status of New Zealand saddle-back *Philesturnus carunculatus*. Bird Conservation International 13:79–95.

Horwitz, P. 1995. An environmental critique of some freshwater captive breeding and reintroduc-tion programmes in Australia. Pp.75–80. In M. Serena, ed., Reintroduction Biology of Australian and New Zealand Fauna. Surrey Beatty and Sons, Chipping Norton, NSW.

Hunter, M. L. 2007. Climate change and moving species: furthering the debate on assisted colonization. Conservation Biology 21:1356–1358.

Huntley, B., P. Barnard, R. Altwegg, L. Chambers, B. W. T. Coetzee, L. Gibson, P. A. R. Hockey, D. G. Hole, G. F. Midgley, L. G. Underhill, and S. G. Willis. 2010. Beyond bioclimatic envelopes: dynamic species' range and abundance modelling in the context of climate change. Ecography 33:621–626

IUCN (International Union for Conservation of Nature). 1987. IUCN Position Statement on the Translocation of Living organisms: Introductions, Re-introductions, and Re-stocking. Available online at: http://www.iucnsscrsg.org/.

IUCN. 1998. Guidelines for Re-introductions. IUCN/SSC Re-introduction Specialist Group, Gland, Switzerland and Cambridge, UK. Available online at: http://www.iucnsscrsg.org/.

IUCN. 2013. Guidelines for Reintroductions and Other Conservation Translocations. 1 IUCN/SSC Re-introduction Specialist Group, Gland, Switzerland/Cambridge, UK. Available online at: http://www.iucnsscrsg.org/

Jackson, S.T., and R.J. Hobbs. 2009. Ecological restoration in the light of ecological history. Science 325:567–569.

Jeong, D.H., D.H. Yang, and B.K. Lee. 2010. Re-introduction of the Asiatic black bear into Jirisan National Park, South Korea. Pp.254–258. In P.S. Soorae, ed., Global Re-introduction Perspectives: 2011. Additional Case Studies from around the Globe. IUCN/SSC Re-introduction Specialist Group, Abu Dhabi, UAE.

Jones, K.E. 2014. From dinosaurs to dodos: who could and should we de-extinct? Frontiers of Biogeography 6:21–24. Available online at: http://escholarship.org/uc/item/9gv7n6d3.

Jorgensen, D. 2013. Reintroduction and De-extinction. BioScience 63:719–720

Kemp, L., G. Norbury, R. West, S. Comer, and R. Groenewegen 2015. The roles of trials and experiments in reintroduction programmes. Pp.73–90. In D.P. Armstrong, M.W. Hayward, D. Moro, and P.J. Seddon, eds., Advances in Reintroduction Biology of Australian and New Zealand Fauna. CSIRO Publishing, Collingwood, VIC.

Kleiman, D.G. 1989. Reintroduction of captive mammals for conservation. Bioscience 39:152–161.

Kleiman, D.G., B.B. Beck, J.M. Dietz, L.A. Dietz, J.D. Ballou, and A.F. Coimbra-Filho.1986. Conservation program for the Golden lion tamarin: captive research and management, ecological studies, educational strategies, and reintroduction. Pp.959–979. In K. Benirschke, ed., Primates: The Road to Self-Sustaining Populations. Proceedings in Life Sciences. Springer, New York.

Larson, B.M.H., and C. Palmer. 2013. Assisted colonization is no panacea, but let's not discount it either. Ethics, Policy and Environment 16:16–18.

Lawton, J.H. 1994. What do species do in ecosystems? Oikos 71:367–374.

Loss, S.R., L.A. Terwilliger, and A.C. Peterson. 2010. Assisted colonization: integrating conservation strategies in the face of climate change. Biological Conservation 144:92–100.

Lyles, A.M. and May, R.M. 1987. Problems in leaving the ark. Nature 326:245–246.

MacDougall, A.S., K.S. McCann, G. Gellner, and R. Tirkington. 2013. Diversity loss with persistent human disturbance increases vulnerability to ecosystem collapse. Nature 494:86–89.

Maddock, A., and M.A. Du Plessis. 1999. Can species data only be appropriately used to conserve biodiversity? Biodiversity and Conservation 8:603–615.

Mark, J. 2013. De-extinction won't make us better conservationists! Available online at: www.salon.com/2013/09/06/de_extinction_wont_make_us_better_conservationists_partner/ (accessed September 19, 2013).

McDonald-Madden, E., M.C. Runge, H.P. Possingham, and T.G. Martin. 2011. Optimal timing for managed relocation of species faced with climate change. Nature Climate Change 1:261–265.

McLachlan, J.S., J.J. Hellmann, and M.W. Schwartz. 2007. A framework for debate of assisted migration in an era of climate change. Conservation Biology 21:297–302.

McNab, J. 1983. Wildlife management as scientific experimentation. Wildlife Society Bulletin 11:397–401.

Mee, A., and L.S. Hall, eds. 2007. California Condors in the 21st Century. Series in Ornithology No. 2. Nuttall Ornithological Club and American Ornithologists' Union, Washington, DC.

Miskelly, C.M., and R.G. Powesland. 2013. Conservation translocations of New Zealand birds, 1863–2012. Notornis 60:3–28.

Mitchell, N., M.R. Hipsey, S. Arnall, G. McGrath, H. Bin Tareque, G. Kuchling, R. Vogwill, M. Sivapalan, W.P. Porter, and M.R. Kearney. 2013. Linking eco-energetics and eco-hydrology to select sites for the assisted colonization of Australia's rarest reptile. Biology 2:1–25.

National Species Reintroduction Forum. 2014. The Scottish Code for Conservation Translocations. Scottish Natural Heritage.

Nichols, J.D., and D.P. Armstrong. 2012. Monitoring for reintroductions. Pp.223–255. In J.G. Ewen, D.P. Armstrong, K.A. Parker, and P.J. Seddon, eds., Reintroduction Biology: Integrating Science and Management. Conservation Science and Practice No. 9. Wiley-Blackwell, Chichester, UK.

Niethammer, G. 1963. Die Einbürgerung von Säugetieren und Vögeln in Europa. Paul Parey, Hamburg-Berlin. [In German]

Nogardy, B. 2013. Devil of a disease. ECOS, February 25. Available online at: www.ecosmagazine.com/?paper=EC13036.

Ormerod, R. 1993. Henry, Richard Treacy: The Dictionary of New Zealand Biography. Te Ara – The Encyclopedia of New Zealand, Wellington, New Zealand.

Osborne P. E., and P. J. Seddon. 2012. Selecting suitable habitats for reintroductions: variation, change and the role of species distribution modelling. Pp.73–104. In J. G. Ewen, D. P. Armstrong, K. A. Parker, and P. J. Seddon, eds., Reintroduction Biology: Integrating Science and Management. Conservation Science and Practice No. 9. Wiley-Blackwell, Chichester, UK.

Parlato, E. H., and D. P. Armstrong. 2012. An integrated approach for predicting fates of reintroductions with demographic data from multiple populations. Conservation Biology 26:97–106.

Peters, R. L., and J. D. S. Darling. 1985. The greenhouse effect and nature reserves. Bioscience 35:707–717.

Pimm, S. 2013. Opinion: the case against species revival. National Geographic News. Available online at: http://news.nationalgeographic.com/news/2013/03/130312--deextincti...ation-animals-science-extinction-biodiversity-habitat-environment/ (accessed September 26 2013).

Ponder, W. F., G. A. Carter, P. Flemons, and R. R. Chapman. 2001. Evaluation of museum collection data for use in biodiversity assessment. Conservation Biology 15:648–657.

Pratolongo, F. A. 2011. Re-introduction of the white-winged guan in Lambayeque, Peru. Pp.141–145. In P. S. Soorae, ed., Global Re-introduction Perspectives: 2011. More Case Studies from around the Globe. IUCN/SSC Re-introduction Specialist Group, Gland, Switzerland/Environment Agency – Abu Dhabi, Abu Dhabi, UAE.

Richardson, D. M., J. J. Hellmann, J. S. McLachlan, D. F. Sax, M. W. Schwartz, P. Gonzaleze, E. J. Brennan et al. 2009. Multidimensional evaluation of managed relocation. Proceedings of the National Academy of Sciences of the United States of America 106:9721–9724.

Richardson, K., I. C. Castro, D. H. Brunton, and D. P. Armstrong. 2015. Not so soft? Delayed release reduces long-term survival in a passerine reintroduction. Oryx 49:446–456.

Ripple, W. J., J. A. Estes, R. L. Beschta, C. C. Wilmers, E. G. Ritchie, M. Hebblewhite, J. Berger et al. 2014. Status and ecological effects of the world's largest carnivores. Science 343. doi:10.1126/science.1241484.

Rout, T. M., C. E. Hauser, and H. P. Possingham. 2009. Optimal adaptive management for the translocation of a threatened species. Ecological Applications 19:515–526.

Sale, J. B., and S. Singh. 1987. Reintroduction of greater Indian rhinoceros into Dudhwa National Park. Oryx 21:81–84.

Sandler, R. 2014. The ethics of reviving long extinct species. Conservation Biology 28:354–360.

Sandom, C., C. J. Donlan, J.-C. Svennin, and D. Hansen. 2013. Rewilding. Pp.430–451. In D. W. MacDonald and K. J. Willis, eds., Key Topics in Conservation Biology 2. John Wiley & Sons, Oxford, UK.

Sanger, A. C. 2013. Extinct habitat, extant species: lessons learned from conservation recovery actions for the Pedder galaxias (*Galaxias pedderensis*) in south-west Tasmania, Australia. Marine and Freshwater Research 64:864–873.

Sarrazin, F. 2007. Introductory remarks: a demographic frame for reintroductions. Ecoscience 14, iv–v. doi:10.2980/1195-6860(2007)14[iv:IR]2.0.CO;2.

Sarrazin, F., and R. Barbault. 1996. Reintroduction: challenges and lessons for basic ecology. Trends in Ecology & Evolution 11:474–478.

Schaub, M., R. Zink, H. Beissmann, F. Sarrazin, and R. Arlettaz. 2009. When to end releases in reintroduction programmes: demographic rates and population viability analysis of bearded vultures in the Alps. Journal of Applied Ecology 46:92–100.

Schneider, J. 2011. Review of reintroduction of Atlantic salmon (*Salmo salar*) in tributaries of the Rhine River in the German Federal States of Rhineland-Palatinate and Hesse. Journal of Applied Ichthyology 27:24–32.

Schwartz, M. W., and T. G. Martin. 2013. Translocation of imperilled species under changing climates. Annals of the New York Academy of Sciences 1286:15–28.

Scientific American. 2013. Why efforts to bring extinct species back from the dead miss the point. Scientific American, June 1.

Seddon, P. J. 2010. From reintroduction to assisted colonization: moving along the conservation translocation spectrum. Restoration Ecology 18:796–802.

Seddon, P. J., D. P. Armstrong, and R. Maloney. 2007. Developing the science of reintroduction biology. Conservation Biology 21:303–312.

Seddon, P. J., C. J. Griffiths, P. S. Soorae, and D. P. Armstrong. 2014a. Reversing defaunation: restoring species in a changing world. Science 345:406–412.

Seddon, P. J., A. Moehrenschlager, and J. Ewen. 2014b. Reintroduction resurrected species: selecting DeExtinction candidates. Trends in Ecology and Evolution 29:140–147.

Seddon, P. J., D. Moro, N. J. Mitchell, A. L. M. Chauvenet, and P. R. Mawson. 2015. Proactive conservation or planned invasion? Past, current and future use of assisted colonisation. Pp.105–126. In D. P. Armstrong, M. W. Hayward, D. Moro, and P. J. Seddon, eds., Advances in Reintroduction Biology of Australian and New Zealand Fauna. CSIRO Publishing, Collingwood, VIC.

Seddon, P. J. and P. S. Soorae. 1999. Guidelines for subspecific substitutions in wildlife restoration projects. Conservation Biology 13:177–184.

Seddon, P. J., P. S. Soorae, and F. Launay. 2005. Taxonomic bias in reintroduction projects. Animal Conservation 8:51–58.

Seddon, P. J. and Y. van Heezik. 2013. Reintroductions to "ratchet up" public perceptions of biodiversity: reversing the extinction of experience through animal restorations. Pp.137–151. In M. Bekoff, ed., Ignoring Nature No More: The Case for Compassionate Conservation. University of Chicago Press, Chicago, IL.

Serena, M., ed. 1995. Reintroduction Biology of Australian and New Zealand fauna. Surrey Beatty & Sons, Chipping Norton, NSW.

Shapiro, B. 2015. How to Clone a Mammoth. Princeton University Press, Princeton, NJ.

Sherkow, J. S., and H. T. Greely. 2013. What if extinction is not forever? Science 340:32–33.

Soulé, M., and R. Noss. 1998. Rewilding and biodiversity: complementary goals for continental conservation. Wild Earth Fall 1998:1–11.

Spalton, J. A., and M. W. Lawrence 1999. Arabian oryx reintroduction in Oman: successes and setbacks. Oryx 33:168–175.

Stanley Price, M. 1989. Animal Re-introductions: The Arabian oryx in Oman, Cambridge University Press, Cambridge, UK.

Stanley Price, M. 2012. The Arabian oryx: saved, yet. . . WAZA magazine 13:15–18.

Stanley Price, M., and P. Soorae. 2003. Re-introductions: whence and wither? International Zoo Yearbook 38:61–75.

Temperton, V. M. 2007. The recent double paradigm shift in restoration ecology. Restoration Ecology 15:344–347.

Temple, S. A., ed. 1978. Endangered Birds: Management Techniques for Preserving Threatened Birds. University of Wisconsin Press, Madison, WI.

Thomas, C. 2011. Translocation of species, climate change, and the end of trying to recreate past ecological communities. Trends in Ecology & Evolution 26:216–221.

Travis, J. M. J., M. Delgado, G. Bocedi, M. Baguette, K. Barton, D. Bonte, I. Boulangeat et al. 2013. Dispersal and species' responses to climate change. Oikos 122:1532–1540.

Viggers, K., D. Lindenmayer, and D. Spratt. 1993. The importance of disease in reintroduction programmes. Wildlife Research 20: 687–698.

Walker, S. F., J. Bosch, T. Y. James, A. P. Litvintseva, J. A. O. Valls, S. Pina, G. Garcia et al. 2008. Invasive pathogens threaten species recovery programs. Current Biology 18:R853–R854.

Walzer, C., P. Kaczensky, W. Zimmermann, and C. Stauffer. 2012. Przewalski's horse reintroduction to Mongolia: status and outlook. World Association of Zoos and Aquariums (WAZA) Magazine 13:3–6.

Williams, B. K. 1997. Logic and science in wildlife biology. Journal of Wildlife Management 61:1007–1015.

Willis, S. G., J. K. Hill, C. D. Thomas, D. B. Roy, R. Fox, D. S. Blakeley, and B. Huntley. 2009. Assisted colonization in a changing climate: a test-study using two U. K. butterflies. Conservation Letters 2:45–51.

Wolf, C. M., T. Garland, Jr., and B. Griffith. 1998. Predictors of avian and mammalian translocation success: reanalysis with phylogenetically independent contrasts. Biological Conservation 86:243–255.

Zimmer, C. 2013. Bringing them back to life: the revival of an extinct species is no longer a fantasy. But is it a good idea? National Geographic Magazine. Available online at: http://ngm.nationalgeographic.com/2013/04/125-species-revival/zimmer-text (accessed September 26, 2013).

THREE

A Conservation Paleobiology Perspective on Reintroduction

CONCEPTS, VARIABLES, AND DISCIPLINARY INTEGRATION

R. Lee Lyman

REINTRODUCTION BIOLOGY (EWEN et al. 2012) occurs within a sociopolitical and economic context (Riley and Sandström, Chapter 4, this volume). This does not mean that plans to reintroduce an organism are not founded on sound biological principles and knowledge (Nygren and Rikoon 2008). Rather it means that the target parameters will likely not be specified solely by biological and ecological variables or principles and without influences of the particular sociopolitical context in which they are discussed (e.g., Ludwig et al. 2001, Holt 2005). Yet, conservation biology is, fundamentally and at heart, a science (e.g., Soulé 1985, Noss et al. 2009, Francis and Goodman 2010, Kareiva and Marvier 2012, or any issue of *Biological Conservation* or *Conservation Biology*). This means choices of conservation applications, including reintroductions, are supposed to be founded on the best available science. A rapidly growing part of the science that might inform conservation biology has over the past few years come to be referred to as conservation paleobiology (Dietl et al. 2015) (see Box 3.1 for a glossary of specialized terms). In short,

conservation paleobiology involves "putting the dead to work" (Dietl and Flessa 2011) as a source of information on ancient biological parameters, processes and outcomes that can inform and influence the plans of conservation biologists. As of yet, this does not happen often.

The uniquely long-term record of biological and ecological (hereafter, biological) statics and dynamics provided by paleontological and archaeological data reveals much that is not perceptible to direct human observation of organisms today interacting with one another and their environments. Even an individual researcher's ~40-year career and the two or three centuries of written records available for many local and regional ecological relationships are temporally limited from a paleobiological perspective. Paleobiological data are readily construed as the results of countless biological experiments (Deevey 1969). Although they often lack temporal resolution similar to the data resulting from observation of modern biological phenomena, paleobiological data more than make up for that in (i) their volume, (ii) the exceptionally long time spans they represent, and (iii) their

BENCHMARK (BASELINE, REFERENCE CONDITION) An ecological (usually static) condition, usually without human influence, to which current condition(s) are compared; often the target of conservation biology applications.

CONSERVATION PALEOBIOLOGY Application of the methods, theories, and data of paleontology to the conservation and restoration of biodiversity and ecosystem services (Dietl et al. 2015).

ETHNOGRAPHY An exhaustive description of a particular ethnic, social, or culturally distinct human group.

FIDELITY STUDIES Assessment of "the quantitative faithfulness of the [paleozoological] record of morphs, age classes, species richness, species abundance, trophic structure, etc. to the original biological signals" (Behrensmeyer et al. 2000, 120).

HISTORY The time when written records describing a human cultural group are available.

NATURAL RANGE OF VARIABILITY The non-anthropogenically influenced variation of ecological and evolutionary patterns and processes over time and across space.

NEOZOOLOGY Study of living animals, or recently (historically) deceased animals remains of which were collected as living organisms for zoological study.

PALEOBIOLOGY Study of ancient plant and animal remains using paleontological and biological methods and theories to learn about ancient biology.

PALEOECOLOGY Study of ancient (not necessarily extinct) plants and animals and their relationships to the ancient environments in which they lived (Fenton 1935, Cloud 1959, Ager 1979).

PALEOZOOLOGY Study of prehistoric animal remains, whether collected from archaeological (with associated artifacts) or paleontological (without associated artifacts) deposits.

PREHISTORY The time when no written records describing a human cultural group are available; >5,000 years everywhere in the world, but extends to more recent times in many areas.

TIME AVERAGING Accumulation of fossil material over a span of time such that material from multiple static states is included.

multiple and varied representations of the effects of biological dynamics.

In this chapter, I examine various aspects of reintroduction biology from a paleobiological perspective. I do so precisely because conservation biology and reintroduction efforts do not take place without biological justification. Because those efforts should be founded on the most complete biological knowledge available and should make biological sense, the unique data of paleobiology should not be ignored. With respect to reintroduction, a paleobiological perspective begs two questions. Can potential conceptual ambiguities of reintroduction as now conceived be identified from paleobiology's uniquely long temporal perspective? Which paleobiologically accessible variables should be consulted when evaluating whether or not an organism should be reintroduced to a particular area? I begin constructing answers to these questions by considering concepts within standard definitions of reintroduction. I then identify and discuss paleobiologically accessible variables that should inform reintroduction efforts. This is followed by a brief review of potential weaknesses of paleobiological data with respect to their relevance to reintroduction. Finally, I outline how paleobiology's contributions to conservation can be more thoroughly and consistently recognized and utilized by identifying means to better integrate the information provided by the former into the latter.

A PALEOBIOLOGICAL PERSPECTIVE ON KEY CONCEPTS

One definition of reintroduction is the artificial "intentional movement of an organism into a part of its native range from which it has disappeared or become extirpated in historic times" (Armstrong and Seddon 2008, 21; see also van Wieren 2012, Jørgensen 2013). Another definition of reintroduction is the IUCN Species Survival Commission's (2013, 3): "the intentional movement and release of an organism inside its indigenous range from which it has disappeared" (Chapter 2, this volume). These definitions provide a means to construct an answer to the question about concepts.

The goal of a reintroduction effort is to establish a viable, free-ranging population of a taxon in an area where it has become extirpated. The prefix *re* indicates that the taxon being released in a particular area occurred in that area sometime during the past. In light of the definitions of reintroduction cited above, the past time is *historical*, and the range to which a species is to be reintroduced is the species' *historic, native,* or *indigenous* range. Implications of the italicized words concern not only a time period, but also the habitat of reintroduction sites, and the inclusiveness of the taxonomic unit under consideration. Examples largely involve mammals, but my discussion applies to all taxa of organisms.

Historic(al) Time

The relevant range of a taxon is the "historical range." The terms "historic" and "historical" both concern a temporal coordinate(s) in the past. The time depth of "historical" could be nearly bottomless, as implied when the term is used as an adjective such as in "historical geology." "History" and "historic" often mean the era when written records are available (e.g., Trigger 1968), though this is not the case in research fields such as "earth history." However, archaeological research shows that writing systems appear in different geographic regions at different times, and as research continues, the date of the origin of particular writing systems extends farther into the past with collections of new data (ignoring for sake of simplicity the slippery question of the neces-

sary and sufficient conditions that make a series of anthropogenically created symbols and icons a writing system). In the Western Hemisphere, 3,000-year-old Mayan archaeological sites in Guatemala have produced evidence of a writing system; there is no such evidence in the contiguous 48 United States. In short, the term "history" and its derivatives (historical, historic) are sufficiently ambiguous as to provide little help with specifying a temporal coordinate. Does the historical era of, say, the state of Wisconsin bottom out at 100, 1,000, or 5,000 years? What about in Spain? Or Borneo?

If "historic" means the era when written records are available, in North America that specifies after 1492 CE (common era) when Christopher Columbus landed on its shores. There are at least three things to keep in mind when using this temporal boundary. First, it ignores the fact that the so-called historic era begins at one time on the East Coast of North America and centuries later on the West Coast. Second, written records can be useful when attempting to determine a historic range, but they may also be incomplete (not indicate a taxon was present in an area when in fact it was), incorrect (indicate a taxon was present in an area when in fact it was not), or ambiguous (be subject to diametrically opposite interpretations [see Lyman 2011b, for an example]). The paleobiological record can help resolve many cases when the historic record is unclear (Box 3.2).

In specifying a historical temporal era, authors may be specifically concerned with postindustrial revolution impacts on biotas (e.g., Houston and Schreiner 1995). What is implied by such a concern is the belief that preindustrial revolution peoples, including prehistoric peoples, had no significant impact on biotas. We have known for some time, based on historic and ethnographic data, that this is incorrect (e.g., Day 1953, Heizer 1955, Krech 1999). Decades of archaeological research indicate that prehistoric Americans also had far-reaching influences on biotas (e.g., Elder 1965,

Grayson 2001; see also the next subsection). In fact, those influences have been used to warrant Pleistocene rewilding (Donlan et al. 2006, Sandom et al. 2013), a proposed form of multiple taxa introduction, despite the weak archaeological evidence that people were in fact responsible for the extinction of North American Pleistocene megafauna (Grayson and Meltzer 2002, 2015, Wolverton 2010, Nagaoka 2012, Meltzer 2015).

We assume two things when choosing a temporal period. First, we assume stasis in an organism's range and physiological tolerances between that temporal period and today. The

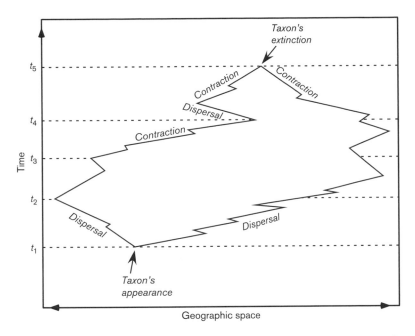

FIGURE 3.1 A simplified model of the spatiotemporal dynamics of a taxon's range. Both axes are ordinal scale. A benchmark range could be chosen at any time after t_1. Time t_3, for example, might signify the beginning of the historic era (when written records begin to be produced). Time t_4 might then represent the beginning of the industrial revolution.

paleozoological record indicates both parts of this assumption might be false (e.g., Faunmap Working Group 1996; Owen et al. 2000). Consider, for example, the model of a species' range over time shown in figure 3.1; this simple model exacerbates all of the problems with defining a temporal period as a benchmark to inform reintroductions. Second, the model in figure 3.1 also highlights that in choosing a temporal era as a benchmark, we assume that the chosen era is somehow more appropriate than another. The basis for deciding on appropriateness is obscure. This latter point is highlighted when we consider the adjectives "native" and "indigenous."

Native and Indigenous Range

Webster's *Third New International Dictionary* (1968) defines "indigenous" as "(i) not introduced directly or indirectly according to historical record or scientific analysis into a particular land or region or environment from the out-side; (ii) originating or developing naturally in a particular land or region or environment." Indigenous and native are synonyms, and I use the former henceforth. IUCN SSC (2013, 2) guidelines indicate the "indigenous range of a species is the known or inferred distribution generated from historical (written or verbal) records, or physical evidence of the species' occurrence." I presume "physical evidence" of a species' occurrence refers to collections of specimens (e.g., skulls, skeletons, and skins of mammals), including records of where the represented organisms were procured, compiled by naturalists over the past several centuries. Importantly, "physical evidence" should also be understood to include paleobiological remains (e.g., Peacock et al. 2012). This point has recently been indirectly, yet eloquently, made by Crees and Turvey (2015). They not only highlight the dynamic nature of species' movements and ranges, but also argue that because of such dynamics the concepts of native and nonnative species actually describe the ends of a continuum. That is, there are different degrees of "nativeness" of an organism that depend on such things as the length of the species' tenure in an area, the means by which the species gained entry to an area, and the like.

A paleobiological perspective highlights that geographic ranges of organisms are dynamic (figure 3.1); thus, we are faced with explaining exactly which indigenous range has been chosen and why. The implicit justification for the adjective indigenous is that it narrows the possible release sites to only those where the species once occurred because such sites increase the probability of (i) a successful reintroduction (a viable population is created) and (ii) avoiding the undesirable impacts of reintroduced species that become invasive (IUCN SSC 2013). Circumstantial evidence of such is found in the IUCN SSC's (2013, 8) caution that a species' "previous indigenous range may no longer be suitable habitat depending on ecological dynamics." Regardless of the correctness of my surmise, the wording presumes that we know not only the location of the boundaries of an organism's indigenous range, but also the causes of that range and those boundaries.

The adjective indigenous assumes there is a range that can be readily delineated; delineation is facilitated from either (presumed) stasis in range boundaries or choice of a range that occurs at some specified time (figure 3.1). The adjective also assumes that the range, because it is native or natural, has not been influenced by anthropogenic processes; the range is neither the function of human introductions nor the result of human-caused extirpations. The first assumption returns us to a point made previously—ranges are dynamic. The ranges of many species of both plants and animals shifted over the most recent geological era—the Quaternary—which only covers the last 2–2.5 million years, and the ranges of many of them were not static even during the Holocene epoch (the last ~11,000 years of the Quaternary) (e.g., Faunmap Working Group 1996, Grayson 2006, Lyman 2007, 2009). So, how do we choose an indigenous range of a taxon? Even narrowing

down the time limits to, say, after 1492 CE, we might, as noted above, find it difficult to determine an indigenous range given contradictory historical documents or contradictory interpretations thereof (Lyman 2011b).

Regarding the second assumption, we must contend with the potential that an organism's range was anthropogenically influenced rather than purely natural, and the anthropogenic influence may be prehistoric, or historic (say, for convenience in the Americas, post-1492 CE), or associated with the industrial age (e.g., Newsome et al. 2007). Paleobiologists can sometimes decipher whether a particular taxon's range was anthropogenically influenced or not, as well as determine when a range changed. Introductions are known to have taken place prehistorically, such as the human translocation of Channel Island foxes (*Urocyon littoralis*) to at least some of California's Channel Islands (Moore and Collins 1995, Rick et al. 2009), the weasel (*Mustela nivalis*) to Majorca Island in the western Mediterranean (Valenzuela and Alcover 2013), red deer (*Cervus elaphus*) to Ireland (Carden et al. 2012), and several mammalian species to islands of the West Indies (Giovas et al. 2012). If a taxon such as these is extirpated from a region where it once occurred as a result of prehistoric human translocation (figure 3.2), how does that alter our concept of an indigenous range? Should it?

What about a range that results only from prehistoric extirpations caused by people? Such might produce a range that is not completely indigenous and thus we overlook areas not occupied today as potential reintroduction loci simply because of our temporally biased modern perceptions of a range (figure 3.2). Paleobiological research can facilitate our efforts to determine the boundaries of an indigenous range, and it can also suggest something about the dynamics of range boundaries—how and when and why they shifted—and also the history of habitats within the range. Reintroduction areas should, as implied by the indigenous adjective, comprise habitats to which the species to be reintroduced is adapted (Osborne and Seddon 2012). A paleo-

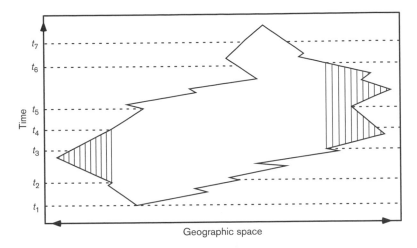

FIGURE 3.2 Examples of how prehistoric anthropogenic introductions and extirpations alter a taxon's range. At t_2 the taxon was anthropogenically introduced to a new area (left cross-hatched area), but at t_4 the taxon was extirpated from that area by non-anthropogenic causes. The taxon was also anthropogenically introduced to a new area (right cross-hatched area) at t_3 but was extirpated in that area by people at t_6. If the historic era begins at t_5, to which of the two cross-hatched areas should the taxon be reintroduced today? What if a benchmark range is chosen at t_3?

biological perspective on this variable is quite pertinent to reintroduction efforts.

Habitat

A suitable habitat is one in which the organism has a good chance to survive, reproduce, and establish a viable population (Dunham et al., Chapter 5, this volume). The area to which a taxon is reintroduced should be ecologically compatible with the physiological tolerances and ecological requirements of the organism (e.g., Brickner et al. 2014). Paleobiological information provides insight into what can be considered a natural habitat and the range of habitat conditions in which a species is likely to survive. The long temporal duration of the paleobiological record emphasizes as well that a suitable habitat is not a simple variable.

As environment (broadly construed) changes, an organism has three options (Gauthreaux 1980). It can be extirpated; it can geographically track favorable habitats; or it can adapt. Paleobiology might reveal the cause of a local extirpation event; knowing the identity of the cause might allow conservation biologists to ensure that that cause is not present in the reintroduction site under consideration. For example, using paleobiological data, Grayson (2005) showed that the range of pika (*Ochotona princeps*) in the Great Basin (Nevada, western Utah) had shrunk over the past 12,000 years or so as isolated populations in shrinking alpine habitats were extirpated. He also showed that the long-term climate-change-driven process of habitat loss had been exacerbated by modern anthropogenic processes. This demonstrated that natural habitat for pikas had been irreversibly depleted, and no good candidate areas for reintroduction remained. Lyman (2004a) made a similar case for an isolated remnant population of pygmy rabbit (*Brachylagus idahoensis*) in Eastern Washington State; based on paleobiological and historical data, minimal satisfactory habitat for potential reintroduction to enhance preservation of this genetically unique population is available.

Geographically tracking a taxon's shifting range over millennia might reveal ecological threats and suggest ecologically favorable conditions for the taxon. Recent trapping in areas trapped a century or so ago reveals upslope movement of some mammals (e.g., Rowe et al. 2010). Paleobiological data extend the time depth of such observations, reveal that taxa respond individualistically (some move upslope, some move downslope as a response to a particular climatic change), and can also reveal responses among taxa that were biogeographically static over the past century or two. For example, Grayson and Delpech (2005) show that the paleobiological record of caribou (*Rangifer tarandus*) indicates that during the late Pleistocene this taxon decreased in abundance in France as summer temperatures rose. Caribou became locally extirpated at the end of the Pleistocene when summer temperatures increased further. Although the range of this ungulate has remained more or less static over the past century or so, the paleobiological record suggests even northern populations might be in jeopardy as global warming continues, and that what seem to be suitable habitats for reintroduction today might not be suitable tomorrow.

Finally, paleobiological data might indicate how particular taxa adapted to past environmental change, thereby revealing how those taxa might respond to future changes or differences in environments faced when they are reintroduced to areas. For example, paleobiological data on black-footed ferrets (*Mustela nigripes*), a taxon thought to be an obligate predator of prairie dogs (*Cynomys* spp.) and thus reintroduced only where colonies of that sciurid are found (Brickner et al. 2014), indicate this intertaxonomic relationship might be a recent adaptation of ferrets (Owen et al. 2000). Another example is found in the fact that several North American mammals went through a period of diminution of body size at the end of the Pleistocene ice age. Taxa include wood rats (*Neotoma* sp.) (Smith and Betancourt 2003, Lyman and O'Brien 2005), bighorn sheep (*Ovis canadensis*) (Lyman 2009), elk (*Cervus elaphus*) (Lyman 2010d), and bison (*Bison* spp.) (Lyman 2004b, Hill et al. 2008). The point here is that reintroducing a taxon

into its former range could result in an unanticipated adaptive response such as diminution. Terminal Pleistocene diminution is thought to be a response to global warming; if in fact we are experiencing a more rapid, anthropogenically driven episode of global warming (e.g., Barnosky 2009), then we must be concerned about how reintroduced populations might respond (Jachowski et al., Chapter 9, this volume). If that were not enough to concern us, consider this: Although not a case of reintroduction, paleozoological data indicate the translocation of small-bodied Rocky Mountain elk (*C. e. nelsoni*) to the southern Cascade Mountains of Washington State in the 1930s led to diminution, likely as a result of hybridization with a small remnant population of resident large-bodied Roosevelt elk (*C. e. roosevelti*) (Lyman 2006a).

Of course, a reintroduced population might not survive. Any number of reasons might be responsible for such an outcome (Gitzen et al., Chapter 12, this volume). These include introduction of a (genotypically or phenotypically) poorly adapted population, or release in a stressful or unfavorable habitat. IUCN guidelines provide suggestions as to how to enhance the success of reintroductions, and some of these center around the biological or taxonomic units selected for reintroduction, yet another variable upon which paleobiological research might shed some light.

Taxonomic Units

The IUCN guidelines use the terms "species" and "organisms" but indicate that these terms signify "species, subspecies or lower taxon" (IUCN SSC 2013, 1). The founder species to be reintroduced "should show characteristics based on genetic provenance, and on morphology, physiology and behavior that are assessed as appropriate through comparison with the original or any remaining wild populations" (IUCN SSC 2013, 8). If the original taxon has become extinct, then "a similar, related species or subspecies can be substituted as an ecological replacement, provided the substitution is based on objective criteria such as phylogenetic closeness, similarity in appearance, ecology and behavior to the extinct form" (IUCN SSC 2013, 9) (Biebach et al., Chapter 8, this volume).

In paleobiology, taxonomic identification is traditionally and usually done by direct morphometric comparison of ancient skeletal remains of unknown taxonomy with homologous remains of known taxonomy curated in a comparative or reference collection (Lyman 2010b). When a fossil is morphometrically matched with a taxonomic known, the former is "identified" as a taxonomic member of the latter (Lyman 2002). More recently, extraction of ancient DNA from prehistoric remains has allowed fine-scale taxonomic (subspecies or population) resolution (e.g., Carden et al. 2012), and in some cases this work has significant implications for reintroduction biology (Box 3.3).

Although the biological units referred to in conservation biology, including the IUCN reintroduction guidelines, often are phrased in terms of biological species, genetics or gene pools, and morphology (phenotype), two aspects of studying the mortal remains of, particularly, prehistoric animals warrant comment. First, those remains sometimes reveal previously undocumented phenotypes that are of potential significance to reintroduction plans (e.g., Lyman 1988, 2006a). Given that natural selection focuses on phenotypes, individuals chosen from a population for reintroduction should not only be taxonomically and genetically similar to the population they are meant to recreate, but also be phenotypically similar to the originals (Box 3.3). The second aspect of studying prehistoric remains is that the paleopathology of animal populations as reflected in skeletal remains could reveal aspects of long-term health within particular populations ancestral to the population under consideration as a source of individuals for translocation. Study of paleopathology ranges from individual skeletal ailments (e.g., Lyman 2010a) to population-wide pathologies (e.g., Byerly 2007).

Both could reveal ecological variables of concern to reintroduction plans, although one must keep in mind the so-called osteological paradox—that an organism must survive an affliction long enough to develop skeletal signatures rather than die shortly after infection, else the proportion of pathological specimens in a collection, and inferred health of a population, will be misleading (Wood et al. 1992, DeWitte and Stojanowski 2015).

Reintroduction and its associated criteria presume we know much about the particular ecological and physiological tolerances of individual taxa that are under consideration for reintroduction (IUCN SSC 2013, 8). We often do not know nearly as much as we might hope about a particular species, a problem long recognized in paleobiology (Harris 1963, Findley 1964) and increasingly recognized among conservation biologists who hope to forecast the likelihood of success of various applications (Mawdsley et al. 2009). Gaps in our knowledge of a taxon's ecology can sometimes be filled with paleobiological data (e.g., Owen et al. 2000, Lyman 2011a).

Using biological units that are genetically, taxonomically, and morphometrically close to the population to be recreated is preferred over substitute or surrogate biological units because the former were built by local natural selection forces and should thus be more successful because they are better adapted to (more fit within) a chosen reintroduction site than a different biological unit. But what if a sufficiently fine-scale biological unit (e.g., subspecies) is not available for reintroduction? Paleobiology might suggest appropriate substitute units, ones that are ecologically equivalent to the unit we hope to reestablish by reintroduction. For example, Hadly (1997) and Hadly et al. (1998) showed that northern pocket gophers (*Thomomys talpoides*) varied phenotypically in response to climate change (small bodied during warm episodes, large bodied during cool episodes), but not genetically over the last 3,000 years. This suggests plans to reintroduce northern pocket gophers to an area should place more

Written records unambiguously indicate sea otters (*Enhydra lutris*) were found historically on coastlines from the Bering Sea south along the eastern Pacific Rim to southern California (Kenyon 1969). Commercial exploitation of this species was one reason that Russian fur traders enslaved native Aleuts; the latter were expert hunters of this valuable fur commodity (Ogden 1933). The species was extirpated from midlatitude portions of its range in the nineteenth century. During the mid-twentieth century there was some dispute as to whether one subspecies occupied northern latitudes of North America and a second subspecies made up a small population in waters adjacent to southern California (Scheffer and Wilke 1950, Roest 1973, 1979, Davis and Lidicker 1975). For a time, consensus was that extant individuals represented a single subspecies (Kenyon 1982). Clinal variation from north to south within the taxon was recognized even then; eventually that variation was studied and it was concluded that an Asian form (*E. l. lutris*), a northern form (*E. l. kenyoni*), and a southern or California (*E. l. nereis*) subspecies exist (Wilson et al. 1991).

A reintroduction effort took place in the late 1960s and early 1970s; it had mixed success (Jameson et al. 1982, 1986; Watson et al. 1997). Northern sea otters from Alaskan waters were translocated to the British Columbia, Washington, and Oregon coasts. Otters transplanted to Vancouver Island and the northern Washington coasts survived and the population(s) grew. The 100 or so otters released on the Oregon coast were gone within a few years. Reasons for the failure of the reintroduction effort along the Oregon and southern Washington coasts were unclear (Jameson et al. 1986).

Paleozoological remains indicate sea otters occurred on the Oregon and Washington coasts and along the northeastern Pacific Rim for several thousand years (Lyman 1988). Prehistoric remains of sea otters also provide information that the historic documents do not. Teeth and mandibles of prehistoric sea otters from the Oregon coast fall morphometrically between historic remains of sea otters from Alaska and from California (Lyman 1988). They fit the cline documented with historic specimens from Alaska and California, but they also fill a significant geographic gap (Roest 1973; Wilson et al. 1991). The morphometric differences suggest one reason for the limited success of the reintroduction efforts may reside in differences in mastication efficiencies (Riley 1985, Lyman 1988).

Recent extraction of ancient DNA from some late prehistoric remains of sea otter recovered from the Oregon coast shows the ancient sea otters were closer genetically to the southern or California sea otter than to the northern form (Valentine et al. 2008). Recall that it was members of the latter subspecies that provided the individuals used in the reintroduction effort. Both the ancient morphometric and the ancient genetic data suggest that the northern sea otter was not the best population from which to capture individuals for the reintroduction effort. This is not to argue that individuals of the California subspecies should be translocated to the Oregon and southern Washington coasts; its population is too small to provide such. The sea otter story indicates that study of paleo-remains—both macroscopically (morphometrically, phenotypically) and microscopically (genetically)—can guide choice of a source population for individuals to be reintroduced.

weight on the correct phenotype than on the extirpated genotype or taxon.

PALEOBIOLOGICAL DATA

The volume of conservation biology literature has grown markedly in the past 20 years. Numerous journal articles, book chapters, and several edited volumes devoted to the topic have been published (e.g., Lauwerier and Plug 2004, Lyman and Cannon 2004, Dietl and Flessa 2009, Louys 2012, Wolverton and Lyman 2012). Together these studies go far toward identifying which biological variables of concern to conservation biologists can be accessed through collection and analysis of paleobiological data (see reviews in Conservation Paleobiology Workshop 2012, Lyman 2012, Dietl et al.

2015, Kidwell 2015). Much of this section is derived from that literature, and the interested reader is referred to it for additional details.

Accessible Variables

Many paleobiological variables are relevant to planning reintroductions. Perhaps the most obvious one is utilizing the paleobiological record to determine which taxonomic units (species, subspecies, morphotypes) were naturally present in a region (Crees and Turvey 2015). In some cases, paleobiological data have shown that species once thought to have been invasive or exotic were in fact part of the indigenous biota. Paleobiological data facilitate identification of a species' range during a particular era. As Gill et al. (2015) note, paleobiology highlights the dynamic nature of ecosystems, landscapes, and biotas, and given its deep-time perspective, it underscores the fragility of the biological world. Both are critically important to reintroduction biology because, as Dietl et al. (2015) note, biologists, including conservationists, seem trapped in the shifting baseline syndrome (Pauly 1995); that is, successive generations perceive different biotas and ecologies. Thus, although shifting baselines demonstrate biological dynamism, they make deciphering which fluctuations are anthropogenic and which are natural difficult (Parmesan 2006). Paleobiological data often can help distinguish these two kinds of causes (but see Pauli et al. 2015). Such data also help define the natural preferred habitat or niche of a species whose modern range is the result of postindustrial anthropogenic causes and thus not completely natural (e.g., Flueck and Smith-Flueck 2012).

The unprecedented time depth of the paleobiological record reveals biological patterns and processes not perceptible in the limited lifespan of a single or even several consecutive researchers (Dietl and Flessa 2011). For example, those long paleo-time spans reveal the normal range of variation in habitats, biotas, and ecosystems; knowing the parameters of that range, we can monitor modern settings and

identify when fluctuation exceeds normal conditions and thus infer that it is time to take conservation action (Hadly and Barnosky 2009).

Another way to phrase the preceding discussion is worthy of consideration here. It originates in a question posed by stratigraphic geologist Derik Ager (1981,55): "Is the present a long enough key to penetrate the deep lock of the past?" Ager was considering the two-century-old principle of uniformitarianism, typically phrased as "the present is the key to the past." He was making the point that ontological uniformitarianism constrained geologists' analytical tool kit to gradualistic change and to the kinds and magnitudes of processes that had been observed in operation in the present. From such a standpoint, consulting the present as a source of actualistic analogs used to interpret extant results in terms of their ancient causes was restrictive for the simple reason that the "present"—however defined—was temporally limited. The significance of Ager's question here is found when it is rephrased to read: "Is the observation of biological phenomena today sufficiently long to more or less accurately predict the biological future?" From Ager's perspective, the answer is "No." This answer can become positive by consulting the extensive temporal record provided by paleobiology from the standpoint of methodological uniformitarianism, that is, our insights into what happened in the past are qualified, high-probability possibilities (Simpson 1963, 1970, Gould 1965, Baker 2014).

This does not mean the paleobiological record is of little value in a conservation context, as some would have it (e.g., White and Walker 1997, Rackham 1998, Morrison 2001). Rather it means due caution is called for because our knowledge of the past is probably, not definitively, correct. This is the so-called paleo conundrum—that knowledge of the past can never be verified or shown to be true (Baker 1999, Cleland 2011, Biondi 2013). Even so, the insights provided by the paleobiological record are valuable in conservation contexts when founded on methodological uniformitarianism—a particular epistemology. For instance, the paleobiological record includes such things as "no-analog" biotas (Williams and Jackson 2007), the likes of which the present fails to have (hence the adjective "no-analog") but which may result in the future from anthropogenic processes (Young 2014). Such biotas were ontologically unexpected until recently, yet their paleoecological significance is readily interpreted from an epistemological perspective (e.g., Graham and Mead 1987, Semken et al. 2010). Paleo-data do have their weaknesses, as do various sorts of analyses of those data, but recognition of those limitations is what prompts thoughtful analyses, particularly in applied cases such as conservation paleobiology (e.g., Jackson et al. 2001).

Paleobiological data reveal how species respond to different kinds and degrees of climatic change and thereby contribute to forecasting the success (or failure) of a planned reintroduction confronted with a high likelihood of environmental flux (Hayward 2009). Such revelations can be used to forecast how a species might respond to future environmental changes of particular kinds (Dietl et al. 2015). Paleobiological data also can provide tests of predicted ecological outcomes of reintroductions and thereby enhance our chances of not releasing a species in an area where it initiates adverse ecological impacts (Conservation Paleobiology Workshop 2012; Dietl et al. 2015). Paleobiological data allow "reverse monitoring" (Jackson and Hobbs 2009) of the ecological consequences of a reintroduction event.

Paleobiological data reveal the variety of ways that particular taxa respond to climatic flux or the immigration of other species. Sometimes a paleobiologically documented immigrant species will become locally extirpated, sometimes it will alter its mean adult body size, and sometimes it will increase in abundance. Importantly, the interdisciplinary exchange between paleobiology and biology is a two-way street; the former uses data from biology to interpret remains from the past and paleobiological data provide tests of and means to calibrate species niche models and species distribution models for the future (McGuire and Davis 2014). Together, all of the facts noted in the preceding paragraphs address Converse et al.'s (2012, 1145) concern that "almost by definition, serious scientific uncertainty is an issue in reintroduction efforts because the species is being reintroduced into an environment that it does not current occupy so the ability to understand fundamental aspects of the ecology and management of a reintroduced species is limited."

Weaknesses

Three potential weaknesses of paleobiological data are particularly worthy of mention with respect to using those data to plan reintroductions. First, paleobiologists cannot always identify the ancient remains of plants or animals to species (let alone subspecies), and sometimes we cannot identify those materials even to the genus level based on morphometric features of bones and teeth (Driver 1992, 2011, Lyman 2002). Further, incorrect identifications have appeared in the literature (e.g., Grayson 1977). Analysis of ancient DNA preserved in prehistoric remains, as noted above, can facilitate correct identification as well as sometimes provide subspecies-level identifications (Box 3.3).

Second, paleobiological data are asymmetrical with respect to biogeography. The presence of ancient remains of a taxon in an area is often good evidence that the taxon was once present there, but the absence of remains of a taxon from an area does not necessarily mean the taxon never occurred there (Grayson 1981). Remains of a taxon might be absent from the known prehistoric record as a result of sampling error, inadequate recovery, poor preservation, or failure to correctly identify the remains (Lyman 2008). Knowing beyond significant doubt on the basis of prehistoric remains that a taxon was present in an area often requires detailed and intensive analysis and study (e.g., Peacock et al. 2012). The same concerns attend the use of historic data, whether written or oral (e.g., Howell and Prevenier 2001).

Finally, paleobiological data might provide distorted reflections of past communities. A collection of remains of organisms might have relatively fine temporal resolution, yet derive from more than one community and thus represent gamma (regional) diversity (e.g., Andrews 2006). Or, a collection of biological remains might represent several communities deposited over a climatically dynamic period of time ranging from a century to several millennia. The term "time averaged" has been used to label this phenomenon in paleoecology (Peterson 1977, Behrensmeyer 1982). Coarse temporal resolution will, of course, translate to coarse paleoecological resolution, so paleontologists have devoted some effort to determining how to analytically deal with time-averaged collections (e.g., Kowalewski 1996, Olszewski 1999, Kidwell and Tomasovych 2013).

The scales of both temporal and spatial sampling of a biota have long been recognized by paleontologists (e.g., Shotwell 1955). One technique paleontologists have used to contend with these and other aspects of collections of ancient remains that render them more or less poor reflections of ecosystems, habitats, communities, and the like involves "fidelity studies" (see Box 3.1) which suggest the limits of paleobiological resolution that might be attained with a particular collection. Fidelity studies involve comparison of the taxonomic composition and abundances of modern communities as reflected by recent remains of organisms, with the remains of prehistoric communities. These and similar methods are regularly used by paleobiologists in general and by paleobiologists interested in conservation biology in particular (e.g., Terry 2010, Miller 2011, Kidwell 2013). Claims that the paleobiological record is unworthy of consideration in the context of conservation biology are ill informed (Lyman 2012), which is not to say that that record is a clear reflection of ancient ecology or that it is easily interpreted. The preceding paragraphs make it clear that it is neither.

Paleobiology Informing Reintroduction

Conservation paleobiology is so new that there are very few instances of a conservation biology application having incorporated paleobiological data from the planning stages through subsequent monitoring. Many instances involve learning that paleobiological data provide a warrant for a particular reintroduction after the reintroduction event has taken place or that such data might have prompted a different plan to be implemented had those data been reviewed during the planning stage. For instance, had paleobiological data been consulted (unlikely in the 1920s as paleobiological research was not at the time often undertaken) prior to the 1930s reintroduction of elk (*Cervus elaphus*) into the southern Cascade Mountains of Washington, we might have larger-bodied individuals than are found there today (Lyman 2006a). The 2011–2012 reintroduction of elk to southern Missouri was not informed by paleobiological data; in fact, the release site was in a county with neither a historic nor a prehistoric record of elk (Harpole 1994). Historic data indicated elk had occupied parts of the state into the nineteenth century. It remains to be seen how this particular reintroduction turns out.

A high-profile instance in which paleobiological data were used as part of the warrant for a reintroduction was the transplanting of wolves (*Canis lupus*) into the Yellowstone ecosystem. This was a particularly complex instance of political ecology (e.g., Fischer 1995) that has played out well in some respects (e.g., Kimble et al. 2011) but not so well in others (e.g., Niemeyer 2007, Bruskotter et al. 2011, Wielgus and Peebles 2014). Because it was so politically charged, all sorts of evidence had to be mustered to build a strong argument that this reintroduction was ecologically, economically, and politically wise (Varley and Brewster 1992). The paleobiological evidence (Cannon 1992) was an important but relatively small part of the warranting argument, and seems to so far have had little role in post-reintroduction monitoring.

A less well-known instance wherein paleo-biological data were used as part of a warrant concerns the recent reintroduction of wood bison (*Bison bison athabascae*) into southern Alaska (Engelking 2015, Schnuer 2015). Extensive paleontological and some zooarchaeological specimens, in conjunction with extensive interviews with native peoples, indicate wood bison were present in much of Alaska until about the end of the nineteenth century, but were rare prior to that time (Stephenson et al. 2001). Paleobiological data are also being consulted in conjunction with other data as conservationists consider reintroducing bison into the backcountry of Banff National Park in the Rocky Mountains of Western Canada (Kay and White 2001, White et al. 2001, Langemann 2004). News releases in March 2015 indicate the Canadian government has announced an initiative to financially invest in this planned reintroduction (http://www.pc.gc.ca/eng/pn-np /ab/banff/plan/gestion-management/bison .aspx). Paleobiologically informed planned reintroductions on the Hawai'ian islands include several species of birds, a crab, snails, and insects (Burney and Burney 2007).

In some cases, it seems that paleobiological data have been consulted after a decision to reintroduce has been made, in which case those data are used in such a way as to suggest they are a post hoc warrant for a particular release area (Box 3.2). Paleobiological data have not been consulted, so far as I know, in several cases of what have been categorized as invasions. These include the immigration of nonnative mountain goats (*Oreamnos americanus*) from Montana to Yellowstone National Park (Laundré 1991, Lemke 2004). Only after the immigration had been documented was the paleobiological record consulted (Schullery and Whittlesey 2001). Similarly, efforts by the US National Park Service to eradicate what they believe to be exotic mountain goats from Olympic National Park did not consult the paleobiological record in any rigorous fashion until the question of what the paleontological record might show was raised (Houston et al. 1994,

Lyman 1998). Paleobiological evidence is limited in boreal and alpine habitats in part because little archaeological and paleontological work has been done in such areas, but as more research is done in such habitats, biogeographic surprises are found, including ancient remains of mountain goats in locations where the written and oral historic records indicate this species did not occur (e.g., Nagorsen and Keddie 2000). It is thus gratifying to occasionally see proactive research, such as the case in which an intensive search of the historic and paleobiological records revealed circumstantial evidence that mountain goats were native to the central Cascade Mountains of Oregon (Matthews and Heath 2008); goats were, literally, *re*introduced to this area in 2010.

INTEGRATING PALEOBIOLOGY INTO CONSERVATION BIOLOGY

Suding and Leger (2012, 282) argue that "historical [and paleobiological] perspectives increase our understanding of the dynamic nature of landscape and provide a frame of reference for assessing modern patterns and processes" (see also Szabó 2010, Szabó and Hédl 2011). As the preceding section makes clear, such statements are not always heeded by conservationists (e.g., Pooley 2013; Szabó and Hédl 2013). Paleobiologists who have argued their data should become regularly used in conservation biology are of mixed opinions as to how commonly and how well those data are in fact integrated with conservation biology. Gillson and Marchant (2014, 318–319) believe that "mainstreaming paleoecological work into adaptive management is still in the early stages." McGuire and Davis (2014) are more positive but consider only the use of paleobiological data to calibrate species niche models such that those models accurately predict species distributions. That paleobiological data should be integrated into all forms of conservation biology seems indisputable. Paleobiologists should be consulted from the start of the planning process for reasons that I have described above; less obviously, they should be included in post-reintroduction monitoring to help assess why successes are successful and why failures are not. The key question therefore is: How should the integration of paleobiological data into conservation biology be encouraged and facilitated?

Gillson and Marchant (2014) outline several means to facilitate integration. Conservation biologists must have transparent management goals and policies (Gillson and Marchant 2014). Goals of a conservation application must be founded on the past range of variability and the best possible scientific data. Paleobiologists must translate their data into forms that are accessible and useful to conservation biologists. Paleobiologists must keep in mind that we might study the paleobiological record to learn *about* the past but, in the context of conservation paleobiology, we study the past in order to learn *for* the future (van der Leeuw 2014). More specifically, as Dietl et al. (2015) indicate, paleobiologists must specify the "deliverables" with which they can provide conservation biologists (e.g., Lyman 1996, 2006b, 2012). They must directly address conservation issues with their data and do so in policy-relevant terms (see various chapters in Lyman and Cannon 2004, Wolverton and Lyman 2012). Paleobiologists "need to find innovative ways of bridging the gap between human well-being, policy, management, and long-term data" (Gillson and Marchant 2014, 321). An obvious way to do this is to publish in appropriate venues; another article on conservation paleobiology in *Journal of Paleontology* or *Paleobiology* will not reach the appropriate audience. Articles should appear in *Conservation Biology* and similar venues where conservation biologists are likely to encounter them (e.g., Wolverton et al. 2007, Peacock et al. 2012). As conservation biologists become more exposed to and thus aware of conservation-oriented paleobiological research, they will come to recognize the value of that research and incorporate it into all stages of all conservation applications.

FUTURE RESEARCH AND DEVELOPMENT

Conservation paleobiology is a relatively new discipline that is a hybrid of biology, paleobiology, and conservation biology. In order for conservation biology to continue to evolve, funding sources must evolve with it. Conservation paleobiologists recently appealed to the National Science Foundation to invest in "infrastructure to further develop analytical techniques and theory" and to support "new cross-disciplinary educational and research opportunities for early-career research scientists and practitioners" (Conservation Paleobiology Workshop 2012, 27). They called for a decade-long initiative. This is clearly a step in the right direction.

Another step that must be taken concerns the structure of advanced education. Given the hybrid nature of conservation paleobiology, cross-disciplinary training far beyond traditional discipline-specific academic departments must occur (Conservation Paleobiology Workshop 2012). This likely will require some rethinking of acceptable courses of study within the typically narrow focus of many departments. Perhaps the greatest hurdle here will be providing sufficient training in both the sciences (biology, ecology, geology, paleontology) and the humanities and social-behavioral sciences (anthropology, history, sociology) while satisfying degree requirements of a student's home department (Szabó and Hédl 2013). Flexibility in accepting a course of study that comprises professional training would, it seem, be mandatory. In my own university, I have colleagues housed in academic departments of anthropology, fisheries and wildlife, geology, and history who have published research that includes mention and even use of conservation paleobiology. Our graduate students interested in conservation paleobiology must fulfill general graduation requirements but also complete sufficient training in several disparate academic departments and colleges. This can require more time (and finances) than many students are willing to commit. This must change if conservation paleobiology is to develop.

SUMMARY

A paleobiological perspective provides a critically important way to explore the implications of some of the variables and concepts attending reintroduction biology. Such a perspective is mandatory to reintroduction biology because both concern ecology and biology, and both concern long durations of time, paleobiology the past, reintroduction biology the future. Biological data recorded in the past couple hundred years might have weaknesses when it comes to using them as the sole foundation on which to base arguments for a particular reintroduction. Paleobiological data also have weaknesses, but in many cases these data on ancient ecosystems, biotas, and habitats are precisely what is needed to better inform reintroduction plans.

Reintroduction biology is a science. As conceived, it entails various conceptual ambiguities that are highlighted by the time depth of a paleobiological perspective. Paleobiological research can provide data on variables important to reintroduction efforts such as an appropriate source population, appropriate release sites, and suitable habitats. Paleobiological data have seldom been used as a strong warrant for a reintroduction; most often they play a role in post hoc justifications. Conservation biologists must learn to consult pertinent paleobiological research beginning in the early planning stages and extending through post-conservation application monitoring. Conservation paleobiology must be more thoroughly integrated into higher education and establish or identify sources of funding for basic research.

MANAGEMENT RECOMMENDATIONS

- Choosing an appropriate taxon (species, subspecies, race, gene pool, phenotype) as an appropriate source for individuals to be reintroduced might be difficult. Paleobiological (morphometric, ancient DNA) data can help determine an appropriate source taxon and population.

- Biogeographic ranges are temporally dynamic. Paleobiological data can help define a range for a selected time period.

- Written records might not exist or be ambiguous as to the historic range of an organism, or the taxon or phenotype present in an area. Paleobiological data can help define a benchmark range and the taxonomic composition of ancient communities.

- Ranges might have been influenced prehistorically by human introductions and extirpations. Paleobiological data can help specify an indigenous range.

- Knowledge of an organism's ecological tolerances might be limited. Paleobiological data can help fill knowledge gaps.

ACKNOWLEDGMENTS

Thanks to Josh Millspaugh for inviting me to contribute to this volume, to David Jachowski for an unmerciful critique of an early draft, and to an anonymous reviewer, Paul Angermeier, Virginia Butler, and D. Jachowski for comments on a second draft. My parents long ago taught me to appreciate and enjoy natural ecosystems. The shoe is now on the other foot; I have grandchildren who I hope can at some distant future time still enjoy the wild places. This chapter is for all generations of my lineage.

LITERATURE CITED

Ager, D. V. 1979. Paleoecology. Pp.530–541. In R. W. Fairbridge and D. Jablonski, eds., Encyclopedia of Paleontology. Dowden, Hutchinson and Ross, Inc., Stroudsburg, PA.

Ager, D. V. 1981. The Nature of the Stratigraphical Record. 2nd ed., John Wiley and Sons, New York.

Andrews, P. 2006. Taphonomic effects of faunal impoverishment and faunal mixing. Palaeogeography, Palaeoclimatology, Palaeoecology 241:572–589.

Armstrong, D. P., and P. J. Seddon. 2008. Directions in reintroduction biology. Trends in Ecology and Evolution 23:20–25.

Baker, V. R. 1999. Geosemiosis. Geological Society of American Bulletin 111:633–645.

Baker, V. R. 2014. Uniformitarianism, earth system science, and geology. Anthropocene 5:76–79.

Barnosky, A. D. 2009. Heatstroke: Nature in an Age of Global Warming. Island Press, Washington, DC.

Behrensmeyer, A. K. 1982. Time resolution in fluvial vertebrate assemblages. Paleobiology 8:211–227.

Behrensmeyer, A. K., S. M. Kidwell, and R. A. Gastaldo. 2000. Taphonomy and paleobiology. Paleobiology 26:S103–S147.

Biondi, F. 2013. The fourth dimension of interdisciplinary modeling. Journal of Contemporary Water Research and Education 152:42–48.

Booth, E. S. 1947. Systematic review of the land mammals of Washington. Dissertation, State College of Washington, Washington State University, Pullman.

Brickner, K. M., M. B. Grenier, A. E. Crosier, and J. N. Pauli. 2014. Foraging plasticity in a highly specialized carnivore, the endangered black-footed ferret. Biological Conservation 169:1–5.

Bruskotter, J. T., S. A. Enzler, and A. Treves. 2011. Rescuing wolves from politics: wildlife as a public trust resource. Science 333:1828–1829.

Burney, D. A., and L. P. Burney. 2007. Paleoecology and "inter-situ" restoration on Kaua'i, Hawai'i. Frontiers in Ecology and Environment 5:483–490.

Byerly, R. M. 2007. Palaeopathology in late Pleistocene and early Holocene Central Plains bison: dental enamel hypoplasia, fluoride toxicosis and the archaeological record. Journal of Archaeological Science 34:1847–1858.

Cannon, K. P. 1992. A review of archeological and paleontological evidence for the prehistoric presence of wolf and related prey species in the Northern and Central Rockies physiographic provinces. Pp.1-175–1-265. In J. D. Varley and W. G. Brewster, eds., Wolves for Yellowstone? A Report to the United States Congress. Vol. IV. Research and Analysis. National Park Service, Yellowstone National Park, Wyoming.

Carden, R. F., A. D. McDevitt, F. E. Zachos, P. C. Woodman, P. O'Toole, H. Rose, N. T. Monaghan, M. G. Campana, D. G. Bradley, and C. J. Edwards. 2012. Phylogeographic, ancient DNA, fossil and morphometric analyses reveal ancient and modern introductions of a large mammal: the complex case of red deer (*Cervus elaphus*) in Ireland. Quaternary Science Reviews 42:74–84.

Cleland, C. E. 2011. Prediction and explanation in historical natural science. British Journal for the Philosophy of Science 62:551–582.

Cloud, P. E., Jr. 1959. Paleoecology: retrospect and prospect. Journal of Paleontology 33:926–962.

Conservation Paleobiology Workshop. 2012. Conservation Paleobiology: Opportunities for the Earth Sciences. Report to the Division of Earth Sciences, National Science Foundation. Paleontological Research Institution, Ithaca, NY.

Converse, S. J., C. T. Moore, M. J. Folk, and M. C. Runge. 2012. A matter of tradeoffs: reintroduction as a multiple objective decision. Journal of Wildlife Management 77:1145–1156.

Crees, J. J., and S. T. Turvey. 2015. What constitutes a "native" species? Insights from the Quaternary faunal record. Biological Conservation 186:143–148.

Davis, J., and W. Z. Lidicker, Jr. 1975. The taxonomic status of the southern sea otter. Proceedings of the California Academy of Sciences 40:429–437.

Day, G. M. 1953. The Indian as an ecological factor in the northeastern forest. Ecology 34:329–346.

Deevey, E. S. 1969. Coaxing history to conduct experiments. Bioscience 19:40–43.

DeWitte, S. N., and C. M. Stojanowski. 2015. The osteological paradox 20 years later: past perspectives, future directions. Journal of Archaeological Research 23:397–450.

Dietl, G. P., and K. W. Flessa, eds. 2009. Conservation Paleobiology: Using the Past to Manage for the Future. Paleontological Society Papers, Vol. 15. Boulder, CO.

Dietl, G. P., and K. W. Flessa. 2011. Conservation paleobiology: putting the dead to work. Trends in Ecology and Evolution 26:30–37.

Dietl, G. P., S. M. Kidwell, M. Brenner, D. A. Burney, K. W. Flessa, S. T. Jackson, and P. L. Koch. 2015. Conservation paleobiology: leveraging knowledge of the past to inform conservation and restoration. Annual Review of Earth and Planetary Sciences 43:79–103.

Donlan, C. J., J. Berger, C. E. Bock, J. H. Bock, D. A. Burney, J. A. Estes, D. Foreman et al. 2006. Pleistocene rewilding: an optimistic agenda for twenty-first century conservation. American Naturalist 168:660–681.

Driver, J. C. 1992. Identification, classification and zooarchaeology. Circaea 9(1):35–47.

Driver, J. C. 2011. Identification, classification and zooarchaeology (and comments). Ethnobiology Letters 2:19–39.

Elder, W. H. 1965. Primeval deer hunting pressures revealed by remains from American Indian middens. Journal of Wildlife Management 29:366–370.

Engelking, C. 2015. Wood bison roam the U. S. for the first time in a century. Discovery Magazine, March. Available online at: http://blogs. discovermagazine.com/d-brief/2015/03/23/ wood-bison-u-s-return/#.VWYmzaYnrBs.

Ewen, J. G., D. P. Armstrong, K. A. Parker, and P. J. Seddon, eds. 2012. Reintroduction Biology: Integrating Science and Management. Wiley-Blackwell, Hoboken, NJ.

Faunmap Working Group. 1996. Spatial response of mammals to late Quaternary environmental fluctuations. Science 272:1601–1606.

Fenton, C. L. 1935. Viewpoints and objectives of paleoecology. Journal of Paleontology 9:63–78.

Findley, J. S. 1964. Paleoecological reconstruction: vertebrate limitations. Pp.23–25. In J. J. Hester and J. Schoenwetter, eds. The Reconstruction of Past Environments: Proceedings. Publication 3. Fort Burgwin Research Center, Taos, NM.

Fischer, H. 1995. Wolf Wars: The Remarkable Inside Story of the Restoration of Wolves to Yellowstone. Falcon Press, Helena, MT.

Flueck, W. T., and J. A. M. Smith-Flueck. 2012. Huemul heresies: beliefs in search of supporting data: 1. Historical and zooarcheological considerations. Animal Production Science 52:685–693.

Francis, R. A., and M. K. Goodman. 2010. Post-normal science and the art of nature conservation. Journal for Nature Conservation 18:89–105.

Gauthreaux, S. A., Jr. 1980. The influences of long-term and short-term climatic changes on the dispersal and migration of organisms. Pp.103–174. In S. A. Gauthreaux, ed., Animal Migration, Orientation and Navigation. Academic Press, New York.

Gill, J. L., J. L. Blois, B. Benito, S. Dobrowski, M. L. Hunter, Jr., and J. L. McGuire. 2015. A 2.5-million-year perspective on coarse-filter strategies for conserving nature's stage. Conservation Biology 29:640–648.

Gillson, L., and R. Marchant. 2014. From myopia to clarity: sharpening the focus of ecosystem management through the lens of palaeoecology. Trends in Ecology and Evolution 29:317–325.

Giovas, C. M., M. J. LeFebvre, and S. M. Fitzpatrick. 2012. New records for prehistoric introduction of neotropical mammals to the West Indies: evidence from Carriacou, Lesser Antilles. Journal of Biogeography 39:476–487.

Gould, S. J. 1965. Is uniformitarianism necessary? American Journal of Science 263:223–228.

Graham, R. W., and J. I. Mead. 1987. Environmental fluctuations and evolution of mammalian faunas during the last deglaciation in North America. Pp.371–402. In W. F. Ruddiman and H. E. Wright, Jr., eds., North America and Adjacent Oceans during the Last Deglaciation. Geology of North America, Vol. K-3. Geological Society of America, Boulder, CO.

Grayson, D. K. 1977. A review of the evidence for early Holocene turkeys in the northern Great Basin. American Antiquity 42:110–114.

Grayson, D. K. 1981. A critical view of the use of archaeological vertebrates in paleoenvironmental reconstruction. Journal of Ethnobiology 1:28–38.

Grayson, D. K. 2001. The archaeological record of human impacts on animal populations. Journal of World Prehistory 15:1–68.

Grayson, D. K. 2005. A brief history of Great Basin pikas. Journal of Biogeography 32:2103–2111.

Grayson, D. K. 2006. The late Quaternary biogeographic histories of some Great Basin mammals (western USA). Quaternary Science Reviews 25:2964–2991.

Grayson, D. K., and F. Delpech. 2005. Pleistocene reindeer and global warming. Conservation Biology 19:557–562.

Grayson, D. K., and D. J. Meltzer. 2002. Clovis hunting and large mammal extinction: a critical review of the evidence. Journal of World Prehistory 16:313–359.

Grayson, D. K., and D. J. Meltzer. 2015. Revisiting Paleoindian exploitation of extinct North American mammals. Journal of Archaeological Science 56:177–193.

Hadly, E. A. 1997. Evolutionary and ecological response of pocket gophers (Thomomys talpoides) to late-Holocene climatic change. Biological Journal of the Linnean Society 60:277–296.

Hadly, E. A., and A. D. Barnosky. 2009. Vertebrate fossils and the future of conservation biology. Pp.39–59. In G. P. Dietl and K. W. Flessa, eds., Conservation Paleobiology: Using the Past to Manage for the Future. Paleontological Society Papers, Vol. 15. Boulder, CO.

Hadly, E. A., M. H. Kohn, J. A. Leonard, and R. K. Wayne. 1998. A genetic record of population isolation in pocket gophers during Holocene climatic change. Proceedings of the National Academy of Sciences of the United States of America 95:6893–6896.

Harpole, J. L. 1994. Zooarchaeological implications for Missouri's elk (Cervus elaphus) reintroduction effort. Pp.103–115. In R. L. Lyman and K. P. Cannon, eds. Zooarchaeology and Conservation Biology. University of Utah Press, Salt Lake City.

Harris, A.H. 1963. Vertebrate remains and past environmental reconstruction in the Navajo Reservoir District. Papers in Anthropology No. 11. Museum of New Mexico, Santa Fe.

Hayward, M.W. 2009. Conservation management for the past, present and future. Biodiversity and Conservation 18:765–775.

Heizer, R.F. 1955. Primitive man as an ecologic factor. Kroeber Anthropological Society Papers 13:1–31.

Hill, M.E., Jr., M.G. Hill, and C.C. Widga. 2008. Late Quaternary *Bison* diminution on the Great Plains of North America: evaluating the role of human hunting versus climate change. Quaternary Science Reviews 27:1752–1771.

Holt, F.L. 2005. The catch-22 of conservation: indigenous peoples, biologists, and cultural change. Human Ecology 33:199–215.

Houston, D.B., and E.G. Schreiner. 1995. Alien species in national parks: drawing lines in space and time. Conservation Biology 9:204–209.

Houston, D.B., E.G. Schreiner, and B.B. Moorhead. 1994. Mountain goats in Olympic National Park: biology and management of an introduced species. Scientific Monographs NPS/NROLYM/NRSM-94/25. National Park Service, Denver, CO.

Howell, M., and W. Prevenier. 2001. From Reliable Sources: An Introduction to Historical Methods. Cornell University Press, Ithaca, NY.

Ingles, L.G. 1965. Mammals of the Pacific States: California, Oregon, Washington. Stanford University Press, Stanford, CA.

IUCN SSC (IUCN Species Survival Commission). 2013. Guidelines for Reintroductions and Other Conservation Translocations, Version 1.0. IUCN Species Survival Commission, Gland, Switzerland.

Jackson, J.B.C., M.X. Kirby, W.H. Berger, K.A. Bjorndal, L.W. Botsford, B.J. Bourque, R.H. Bradbury et al. 2001. Historical overfishing and the recent collapse of coastal ecosystems. Science 293:629–637.

Jackson, S.T., and R.J. Hobbs. 2009. Ecological restoration in the light of ecological history. Science 325:567–569.

Jameson, R.J., K.W. Kenyon, S. Jeffries, and G.R. VanBlaricom. 1986. Status of a translocated sea otter population and its habitat in Washington. The Murrelet 67:84–87.

Jameson, R.J., K.W. Kenyon, A.M. Johnson, and H.M. Wight. 1982. History and status of translocated sea otter populations in North America. Wildlife Society Bulletin 10:100–107.

Jørgensen, D. 2013. Reintroduction and de-extinction. Bioscience 63:719–720.

Kareiva, P., and M. Marvier. 2012. What is conservation science? BioScience 62:962–969.

Kay, C.E., and C.A. White. 2001. Reintroduction of bison into the Rocky Mountain parks of Canada: historical and archaeological evidence. Pp.143–151. In D. Harmon, ed., Crossing Boundaries in Park Management. Proceedings of the 11th Conference on Research and Resource Management in Parks and on Public Lands. George Wright Society, Hancock, MI.

Kenyon, K.W. 1969. The Sea Otter in the Eastern Pacific Ocean. North American Fauna No. 68. US Fish and Wildlife Service, Fish and Wildlife Service, Washington, DC.

Kenyon, K.W. 1982. Sea otter, *Enhydra lutris*. Pp.704–710. In J.A. Chapman and G.A. Feldhamer, eds. Wild Mammals of North America: Biology, Management, Economics. Johns Hopkins University Press, Baltimore, MD.

Kidwell, S.M. 2013. Time-averaging and fidelity of modern death assemblages: building a taphonomic foundation for conservation palaeobiology. Palaeontology 56:487–522.

Kidwell, S.M. 2015. Biology in the Anthropocene: challenges and insights from young fossil records. Proceedings of the National Academy of Sciences of the United States of America 112:4922–4929.

Kidwell, S.M., and A. Tomasovych. 2013. Implications of time-averaged death assemblages for ecology and conservation biology. Annual Review of Ecology, Evolution and Systematics 44:539–563.

Kimble, D.S., D.B. Tyers, J. Robison-Cox, and B.F. Sowell. 2011. Aspen recovery since wolf reintroduction on the Northern Yellowstone winter range. Rangeland Ecology and Management 64:119–130.

Kowalewski, M. 1996. Time-averaging, overcompleteness, and the geological record. Journal of Geology 104:317–326.

Krech, S., III. 1999. The Ecological Indian: Myth and History. Norton, New York.

Langemann, E.G. 2004. Zooarchaeological research in support of a reintroduction of bison to Banff National Park, Canada. Pp.79–89. In R.C.G.M. Lauwerier and I. Plug, eds., The Future from the Past: Archaeozoology in Wildlife Conservation and Heritage Management. Oxbow Books, Oxford, UK.

Laundré, J.W. 1991. Mountain goats in Yellowstone: the horns of a dilemma? Park Science 11(3):8–9.

Lauwerier, R. C. G. M., and I. Plug, eds. 2004. The Future from the Past: Archaeozoology in Wildlife Conservation and Heritage Management. Oxbow Books, Oxford.

Lemke, T. O. 2004. Origin, expansion, and status of mountain goats in Yellowstone National Park. Wildlife Society Bulletin 32:532–541.

Louys, J., ed. 2012. Paleontology in Ecology and Conservation. Springer-Verlag, Berlin.

Ludwig, D., M. Mangel, and B. Haddad. 2001. Ecology, conservation, and public policy. Annual Review of Ecology and Systematics 32:481–517.

Lyman, R. L. 1988. Zoogeography of Oregon coast marine mammals: the last 3000 years. Marine Mammal Science 4:247–264.

Lyman, R. L. 1996. Applied zooarchaeology: the relevance of faunal analysis to wildlife management. World Archaeology 28:110–125.

Lyman, R. L. 1998. White Goats, White Lies: The Misuse of Science in Olympic National Park. University of Utah Press, Salt Lake City.

Lyman, R. L. 2002. Taxonomic identification of zooarchaeological remains. The Review of Archaeology 23(2):13–20.

Lyman, R. L. 2004a. Biogeographic and paleoenvironmental implications of late Quaternary pygmy rabbits (Brachylagus idahoensis) in eastern Washington. Western North American Naturalist 64:1–6.

Lyman, R. L. 2004b. Late-Quaternary diminution and abundance of prehistoric bison (Bison sp.) in eastern Washington state, USA. Quaternary Research 62:76–85.

Lyman, R. L. 2006a. Archaeological evidence of anthropogenically induced twentieth-century diminution of North American wapiti (Cervus elaphus). American Midland Naturalist 156:88–98.

Lyman, R. L. 2006b. Paleozoology in the service of conservation biology. Evolutionary Anthropology 15:11–19.

Lyman, R. L. 2007. The Holocene history of pronghorn (Antilocapra americana) in eastern Washington State. Northwest Science 81:104–111.

Lyman, R. L. 2008. Estimating the magnitude of data asymmetry in paleozoological biogeography. International Journal of Osteoarchaeology 18:85–94.

Lyman, R. L. 2009. The Holocene history of bighorn sheep (Ovis canadensis) in eastern Washington State, northwestern USA. The Holocene 19:143–150.

Lyman, R. L. 2010a. Mandibular hypodontia and osteoarthritis in prehistoric bighorn sheep (Ovis canadensis) in eastern Washington State, USA.

International Journal of Osteoarchaeology 20:396–404.

Lyman, R. L. 2010b. Paleozoology's dependence on natural history collections. Journal of Ethnobiology 30:126–136.

Lyman, R. L. 2010c. Prehistoric anthropogenic impacts to local and regional faunas are not ubiquitous. Pp.204–224. In R. M. Dean, ed., The Archaeology of Anthropogenic Environments. Occasional Paper No. 37. Center for Archaeological Investigations, Southern Illinois University, Carbondale.

Lyman, R. L. 2010d. Taphonomy, pathology and paleoecology of the terminal Pleistocene Marmes Rockshelter (45FR50) "big elk" (Cervus elaphus), southeastern Washington State, USA. Canadian Journal of Earth Sciences 47:1367–1382.

Lyman, R. L. 2011a. Paleoecological and biogeographical implications of late Pleistocene noble marten (Martes americana nobilis) in eastern Washington State, USA. Quaternary Research 75:176–182.

Lyman, R. L. 2011b. Paleozoological data suggest Euroamerican settlement did not displace ursids and North American elk from lowlands to highlands. Environmental Management 47:899–906.

Lyman, R. L. 2012. A warrant for applied paleozoology. Biological Reviews 87:513–525.

Lyman, R. L., and K. P. Cannon, eds. 2004. Zooarchaeology and Conservation Biology. University of Utah Press, Salt Lake City.

Lyman, R. L., and M. J. O'Brien. 2005. Within-taxon morphological diversity in late-Quaternary Neotoma as a paleoenvironmental indicator, Bonneville Basin, northwestern Utah, USA. Quaternary Research 63:274–282.

Lyman, R. L., and S. Wolverton. 2002. The late prehistoric–early historic game sink in the northwestern United States. Conservation Biology 16:73–85.

Matthews, P. E., and A. C. Heath. 2008. Evaluating historical evidence for occurrence of mountain goats in Oregon. Northwest Science 82:286–298.

Mawdsley, J. R., R. O'Malley, and D. S. Ojima. 2009. A review of climate-change adaptation strategies for wildlife management and biodiversity conservation. Conservation Biology 23:1080–1089.

McCabe, R. E., B. W. O'Gara, and H. M. Reeves. 2004. Prairie Ghost: Pronghorn and Human Interaction in Early America. University Press of Colorado, Boulder.

McGuire, J. L., and E. B. Davis. 2014. Conservation paleobiogeography: the past, present and future of species distributions. Ecography 37:1092–1094.

Meltzer, D. J. 2015. Pleistocene overkill and North American mammalian extinctions. Annual Review of Anthropology 44:33–53.

Miller, J. H. 2011. Ghosts of Yellowstone: multidecadal histories of wildlife populations captured by bones on a modern landscape. PLoS ONE 6:e18057.

Moore, C. M., and P. W. Collins. 1995. *Urocyon littoralis*. Mammalian Species 489:1–7.

Morrison, M. L. 2001. Techniques for discovering historic animal assemblages. Pp.295–315. In D. Egan and E. A. Howell, eds., The Historical Ecology Handbook: A Restorationist's Guide to Reference Ecosystems. Island Press, Washington, DC.

Nagaoka, L. 2012. The overkill hypothesis and conservation biology. Pp.110–138. In S. Wolverton and R. L. Lyman, eds., Conservation Biology and Applied Zooarchaeology. University of Arizona Press, Tucson.

Nagorsen, D. W., and G. Keddie. 2000. Late Pleistocene mountain goats (*Oreamnos americanus*) from Vancouver Island: biogeographic implications. Journal of Mammalogy 81:666–675.

Newsome, S. D., M. A. Etnier, D. Gifford-Gonzalez, D. L. Phillips, M. van Tuinen, E. A. Hadly, D. P. Costa, D. J. Kennett, T. P. Guilderson, and P. L. Koch. 2007. The shifting baseline of northern fur seal ecology in the northeast Pacific Ocean. Proceedings of the National Academy of Sciences of the United States of America 104:9709–9714.

Niemeyer, C. C. 2007. The good, bad and ugly, depending on your perspective. Pp.287–296. In Wildlife Management Institute Publications Department, ed., Transactions of the 72nd North American Wildlife and Natural Resources Conference, Wildlife Management Institute, Washington, DC.

Noss, R. F., E. Fleishman, D. A. Dellasala, J. M. Fitzgerald, M. R. Gross, M. B. Main, F. Nagle, S. L. O'Malley, and J. Rosales. 2009. Priorities for improving the scientific foundation of conservation policy in North America. Conservation Biology 23:825–833.

Nygren, A., and S. Rikoon. 2008. Political ecology revisited: integration of politics and ecology does matter. Society and Natural Resources 21:767–782.

O'Gara, B. W. 1978. *Antilocapra americana*. Mammalian Species 90:1–7.

Ogden, A. 1933. Russian sea otter and seal hunting on the California coast. California Historical Society Quarterly 12:29–51.

Olszewski, T. D. 1999. Taking advantage of time averaging. Paleobiology 25:226–238.

Osborne, D. 1953. Archaeological occurrences of pronghorn antelope, bison, and horse in the Columbia Plateau. Scientific Monthly 77:260–269.

Osborne, P. E., and P. J. Seddon. 2012. Selecting suitable habitats for reintroductions: variation, change and the role of species distribution modelling. Pp.73–104. In J. G. Ewen, D. P. Armstrong, K. A. Parker, and P. J. Seddon, eds., Reintroduction Biology: Integrating Science and Management. Wiley-Blackwell, Hoboken, NJ.

Owen, P. R., C. J. Bell, and E. M. Mead. 2000. Fossils, diet, and conservation of black-footed ferrets (*Mustela nigripes*). Journal of Mammalogy 81:422–433.

Parmesan, C. 2006. Ecological and evolutionary responses to recent climate change. Annual Review of Ecology, Evolution and Systematics 37:637–669.

Pauli, J. N., W. E. Moss, P. J. Manlick, E. D. Fountain, R. Kirby, S. M. Sultaire, P. L. Perrig, J. E. Mendoza, J. W. Pokallus, and T. H. Heaton. 2015. Examining the uncertain origin and management role of martens on Prince of Wales Island, Alaska. Conservation Biology 29:1257–1267.

Pauly, D. 1995. Anecdotes and the shifting baseline syndrome of fisheries. Trends in Ecology and Evolution 10:430.

Peacock, E., C. R. Randklev, S. Wolverton, R. A. Palmer, and S. Zaleski. 2012. The "cultural filter," human transport of mussel shell, and the applied potential of zooarchaeological data. Ecological Applications 22:1446–1459.

Peterson, C. H. 1977. The paleoecological significance of undetected short-term temporal variability. Journal of Paleontology 51:976–981.

Pooley, S. 2013. Historians are from Venus, ecologists are from Mars. Conservation Biology 27:1481–1483.

Rackham, O. 1998. Implications of historical ecology for conservation. Pp.152–175. In W. J. Sutherland, ed. Conservation Science and Action. Blackwell Scientific, Oxford, UK.

Rick, T. C., J. M. Erlandson, R. L. Vellanoweth, T. J. Braje, P. W. Collins, D. A. Guthrie, and T. W. Stafford, Jr. 2009. Origins and antiquity of the island fox (*Urocyon littoralis*) on California's Channel Islands. Quaternary Research 71:93–98.

Riley, M. A. 1985. An analysis of masticatory form and function in three mustelids (*Martes americana, Lutra canadensis, Enhydra lutris*). Journal of Mammalogy 66:519–528.

Roest, A. I. 1973. Subspecies of sea otter, *Enhydra lutris*. Los Angeles County Natural History Museum, Contributions in Science No. 252.

Roest, A. I. 1979. A reevaluation of sea otter taxonomy. Paper presented at the Abstracts of the Sea Otter Workshop, Santa Barbara, CA, August 23–25, 14pp.

Rowe, R. J., J. A. Finarelli, and E. A. Rickart. 2010. Range dynamics of small mammals along an elevational gradient over an 80-year interval. Global Change Biology 16:2930–2943.

Sandom, C. J., C. J. Donlan, J.-C. Svenning, and D. Hansen. 2013. Rewilding. Pp.430–451. In D. W. Macdonald and K. J. Willis, eds., Key Topics in Conservation Biology 2. John Wiley & Sons, New York.

Scheffer, V. B., and F. Wilke. 1950. Validity of the subspecies *Enhydra lutris nereis*, the southern sea otter. Journal of the Washington Academy of Sciences 40:269–272.

Schnuer, J. 2015. Bringing the wood bison back to Alaska. Smithsonian Magazine, March. Available online at: http://www.smithsonianmag.com /science-nature/bringing-wood-bison-back-to-alaska-180954326/.

Schullery, P., and L. Whittlesey. 2001. Mountain goats in the Greater Yellowstone Ecosystem: a prehistoric and historical context. Western North American Naturalist 61:289–307.

Semken, H. A., Jr., R. W. Graham, and T. W. Stafford, Jr. 2010. AMS ^{14}C analysis of late Pleistocene non-analog faunal components from 21 cave deposits in southeastern North America. Quaternary International 217:240–255.

Shotwell, J. A. 1955. An approach to the paleoecology of mammals. Ecology 36:327–337.

Simpson, G. G. 1963. Historical science. Pp.24–48. In C. C. Albritton, Jr., ed., The Fabric of Geology. Freeman, Cooper, and Co., Stanford, CA.

Simpson, G. G. 1970. Uniformitarianism: an inquiry into principle, theory, and method in geohistory and biohistory. Pp.43–96. In M. K. Hecht and W. C. Steere, eds., Essays in Evolution and Genetics in Honor of Theodosius Dobzhansky. Appleton, New York.

Smith, F. A., and J. L. Betancourt. 2003. The effect of Holocene temperature fluctuations on the evolution and ecology of *Neotoma* (wood-rats) in Idaho and northwestern Utah. Quaternary Research 59:160–171.

Soulé, M. E. 1985. What is conservation biology? Bioscience 35:727–734.

Stephenson, R. O., S. C. Gerlach, R. D. Guthrie, C. R. Harington, R. O. Mills, and G. Hare. 2001. Wood bison in late Holocene Alaska and adjacent Canada: paleontological, archaeological and historical records. Pp.124–158. In S. C. Gerlach and M. S. Murray, eds., People and Wildlife in Northern North America. British Archaeological Reports International Series 944. Archaeopress, Oxford, UK.

Suding, K., and E. Leger. 2012. Shifting baselines: dynamics of evolution and community change in a changing world. Pp.281–292. In J. van Andel and J. Aronson, eds. Restoration Ecology: The New Frontier, 2nd ed., Blackwell, Oxford, UK.

Szabó, P. 2010. Why history matters in ecology: an interdisciplinary perspective. Environmental Conservation 37:380–387.

Szabó, P., and R. Hédl. 2011. Advancing the integration of history and ecology for conservation. Conservation Biology 25:680–687.

Szabó, P., and R. Hédl. 2013. Grappling with interdisciplinary research: response to Pooley. Conservation Biology 27:1484–1486.

Terry, R. C. 2010. The dead do not lie: using skeletal remains for rapid assessment of historical small-mammal community baselines. Proceedings of the Royal Society B 277:1193–1201.

Trigger, B. G. 1968. Beyond History: The Methods of Prehistory. Holt, Rinehart and Winston, New York.

Valentine, K., D. A. Duffield, L. E. Patrick, D. R. Hatch, V. L. Butler, R. L. Hall, and N. Lehman. 2008. Ancient DNA reveals genotypic relationships among Oregon populations of the sea otter (*Enhydra lutris*). Conservation Genetics 9:933–938.

Valenzuela, A., and J. A. Alcover. 2013. Radiocarbon evidence for a prehistoric deliberate translocation: the weasel (*Mustela nivalis*) of Mallorca. Biological Invasions 15:717–722.

van der Leeuw, S. E. 2014. Transforming lessons from the past into lessons for the future. Pp.215–231. In A. F. Chase and V. L. Scarborough, eds., The Resilience and Vulnerability of Ancient Landscapes: Transforming Maya Archaeology through IHOPE. Archaeological Papers of the American Anthropological Association 24. Wiley, New York.

van Wieren, S. E. 2012. Reintroductions: learning from successes and failures. Pp.87–100. In J. van Andel and J. Aronson, eds. Restoration Ecology: The New Frontier. 2nd ed., Blackwell, Oxford, UK.

Varley, J. D., and W. G. Brewster, eds. 1992. Wolves for Yellowstone? A Report to the United States Congress, Vol. IV: Research and Analysis. National Park Service, Yellowstone National Park, Wyoming.

Walgamott, A. 2011. How pronghorns returned to Washington. Northwest Sportsman 3(6):14–20, 126–128.

Watson, J.C., G.M. Ellis, T.G. Smith, and J.K.B. Ford. 1997. Updated status of the sea otter, *Enhydra lutris*, in Canada. Canadian Field Naturalist 111:277–286.

White, C.A., E.G. Langemann, C.C. Gates, C.E. Kay, T. Shury, and T.E. Hurd. 2001. Plains bison restoration in the Canadian Rocky Mountains? Ecological and management considerations. Pp.152–160. In D. Harmon, ed., Crossing Boundaries in Park Management: Proceedings of the 11th Conference on Research and Resource Management in Parks and on Public Lands. George Wright Society, Hancock, MI.

White, P.S., and J.L. Walker. 1997. Approximating nature's variation: selecting and using reference information in restoration ecology. Restoration Ecology 5:338–349.

Wielgus, R.B., and K.A. Peebles. 2014. Effects of wolf mortality on livestock depredations. PLoS ONE 9(12):e113505.

Williams, J.W., and S.T. Jackson. 2007. Novel climates, no-analog communities, and ecological surprises. Frontiers in Ecology and Environment 5:475–482.

Wilson, D.E., M.A. Bogan, R.L. Brownell, Jr., A.M. Burdin, and M.K. Maminov. 1991. Geographic variation in sea otters, *Enhydra lutris*. Journal of Mammalogy 72:22–36.

Wolverton, S. 2010. The North American Pleistocene overkill hypothesis and the re-wilding debate. Diversity and Distributions 16:874–876.

Wolverton, S., J.H. Kennedy, and J.D. Cornelius. 2007. A paleozoological perspective on white-tailed deer (*Odocoileus virginianus texana*) population density and body size in central Texas. Environmental Management 39:545–552.

Wolverton, S., and R.L. Lyman, eds. 2012. Conservation Biology and Applied Zooarchaeology. University of Arizona Press, Tucson.

Wood, J.W., G.R. Milner, H.C. Harpending, and K.M. Weiss. 1992. The osteological paradox: problems of inferring prehistoric health from skeletal samples. Current Anthropology 33:343–370.

Young, K.R. 2014. Biogeography of the anthropocene: novel species assemblages. Progress in Physical Geography 38:664–673.

PART TWO

Setting Goals

FOUR

Human Dimensions Insights for Reintroductions of Fish and Wildlife Populations

Shawn J. Riley and Camilla Sandström

As DISCUSSED THROUGHOUT THIS book and elsewhere, biological and environmental considerations associated with reintroduction of fish and wildlife populations are numerous, complex, and accompanied by uncertainty (Converse et al. 2013). Human dimensions aspects of reintroductions, the decision space not directly about the organism and its habitat (Decker et al. 2012a), are characterized by additional complexities and uncertainties. Human dimensions aspects of reintroduction are associated with cultural, political, economic, sociological, and psychological considerations. We use the term "sociocultural" throughout this chapter to mean relating to, or involving a combination of, these social and cultural factors. Public fisheries and wildlife management, for which reintroduction is one sort of intervention, is an endeavor of contemporary democratic decision-making within governance of natural systems (Rudolph et al. 2012). Sustainable reintroductions of fish and wildlife populations require alignment of environmental and social considerations (Macdonald 2009). Reintroductions as management interventions are likely to be more socially sustainable if distinguished from traditional models of conservation, which are often based on implicit, deeply rooted assumptions that humans occur apart from nature (Mangel et al. 1996, Botkin 2012).

Even if biological and environmental uncertainties associated with a reintroduction are low, which seldom occurs (Converse et al. 2013), achieving social acceptance or political support in terms of legitimacy is antecedent to sustainable reintroductions (Kleiman et al. 1994, Clark et al. 2002). In its generic form, legitimacy is defined as the right to govern according to justifiable rules (Beetham 1991). In relation to governance of reintroductions, however, the concept describes intervening roles of institutions, such as rules and norms. Legitimacy also is assessed through institutional arrangements, through public participation, and the extent to which these arrangements are perceived as legitimate (Pierre and Peters 2005). To include both of these aspects, we use legitimacy to mean "the acceptance and justification of shared rule by a community" (Bernstein 2005, 142).

We consider legitimacy through acceptance capacity of actors by either assessments of stakeholder belief systems or articulation of reasons why stakeholders find reintroductions to be legitimate or illegitimate (Johansson 2013). Stakeholders are used in this chapter to mean any person or organization that is significantly (inherently subjective in some cases) affected by, or that significantly affects, reintroduction of fish or wildlife populations (Decker et al. 2012a). This reasoning, in turn, is linked to a concept of stakeholder acceptance capacity (Decker and Purdy 1988, Carpenter et al. 2000), which has been investigated in a variety of situations with species pertinent to reintroductions generally (Riley and Decker 2000, Morzillo et al. 2007, Carter et al. 2012, Zajac et al. 2012). Results from these investigations repeatedly stressed a need to anticipate factors affecting society's acceptance capacity for the particular organism of interest (Kleiven et al. 2004, Smith et al. 2014) and the quality of public decision-making processes. The quality of those processes is defined/measured through "input legitimacy" in terms of which various publics are involved and of the power that those publics exercise in decision-making, "throughput legitimacy" such as efficacy, accountability, and transparency of the decision-making process, and "output legitimacy" such as the effectiveness of the policy outcome (Lauber and Knuth 1999, Scharpf 1999, Arts et al. 2014, Smith et al. 2014).

Our objective in this chapter is to offer a framework that guides human-assisted reintroduction of fish or wildlife populations as well as recolonization (those reintroductions that occur without active human intervention). Similarities between these two circumstances exist because in a sense all reintroductions are a form of "forced dispersal" (Morell 2008) or human-induced recolonization (Jørgensen 2011). Macdonald (2009) emphasized that generic guidelines are more easily drafted than followed. Our intent is to offer insights for consideration rather than a cookbook for the human dimensions of reintroductions. We draw on research and case

studies that provide descriptive assessments of past reintroduction efforts in the context of adaptive decision-making and contemporary governance. The outcome we seek is that more careful, comprehensive, and inclusive deliberations about reintroductions will create greater likelihood of meeting the criteria for successful reintroductions.

A PROPOSED HUMAN DIMENSIONS FRAMEWORK FOR REINTRODUCTIONS

Strategic reintroduction processes follow predictable steps or stages (Sarrazin and Barbault 1996, Arts et al. 2014). We propose an adaptive governance framework that includes four stages: conceptualization, feasibility, implementation, and learning (figure 4.1). Our framework encompasses similar stages to other frameworks to guide reintroductions such as those depicted in the 2013 IUCN Species Survival Commission guidelines (IUCN/SSC 2013), yet we emphasize a more acute focus on when human dimensions considerations play the most important role. All stages are important, yet in this chapter we focus most intensely on stages 1 and 2 (input legitimacy and throughput legitimacy) because it is at these stages that human dimensions insights are most crucial in their contribution to sustained (Kleiman et al. 1994) and ethical (Bekoff 1999) reintroductions. Reintroductions are most effective when viewed as long-term commitments and processes that are monitored for success, by whatever criteria success is defined early in the process, and then managed within an adaptive process (Nichols and Armstrong 2012, Chauvenet et al., Chapter 6, this volume). As a note of caution, the danger in deeming a reintroduction as "successful" is that it implies there is an end point in what is really an ongoing process. To establish or maintain legitimacy, it is crucial to include the explicit possibility that a reintroduction might not occur, or at a minimum, a step back to a previous stage is possible. This is depicted in our diagram as "no" between stages at which time the reintro-

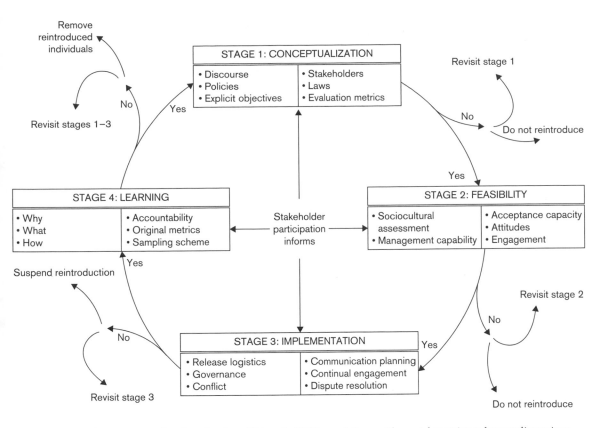

FIGURE 4.1 An adaptive process for reintroduction of fish and wildlife populations, with special attention to human dimensions considerations.

duction may be stopped or the previous stages may be reconsidered.

Stage 1: Conceptualization

Species reintroductions are a prominent feature of global conservation discourses and practical efforts through the International Union for Conservation of Nature (IUCN) encouraging responsible reintroductions. These broad policies are apparent throughout the world. The 1992 Convention on Biological Diversity created an obligation for contracting parties (9c) to adopt measures for recovery and rehabilitation of threatened species and for their reintroduction into their natural habitats under appropriate conditions (McCarthy et al. 2012). In Europe, the 1979 Convention on the Conservation of European Wildlife and Natural Habitats encouraged its "Parties" to reintroduce native

species as a distinct method of conservation. Furthermore, under European Legislation, the European Commission's Habitats Directives 92/43/EEC direct member states to consider feasibility of restoring species that have been locally extinct. These international commitments, in turn, have been introduced in national policy documents and legislations throughout the various EU nation-states.

In adoption of similar broad policy within the United States, the Endangered Species Act (ESA) requires federal agencies, such as the US Fish and Wildlife Service and the US National Marine Fisheries Service to develop and implement recovery plans for species listed under the ESA. Recovery plans identify actions needed to restore threatened and endangered species to the point that they are again self-sustaining elements of their ecosystems and no longer need the protections of the ESA. Although recovery plans offer guidance, and are not regulatory documents per se, the ESA clearly envisions recovery plans as the central organizing tool for guiding and coordinating recovery efforts across a wide spectrum of federal, state, tribal, local, and private entities. Reintroductions of native fauna are often a component of recovery plans.

Reintroductions of wildlife populations thus have become a common element of global and national conservation policies. Since reintroductions of fish or wildlife populations almost always involve public decisions about public resources, politics in its various forms are involved despite efforts of scientists and experts to maintain objective rigor (Arts et al. 2012, 2014). Kleiman et al. (1994) counseled that successful reintroductions might be more closely linked to political conditions than to the rigor of scientific reasoning. Those conclusions were reinforced during a global assessment of reintroductions 20 years later, using case studies ranging from monk seals (*Monachus schauinslandi*) in Hawaii to chimpanzees (*Pan troglodytes*) in Guinea, West Africa, to Eagle Lake rainbow trout (*Oncorhynchus mykiss aquilarum*) in California (Soorae 2008, 2013).

Given the political aspects, both the framing (discourse) of the issue (Arts et al. 2012) and the input legitimacy become vital considerations in reintroduction processes. How reintroductions are framed in discourse during the conceptualization stage influences the feasibility of a particular reintroduction. Discourse, used in this context to mean the lens though which reintroductions are viewed, reinforces norms and is thus embedded in broader environmental discourses (Svarstad et al. 2008, Arts et al. 2014). The framing discourse might take on a biocentric or anthropocentric focus or a management perspective, which is more or less an interventionist approach (Hajer 1995, Adger et al. 2001). Pro-reintroduction stakeholders might evoke scientific evidence for why a reintroduction is needed that induces a cognitive response (see Box 4.1). For instance, reintroducing lynx to keep numbers of animals

high enough to in turn keep the federal government from intervening under the ESA was suggested as a motivation for reintroductions in Colorado and Idaho (Bekoff 1999). Anti-reintroduction sides, on the other hand, might use imagery that evokes emotional or affective responses (Reading and Kellert 1993). Discourses are thus formative to people's attitudes toward reintroductions and the debate can become stalled or "frozen" at this stage if the discursive struggle between pro- and anti-reintroduction is not facilitated and managed. By understanding discourses and related attitudes early in a process, along with stakeholder engagement, the conceptualization of a reintroduction has a greater probability of matching societal needs.

Discourses related to reintroductions also consider broader philosophical meanings of what is native, natural or invasive and unnatural, and societal obligations to reintroduce wildlife populations (Rees 2001). Global policies typically are ambiguous and do not provide precise reference points in time or space of when species existed that might justify reintroduction as biologically or sociologically appropriate. These discourses occur commonly and broadly in science as well as in society (Seddon 2010, Jørgensen, 2011, Dalrymple and Moehrenschlager 2013). Although the debate has semantic connotations, if the different meanings associated with concepts such as "presence of species in historic times or historic range," or use of "native species" are made without clarifying of what constitutes "nativeness," this leaves interpretation up to persons involved in setting policy, which can lead to conflicts (Jørgensen 2011). Different discourses and motives of the public as well as stakeholder opinions may be channelized and aggregated through deliberative and participatory procedures.

A deliberative turn in environmental governance, where citizens and stakeholders gain access to policy arenas for the purpose of influencing political decision-making during early stages of reintroductions, usually requires new procedures in parallel with traditional repre-

BOX 4.1 · The Writing on the Wall: Discourses of Reintroduction

Koen Arts and Anke Fischer

4a The white-tailed eagle (*Haliaeetus albicilla*) has been the subject of ongoing reintroduction attempts in Scotland for over the past 55 years (credit: Koen Arts).

Decision-making on the reintroduction of wildlife populations is increasingly required to be evidence based. This often leads to production of numerous expert documents that deal with a broad range of socio-ecological dimensions of reintroduction, including case studies from examples elsewhere, public and stakeholder consultations, risk and impact assessments, and legal examinations. At face value, such expert documents—with their scientific language and logical reasoning—deal with technical aspects of a reintroduction. Because they are usually authored and commissioned by proponents of introductions, it is conceivable that they are also employed, intentionally or not, to influence public opinion and persuade decision-makers. We scrutinized such expert documents to better understand the roles of rhetoric and argumentation in the discursive practices around animal reintroduction. Here, we review findings from our review of 43 expert documents regarding reintroduction of the white-tailed eagle (*Haliaeetus albicilla*), to Scotland (figure 4a).

The white-tailed eagle became extinct in Scotland in 1918, supposedly as a result of prolonged human-caused mortality. Two small-scale and unsuccessful reintroduction attempts were made by ecologists in 1959 and 1968 with three and four birds, respectively. Between 1975 and 1985, a UK government-funded body released 85 juvenile eagles of Norwegian origin on an island off the west coast of Scotland. In a second phase, 58 juveniles were released between 1993 and 1998, and in 2007, a third phase commenced, which brought about 100 birds to the eastern Scottish mainland. By the beginning of 2008, the entire reintroduction program had resulted in >250 fledged chicks in Scotland.

Storyline 1: The Necessary Reintroduction of a Native Bird

Our analysis was based on the identification of storylines in expert documents: fundamental statements (whether explicit or implicit) that together constitute a discourse. The first storyline, which permeated the discourse from 1960 to 2010, could be described as: "The necessary reintroduction of a native bird." This storyline suggested that active progress toward white-tailed eagle conservation was an imperative and Scotland had to proactively provide its share of the responsibility. Seemingly isolated sub-storylines (human cause of extinction, time of extinction, once common, slow recolonizer, globally endangered), from different angles (historical, ecological, ethical), and intensified with rhetorical techniques (comparison, time pressure, personification), were tied together into an argumentatively inescapable rationale.

Storyline 2: Foremost a Scavenger, with Importance for Scotland's Tourism

From the early 2000s a second storyline emerged due to the rise of a serious issue for reintroduction proponents: farmers accused the eagles of lamb predation. Following the allegations, public authority leading the reintroduction commissioned extensive research on the eagles' diet. The resulting documents revealed that, at least during late April and May, some of the 15 studied breeding pairs scavenged on lamb carcasses. More so, despite the researchers' observation that most of those lambs were nonviable, there was evidence that one pair actively killed lambs. In response, many expert documents now suggested that the white-tailed eagle *was foremost a scavenger* and often drew on two sub-storylines (the white-tailed eagle occasionally preys mainly nonviable lambs and has minimal impact on Scottish sheep farming) to support this claim. Importantly, though, the conclusions of the commissioned research were also used by opponents. The possibility of eagles causing economic loss to individual farms contributed directly to the setup of a permanent "management scheme" that offered at least 30 farmers financial aid for "positive" and "friendly" co-management of the eagles. Despite these mitigation measures, farmers continued to blame white-tailed eagles for large-scale lamb killing. Possibly because of this prolonged controversy, an older notion, namely, that the eagles attracted visitors, became now explicitly connected to financial benefit. One of the analyzed documents presented calculations that eagle tourism was worth about £1.5 million per year on the Isle of Mull. The storyline of the white-tailed eagle being "foremost a scavenger, with importance for Scotland's tourism" can thus be interpreted as a discursive tool that played down predation on lambs on the basis of ecological science and an overall positive economic balance.

Storyline 3: A Slow Reproducer, but Not Only a Bird of Wild Coasts

The failure of the first two reintroduction attempts allegedly had manifold causes, including the small number of released birds, poor habitat, and lack of secluded cliffs. But Scotland's coastal habitat in general was still seen as suitable, resulting in a renewed attempt on the Isle of Rum with many more birds (Phase 1). Nonetheless, serious issues with regard to breeding success were reported soon after release. The sub-storyline "lack of breeding success is the main factor for insufficient population growth" came to underpin a broader storyline that "white-tailed eagles are slow reproducers." It was calculated the existing population would become extinct within 100 years. Phase 2 was thus implemented, which helped to increase the population to 19 breeding pairs. Still, a third phase was deemed necessary. By 2003, an important change had taken place in the ecological discourse. The initial question of suitable habitat, such as remote islands and wild coasts, was reconsidered, and areas previously thought unsuitable for release, such as rivers, estuaries, and wetlands, came to be regarded as the most productive areas. Thus, with a changing focus from an "internal" (slow reproduction) to an "external" (released in suboptimal habitats) impediment to population growth, the first and second parts of the storyline merged.

The Writing on the Wall

Comparing the pro-reintroduction discourse of the white-tailed eagle with the discourses of beaver and lynx, clear similarities at the level of storylines are apparent. These include historical or ethical justifications for intervention, and an emphasis on the positive economics or the ecological benefits of a reintroduction. These similarities reflect a "zeitgeist." For each species, the first storyline focused on history and a need to establish the human cause of extinction, which mirrors the importance of the IUCN guidelines of 1987 and 1998. More recently, rhetorical and argumentative complexity of the discourses increased, suggesting a growing reflexivity. For example, storylines adjusted their own position in relation to counter-argumentation, and they imbued awareness of the importance of perceptions and support of stakeholders and the wider public. A final reflexive element was that many documents feeding into storylines seemed "self-aware" of their role in the process of advancing reintroductions. Through cross-referencing to and anticipating future research, a resolute, goal-oriented chain of publications was created. Implications revealed by storylines are worthy of consideration for at least three reasons. First, reflexive storylines may conceal a discrepancy between "good language" and "good practice." Second, as storylines become discursively armored, rhetorically advanced, and increasingly aim to preempt anticipated problems, debates get frozen at an early stage. This, in turn, might drive proponents and opponents of reintroductions farther apart and increase controversy. Third, reflexive storylines might also restrict perspectives and narrow scientific inquiry. Our study calls for caution. Expert documents are most useful when they do not furnish a foregone conclusion, so they are merely viewed as the "writing on the wall" to opponents. The litmus test for a democratic discourse is that a possibility of no reintroduction is openly and honestly considered.

sentative institutions. These procedures might include elections and party competition to link political decisions with citizens' or stakeholders' preferences (Sandström et al. 2009, Bäckstrand et al. 2010). Because traditional mechanisms, such as elections, do not provide genuine or well-informed opportunities for stakeholders to influence policy-making, deliberative processes through public participation are an essential way to link political decisions with stakeholder needs or preferences (Lauber et al. 2012).

Interactive modes of governance attempt to improve input and throughput legitimacy and induce learning between stakeholders and policy-makers (Dryzek 2012). Experience in how to design these processes, how to identify relevant stakeholders (stakeholder analysis), and what decision support tools to use is extensive (Lauber et al. 2012). Nonetheless, the most successful processes in terms of perceived legitimacy are those that are open and transparent; solicit and evaluate alternatives; efficient but do not take short cuts; and are not guided by a "hidden agenda" (Keohane 2011, Lauber et al. 2012). Based on research into what stakeholders are wanting from participation processes, it can be predicted that the quality of a reintroduction process can be improved when scientific information is provided, the participation has genuine influence on decisions, stakeholders are treated fairly, and the participation promotes informative communication and learning among all participants (Chase et al. 2004). Deliberative processes that fail or become prolonged are usually those enacted in later stages to resolve conflicts or otherwise used to "convince" people (e.g., attain "buy in") of a predetermined outcome. An option to stop a reintroduction process if it can be demonstrated that costs outweigh the benefits will bolster confidence of stakeholders to continue into the next stages (Lauber and Knuth 1997).

Stage 2: Feasibility

Once a reintroduction idea passes the conceptualization stage, emphasis is on feasibility. Biological feasibility studies for a reintroduction focus on the capacity of the species to thrive and sustain itself through time (see Chapters 8–10, this volume). Similarly, from a human dimensions perspective, the capacities of society to allow, support, and sustain a reintroduction effort are crucial considerations in gauging social feasibility (Macdonald 2009). The feasibility stage considers sociocultural effects of a reintroduction as well as legal requirements of a specific reintroduction site. Important considerations include the assessment of trade-offs of reintroduction versus recolonization (Converse et al. 2013), and if feasibility can be demonstrated, the development of appropriate stakeholder participation processes matched to the scale of expected effects created by a reintroduction (Lauber et al. 2012).

Feasibility is linked to characteristics of the species (e.g., phylogeny, morphology, behavior), as well as human beliefs (e.g., perceived worth, symbolic qualities, or risk perceptions) attributable to the species (Kellert and Berry 1980, Skogen and Krange 2003, Rice et al. 2007). Feasibility is further influenced by political issues of legitimacy and conflict over larger issues of sociopolitical power and control (Wilson 1997, Dak 2015). A proposal to reintroduce species such as burbot (*Lota lota*), for which there is sparse public awareness and that is generally viewed as innocuous, likely will encounter minimal public opposition (Worthington et al. 2010, Jørgensen 2013). Proposals to reintroduce prominent carnivores, such as wolves (*Canis lupus*; Williams et al. 2002), grizzly bears (*Ursus arctos*; Clark et al. 2002), or eagles (Arts et al. 2012), however, frequently are met with resistance from influential stakeholders (e.g., agriculturalists or hunters). Even reintroductions of smaller, less threatening species such as black-footed ferrets (*Mustela nigripes*) can be resisted when they are symbolic of other underlying issues such as threats associated with ESA control of land use (Reading and Kellert 1993, Skogen and Krange 2003) or loss of loci of control in a community's way of life (Bjerke et al. 2000).

Social feasibility assessments (Yohe and Tol 2002) for reintroductions are sparse; most are post hoc reviews or analyses. More commonly, assessments are conducted of attitudes toward the species being considered, or in some cases, attitudes toward the reintroduction are measured (e.g., Browne-Nuñex and Taylor 2002). Because the present volume targets readers from myriad backgrounds with a common interest in reintroduction of wildlife populations, it might be helpful to gain a little perspective on attitudes, what they do and do not represent, and how attitudes have been used in regard to reintroductions.

Attitudes

Attitudes provide one measure of the "political conditions" when it comes to a well-known species such as wolves, white-tailed eagles (*Haliaeetus albicilla*), or black-footed ferrets, but are less reliable when it comes to other lesser known species (Heberlein 2012, 22) such as monk seals, Natterjack toads (*Bufo calamita*; Denton et al. 1997), or burbot. An assumed positive link between an expressed attitude and

human behavior (Ajzen and Gilbert 2008, Vaske and Manfredo 2012) is one reason why so much focus on attitudes has occurred in human dimensions research generally and particularly when it comes to reintroductions (Browne-Nunez and Taylor 2002).

An attitude is "a psychological tendency that is expressed by evaluating a particular entity with some degree of favor or disfavor" (Eagly and Chaiken 1993). An attitude is a tendency in that it lasts at least a short time, yet is subject to change. Attitudes are composed of evaluative and a cognitive dimensions. The evaluative component refers to whether a person perceives the attitude object (e.g., reintroduction of black-footed ferrets to Montana, USA) in a positive or negative manner. The cognitive component refers to beliefs—what people believe to be true, which is not necessarily accepted objective fact—that influence formation of attitudes. For example, one person might believe reintroduction of black-footed ferrets could, through actions of regulatory agencies, prevent other land uses such as oil and gas development, whereas another person might believe reintroduction of ferrets could restore a balance of nature (Reading and Kellert 1993). Attitudes have their strongest predictive capability when measured specifically rather than generally. For example, if managers wish to know whether people will support reintroduction of black-footed ferrets, the attitudinal questions asked should be about ferret reintroductions and not about ferrets (Whittaker et al. 2006).

Attitudes matter, but more importantly to inform decisions about reintroductions, it matters that authorities be absolutely clear about what they are trying to change with a reintroduction, and then seek insights about attitudes related to that reintroduction (Pate et al. 1996). A hierarchy is proposed (figure 4.2) that reasonably predicts how perceived threats (risks) affect the strength and valence (positive or negative) of human-wildlife interactions. Interactions that affect human health and safety (e.g., attacks on people, collisions with vehicles)

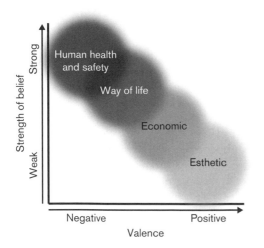

FIGURE 4.2 A proposed hierarchy of response (valence and strength) by people in interactions with wildlife, depending on what is being affected by the interaction.

this volume), is an attempt to influence human thought and behavior through information.

A common assumption of a cognitive fix is that the more information people are provided, the more they will agree with experts about the appropriateness or consequences of reintroducing a species. Yet, little evidence exists to support the notion that providing information alone will change behaviors as intended (Ajzen 1991, Gregory and Miller 2000, Heberlein 2012). The most effective communication efforts will be those that are coordinated and focused on the anticipated impacts depicted in figure 4.2 during each step of a reintroduction process vis-à-vis figure 4.1. Each step of a reintroduction process will have its own unique objectives for communication (George and Sandhaus, Chapter 14, this volume).

are likely to be resisted more than threats to economic well-being (e.g., damage to hay crops, loss of livestock), which in turn are likely to create greater resistance than threats to esthetic conditions (browsing damage to household plantings, animals tipping over garbage cans). The types of threats are completely consistent with the notion of impacts (Riley et al. 2002) advocated for throughout this chapter.

To illustrate subtle but important differences in the role of attitudes, Heberlein (2012, 8) distinguishes three types of fixes for environmental problems: technological, structural, and cognitive. Reintroductions are a classic example of a technological fix to extinction or low abundance of a species, where managers attempt to actively influence the environment. The role of attitudinal assessment in such situations is to gauge whether the technology (e.g., reintroduction) is consistent with dominant public beliefs and values (Bright and Manfredo 1996). A structural fix is when the situation influences human behavior. Structural fixes require changing behaviors of people to ensure survival and sustainability of a reintroduction. A cognitive fix, the most popular approach and the goal of most communication and education campaigns (George and Sandhaus, Chapter 14,

INSIGHTS FROM ATTITUDES ABOUT REINTRODUCTIONS Attitudes have been measured mostly in reintroduction efforts with high-profile species. For example, Browne-Nunez and Taylor (2002) described ≥50 reports on attitudes toward reintroduction of wolves in the United States, of which 36% were just from the Greater Yellowstone Ecosystem. This is to be expected because resistance to reintroductions is most likely to occur with species perceived as a threat to human safety, economics, or way of life. A meta-analysis, including 105 quantitative surveys conducted in 24 countries in Europe from 1976 to 2012, suggests that people's attitudes were more positive toward bears than wolves (Dressel et al. 2015). Attitudes toward bears became more positive over time, but attitudes toward wolves seemed to become less favorable the longer people coexisted with them.

There has been a tendency for human dimensions research to be employed once conflict arises rather than proactively in the planning process as we are advocating (Decker et al. 2012b). Inquiry early rather than late in the process establishes legitimacy, enables proactive conflict mitigation, and provides reference points for monitoring. Attitudinal inquiry, in combination with qualitative analysis through methods such as frame or discourse analysis, also promotes learning about how people and communities adapt, or not, to reintroduced species. This sort of learning strengthens guidance for future reintroductions. Support for moose restoration in the Adirondacks occurred in part because of early inquiry into factors affecting acceptance of moose, which led to a decision to promote unassisted recolonization by moose as opposed to active reintroduction (see Box 4.2). This sort of action represents how it is possible to step back to a previous stage in the reintroduction process, or return to the conceptualization stage to move forward again with what might be the most probable way to ensure conservation of a species.

Norms

Social psychologists define concepts of norms in the context of personal (individual) and social (shared norms) norms (see Vaske and Manfredo 2012). Norms, which are more stable than attitudes, help explain why people behave the way they do. Internal sanctions, beliefs about positive or negative consequences of a behavior, and feelings of personal responsibility are all norms that affect behavior. At another level, norms widely shared by a society (e.g., not killing wild animals outside of established hunting seasons) sometimes become codified into legal mandates (laws) with external sanctions. Norms influence social feasibility of any management action and are especially important to reintroductions (Zinn et al. 1998). Attempts to change norms are one form of a structural fix suggested by Heberlein (2012).

Acceptance Capacity

Wildlife acceptance capacity was originally defined analogous to biological carrying capacity. That is, capacity was defined in terms of the maximum population of a species acceptable to people in an area at any given time (Carpenter et al. 2000). Wildlife stakeholder acceptance capacity has been refined as the ability of a human population to live with the frequency and quality of human-wildlife interactions

(Carter et al. 2012, Zajac et al. 2012). Acceptance capacity is based on stakeholders' perceptions of a species' impacts on human land use, environments, other species, or people's well-being (Zinn et al. 2000). Personal values, beliefs, and attitudes, however, are part of a much broader sociopolitical complex within society, which affects attitudinal formation (Kleiven et al. 2004). Larger sociopolitical or cultural conflicts occur, which might override biological feasibility. This was observed in stage 2 (figure 4.1) of a potential reintroduction of grizzly bears into the Selway-Bitterroot area of the Rocky Mountains (Dak 2015) as well in the case of white-tailed eagles in Scotland (Arts et al. 2014). To achieve legitimacy for reintroductions it is necessary to take these broader aspects into consideration of what factors affect the acceptance capacity.

Investigations of factors affecting wildlife acceptance capacity suggest a general hierarchy of interactions that affect tolerance for interactions with wildlife. Interactions that threaten human health or safety elicit a stronger negative reaction than those that threaten economic well-being, which in turn create more negative influence on acceptance than interactions of a more esthetic nature. For instance, in Norway, brown bears and wolves were found to be less acceptable than lynx (*Lynx lynx*) or wolverine (*Gulo gulo*), especially when in close proximity to people (Kleiven et al. 2004). Similar results have been reported in Sweden (Dressel et al. 2015) and in a meta-study of attitudes toward wolves and brown bears in Europe (Dressel et al. 2015). Perceptions of risk—risk to human health and safety, economics, way of life—play a role in the social feasibility of reintroductions or restoration efforts (Morzillo et al. 2007, Rice et al. 2007, Carter et al. 2012), although these risk perceptions are usually context specific (Riley and Decker 2000, Kleiven et al. 2004). Under this hierarchical scheme, species such as burbot, for which no harmful effects to humans are perceived (Worthington et al. 2010), are not expected to create public resistance to reintroduction.

BOX 4.2 · Stakeholder Participation and Moose in the Adirondacks

Bruce Lauber

4b A moose (*Alces alces*) being transported by helicopter in Michigan
(credit: Michigan Department of Natural Resources).

In the early 1990s, the New York State Department of Environmental Conservation (DEC) was considering the reintroduction of moose to northern New York State (figure 4b). Moose are native to New York, but were extirpated by the early twentieth century because of excessive hunting and habitat degradation. Occasional stray moose entered New York from Canada or Vermont over the years, and by about 1980 a small permanent moose population took up residence in the Adirondack region of northern New York again. By the 1990s, estimates of the size of the population ranged from 25 to 50 animals.

Biologists expected this population to continue to increase naturally to a population of 1,300 animals within 40 years. Because of public interest in moose, however, DEC developed a proposal to restore moose by accelerating the process of moose's natural return through reintroduction of animals to unoccupied native habitats. In specific, they proposed transporting approximately 100 additional moose into the state over a period of several years, with the expectation that the population would increase to 1,300 animals within 20 years, half the time that could be expected otherwise. The estimated cost of the reintroduction was $1.3 million. DEC knew the proposal had potential

to be controversial. The relationship between state government and Adirondack residents historically had been strained for more than 100 years. Conflict erupted periodically about the degree to which natural resources in the region should be protected by land acquisition and restrictions on development, with local residents concerned about the effects of resource protection on economic development. The early 1990s was a particularly tense period because a governor's commission released a report calling for significant additional land acquisition in the region.

In addition to the potential for public controversy, the DEC anticipated specific concerns about human-moose interactions. Moose are such large animals that moose-vehicle collisions in areas with sizable moose populations can be a danger. Because DEC did not have the authority to allow moose hunting, they were concerned that the moose population could grow out of control, leading to an unacceptable level of moose-vehicle collisions. The DEC also believed that some residents would oppose the proposal on the basis of cost.

DEC officials decided to proceed with the reintroduction only if they were granted the legal authority to initiate a moose-hunting season when northern New Yorkers became concerned about the number of moose-related problems. They also proposed to fund the reintroduction with voluntary donations. The DEC recognized, however, that it was not just the proposal itself that would determine the acceptability of moose reintroduction—the process used to make a decision about moose reintroductions would have a big influence on public perceptions. People can support a decision, but dislike the process used to make that decision. Conversely, people may dislike a decision, but think that it was reached through a fair process.

Consequently, the DEC provided numerous opportunities for citizen participation in the decision-making process. DEC staff believed that the final decision to introduce moose should be based on the desires of New York State citizens, particularly citizens living in northern New York. Moose were not an endangered species, and the DEC did not think the restoration of moose in New York was a critical issue. Staff members involved in the process tried to play a neutral role in regard to the reintroduction proposal.

Early in the process, four scoping meetings were held to evaluate potential citizen concerns about moose. Based in part on input from these sessions, the DEC developed a draft environmental impact statement (EIS) describing the reintroduction proposal, several alternative moose management options, including the preferred alternative of accelerating the reintroduction of moose, and the anticipated consequences of each option. The draft EIS was released to the public in June 1992.

A public comment period followed the release of the EIS. The DEC held 15 public meetings throughout New York State to explain the reintroduction proposal and accept citizen comments. About half of these meetings were held in northern New York, the region where moose were to be reintroduced. The DEC also solicited letters from interested citizens. Finally, in December 1992, the DEC conducted a telephone survey of randomly selected residents of northern New York to determine what a representative sample of northern New Yorkers thought about the issue.

Based on the input received during the public comment period, the DEC announced in March 1993 its decision to abandon the active reintroduction proposal and, instead, allow moose to continue to return naturally to New York. The reaction to the decision was largely favorable. Most people did not oppose having moose in northern New York, but they believed that allowing moose to return on their own was different than actively accelerating the return of moose. Transporting moose into northern New York was viewed as "unnatural" by some, and many thought it was wrong to accelerate a reintroduction if that would accelerate the time at which moose-related problems began to surface.

But perceptions of the decision were also heavily influenced by perceptions of the decision-making process. The extensive public involvement period was viewed as a fair and appropriate way to make the decision. DEC's receptivity to citizen input had a big influence on the perceived fairness of the process. This perceived receptivity was reinforced by the fact the DEC decided not to adopt its preferred alternative based on the public input it received. People who were most supportive of DEC's decision-making process were those who attended at least 1 of the 15 public meetings DEC held—people who only wrote letters to DEC or participated in the telephone survey were less supportive. These meetings allowed for more interaction between DEC and citizens, allowing DEC to communicate its reasoning more clearly and facilitating opportunities for relationship building between the agency and various publics.

Enck and Brown (2000) warned against using the simple measure of positive or negative attitudes toward reintroduction as a sole input rather than a more comprehensive assessment of the capacity of the community to sustain a reintroduction. Attitudes toward the species do not necessarily correspond perfectly, or at all, with attitudes toward a reintroduction. The important point is that when wanting to know about attitudes toward a reintroduction, an accurate estimate will be achieved only when the referent of the attitude—the reintroduction—is the focus of inquiry. Nonetheless, identifying at a local level the antecedents of tolerance (or intolerance) for a species being reintroduced is instrumental in achieving goals for any specific reintroduction. Insights into factors affecting stakeholder wildlife acceptance capacity can help managers develop reintroduction programs to foster benefits and mitigate costs and to consider wider aspects of throughput legitimacy such as efficacy, transparency, and accountability.

Social Feasibility Assessments

As mentioned earlier, few a priori social feasibility assessments of reintroductions exist; most are post hoc reviews. We review here published attempts to demonstrate different possible approaches that have been taken, which might reveal some lessons for future reintroductions. McClafferty and Parkhurst (2001) reported on an effort to integrate habitat data with estimates of socioeconomic costs and benefits of reintroducing elk (*Cervus elaphus*) in Virginia. They used surveys and workshop venues to identify stakeholder reasons for support and opposition to reintroductions of elk. Chief social reasons for reintroductions were related to restoration of elk as part of the natural heritage of Virginia, educational values, and to increase opportunities to view elk. Yet, three of the top five reasons for reintroductions of elk were more biocentric: elk were present historically; reintroduction would benefit biodiversity; and the belief that elk have a right to exist in native range. Arguments against reintroduction were related to human safety from automobile collisions, agricultural damage, disease transmission to other wildlife as well as livestock, and a belief that reintroduction efforts were a poor use of scarce financial resources. The process resulted in a cost-benefit analysis that ranked feasibility along a gradient of high to low, and identified specific areas of the state where societal support for elk existed.

A similar social feasibility assessment was conducted for elk in New York (Enck and Decker 1999), but focused more on capacity of human communities to identify opportunities and achieve benefits, as well as mitigate negative consequences, of reintroductions. A greater effort in this study was on how restoration of elk through a reintroduction aligned with existing community planning. High social feasibility, based on cost-benefit analyses, did not nec-

essarily translate into community support for reintroduction of elk. Rather, this study's findings identified the social context within which reintroductions could be deliberated.

Another attempt at a social feasibility assessment for reintroducing wolves to the Adirondack Mountain region of New York outlined possible procedures to include in such an assessment (Enck and Brown 2000). The main focus was on describing the context, including social and economic data, to estimate capacity of communities to respond to changes such as the presence of wolves. Otherwise, this Adirondack study was a variation on the cost-benefit approach, embellished by consideration of impacts—the important effects on human values (Riley et al. 2002)—created by wolves on agriculture, deer hunters, and tourism. The researchers also measured attitudes toward wolves locally and regionally. Attitudes toward wolves were generally positive—most residents reported beliefs that effects of wolves would be positive or the potential negative effects were not likely to happen—and yet, stakeholders were almost evenly split in terms of whether they approved or disapproved of wolf reintroductions to the Adirondacks.

Although no single feasibility assessment method specific to reintroductions has been tested, the main considerations are the local context, and assessments of benefits and costs of reintroductions—all the benefits and costs, not just those that can be monetized. A variety of social impact assessment techniques exist (see Becker 2001) and provide useful guidance. In some cases, especially in the United States, a formal environmental impact statement or environmental impact assessment must be followed to meet legal obligations.

An impacts approach, introduced by Riley et al. (2002, 2003) and expanded on by Decker et al. (2014a, b), provides methods specific to wildlife that considers stakeholder-defined outcomes (desired impacts) through reliance on stakeholders' input for identifying and weighing competing desired outcomes. The main metric is impacts, which are the important (importance defined by stakeholders) effects of events such as reintroductions on human values (Riley et al. 2002). The number of animals in a population (a more typical metric) is a relevant metric, but overly narrow and generally inaccurate measure of sustainability in reintroductions (Macdonald 2009). Numbers, obviously, are an important consideration, yet population estimates cannot solely predict or infer acceptance capacity and thus legitimacy of reintroductions in a democratic society. An impact-oriented approach during Stage 2 encourages participants to concentrate on social as well as biological feasibility.

Governing Reintroductions

If a reintroduction appears feasible from environmental and sociocultural perspectives, a necessary step toward success is design of the governance and management of reintroductions. Since governance increasingly is a shared responsibility of state, market, and civil components of society (Pierre and Peters 2005, Rudolph et. al. 2012), there are varied coexisting modes of governance that consider aspects or dimensions of legitimacy (input, throughput, and output). These can be arranged along a continuum with centralized top-down governance at one end and bottom-up or self-governance at the other end (Lange et al. 2013). Various forms of co-governance, such as decentralized governance, public-private partnerships, and interactive governance are classified in the range between top-down and bottom-up. Similarities exist among the forms of co-governance, yet those tend to emphasize different aspects of legitimacy, which in turn implies that they have different potentials and limitations in the "reintroduction landscape" (Dunham et al., Chapter 5, this volume).

Top-down approaches emphasize the input dimension and decision-making through democratic participation or representation with an emphasis on inclusiveness or consent by affected stakeholders. Bottom-up approaches emphasize efficiency, which often is linked with output legitimacy (Kronsell and Bäckstrand 2010).

Various forms of co-governance, including both public and private actors, emphasize throughput legitimacy and aim at including both stakeholder-oriented input legitimacy and results-oriented output legitimacy. This is achieved by efforts to create interaction, integrative learning, and structured processes and procedures (Schmidt 2013). Depending on objectives, feasibility, and initial reasoning for a reintroduction, it is possible to choose between different modes of governance or combinations thereof.

Stage 3: Implementation

Considerable attention is given in this book about various ways to increase the probability of success in the implementation stage (see chapters in Part 3, this volume). From a human dimensions standpoint, communication and continual involvement on the part of stakeholders is important in maintaining transparency, inclusiveness, and fairness from which legitimacy and support for reintroductions is sustained. Most reintroductions require multiple releases over time (see Chapters 12 and 13). This sets up a potential situation for learning about communication through adaptive management (McCarthy et al. 2012), but also necessitates formation of a sustained communication planning effort (Shanahan et al. 2012, George and Sandhaus, Chapter 14, this volume). Human dimensions insights from past reintroduction efforts, and inquiry during the conceptualization and feasibility stages, can greatly contribute to more thoughtful communication leading up to and during actual reintroductions.

At the implementation stage, however, it is important to remember Heberlein's (2012) distinction between technological, cognitive, and structural fixes. Experts tend to focus on the cognitive fix (i.e., provide information) based on the assumption that more information will change human attitudes and behavior about the technological fix (i.e., reintroduction). The correlated belief is that simply providing information equates to communication, but it is more complex if the intended outcome is a behavioral change (Shanahan et al. 2012). Jacobson (2009) and Shanahan et al. (2012) provide thorough reviews of techniques and skills needed for effective communication. It is difficult to overemphasize the importance of skilled communication. Sustainable governance or reintroductions seldom occur in the absence of communication with stakeholders (George and Sandhaus, Chapter 14, this volume).

Stakeholder engagement is the process of involving and interacting with stakeholders in making, understanding, implementing, or evaluating management decisions (Lauber et al. 2012). An engagement approach as opposed to merely an audience for communication outward, if initiated early in the reintroduction process—Stage 1: Conceptualization—and sustained, will aid in minimizing "surprises" during the implementation stage. These surprises may come in the form of lawsuits, political interference, or abrupt change in the decision by policy-makers (Dak 2015). Engagement, especially when started early, can serve as more than a communication effort by facilitating coordination among agencies, nongovernmental organizations, and interested stakeholders.

A benefit of collaborative governance schemes arising from stakeholder engagement is that they create opportunities to pool resources and expertise, among which are capabilities in communication. Communication, when it comes from a source viewed as independent, builds transparency and thus legitimacy to management efforts such as reintroductions (Jacobson 2009). Resistance to lake trout reintroductions in the Great Lakes brought those efforts to a halt before a concerted engagement process was developed (see Box 4.3). Use of communication technologies, such as webcams, GIS mapping, and social media might not be resources available to all governmental agencies around the world, yet where available they can contribute to transparency and communicate progress.

Presumably, if the processes and insights promoted in this chapter are followed, conflict will be less than it might otherwise be. None-theless, sustainable reintroduction planning includes strategies for how to manage conflict and disputes. There are numerous methods to draw upon (Woodroffe et al. 2005). Alternative dispute resolution techniques (Maguire and Boiney 1994) typically include mediation, arbitration, neutral evaluation, consensus building, or ombudsmen. One thing dispute resolution processes almost always have in common is an objective third party, often referred to as a "neutral." The neutral party, who has no substantive stake in the outcome, orchestrates disputes and ensures that the chosen course of action, whatever it might be, is decided upon inclusively and implemented fairly.

Stage 4: Learning

Monitoring is an important stage in learning related to reintroductions (Gitzen et al., Chapter 12, this volume). Nichols and Armstrong (2012) describe three essential questions to monitoring that are equally pertinent to human dimensions aspects of reintroductions: why, what, and how? "Why" relates to the role of monitoring, which is evaluation for accountability, yet also for learning about reintroduction as a management intervention or technical fix. "What" is meant to address which "state variables" to measure—quantities that provide insight into the health or success of the reintroduction. "How" refers to sampling schemes used to monitor and the inferences made from that sampling. Nichols and Armstrong (2012) and Gitzen et al. (Chapter 12, this volume) provide a thorough review of typical biological state variables and monitoring techniques used in reintroductions.

Monitoring of the sociocultural attributes is as essential as monitoring of the reintroduced species' situation. Objectives formed for sociocultural outcomes related to proposed reintroduction during the conceptualization and feasibility stages form a basis for monitoring that supplements monitoring of biological state variables. The state variables monitored should provide a basis for assessing whether objectives

BOX 4.3 · Lake Trout Reintroductions in the Laurentian Great Lakes

C. C. Krueger and T. J. Newcomb

4c Reintroduction has been a key but also controversial tool for restoring lake trout (*Salvelinus namaycush*) populations in the Great Lakes of North America (credit: Andre Muir).

Lake trout (*Salvelinus namaycush*) were the dominant native predator in the Laurentian Great Lakes when Europeans first occupied North America. The species showed a diversity of life forms, which function much like distinct species, each with unique habitat needs for feeding and spawning. Louis Agassiz, an early naturalist, noted use of Lake Superior lake trout for food and barter by aboriginal peoples and their recognition of different lake trout types, especially the deep-water fat form called siscowet. With European colonization during the latter half of the 1800s, commercial fisheries flourished, along with widespread environmental changes in the Great Lakes Basin that came with shipping, agriculture, exotic species, and industrial pollution. The population response by lake trout was catastrophic. With the exception of Lake Superior and two small remnant populations in Iroquois Bay and Parry Sound in Lake Huron, lake trout became extinct from the Great Lakes. This story repeated itself in each of the other four Great Lakes.

Stocking of nonnative Pacific salmon in the 1960s, in part to control exotic nuisance alewife (*Alosa pseudoharengus*) that flourished in absence of lake trout, changed the direction of the Great Lakes fisheries from a commercially dominated native species fisheries to a spectacular recreational fisheries for nonnative chinook (*Oncorhynchus tshawytscha*) and coho salmon (*O. kisutch*). Thus began a long-running conflict among anglers, environmentalists, and natural resources agencies over lake trout restoration: whether or not to pursue restoration of a native species through reintroduction of original life forms of lake trout. Some stakeholders desired lake trout to be restored and others did not. Reintroductions as part of fisheries management within the Great Lakes are especially difficult due to the complexity of the ecosystem and its governance.

Remarkable in the governance structure is that the shared waters between two countries is predominately managed by subnational governments rather than federal governments, and that interjurisdictional management by subnational governments is achieved voluntarily and is not mandated by legislation or treaty. Governance of Great Lakes fisheries involves eight US states, one Canadian province, and three tribal organizations. Lake trout stocking and biological assessments of the lakes are conducted by federal and the state and provincial agencies (figure 4c). Tribal fishery jurisdictions have been affirmed by courts over treaty-ceded areas of US waters of Lakes Superior, Huron, and Michigan. The Great Lakes Fishery Commission, a transnational organization, was established in 1955 and charged with coordinating management and implementing a program of sea lamprey control. Furthermore, state jurisdictions are funded primarily through fishing license sales and excise taxes on the sale of fishing equipment. Thus, stakeholder support for the direction of the fishery is crucial to the agencies' quest for lake trout restoration.

The first rounds of plans to restore native lake trout, however, were developed with little public involvement. Reintroduction of lake trout, fishery regulation, and sea lamprey control were the primary management actions used. Unfortunately, the establishment of self-sustaining, naturally reproducing populations showed little progress over the first 20 years. Stocked hatchery lake trout survived to maturity, but little detectable natural recruitment occurred. At the same time, management agencies recognized that single-species plans needed to be consistent with the public's desires for the fish community as a whole, and consistent with the overall ecology of the systems. Thus, after the first lake trout restoration plans were developed, the agencies began development of "fish community objectives" (FCOs) and married FCOs to a planned review of progress every 5 years via "state of the lake reports." As programs moved into the late 1990s and 2000s, the lakes further changed with new invasions of nonnative species, both fishes and invertebrates, which fundamentally changed the ecology of the lakes. The strategies to achieve lake trout restoration and the FCOs had to be revised and updated. By then, the sociocultural situation had also changed with legal requirements of more public participation and formal consultation bodies, such as Citizen's Fishery Advisory Councils in Michigan, were in place for coordinated input. The charge of the Lake Huron Citizens Fishery Advisory Committee, for example, was to review and provide recom-

mendations and direction for the fishery goals, objectives, and management plans for Lake Huron. These councils were organized to represent a diverse set of stakeholder interests in lake management.

In the late 1990s, tensions among stakeholders related to salmon versus lake trout intensified as assessment data in Lake Huron once again suggested predation was exceeding capacity of the available forage base, particularly alewife. As agencies moved to reduce stocking of salmon, various public protested vociferously. The public focused on reduction of all predators, lake trout included, thus potentially jeopardizing the restoration program.

To move forward with public discourse in 2003, a portfolio of science communication tools was developed. A targeted communication plan for stakeholders, the governor's office, the legislature, and the general public was implemented. A transparent path to a final decision was an essential message. Although the process took 12 months to accomplish, agencies were able to reduce Chinook salmon stocking by 50%, retain the lake trout reintroductions, and a framework was established for future reintroduction. The success of this public approach was evidenced in 2013, when another Chinook salmon stocking reduction was successfully implemented as the species was reproducing in the wild in Ontario tributaries and the forage base remained suppressed.

Changes in the ecology of Lake Huron during the past 10 years favored lake trout and other native species; these species are rebounding! Evidence of natural reproduction of lake trout has been found throughout Lake Huron, with some areas having more than 50% of spawning lake trout being wild fish. Fisheries agencies on the lake are discussing a reduction of lake trout reintroductions over concerns not to suppress wild lake trout production through a density-dependent process. The lake trout fishery rehabilitation success was predicated on gaining support from the recreational fishery, on making information widely available, and on creating transparent processes of how reintroductions were decided upon and conducted.

have been met. These might include costs and benefits predicted during the feasibility stage, such as predation on livestock from reintroduced predators (cost), or tourist revenue from increased visitation related to reintroduced species (benefits). The state variable might also be human attitudes or behaviors that provide insight into acceptance capacity for the reintroduced species.

Answers to why and what questions related to monitoring depend on the situation (Nichols and Armstrong 2012). The answer to "how" depends on accepted statistical procedures and ability to draw cause-and-effect inferences. There is an array of procedures on which to draw (Bickman and Rog 2008). The appropriateness of techniques used depends again on the specifics of the reintroduction situation. Many techniques, such as survey research, are not feasible in places without the infrastructure to conduct reliable surveys (Carter et al. 2012). If assessments were conducted during the feasibility stage, learning can take place by revisiting all or a sample of the population measured previously (e.g., attitudes toward reintroduction before and after). There is no single best way to monitor. Different schemes may be used for different species and for different situations. Techniques that reveal the most about the reintroduction and focus on the outcomes of objectives established prior to reintroduction are likely to generate the most learning.

An often overlooked issue is the target of learning from monitoring. Most of the suggested approaches focus on learning by experts (Nichols and Armstrong 2012). Although expert knowledge is needed, approaches that develop social learning simultaneously contribute to evaluation and social acceptance (Schusler et al. 2003). Those authors list eight characteristics that foster social learning: open communication, diverse participation, unrestricted thinking, constructive conflict, democratic structure, multiple sources of knowledge, extended engagement, and facilitation. Many of these characteristics are discussed within the present chapter for inclusion at all stages of reintroduction conducted under a framework of collaborative governance (Rudolph et al. 2012). Although involvement of trained scientists is needed to maintain reliability, the nascent field of citizen science or the involvement of volunteers in research and monitoring is one form of stakeholder engagement that potentially helps with inherent costs and logistical difficulties of monitoring (Dickinson et al. 2010, Gitzen et al., Chapter 12, this volume).

Monitoring presents challenges because it is difficult to do well, logistics are often difficult in the far-flung places where many animals are being reintroduced, and once the animals are reintroduced (normally an expensive undertaking), the incentives for continued investment diminish ("couldn't money be better spent on another reintroduction?"). Challenges of monitoring (Gitzen et al., Chapter 12, this volume) frequently lead to a lack of monitoring, which only serves to slow the acquisition of knowledge needed for sustained reintroductions of fish and wildlife populations.

This chapter (and several others in this volume) mostly focuses on how insights from human dimensions of wildlife research and other social sciences can help inform decisions about reintroductions. It really is a call, however, for a comprehensive approach to conceptualizing, assessing feasibility, implementing, and evaluating reintroduction like any other management intervention. There is a history of calling on human dimensions insights to resolve conflict or administer an opinion poll late in the management process—too late to be effective (Lauber et al. 2012). The sustainability of reintroductions will not depend on political opinion polls. Rather, it will be influenced by stakeholder engagement that focuses on factors affecting legitimacy of the process: inclusiveness, transparency, and accountability as well as an effective outcome.

FUTURE RESEARCH AND DEVELOPMENT

Uncertainties associated with reintroductions are prevalent for human as well as environmental dimensions of reintroductions. Those uncertainties can be reduced through adaptive approaches complemented by social science (see also George and Sandhaus, Chapter 14, this volume) about the need to learn how to better integrate outreach and engagement in the reintroduction process). Research to develop stronger social feasibility assessments that take into account more than mere attitudes toward the species, or even attitudes toward reintroductions, can be expected to aid sustainability of reintroductions. Evaluation of context-specific attributes associated with success and failure of reintroductions will contribute to developing stronger feasibility assessments. A focus on capacity of communities to mitigate costs while bolstering benefits to society from reintroductions will provide insights into how to better address long-term sustainability. Inclusion of more political and government science is especially needed because nearly all reintroductions are done by governments, with governments, or with delegated authority from government (see also Dunham et al., Chapter 5, this volume). To be sustainable, the complex nature of reintroductions requires more governance than is provided by a managerial, command-and-control approach. What sorts of governance structures lead to changes in capacity of communities to adapt to the myriad changes created by reintroductions? Ideally, time-series investigations would provide insights into longer-term costs and benefits of reintroductions to society, yet funding for such efforts is generally episodic around political crises or specific, one-time events of a reintroduction.

SUMMARY

Human dimensions aspects of reintroductions are those attributes not directly about the organism and its habitat. Our objective is to offer a framework that guides human-assisted reintroduction of fish or wildlife populations as well as recolonization (those reintroductions that occur without active human intervention). Sustainable reintroductions of fish and wildlife

populations, a technological fix to an environmental problem, are more likely to occur if those reintroductions are approached through a deliberate and multidisciplinary process. We outline a four-step process that includes human dimensions considerations while advocating an adaptive process. That adaptive process is initiated with conceptualization that includes well-articulated, stakeholder-informed objectives for the reintroduction—sociocultural as well as biological—formed through public discourse. Feasibility of reintroductions is assessed using methods of social science inquiry that assess anticipated benefits weighed against likely costs. Species that pose risks—assessed or perceived—to human communities are likely to meet greater resistance and require increased capacity of those communities to cope. A hierarchy of risks exists that reasonably predicts threats to human health and safety are likely to be resisted more than threats to economic well-being, which in turn are likely to create greater resistance than threats to esthetic conditions. Although no single best social feasibility assessment technique exists, we suggest an impacts-oriented approach to complement biological feasibility assessments. Once feasibility is assessed, the next stage is the reintroduction, which is likely to be more successful when accompanied by a well-thought-out communication plan that addresses concerns revealed in the feasibility assessment as well as progress achieved in the reintroduction. A dedicated learning and monitoring stage assures that knowledge gained from a reintroduction effort will be available to reduce uncertainties associated with future endeavors.

MANAGEMENT RECOMMENDATIONS

The probability of sustainable reintroductions is increased when structured, deliberative, and adaptive processes are articulated and followed (figure 4.1). Human dimensions inquiry and insights inform each step of those processes to ensure alignment of environmental and sociocultural considerations. Human dimension insights and processes are most useful when employed early in conceptualization of reintroductions and feasibility analyses rather than brought in late to understand or mitigate conflict. The following are suggestions to help improve the probability that human dimensions inquiry and insights contribute effectively to sustained reintroductions:

- Reintroduction programs that are adaptive—focused on learning—are most effective when they follow a process of conceptualization that is inclusive of stakeholders and defines specific objectives with measurable metrics of progress; feasibility assessment that includes social feasibility to complement traditional environmental feasibility assessments; implementation efforts supported by communication planning; learning through monitoring of variables established in conceptualization stage.

- The explicit possibility that a reintroduction might not occur or that the process can go back a stage if perceived costs exceed benefits will build trust and legitimacy.

- Feasibility assessments focused on the capacity of the system—human and natural—to adapt to population relocations will be more informative than merely measuring attitudes. Attitudes, when measured, should focus on attitudes toward the reintroduction rather than the species. Acceptance capacity is a useful framework for assessing attributes of feasibility.

- Simply providing more information cannot be expected to change attitudes (cognitive change) toward a reintroduction (technological change). Rather, structural change affected by participatory processes and legitimacy will increase the chances of human behaviors conducive to sustainable reintroductions.

- Monitoring and evaluation of reintroductions that share learning among experts as well as stakeholders contribute to current and future successful reintroductions.

LITERATURE CITED

Adger, W. N., T. A. Benjaminsen, K. Brown, and H. Svarstad. 2001. Advancing a political ecology of global environmental discourses. Development and Change 32:681–715.

Ajzen, I. 1991. The theory of planned behavior. Organizational Behavior and Human Decision Processes 50:179–211.

Ajzen, I., and C. N. Gilbert. 2008. Attitudes and the prediction of behavior. Pp.289–311. In W. D. Crano and R. Prislin, eds., Attitudes and Attitude Change. Psychology Press, New York.

Arts, K., A. Fisher, and R. van de Wal. 2012. Common stories of reintroduction: a discourse analysis of documents supporting animal reintroduction to Scotland. Land Use Policy 29:911–920.

Arts, K., A. Fisher, and R. van de Wal. 2014. Political decision making, governance shifts and Scottish animal reintroductions: are democratic principles at stake? Journal of Environmental Planning and Management 4:612–628.

Bäckstrand, K., J. Khan, A. Kronsell, and E. Lövbrand. 2010. Environmental politics after the deliberative turn. Pp.217–234. In K. Bäckstrand, J. Khan, A. Kronsell, and E. Lövbrand, eds., Examining the Promise of New Modes of Governance. Edward Elgar, Cheltenham, UK.

Becker, H. A. 2001. Social impact assessment. European Journal of Operational Research 128:311–321.

Beetham, D. 1991. The Legitimation of Power. Macmillan, Basingstoke, UK.

Bekoff, M. 1999. Jinxed lynx? Journal of Applied Animal Welfare Science 2:239–242.

Bernstein, S. 2005. Legitimacy in global environmental governance. Journal of International Law and International Relations 1:139–166.

Bickman, L., and D. J. Rog. 2008. The Sage Handbook of Applied Social Research Methods. Sage Publications, Thousand Oaks, CA.

Bjerke, T., J. Vitterso, and B. P. Kaltenborn. 2000. Locus of control and attitudes toward large carnivores. Psychological Reports 86:37–46.

Botkin, D. B. 2012. The Moon in the Nautilus Shell: Discordant Harmonies Reconsidered. Oxford University Press, New York.

Bright, A. D., and M. J. Manfredo. 1996. A conceptual model of attitudes toward natural resource issues: a case study of wolf reintroduction. Human Dimensions of Wildlife 1:1–21.

Browne-Nuñez, C., and J. G. Taylor. 2002. American's Attitudes toward Wolves and Wolf Reintroduction: An Annotated Bibliography. Information

Technology Report USGS/BRD/ITR—2002-0002. US Government Printing Office, Denver, CO.

Carpenter, L. H., D. J. Decker, and J. F. Lipscomb. 2000. Stakeholder acceptance capacity in wildlife management. Human Dimensions of Wildlife 5:5–19.

Carter, N. H., S. J. Riley, and J. Liu. 2012. Utility of a psychological framework for carnivore conservation. Oryx 46:525–535.

Chase, L. C., D. J. Decker, and T. B. Lauber. 2004. Public participation in wildlife management: what do stakeholders want? Society and Natural Resources 17:629–639.

Clark, J. D., D. Huber, and C. Servheen. 2002. Bear reintroductions: lessons and challenges. Ursus 13:335–345.

Converse, S. J., C. T. Moore, M. J. Folk, and M. C. Runge. 2013. A matter of tradeoffs: reintroduction as a multiple objective decision. Journal of Wildlife Management 77:1145–1156.

Dak, M. J. 2015. Grizzly West: A Failed Attempt to Reintroduce Grizzly Bears in the Mountain West. University of Nebraska Press, Lincoln.

Dalrymple, S. E., and A. Moehrenschlager. 2013. Words matter: a response to Jørgensen's treatment of historic range and definitions of reintroduction. Restoration Ecology 21:156–158.

Decker, D. J., A. B. Forstchen, J. F. Organ, C. A. Smith, S. J. Riley, C. A. Jacobson, G. R. Batcheller, and W. F. Siemer. 2014a. Impact management: an approach to fulfilling public trust responsibilities of wildlife agencies. The Wildlife Society Bulletin 38:2–8.

Decker, D. J., and K. G. Purdy. 1988. Toward a concept of wildlife acceptance capacity in wildlife management. Wildlife Society Bulletin 16:53–57.

Decker, D. J., S. J. Riley, J. F. Organ, W. F. Siemer, and L. H. Carpenter. 2014b. Applying Impact Management: A Leaders' Guide, 3rd ed. Human Dimensions Research Unit, Department of Natural Resources, Cornell University, Ithaca, NY.

Decker, D. J., S. J. Riley, and W. F. Siemer. 2012a. Human dimensions of wildlife management. Pp.3–14. In Decker, D. J., S. J. Riley, and W. F. Siemer, eds., Human Dimensions of Wildlife Management. Johns Hopkins University, Baltimore, MD.

Decker, D. J., S. J. Riley, and W. F. Siemer. 2012b. Adaptive value of human dimensions for wildlife management. Pp.248–255. In D. J. Decker, S. J. Riley, and W. F. Siemer, eds., Human Dimensions of Wildlife Management. Johns Hopkins University, Baltimore, MD.

Denton, J. S., S. P. Hitchings, T. J. C. Beebee, and A. Gent. 1997. A recovery program for the natterjack toad (*Bufo calamita*) in Britain. Conservation Biology 11:1329–1338.

Dickinson, J. L., B. Zuckerberg, and D. N. Bonter. 2010. Citizen science as an ecological research tool: challenges and benefits. Annual Review of Ecology, Evolution, and Systematics 41:149–172.

Dressel, S., C. Sandström, and G. Ericsson. 2015. A meta-analysis of studies on attitudes toward bears and wolves across Europe 1976–2012. Conservation Biology 29:565–574.

Dryzek, J. S. 2012. The Politics of the Earth: Environmental Discourses. Oxford University Press, Oxford, UK.

Eagly, A. H., and S. Chaiken. 1993. The Psychology of Attitudes. Harcourt Brace Jovanovich College Publishers, Fort Worth, TX.

Enck, J. W., and T. L. Brown. 2000. Preliminary assessment of social feasibility for reintroducing gray wolves to the Adirondack Park in northern New York. Human Dimensions Research Unit Series 00-03. Cornell University, Ithaca, NY.

Enck, J. W., and D. J. Decker. 1999. Assessing the social feasibility of restoring elk. Human Dimensions of Wildlife 4:68–69.

Gregory, J., and S. Miller. 2000. Science in Public: Communication, Culture and Credibility. Perseus, London, UK.

Hajer, M. 1995. The Politics of Environmental Discourse: Ecological Modernization and the Policy Process. Oxford University Press, Oxford, UK.

Heberlein, T. A. 2012. Navigating Environmental Attitudes. Oxford University Press, New York.

IUCN/SSC (International Union for Conservation of Nature Species Survival Commission). 2013. Guidelines for Reintroductions and Other Conservation Translocations, Version 1.0. IUCN Species Survival Commission, Gland, Switzerland.

Jacobson, S. K. 2009. Communication Skills for Conservation Professionals, 2nd ed. Island Press, Washington, DC.

Johansson, J. 2013. Constructing and Contesting the Legitimacy of Private Forest Governance: The Case of Forest Certification in Sweden. Department of Political Science, Umeå University, Umeå, Sweden.

Jørgensen, D. 2011. What's history got to do with it? A response to Seddon's definition of reintroduction. Restoration Ecology 19:705–708.

Jørgensen, D. 2013. Reintroduction and de-extinction. BioScience 63:719–720.

Kellert, S. R., and J. K. Berry. 1980. Knowledge, Affection and Basic Attitudes toward Animals in

American Society. US Fish and Wildlife Service, Washington, DC.

Keohane, R. O. 2011. Global governance and legitimacy. Review of International Political Economy 18:99–109.

Kleiman, D., G. Stanley Price, and B. B. Beck. 1994. Criteria for reintroductions. Pp.283–303. In P. J. S. Olney, G. M. Mace, and A. T. C. Feistner, eds., Creative Conservation: Interactive Management of Wild and Captive Animals. Chapman and Hall, London, UK.

Kleiven, J., T. Bjerke, and B. Kaltenborn. 2004. Factors influencing the social acceptability of large carnivore behaviors. Biodiversity and Conservation 13:1647–1658.

Kronsell, A., and K. Bäckstrand. 2010. Rationalities and forms of governance: a framework for analyzing the legitimacy of new modes of governance. Pp.28–46. In K. Bäckstrand, J. Khan, A. Kronsell, and E. Lövbrand, eds., Environmental Politics and Deliberative Democracy: Examining the Promise of New Modes of Governance. Edward Elgar, Cheltenham, UK.

Lange, P., P. J. Driessen, A. Sauer, B. Bornemann, and P. Burger. 2013. Governing towards sustainability: conceptualizing modes of governance. Journal of Environmental Policy and Planning 15:403–425.

Lauber, T. B., and B. A. Knuth. 1997. Fairness in moose management decision-making: the citizens' perspective. Wildlife Society Bulletin 25:776–787.

Lauber, T. B., and B. A. Knuth. 1999. Measuring fairness in citizen participation: a case study of moose management. Society and Natural Resources 11:19–37.

Lauber, T. B., D. J. Decker, K. M. Leong, L. C. Chase, and T. M. Shusler. 2012. Stakeholder engagement in wildlife management. Pp.139–156. In D. J. Decker, S. J. Riley, and W. F. Siemer, eds., Human Dimensions of Wildlife Management. Johns Hopkins University, Baltimore, MD.

Macdonald, D. W. 2009. Lessons learnt and plans laid: seven awkward questions for the future of reintroductions. Pp.411–448. In M. W. Hayward and M. J. Somers, eds., Reintroduction of Top-Order Predators. Blackwell Publishing, Oxford, UK.

Maguire, L. A., and L. G. Boiney. 1994. Resolving environmental disputes: a framework incorporating decision analysis and dispute resolution techniques. Journal of Environmental Management 42:31–48.

Mangel, M., L. M. Talbot, M. Meffe, G. K. Agardy, M. Tundi, D. L. Alverson, J. Barlow et al. 1996.

Principles for the conservation of wild living resources. Ecological Applications 6:338–362.

McCarthy, M. A., D. P. Armstrong, and M. C. Runge. 2012. Adaptive management of reintroductions. Pp.284–288. In J. G. Ewen, D. P. Armstrong, and K. A. Parker, eds., Reintroduction Biology: Integrating Science and Management. Blackwell, Chichester, UK.

McClafferty, J. A., and J. A. Parkhurst. 2001. Using public surveys and GIS to determine the feasibility of restoring elk to Virginia. Pp.83–100. In D. S. Maehr, R. F. Noss, and J. L. Larkin, eds., Large Mammal Restoration: Ecological and Sociological Challenges in the 21st Century. Island Press, Washington, DC.

Morell, V. 2008. Into the wild: reintroduced animals face daunting odds. Science 320:742–743.

Morzillo, A. T., A. G. Mertig, N. Garner, and J. Liu. 2007. Resident attitudes toward black bears and population recovery in East Texas. Human Dimensions of Wildlife 12:417–428.

Nichols, J. D., and D. P. Armstrong. 2012. Monitoring for reintroductions. Ppp.223–255. In J. G. Ewen, D. P. Armstrong, and K. A. Parker, eds., Reintroduction Biology: Integrating Science and Management. Blackwell, Chichester, UK.

Pate, J., M. J. Manfredo, A. D. Bright, and G. Tischbein. 1996. Coloradans' attitudes toward reintroducing the gray wolf into Colorado. Wildlife Society Bulletin 24:421–428.

Pierre, J., and B. G. Peters. 2005. Governing Complex Societies: Trajectories and Scenarios. Palgrave MacMillan, London, UK.

Reading, R. P., and S. R. Kellert. 1993. Attitudes toward a proposed reintroduction of black-footed ferrets (Mustela nigripes). Conservation Biology 7:569–580.

Rees, P. A. 2001. Is there a legal obligation to reintroduce animal species into their former habitats? Oryx 35:216–223.

Rice, M. B., W. B. Ballard, E. B. Fish, D. B. Webster, and D. Holdermann. 2007. Predicting private landowner support toward recolonizing black bears in the Trans-Pecos region of Texas. Human Dimensions of Wildlife 12:405–145.

Riley, S. J., and D. J. Decker. 2000. Wildlife stakeholder acceptance capacity for cougars in Montana. Wildlife Society Bulletin 28:931–939.

Riley, S. J., D. J. Decker, L. H. Carpenter, J. F. Organ, W. F Siemer, G. F. Mattfeld, and G. Parsons. 2002. The essence of wildlife management. Wildlife Society Bulletin 30:585–593.

Riley, S. J., W. F. Siemer, D. J. Decker, L. H. Carpenter, J. F. Organ, and L. T. Berchielli. 2003.

Adaptive impact management: an integrative approach to wildlife management. Human Dimensions of Wildlife 8:81–95.

Rudolph, B. A., M. G. Schechter, and S. J. Riley. 2012. Governance and the human dimensions of wildlife management. Pp.15–25. In D. J. Decker, S. J. Riley, and W. F. Siemer, eds., Human Dimensions of Wildlife Management. Johns Hopkins University, Baltimore, MD.

Sandström, C., J. Pellilekka, and O. Ratamäki. 2009. Management of large carnivores in Fennoscandia: new patterns of regional participation. Human Dimensions of Wildlife 14:37–50.

Sarrazin, F., and R. Barbault. 1996. Reintroduction: challenges and lessons for basic ecology. Trends in Ecology and Evolution 11:474–478.

Scharpf, F. W. 1999. Governing in Europe: Effective and Democratic? Oxford University Press, New York.

Schmidt, V. 2013. Democracy and legitimacy in the European Union revisited: input, output and "throughput." Political Studies 61:2–22.

Schusler, T. M., D. J. Decker, and M. J. Pfeffer. 2003. Social learning for collaborative natural resource management. Society and Natural Resources 16:309–326.

Seddon, P. 2010. From reintroduction to assisted colonization: moving along the conservation translocation spectrum. Restoration Ecology 18:796–802.

Shanahan, J. E., M. L. Gore, and D. J. Decker. 2012. Communication for effective wildlife management. Pp.157–173. In D. J. Decker, S. J. Riley, and W. F. Siemer, eds., Human Dimensions of Wildlife Management. Johns Hopkins University Press, Baltimore, MD.

Skogen, K., and O. Krange. 2003. A wolf at the gate: the anti-carnivore alliance and the symbolic construction of community. Sociologia Ruralis 43:309–325.

Smith, J. B., C. K. Nielsen, and E. C. Hellgren. 2014. Illinois resident attitudes toward recolonizing large carnivores. Journal of Wildlife Management 78:930–943.

Soorae, P. S. 2008. Global Re-introduction Perspectives: Re-introduction Case Studies from around the Globe. IUCN, Abu Dhabi.

Soorae, P. S. 2013. Global Re-introduction Perspectives: Further Case Studies from around the Globe. IUCN/SSC Re-introduction Specialist Group, Gland, Switzerland/Environment Agency, Abu Dhabi, UAE.

Svarstad, H., L. K. Petersen, D. Rothman, H. Siepel, and F. Wätzold. 2008. Discursive biases of the environmental research framework DPSIR. Land Use Policy 25:116–125.

Vaske, J. J., and M. J. Manfredo. 2012. Social psychological considerations in wildlife management. Pp.43–57. In D. J. Decker, S. J. Riley, and W. F. Siemer, eds., Human Dimensions of Wildlife Management. Johns Hopkins University Press, Baltimore, MD.

Whittaker, D. J., J. J. Vaske, and M. J. Manfredo. 2006. Specificity and the cognitive hierarchy: value orientations and acceptability of urban wildlife management actions. Society and Natural Resources 19:515–530.

Williams, C. K., G. Ericsson, and T. A. Heberlein. 2002. A quantitative summary of attitudes toward wolves and their reintroduction (1972–2000). Wildlife Society Bulletin 30:575–584.

Wilson, M. A. 1997. The wolf in Yellowstone: science, symbol, or politics? Deconstructing the conflict between environmentalism and wise use. Society and Natural Resources 10:453–468.

Woodroffe, R., S. Thirgood, and A. Rabinowitz. 2005. People and Wildlife, Conflict or Co-existence? Cambridge University Press, Cambridge, UK.

Worthington, T., J. Tisdale, P. Kemp, I. Williams, and P. E. Osborne. 2010. Public and stakeholder attitudes to the reintroduction of the burbot, Lota lota. Fisheries Management and Ecology 17:465–472.

Yohe, G, and R. S. J. Tol. 2002. Indicators for social and economic coping capacity: moving toward a working definition of adaptive capacity. Global Environmental Change 12:25–40.

Zajac, R. M., J. T. Bruskotter, R. S. Wilson, and S. Prange. 2012. Learning to live with black bears: a psychological model of acceptance. Journal of Wildlife Management 76:1331–1340.

Zinn, H. C., M. J. Manfredo, and J. J. Vaske. 2000. Social psychological bases for stakeholder acceptance capacity. Human Dimensions of Wildlife 5:20–33.

Zinn, H. C., M. J. Manfredo, J. J. Vaske, and K. Wittman. 1998. Using normative beliefs to determine the acceptability of wildlife management actions. Society and Natural Resources 11:649–662.

FIVE

The Reintroduction Landscape

FINDING SUCCESS AT THE INTERSECTION OF ECOLOGICAL, SOCIAL, AND INSTITUTIONAL DIMENSIONS

Jason B. Dunham, Rollie White, Chris S. Allen,
Bruce G. Marcot, and Dan Shively

FROM AN ECOLOGICAL PERSPECTIVE, an ideal site for a species reintroduction meets all of the conditions necessary to ensure establishment, long-term viability, and perhaps even spread of reintroduced species to other sites. Recently Osborne and Seddon (2012) described a useful and comprehensive list of considerations for the suitability of conditions in sites proposed for reintroductions (Box 5.1). The authors also emphasized the role of a variety of habitat models in providing a foundation for evaluating the biological suitability of sites for a reintroduction. In brief (see Osborne and Seddon 2012, for details), it is useful to consider the past, present, and future of habitat conditions, how this variability relates to the species of interest, and how the species itself may vary in space or through time with respect to its habitat requirements. This might seem to be a very tall order, but given the costs of conducting reintroductions and their limited record of success (Griffith et al. 1989, Wolf et al. 1996, Pérez et al. 2012) it is worth at least considering each specific point (Box 5.1) and explicitly acknowledging uncertainties where they exist, rather than taking hasty actions without fully considering the consequences.

Whereas the question of biological suitability of a given site can involve a rather complex suite of considerations (Box 5.1), definition of a site itself often is not clear. Accordingly, recent guidelines issued by the International Union for Conservation of Nature (IUCN) have distinguished a release "site" from a release "area" for reintroductions (IUCN/SSC 2013). According to the IUCN, a release site refers to the immediate vicinity in which a species is released for a reintroduction, whereas a release area comprises a much larger area into which released animals might disperse. Specific aspects of the release area listed by the IUCN/SSC (2013) include an area that will:

> Meet all the species' biotic and abiotic requirements, be appropriate habitat for the life stage released and all life stages of the species, be adequate for all seasonal habitat needs, be large enough to meet the required conservation benefit, have adequate connectivity to suitable habitat if that habitat is fragmented, and be adequately isolated from suboptimal or non-habitat areas which might be sink areas for the population.

1. Historical locations of a species' presence may not indicate a present-day suitable habitat.

2. Present-day locations of a species' presence may not indicate a currently suitable habitat.

3. Present-day locations where a species is absent may not indicate an unsuitable habitat.

4. Present-day locations of a species' presence may not indicate a future suitable habitat.

5. Not all suitable habitat patches may be colonized because landscape components may be missing.

6. A habitat's suitability and its characteristics likely vary across a species' range.

7. Individuals from across a species' range may not all be equally suited to the chosen release site.

8. A suitable habitat may need to be engineered (restored or created) to aid colonization and then managed to maintain its perceived value.

It is clear that a release "site" (as defined by the IUCN) is potentially necessary but might not be sufficient to ensure the long-term success of a reintroduction, which can ultimately depend on characteristics of the surrounding release area and landscape (see also Seddon 2013). Osborne and Seddon (2012) similarly emphasized the role of "landscape components" around reintroduction sites but neither their chapter nor IUCN guidelines provide more than a brief discussion of the topic. IUCN guidelines do highlight the importance of social and institutional factors, however, which influence landscapes and are important considerations in reintroductions (Reading et al. 2002; Chauvenet et al., Chapter 6, this volume; Riley and Sandström, Chapter 4, this volume).

Our objective in this chapter is to elaborate on the utility of applying landscape principles when planning species reintroductions. In doing so, we define the *reintroduction landscape*, which comprises ecological, social, and institutional dimensions, and encompasses release sites and areas described by the IUCN/SSC (2013). With this definition in hand, we review each dimension (ecological, social, and institutional) of the reintroduction landscape in further detail, beginning with consideration of salient ecological landscape processes. We follow this with a discussion of how social and institutional processes interact in landscapes to determine the success of species reintroductions. To illustrate the relevance of viewing reintroductions in a landscape context, we draw on a range of case studies involving recent reintroductions to illustrate key points. With this, we conclude with a concise set of guidelines for applying landscape principles to reintroductions.

THE REINTRODUCTION LANDSCAPE

We define the reintroduction landscape in terms of ecological, social, and institutional dimensions (figure 5.1). Ecological dimensions of reintroductions include the biotic and abiotic factors that allow a species to successfully establish a self-sustaining population in response to a reintroduction. Social dimensions include the values accorded to reintroduced species and associated political and social interactions driven by stakeholders (e.g., nongovernmental organizations or individuals with an interest in reintroductions). Institutional dimensions include formal governance structures (e.g., government agencies) and regulations that provide a framework within which social factors operate. Each of these dimensions has been emphasized in the literature on reintroductions, but they are infrequently considered together (but see IUCN/SSC

FIGURE 5.1 An illustration of relationships among the ecological, social, and institutional dimensions of the reintroduction landscape, with specific factors listed within each. Intersections among circles corresponding to only two of the three dimensions indicate how the potential success of a reintroduction might be limited by the third. Reintroduction landscapes are only considered to be fully suitable or appropriate when specified conditions for each dimension are met.

2013). The degree to which factors within each dimension must be addressed will vary on a case-specific basis, but it is clear in our experience that reintroductions that fail to consider each of these three dimensions are much less likely to succeed. Analyses of factors contributing to the success or failure of reintroductions typically focus on the immediate site of a reintroduction (Griffith et al. 1989, Seddon et al. 2007, Cochran-Biederman et al. 2014) or focus solely on biological or ecological factors (Ewen et al. 2012). It is well known, however, that social and institutional factors can play a significant, if not dominant role in determining success (Reading et al. 2002). In many cases, it may be most efficient to evaluate biological or ecological factors first to determine the potential feasibility of a reintroduction (e.g., Dunham et al. 2011, Pérez et al.

2012). If such an assessment indicates a reintroduction is ecologically feasible, then it is worth investing effort to evaluate social and institutional feasibility. This order of events need not be strictly followed, particularly in the case of species associated with strong social values. These may include reintroductions of species that could pose a risk to human welfare or those with strong cultural significance (Fischer and Lindenmayer 2000). Decisions regarding such species might justifiably merit considering social and institutional factors first (see Riley and Sandström, Chapter 4, this volume), so long as the ecological feasibility of a reintroduction is not deemed out of the question.

Although not illustrated by our framework (figure 5.1), it is worth highlighting that the ecological, social, and institutional factors may not be independent of each other, and may covary through time. For example, if an evaluation of ecological factors suggests a reintroduction is feasible, such a finding may influence social perceptions and institutional behaviors. Viewing the intersections among these three factors as the conceptual space that defines an appropriate reintroduction landscape, one can envision that over time the degree of overlap among them may increase or decrease, thus improving or degrading the chances of a successful reintroduction. The rate at which each of these factors can change may vary considerably. Of the three, ecological factors are often, though not always, inherently the least variable. Ecological aspects of landscapes generally change slowly though time, although extreme physical events (e.g., floods, fires, droughts) or rapid biological changes (e.g., outbreaks of pests or pathogens, rapid invasions) can act on short time frames (<1 year). Institutional factors generally vary over moderate time frames, as funding, regulations, bureaucratic processes, and governance can be changed over time, but may take years to do so. In contrast, social factors can shift rapidly, especially as capabilities to rapidly disseminate information have grown (e.g., social media, 24-hour news cycles, mobile technology). A wide range of temporal variabil-

ity in these factors should be expected, and can work against, or in favor of, any given reintroduction being considered. For example, savvy practitioners might be able to pursue a reintroduction opportunity where ecological and institutional factors are well aligned, but social factors are not. By explicitly acknowledging and targeting communication and outreach to address social barriers to implementation (George and Sandhaus, Chapter 14, this volume; Riley and Sandström, Chapter 4, this volume), a project can go from a low chance of success (i.e., socially unacceptable) to a high chance of success (figure 5.1).

THE ECOLOGICAL DIMENSIONS OF REINTRODUCTION LANDSCAPES

The ecological dimension of the reintroduction landscape (figure 5.1) is often the first consideration in the minds of practitioners. As with most disciplines, the lexicon of landscape ecology is populated with an ever-increasing list of variably defined terms and concepts, which we define here for the sake of clarity (Boxes 5.2–5.4). Definitions of landscapes and landscape ecology (Box 5.2) specify that landscapes are characterized by spatial and temporal heterogeneity. Although the focus is often on ecological processes, human dimensions of variation (i.e., social and institutional) are integral to landscape ecology as a discipline, as they are in the concept of the reintroduction landscape. In other words, landscape ecology is not considered to be a strictly ecological discipline, but rather a discipline that explicitly integrates human dimensions of landscapes. Key features of landscapes (Box 5.3) include what landscapes are composed of (composition), how these components are arrayed (configuration), how they are linked (connectivity), and how all three of these vary through time. These features are not independent, of course, but interact to influence spatial and temporal availability of resources that drive species' persistence (Box 5.4). With these concepts in hand, we move here to address key questions about ecological processes in the reintroduction landscape. We then turn to address social and institutional dimensions.

Key Questions about Ecological Processes in the Reintroduction Landscape

Here we discuss ecological landscape processes outlined above in the context of four basic questions that might be addressed when selecting reintroduction landscapes:

1. How large should the reintroduction landscape be?

2. How might spatial processes influence the success of a reintroduction?

3. How does connectivity play into selection of reintroduction landscapes?

4. "How should temporal dynamics factor into selection of a reintroduction landscape?"

We consider each in turn.

Question 1: How Large Should the Reintroduction Landscape Be?

This is one of the simplest questions we can ask, but the area or extent required for a species' reintroduction can depend on a broad range of factors, such as the degree of habitat specificity exhibited by a species, their capacity to move, and per capita resource requirements (e.g., body size). Among these, most fundamental are individual or per capita resource requirements and how they translate into the total number of individuals a given area can support, which ultimately drives species persistence and viability (Gilpin and Soulé 1988, Groves et al. 2002; Converse and Armstrong, Chapter 7, this volume). Generally speaking, larger-bodied species are expected to have much greater area requirements than smaller species (Marcot et al. 2001) by virtue of their greater per capita resource requirements and lower corresponding population densities (Peters 1986, Brown et al. 2004). A recent synthesis of minimum area requirements focused on birds and mammals revealed that, indeed, body size, as well as feeding guild, can explain much of the variation in the area required for a species (Pe'er et al. 2014), and by implication the area required for a reintroduction. Although simple in concept, minimum area requirements are generated by approaches that suffer from a host of widely debated caveats (Scott et al. 2002, Pe'er et al. 2014). Accordingly, such estimates should not be taken strictly at face value by practitioners, but rather evaluated in light of key assumptions, limitations, and uncertainties. In particular, several quantitative tools such as spatial occupancy modeling can be used to help predict potential response of reintroduced populations to specific landscape configurations (Chandler et al. 2015).

Addressing the question of minimum area requirements would seem fundamental to the success of a reintroduction; however, assessments of the success of reintroductions for various taxa (Cochran-Biederman 2014, Fischer and Lindenmayer 2000, Germano and Bishop 2009, Griffith et al. 1989) did not provide evidence in support of area as a key factor. If valid, the findings of current assessments of reintroductions run counter to the general observation that minimum area requirements strongly drive species persistence (Whittaker

BOX 5.3 · Ecological Processes in Landscapes

Ecological processes in landscapes that might be critical to the success of a reintroduction can be divided into three broad spatial components: *composition*, *configuration*, and *connectivity* (Cushman et al. 2010). We provide some definitions and discussions of each of these components, followed with a consideration of temporal dynamics on landscapes.

The Composition of landscapes, by definition (Turner 2005), is heterogeneous. Heterogeneity can arise from continuous gradients in physical or biological conditions and be manifested in more discontinuous or discrete forms. Often the term *patch* is applied to describe discrete components of a landscape. As with all terms used here, "patch" is variously defined in the literature, but can be simply thought of as a relatively homogeneous area with discrete boundaries that distinguish it from the landscape at large (see Kotliar and Wiens 1990). Examples include patches of host plants for butterflies (Hanski 1999), patches of physical conditions suitable for supporting fish in streams (Dunham et al. 2002), and patches of vegetation that support forest-dependent species (Zuckerberg and Porter 2010). Patches can be viewed as being embedded within a matrix of landscape conditions that cannot independently sustain patch-dependent species. The quality and quantity of patch and matrix habitats in a landscape can independently or interactively drive dynamics of species populations (Driscoll et al. 2013). In addition to cases of discrete variability, landscapes can be composed of more continuous gradients linked to variation in physical (e.g., temperature, moisture, geology) or biotic conditions (e.g., distribution of interacting predators, prey, competitors, parasites, or pathogens).

Configuration refers to the spatial arrangement of landscape components (i.e., what a landscape is composed of). Landscape configuration is tied to the locations, shapes, and orientations of landscape components. The configuration of landscapes can be described in many ways (e.g., spatial patterns of fragmentation, landscape geometry), and a large arsenal of analytical methods and software packages have emerged for this purpose (Beyer et al. 2010). Habitat fragmentation is a familiar change in the configuration of landscapes involving the breakup of continuous habitats into smaller pieces (Wu 2009).

Connectivity refers to the interactions among composition, configuration, and physical and biotic fluxes (e.g., movements of individuals) across a landscape (Taylor et al. 1993, Baguette et al. 2013). In the context of reintroductions, connectivity is fundamentally a product of movement by individuals, which is influenced by a variety of constraints, including external factors influencing movement (e.g., corridors that enable movement or external barriers or other factors that pose resistance to movement), motion capacity, capacity to navigate, and an individual's internal state (e.g., maturity, condition; Nathan et al. 2008). Landscape resistance refers to a host of external factors that may restrict movement, including distance, conditions that limit movement within different pathways, or barriers that unconditionally prevent movement (Rudnick et al. 2012, Wilkerson 2013, Graves et al. 2014). Motion capacity refers to the physical potential for a species to move, which is a function of its biomechanical design and mode of movement (Nathan et al. 2008). Even if an individual is potentially capable of moving long distances, a host of behavioral factors can modify realized patterns of movement (Reed and Dobson 1993). Genetic connectivity, referring to the degree to which alleles are shared and can spread throughout a landscape, might be particularly pertinent to more sessile or less vagile organisms (e.g., Vandergast et al. 2009) but can still be a critical consideration for wide-ranging species such as ungulates and carnivores (Reding et al. 2013, Creech et al. 2014). The discipline of landscape genetics (Manel et al. 2003, Epps et al. 2007, van Strien et al. 2014) provides tools and methods for studying genetic connectivity influenced by environmental variability.

Temporal variability of landscape composition, configuration, and connectivity is relevant because most landscapes have the potential to be highly dynamic (Pickett and Thompson 1978; Pickett and White 1985), and face the prospect of altered dynamics in the future due to changing climate and land use change (Peters 1988, Verboom et al. 2010), including efforts to restore natural ecological functions or ecosystem services (Millar et al. 2007). Temporal variability can emerge in a variety of forms related to the timing, duration, magnitude, variability, and predictability of events (Gaines and Denny 1993, Landres et al. 1999; Carpenter and Brock 2006), all of which might be important biologically. For example, the timing of events influences the life histories of many species because they are strongly linked to seasonal cycles, particularly climatic cycles, which are changing across the planet (Denny et al. 2014). The duration and magnitude of events (e.g., durations of floods, droughts, thermal events, human disturbances) are clearly important, as well as are changes in variability (Carpenter and Brock 2006). The predictability of events also matters. In situations where events are highly predictable, species may be adapted to them, even in the case of disturbance (Lytle and Poff 2004).

and Fernandez-Palacios 2007). Within the literature on reintroductions, however, there are examples of the roles of habitat size and quality as drivers of success (e.g., Jachowski et al. 2011). We suspect the lack of a connection between species' area requirements and the success of reintroductions is partly a consequence of the challenge of consistently defining area requirements (Pe'er et al. 2014), and the observation that reintroduced individuals might exhibit novel patterns of habitat selection or movement (Bennett et al. 2013; Jachowski et al., Chapter 9, this volume). In the latter case, it would be difficult to specify minimum area requirements in advance of a reintroduction, as the habitat requirements of the species might change unexpectedly. Finally, frequent observations of animals moving outside of reintroduction sites (Germano and Bishop 2009, Le Gouar et al. 2012) might partly be attributed to insufficient area of the site to support a reintroduction.

Question 2: How Might Spatial Processes Play into the Success of a Reintroduction?

Our review of the reintroduction literature found almost no examples explicitly citing specific landscape spatial processes. Literature search terms such as "landscape complementation," "landscape supplementation," "neighborhood effects," or "source-sink dynamics" (Box 5.4) were very infrequently cited in journal articles involving reintroductions. Although these well-known processes might be implicitly understood by researchers and practitioners, lack of explicit reference to these terms in the literature on reintroductions signifies a gap in applying some of the core concepts from landscape ecology.

One recent example of the role of source-sink dynamics in reintroductions is work by Marcot et al. (2013), who simulated spatial dynamics of northern spotted owls (*Strix occidentalis caurina*) in Pacific Northwest United States. They found that the proportion of the population of unpaired, nonreproductive, "floater" owls in sink habitat and the proportion of sink habitat throughout the landscape both correlated with increasing rates of population decline. In this study, landscapes with at least 40% source habitat provided opportunities for stable populations, but configurations of source and sink habitats also had an effect on population trend and persistence. Such findings can provide guidance for selecting reintroduction landscapes with appropriate amounts and configurations of source habitat, such as with the northern spotted owl captive breeding and reintroduction program of the Ministry of Lands and Forests, British Columbia, Canada (Fenger et al. 2007).

Within the literature on reintroductions, we found an illustrative example of the role of ecological traps (Box 5.4) involving gray partridge (*Perdix perdix*) reintroductions in the United Kingdom (Rantanen et al. 2010). In this case, captive-bred animals were released as family groups (coveys) or as pairs. Partridges released as pairs selected crop and field margin habitats. Birds in crop habitats experienced increased

BOX 5.4 · Critical Landscape Processes

Landscape composition, configuration, and connectivity interact with temporal variability (Box 5.3) to drive species persistence. Here we highlight five key interactions: *landscape complementation, supplementation, neighborhood effects, source-sink dynamics* (Dunning et al. 1992), and *ecological traps* (Battin 2004).

Landscape complementation is relevant when resource requirements for a species cannot be satisfied in a single site. In such cases, availability of complementary habitats that satisfy unique resource requirements within a landscape is critical. There are innumerable examples of the need for complementation for a host of species (Dunning et al. 1992), ranging from well-known requirements of amphibians for aquatic and terrestrial habitats (Pope et al. 2000, Ficetola et al. 2011) to the juxtaposition of nesting and foraging locations for birds (Rosenberg and McKelvey 1999, D'Elia et al. 2015).

Landscape supplementation is relevant when species depend on more than one location to meet the same resource requirement. In this case, resources in different locations are substitutable in terms of meeting resource needs. This contrasts with the case of complementation, where resources are not substitutable (Dunning et al. 1992).

Neighborhood effects refer to the influences of environmental conditions directly adjacent to a given site within a landscape (Addicott et al. 1987). Few sites in a landscape are completely unaffected by conditions immediately surrounding them. Neighborhood effects depend on the degree to which physical and biological fluxes can occur (boundary conditions or permeability), the contrast or environmental differences between a site and the ecological neighborhood in which it is embedded, and types of interactions that can occur. In practice, neighborhood effects might be difficult to distinguish from other landscape processes and in fact not be independent of them, but

like all of the processes described herein they are pervasive and often scale-dependent (Nash et al. 2014).

Source-sink dynamics refer to cases in which some portions of the landscape (e.g., patches or "habitats"), referred to as "sinks," cannot sustain local populations without demographic support from more productive source populations (Pulliam 1988). In other words, sinks are locations where mortality exceeds reproduction, and thus the persistence of populations in sinks must depend on demographic sources within the landscape (Pulliam 1988). If individuals are able to perceive these differences or gradients in conditions and preferentially use source locations, the presence of sinks can increase the total population size of a species if such locations can support individuals emigrating from source locations when motivated to do so (e.g., density-dependent emigration from sources). Simulations show, however, that as landscape composition shifts to a high prevalence of sink habitats, species abundance can decrease (Pulliam and Danielson 1991). This is because individuals may have difficulty locating a small number of higher-quality source habitats embedded in landscape primarily composed of sink habitats.

Ecological traps are situations in which individuals actually select sink habitats (Schlaepfer et al. 2002). In such cases, populations can go extinct rapidly, particularly when initial population sizes are low (Battin 2004). Informally, ecological traps may be viewed as cases of "fatal attraction," whereby individuals are attracted to locations with negative fitness consequences. Identifying sink habitats or ecological traps can be difficult, but they represent relevant conservation considerations (Battin 2004). A more recent variation on this theme is a "perceptual trap" whereby individuals fail to recognize and select high-quality habitats, with similar consequences (Patten and Kelly 2010).

survival, whereas those in field margins suffered lower survival—likely due to increased activity of predators in such habitats. Birds released as coveys selected habitats with greater cover and experienced higher survival. The fact that in some cases birds actually selected lower-quality habitats (e.g., sinks) that reduced their survival provides evidence for the influence of ecological traps. Overall, this case study emphasizes the need to connect behaviors (e.g., habitat selection) to survival, and to consider the effect of release strategies on both (Moehrenschlager and Lloyd, Chapter 11, this volume). In the case of the gray partridge (Rantanen et al. 2010), the presence of ecological traps might be linked to naivety that could be attributed to captive rearing (Griffith et al. 1989, Fischer and Lindenmayer 2000, but see Roe et al. 2015). In any case, although we might expect habitat models to faithfully represent the needs of animals and provide useful guidance for site selection, the influences of landscape processes such as source-sink dynamics and ecological traps might dictate otherwise (Suvorov and Svobodová 2012).

A recent analysis of activity-specific ecological niche models for reintroduction of California condors (*Gymnogyps californianus*) considered the complementary requirements of the species for nesting, roosting, and feeding (D'Elia et al. 2015). This effort produced maps depicting where these three requirements were satisfied across a very extensive area, extending over the states of Washington, Oregon, and California in the western United States. Locations supporting the coincidence of these conditions corresponded well with locations currently occupied by California condors. Furthermore, resulting maps identified a number of presently unoccupied locations that may be capable of supporting the species. The latter areas could be suitable candidates for a more detailed analysis of the feasibility of a reintroduction. In the context of a landscape perspective, a key feature of this work was to explicitly consider the reintroduction landscape with specific reference to complementary habitat needs of the species.

Although these spatial landscape characteristics are critical, they are part of a broader consideration of stressors that could determine the success of a reintroduction (e.g., lead poisoning in California condors; Finkelstein et al. 2012).

Question 3: How Does Connectivity Influence Selection of Reintroduction Landscapes?

Throughout this chapter we emphasize it is not unusual for reintroduced individuals to disperse from the original site of release (Le Gouar et al. 2012). From the perspective of the individuals being reintroduced, a reintroduction landscape can represent a completely novel situation, where individuals are not bound to a location by natal philopatry (the propensity for individuals to reside in or near their natal location) or have prior information about the landscape into which they have been released. Given these conditions, perhaps it is no surprise that reintroduced individuals can exhibit novel patterns of habitat selection, which often involve extensive movements (Rantanen et al. 2010, Bennett et al. 2012). This naturally brings the question of connectivity into play. Connectivity, as considered here, is a function of the organism's ability to move and the degree of resistance to movement imposed by the landscape. The role of connectivity in conservation is widely recognized (Crooks and Sanjayan 2006, Jachowski et al., Chapter 9, this volume; Seddon and Armstrong, Chapter 2, this volume), but it is becoming increasingly apparent that the opposite—intentional isolation—may be warranted in some cases (Fausch et al. 2009, Rahel 2013).

To discuss the issue of connectivity more specifically, it may be worth reframing our question to ask "When is connectivity more beneficial than intentional isolation?" Connectivity may be necessary for reintroduced species that do not pose threats to other valued attributes, e.g., other valued ecosystem components or human welfare (Riley and Sandström, Chapter 4, this volume), cases in which we might expect individuals to move extensively without seriously compromising the probability of establishment, or cases where connectivity is an obligate requirement. Many species naturally thrive in patchy and dynamic environments, and depend strongly on connectivity for persistence (e.g., Pickett and White 1985, Hanski 1999). Examples include species that occupy habitats at a specific seral stage (Sousa 1984) or depend on patch dynamics for coexistence with natural enemies (Baggio et al. 2011). More obviously, obligate migratory and partially migratory species depend strongly on connectivity (Chapman et al. 2011, Seddon and Armstrong, Chapter 2, this volume). In contrast to these examples, there may be cases where limited connectivity or even intentional isolation is more desirable (Fausch et al. 2009, Rahel 2013). These include cases where invasion of nonnative species might adversely impact the reintroduced species (e.g., Peterson et al. 2008, van Heezik et al. 2009, Muhlfeld et al. 2012), or there are concerns regarding the spread of pathogens by a reintroduction (Muths and McCallum, Chapter 10, this volume), attraction of reintroduced individuals to sinks (Rantanen et al. 2010, Suvorov and Svobodová 2012), impacts of the reintroduced species on other valued species (Marcot et al. 2012b), or human-wildlife conflicts (Treves et al. 2009). The latter is a particularly relevant consideration as translocations (including reintroductions) of species to mitigate human-wildlife conflicts may be less successful than translocations initiated for other reasons (e.g., Germano and Bishop 2009, Germano et al. 2015).

Question 4: How Should Temporal Dynamics Factor into Selection of a Reintroduction Landscape?

Although the natural dynamics of landscapes has long been appreciated, growing awareness of the extent and magnitude of human-caused changes has fundamentally changed our view of how natural ecosystems function now and into the foreseeable future (Turner 2010). Such changes may challenge the conventional wisdom behind classic reserve design concepts and their implications for selecting landscapes for species reintroductions (Pickett and

Thompson 1978, Pickett and White 1988). Poleward shifts in the distributions documented for dozens of species in the face of warming climates (Parmesan and Yohe 2003) and widespread human domination of a host of ecosystem processes is now the new "normal" (Ehlers and Krafft 2006, Jachowski et al., Chapter 9, this volume). In the face of these changes, the view of species translocations in landscapes has been expanded beyond reintroductions to include consideration of "assisted colonization," or the deliberate translocation[1] by humans of a founder population to a location not historically occupied by the species of interest but that is anticipated to provide suitable (source) habitat under a new environmental regime (Grady et al. 2011, Catford et al. 2013, IUCN/SSC 2013). Such conservation tactics are not without controversy (Minteer and Collins 2010, Seddon 2010, Lawler and Olden 2011, Hagerman and Satterfield 2014), however, and we expect much further ethical debate on these concepts in scientific, management, and social realms. Even if ecosystem changes are highly localized, as local conditions change, so will a host of linked processes as the corresponding composition, configuration, and connectivity of whole landscapes. As with assisted colonization, the idea of actively engineering or selecting these features of landscapes to ensure that species are more resilient in the face of extreme events and change is a topic of growing discussion and application (Verboom et al. 2010). Along these lines, Crees and Turvey (2015) raised the question of defining species as

---

1 The IUCN defines translocation as "the deliberate movement of organisms from one site for release in another" with intention to "yield a measurable conservation benefit" (IUCN/SSC 2013, VIII). Reintroductions are defined by the IUCN as "the intentional movement and release of an organism inside its indigenous range from which it has disappeared" with the intention to "re-establish a viable population of the focal species within its indigenous range" (IUCN/SSC 2013, 3). Assisted colonization is defined by the IUCN as "the intentional movement and release of an organism outside its indigenous range to avoid extinction of populations of the focal species" (IUCN/SSC 2013, 3).

"native" in a paleohistorical context, which suggests that as future "novel ecosystems" are formed from climate change dynamics, using historical ranges of native species will have limited utility and success for reintroductions.

It is clear that the field of translocations (Armstrong and Seddon 2007, Ewen et al. 2012, Seddon and Armstrong, Chapter 2, this volume) will need to remain flexible and strive for innovative solutions not yet conceived to conditions not yet manifested. These might result from influences such as climate change, continuing introduction and spread of nonnative species, and the continued disruption of native ecosystems from burgeoning human populations (Hobbs et al. 2009). An example of adopting assisted colonization in the face of recent and rapid changes is that of the buff weka (*Gallirallus australis hectori*), a large flightless rail and one of four subspecies of weka that was extirpated from its natural range in eastern South Island, New Zealand, but "was translocated to the Chatham Islands where it is common in many areas and where limited harvesting of food is permitted" (Gill 1999, 19). This is a case of a subspecies extirpated within, but successfully translocated outside and adjacent to, its native range. Finally, it is possible that a reintroduction itself (or any other form of translocation) might result in substantial changes to landscapes (Hayward and Slotow, Chapter 13, this volume). For example, reintroductions of top mammalian carnivores have been documented to precipitate trophic cascades in terrestrial and marine ecosystems (Ripple et al. 2014). Reintroductions of ecosystem engineers, such as beaver (*Castor canadensis*), can also lead to substantial modifications of aquatic and riparian ecosystems (Pillai and Heptinstall 2013, Pollock et al. 2014). Reintroductions can induce unforeseen trophic cascade effects in ecosystems well beyond the predicted outcome of the reintroduction itself (Painter et al. 2015), thus suggesting that surprises may arise and monitoring may be warranted. These examples highlight the value of considering temporal dynamics from multiple change agents, including natural disturbance, succession, and patch dynamics; influences of climate and land use change; our ability to engineer ecological landscapes for conservation; and influences of the species themselves on the landscape.

SOCIAL AND INSTITUTIONAL LANDSCAPES: REGULATION AND COLLABORATION

Social and institutional factors are integral to landscape ecology and species reintroductions. This notion is not new, and dates back to the original motivation to develop landscape ecology as a unique discipline in the mid-twentieth century (Turner 2005). Though the term "landscape ecology" implies only ecological considerations, the discipline is explicitly inclusive of human dimensions. Similarly, species reintroductions are fundamentally motivated by a host of societal values, ranging from purely moralistic to practical (Kellert 1997, Riley and Sandström, Chapter 4, this volume), and these values in turn are often embodied in the form of laws or regulations implemented by governing bodies or institutions. There are many ways to approach the topic of socio-institutional landscapes. We follow the perspective offered by Barrett et al. (2001, 399):

> The institutional landscape should be approached as carefully as the ecological if biodiversity conservation is to be successful. It makes no more sense to valorize the community as the best defender of conservation in all cases than it does to claim that national governments are always in the best position to protect nature. Discussion about which institutions are appropriate to govern biodiversity conservation must move beyond the false dichotomy of community versus central government. Instead, scholars and policymakers must focus first on how institutions work at multiple levels and explore which configurations appear best for different types of biodiversity conservation.

This view acknowledges the role of constraints imposed by centralized regulations or governance on the actions of local stakeholders or communities, the role of leadership and engagement at the local level (Chauvenet et al.,

Chapter 6, this volume), and the configuration of social and institutional factors at both levels. It is worth noting that this simplified characterization, although with heuristic value, does not capture the full complexity of how politics, property, and policy play out in reality across landscapes (Brunckhorst 2011). Furthermore, it is worth pointing out that characteristics of the players themselves (e.g., organizations and individuals) can be key factors driving the success of reintroductions (Westrum 1994). These factors are at the heart of the emerging science of structured decision-making (Conroy and Peterson 2013), which has become rapidly adopted as a standard practice in natural resource management (e.g., Marcot et al. 2012a, Williams and Brown 2012). In the context of landscapes, it is clear that the configurations of these social and institutional factors vary in relation to complex patterns of land ownership and institutional or geo-political boundaries (Reading et al. 2002). These landscape properties are arguably just as important as purely ecological patterns and processes on landscapes.

To illustrate interactions between social and institutional settings in the context of reintroductions, below we examine several case studies of reintroductions proposed or conducted under the regulatory authority of the US Endangered Species Act (ESA). These examples highlight a wide range of social and institutional landscape contexts and demonstrate the regulatory flexibility needed to effectively implement successful reintroduction projects. We hope that insights from our discussion of the ESA can apply to a host of other situations where similar factors are in play.

The US Endangered Species Act: A Brief Primer

Before turning to a discussion of applications, we provide a very brief overview of the ESA of 1973 (16 U.S.C. §§ 1531–1544).[2] The ESA was enacted to prevent the extinction of species and to "provide a means whereby the ecosystems upon which endangered species and threatened species depend may be conserved." Passage of the ESA was the culmination of a series of wildlife protection treaties and laws, including the Endangered Species Preservation Act of 1966 (P.L. 89-669, 80 Stat. 926), the Endangered Species Conservation Act of 1969 (P.L. 91-135, 83 Stat. 275), and going as far back as the 1918 passage of the Migratory Bird Treaty Act (16 U.S.C. 703-712; Ch. 128; July 13, 1918; 40 Stat. 755, as amended). The ESA is generally regarded as one of the strongest environmental laws in existence for two reasons. First, the decision to list a species as threatened or endangered must be made without consideration of the potential economic impacts of such a decision. Second, all federal agencies of the US government have an affirmative duty to conserve listed species and must avoid jeopardizing the continued existence of listed species. Although it can be argued that the ESA is inherently reactive in the sense that action is not taken until a species is listed, in practice the prospect of listing a species can result in significant conservation actions meant to avoid further imperilment (Belton and Jackson-Smith 2010).

The ESA has been amended substantially three times, in 1978, 1982, and 1988. An additional minor amendment in 2004 addressed issues around military installations and how the ESA affects national defense training and operations. In its current state, the ESA protects listed species by requiring all federal agencies to utilize their authorities to carry out programs for listed species conservation, by

- requiring consultation with regulatory federal agencies (US Fish and Wildlife Service [USFWS] or National Marine Fisheries Service) under Section 7 of the ESA to ensure Federal actions do not jeopardize the continued existence of listed species or destroy designated critical habitat;

2 For details, please see http://www.fws.gov/endangered/laws-policies/timeline.html.

- prohibiting the take[3] of individuals of listed species; and
- requiring the development and implementation of recovery plans intended to bring the species to the point where the protections of the ESA are no longer required.

Prohibitions against take of listed animals apply to all entities and individuals: no take may occur for the species where take is prohibited without either a permit or an exemption. In the context of reintroducing listed species, this creates a potential disincentive: species that would benefit from reintroductions are often not wanted by landowners who may, in effect, be importing significant regulatory prohibitions along with the listed species (Wilcove et al. 2004). In acknowledgment of this issue, the ESA includes a number of provisions that can be applied to reduce the effect of regulatory prohibitions while still promoting species recovery (Table 5.1). Each of these provisions has advantages and limitations that should be carefully considered in light of the social and institutional dimensions of the reintroduction landscape (figure 5.1).

Adapting the US Endangered Species Act to Reintroduction Landscapes

How can the diverse provisions of the ESA (Table 5.1) be adapted to find the appropriate space (figure 5.1) among the diverse ecological, social, and institutional factors in play within reintroduction landscapes? In the context of reintroductions Section 10(j) "nonessential experimental populations" rules can significantly modify the regulatory prohibitions of the ESA, which can be useful in removing social and institutional barriers. Section 10(j) populations can only be designated through a federal rulemaking process, which can be lengthy. In addition, experimental populations must be wholly separate from other populations of the species so as not to create con-

fusion around enforcement of the ESA. As formulated, Section 10(j) may be particularly effective if the reintroduction landscape is composed of a multitude of public and private landowners that share common concerns about the implications of a listed species on their lands.[4] An alternative to the 10(j) provision is Section 4(d) of the ESA. This allows USFWS (Table 5.1) to establish special regulations (called "4(d) rules") for threatened (but not endangered) species in cases where such changes benefit species conservation. Among these are cases in which conservation of threatened species conflicts with other human values. If modifications of normal ESA protections can minimize conflicts and not slow species recovery, provisions within section 4(d) may be useful. Conversely, if the reintroduction landscape has only one or two private landowners with a low level of concern about the economic impacts of reintroducing a protected species, more specific tools like a Section 10(a)1(A) Safe Harbor Agreement might be most effective (Table 5.1). Safe harbor agreements encourage landowners to manage for, and conserve, listed species on their land by offering a return to the original baseline status of those lands at a later date in exchange for their interim conservation. This "return to baseline" option can allow for removal of the listed species entirely (in cases of a "zero baseline"), but is rarely used and offers landowners significant peace of mind. Additional provisions include Section 7 provisions that involve agreements that require internal regulatory consultation and approval, and Section 6 provisions that authorize federal regulatory agencies to provide exemptions to state agencies or state-delegated "agents of the state" for the purpose of conservation of listed species, including support of reintroductions.

From the preceding discussion, it is clear that the ESA has a number of provisions that

---

3 Take is broadly defined under the ESA as: "To harass, harm, pursue, hunt, shoot, wound, kill, trap, capture or collect, or to attempt to engage in any such conduct."

4 It is also worth noting that Section 10(j)(2)(A) also allows the Secretary of Interior to authorize the release of a population of an endangered or threatened species outside the current range of the species if such a release will further the conservation of the species.

TABLE 5.1

Regulatory Provisions of the US Endangered Species Act That Can Provide Social Incentives to Implementing Reintroductions

ESA Provision	Description
Section 10(j) "nonessential experimental population"	Rulemaking that designates a population as "experimental." "Take" prohibitions can be tailored to the specific circumstances; most agencies do not have to consult under Section 7 and critical habitat is not designated for such populations.
Section 4(d) Rule	A "protective regulation" that exempts certain forms of take from the prohibitions of Section 9. Carefully crafted exemptions can be tailored to specific circumstances in ways that advance conservation and reintroductions. Because the USFWS published a blanket 4(d) rule extending full ESA take protections to all listed animals (43 Federal Register 18181, April 28, 1978), USFWS 4(d) rules eliminate take prohibitions for threatened species when the conservation of the species is best served by reduced take protections. Conversely, NOAA-Fisheries, which administers the ESA for marine species and many anadromous fish, has not published such a blanket 4(d) rule. As a result, NOAA-Fisheries 4(d) rules apply take prohibitions to threatened species that otherwise would have none, in accordance with the language of the statute.
Section 6 Agreement	Agreement between USFWS and the relevant state fish and wildlife agency with allowances for take for recovery purposes. The state agency may further designate "agents of the state," including local public and private landowners, to extend the take exemption to these parties.
Section 10(a)1(A) Safe Harbor Agreement (SHA)	Agreement between USFWS or a state fish and wildlife agency and private landowners to provide habitat for species for an extended period of time. In exchange, landowners are allowed to return habitat condition to a previously identified "baseline condition," even if that means individuals of the listed species are taken or removed.
Section 7 Memorandum of Agreement (MOA) or Conservation Agreement (CA)	Agreement signed by USFWS and other entities. USFWS then conducts internal Section 7 consultation on the MOA/CA, which provides an exemption from the take prohibition in the "incidental take statement."

can be adapted to a wide range of conditions that may be encountered in reintroduction landscapes (figure 5.1). These provisions address both social and institutional dimensions of reintroduction landscapes. To illustrate how these interact with ecological dimensions of the reintroduction landscape, we turn to briefly discuss a range of case studies of reintroductions involving ESA-listed species. It is worth noting that all of the examples used herein represent relatively new applications (within the past 5 years), and include both relatively new and long-established provisions of the ESA.

The case of the Oregon chub (*Oregonichthys crameri*) involved an effort that was distributed across the landscape of the Willamette River basin in western Oregon (Baker et al. 2004). Due to a variety of influences, the distribution of the Oregon chub within the Willamette River basin was severely contracted at the time the species was initially listed under the ESA in 1993 (USFWS 1998). The species is a small-bodied (figure 5.2), relatively short-lived fish (Scheerer and McDonald 2003), and capable of establishing self-sustaining populations in small, localized habitats (Scheerer 2002): characteristics that lend themselves well to translo-

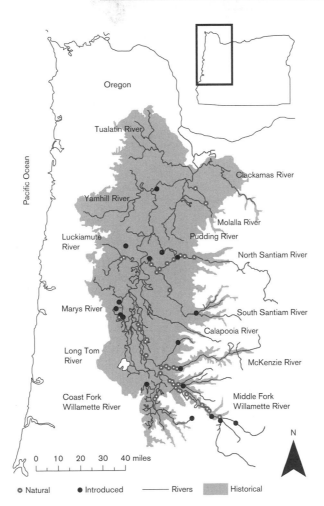

FIGURE 5.2 Map of the Willamette River Basin in Western Oregon (bottom) with the historical range of the Oregon chub (*Oregonichthys crameri*) indicated by gray shading. Points on the map correspond to known instances of species presence in natural habitats (dark gray and white dots) and introductions (black dots). On the top right panel a location on private land where Oregon chub were introduced is pictured, along with two biologists from the Oregon Department of Fish and Wildlife (Brian Bangs, left; Paul Scheerer, right) who led the recovery effort for the state of Oregon. An Oregon chub is pictured in the upper left panel.

SOURCE: Images provided with permission from the Oregon Department of Fish and Wildlife (B. Bangs and P. Scheerer, pers. comm.).

cations (Hendrickson and Brooks 1991). Consequently, recovery of the Oregon chub involved a series of translocations, including reintroduction into historically occupied localities and introductions into novel (sometimes newly created) locations within the Willamette basin (figure 5.2). These actions involved active engagement of diverse landowners across the Willamette basin. This effort resulted from the USFWS's Section 7(a)(1) Recovery implementation program for the chub, and required use of Safe Harbor Agreements (Section 10(a)1(A) of the ESA; Table 1) with private entities, ESA consultation between federal agencies (between USFWS and the US Army Corps of Engineers under Section 7(a)(2) of the ESA), and cooperative agreements between USFWS and the state of Oregon (ESA Section 6) to provide the state with the legal and financial means to conduct translocations and other actions needed to recover the species. Private landowners in some cases also entered into conservation agreements or easements with the US Department of Agriculture's Natural Resources Conservation Service (NRCS) to receive payments for setting aside lands that assisted Oregon chub conservation. The Oregon chub has been delisted under the ESA (80 Federal Register 9125, February 19, 2015), and is the first fish to be recovered under the ESA. In this case, judicious application of several provisions of the ESA fostered effective partnerships among diverse parties, removing barriers to implementation, and ultimately resulting in a highly successful outcome.

Unlike the Oregon chub, many species that could benefit from reintroductions have a high potential for adverse human-wildlife conflicts, or in some cases conflicts with other species of concern. Perhaps the most well-known case of reintroductions and human-wildlife conflicts is the case of the gray wolf (*Canis lupus*, figure 5.3; Smith and Bangs 2009). Diverse public perceptions and effects of wolves on human welfare (Treves et al. 2013) and their capacity to disperse widely conspire to create great potential for conflicts. Gray wolves were reintroduced to

Yellowstone National Park and central Idaho in 1995 and have spread across the Western United States (Smith et al. 2010). To manage institutional and social factors, reintroduced populations were designated under ESA Section 10(j) as nonessential experimental populations (59 Federal Register 60252 and 60266, respectively, November 22, 1994). The reintroduction of the gray wolf has been highly successful in reestablishing the species, and it was delisted in the Northern Rocky Mountain population in two stages (2009: 74 Federal Register 151123, April 2, 2009, and 2012: 77 Federal Register 55530, September 10, 2012). At the time of this writing some of these decisions are still being legally challenged,[5] but there is little question that the success that has been realized to date in recovery of the gray wolf would not have been possible without allowing for some management flexibility (e.g., killing wolves that depredate livestock) via the 10(j) provision of the ESA.

Although the ESA is focused on recovery of individual species, the narrative in the original act explicitly acknowledges the role of ecosystem processes. In this regard, it is becoming increasingly likely that reintroduction of one imperiled species may impact another (e.g., Soulé et al. 2005). The case of reintroduction of threatened bull trout (*Salvelinus confluentus*) into the Clackamas River, Oregon, provides an interesting example. Although reintroduction of bull trout into this system was deemed biologically feasible (Dunham et al. 2011), concerns about conflicts with ESA-listed Pacific salmon and steelhead (*Oncorhynchus*, spp.) posed a potential barrier to implementation. Specifically, concerns about impacts of reintroduced bull trout (an apex predator) on survival of juvenile salmon and steelhead posed a potential conflict between two federal agencies responsible for recovery of bull trout (USFWS)

---

5 As of the date this was authored, the most recent updates on these legal actions were posted on the USFWS web page: http://www.fws.gov/mountain-prairie/species/mammals/wolf/Document68_NRM_WYOpinion.pdf.

FIGURE 5.3. A male gray wolf (*Canis lupus*), named OR-7 photographed at a camera trap in southern Oregon (eastern Jackson County; photo courtesy of US Fish and Wildlife Service). This individual dispersed over 1,600 km from its birthplace in the Wallowa Mountains in northeast Oregon to range through northern California and southern Oregon. This individual is believed to have paired with a female wolf in southern Oregon and they have produced at least three pups in 2014 and at least two more in 2015 (see http://www.fws.gov/oregonfwo/Species/Data/GrayWolf/PhotoGallery.asp and http://www.fws.gov/oregonfwo/OR7_YearlingAndPupGallery.cfm). The example of the gray wolf illustrates just how extensive a reintroduction landscape can be for highly mobile ecological generalists.

and Pacific salmon and steelhead (National Marine Fisheries Service for Pacific salmon). To address this issue, an expert panel was convened to evaluate threats posed by bull trout to listed Pacific salmon and steelhead (Marcot et al. 2012b). Based on results of this and subsequent negotiations with federal, state, tribal, and private stakeholders, a rule making process using the ESA 10(j) provision was used to develop a systematic plan for monitoring and mitigating the potential adverse impacts of bull trout on listed Pacific salmon and steelhead, including termination and reversal of the reintroduction if impacts were unacceptable (76 Federal Register 35979, June 21, 2011). With this provision in place, bull trout were reintroduced into the Clackamas River in 2011. As of the drafting of this chapter, introduced fish have survived with evidence of attempts to reproduce and no evidence to suggest adverse

impacts to listed Pacific salmon and steelhead (Barry et al. 2014; work supported in part by Section 6 provisions of the ESA). Although the ecological success of the Clackamas bull trout reintroduction has yet to be confirmed, development of innovative approaches to enable it to occur in the first place as well as commitments to monitoring of the reintroduction represent social and institutional successes.

Reintroduction case studies considered thus far involve potential conflicts or trade-offs in the reintroduction landscape and how they can be addressed in a flexible regulatory environment that can adapt to local social and institutional contexts. Such complexities are not always the case, however. In the case of Fender's blue butterfly (*Icaricia icarioides fenderi*, figure 5.4), limited mobility of the species, small area requirements, limited potential for human-wildlife conflicts, and the relative ease

FIGURE 5.4. Fender's blue butterfly (*Icaricia icarioides fenderi*, right photo), a threatened ecological specialist with limited dispersal. The left photo is of Kincaid's lupine (*Lupinus sulphureus kincaidii*), one of the few host plants used by Fender's blue butterfly (Schultz and Dlugosch 1999). (Images provided with permission from US Fish and Wildlife Service). The example of Fender's blue butterfly illustrates a reintroduction landscape that is relatively limited in extent, dependent on the presence of specific components (host plants), and inhabited by a species with limited dispersal capability (Schultz and Crone 2005).

of restoring habitat on the federally owned land (William L. Finley National Wildlife Refuge, located in west-central Oregon) posed few barriers to conducting a reintroduction. In this case, social and institutional dimensions of the reintroduction landscape were aligned in favor of supporting a reintroduction, and the ecological dimension required only minimal site preparation to bring these dimensions into alignment. As a result, the species was reintroduced to the refuge in 2014. We highlight this case to acknowledge that many reintroductions may not require extensive analyses of the three dimensions of the reintroduction landscape, nor exhaustive efforts to bring all three into alignment (overlap; figure 5.1). The extensive list of considerations offered for reintroductions in this volume or other recent references on reintroductions (Ewen et al. 2012) merit serious attention, but the process of evaluation itself should not become a barrier to implementation.

Although our discussion of socio-institutional landscapes focuses on species protected under the ESA and cases drawn from the western United States, the lessons learned illustrate how the more generalized framework of the reintroduction landscape can be applied. A more generalized presentation of many of these ideas that can be extended to other contexts can be found in the literature on structured decision-making (Conroy and Peterson 2013). Even from the small number of examples we examined within a single regulatory framework, it is clear that intersecting all three elements of the reintroduction landscape can involve a diversity of solutions to find the necessary space defined by an appropriate reintroduction landscape (figure 5.1).

SUMMARY

Based on our review of past assessments of factors contributing to the success or failure of reintroductions (e.g., Griffith et al. 1989,

Germano and Bishop 2009, Fischer and Lindenmayer 2010, Cochran-Biederman 2014), we find that the principles of landscape ecology are often not explicitly considered. In this chapter, we return to the original view of landscape ecology as the integration of human and ecological factors to coin a new phrase—the *reintroduction landscape*—that we use to describe an integrated framework that can be readily applied to reintroductions (see also IUCN/SSC 2013). Our framework emphasizes the inextricable nature of social, institutional, and ecological dimensions, as well as interactions among them in determining the success of reintroductions (or other types of translocations). For each dimension of the framework, we provide examples to illustrate their relevance and application (where examples exist) to reintroductions. Based on these examples and the lessons learned, we conclude with a list of 10 management recommendations for adopting the reintroduction landscape as a framework for explicitly considering landscape processes before, during, and after reintroductions are attempted.

MANAGEMENT RECOMMENDATIONS

Based on our framework (figure 5.1) and exploration of the literature on reintroductions, as well as practical examples, we have identified 10 key considerations for reintroduction landscapes that are listed below.

- The reintroduction landscape is large enough to allow establishment of a population of sufficient size to ensure persistence for a desired time frame.

- The reintroduction landscape supports processes that represent the anticipated composition, configuration, and connectivity of resources required to allow the species to successfully complete its life cycle.[6]

- Connectivity within the reintroduction landscape serves the species' need to move within it while considering the potential for adverse ecological impacts or impacts to human welfare.

- The future dynamics of the reintroduction landscape are considered in terms of how they influence landscape processes that affect the probability of establishment or persistence of the reintroduced species.

- The likelihood that the reintroduced species may exhibit novel patterns of habitat use or selection is considered in selecting the reintroduction landscape.

- Influences of a reintroduction on potentially conflicting values held by stakeholders in the reintroduction landscape are explicitly considered.

- Regulatory alternatives or flexibilities to surmount barriers to implementation within the reintroduction landscape are explored.

- Critical contingencies or decision thresholds are identified and linked to specific action alternatives, including termination of the reintroduction.

- The fact that (sometimes rapidly) changing relationships among ecological, social, and institutional landscape factors can drive the success or failure of a reintroduction is acknowledged.

- Cases in which ecological, social, and institutional landscape factors are clearly in or out of alignment are quickly recognized and addressed; if necessary, alternative reintroduction landscapes are identified without investing excessive amounts of time or resources in trying to force implementation in cases where reintroductions are not feasible.

ACKNOWLEDGMENTS

We would like to thank Paul Angermeier, Bill Brignon, Jesse D'Elia, Martin Fitzpatrick, David Jachowski, Philip Seddon, and Jack

---

6 For species with obligate migratory life histories, the release area should at least satisfy the needs for the targeted migratory use (e.g., feeding, reproductive, or refuge).

Williams for their constructive comments on this manuscript. This chapter has been peer reviewed and approved for publication consistent with US Geological Survey Fundamental Science Practices (http://pubs.usgs.gov.circ /1367). The findings and conclusions in this chapter are those of the author(s) and do not necessarily represent the views of the US Forest Service or USFWS. Use of trade or firm names is for reader information only and does not constitute endorsement of any product or service by the US government.

LITERATURE CITED

Addicott, J. F., J. M. Aho, M. F. Antolin, D. K. Padilla, J. S. Richardson, and D. A. Soluk. 1987. Ecological neighborhoods: scaling environmental patterns. Oikos 49:340–346.

Armstrong, D. P., and P. J. Seddon. 2007. Directions in reintroduction biology. Trends in Ecology and Evolution 23:20–25.

Baggio, J. A., K. Salau, M. A., Janssen, M. L. Schoon, and Ö. Bodin. 2011. Landscape connectivity and predator–prey population dynamics. Landscape Ecology 26:33–45.

Baguette, M., S. Blanchet, D. Legrand, V. M. Stevens, and C. Turlure. 2013. Individual dispersal, landscape connectivity and ecological networks. Biological Reviews 88:310–326.

Baker, J. P., D. W. Hulse, S. V. Gregory, D. White, J. Van Sickle, P. A. Berger, D. Dole, and N. H. Schumaker. 2004. Alternative futures for the Willamette River basin, Oregon. Ecological Applications 14:313–324.

Barrett, C. B., K. Brandon, C. Gibson, and H. Gjertsen. 2001. Conserving tropical biodiversity amid weak institutions. BioScience 51:497–502.

Barry, P. M., J. M. Hudson, J. D. Williamson, M. L. Koski, and S. P. Clements. 2014. Clackamas River Bull Trout Reintroduction Project, 2013 Annual Report. Oregon Department of Fish and Wildlife and US Fish and Wildlife Service, Portland, OR.

Battin, J. 2004. When good animals love bad habitats: ecological traps and the conservation of animal populations. Conservation Biology 18:1482–1491.

Belton, L. R., D. Jackson-Smith. 2010. Factors influencing success among collaborative sage-grouse management groups in the western United States. Environmental Conservation 37:250–260.

Bennett, V. A., V. A. Doerr, E. D. Doerr, A. D. Manning, D. B. Lindenmayer, and H. J. Yoon. 2012. Habitat selection and post-release movement of reintroduced brown treecreeper individuals in restored temperate woodland. PloS ONE 7:e50612.

Beyer, H. L., J. Jenness, and S. A. Cushman. 2010. Components of spatial information management in wildlife ecology: software for statistical and modeling analysis. Pp.245–253. In S. A. Cushman and F. Huettmann, eds., Spatial Complexity, Informatics, and Wildlife Conservation. Springer, New York.

Brown, J. H., J. F. Gillooly, A. P. Allen, V. M. Savage, and G. B. West. 2004. Toward a metabolic theory of ecology. Ecology 85:1771–1789.

Brunckhorst, D. 2011. Ecological restoration across landscapes of politics, policy, and property. Pp.149–161. In D. Egan, E. E. Hjerpe, and J. Abrams, eds., Human Dimensions of Ecological Restoration: Integrating Science, Nature, and Culture. Island Press, Washington, DC.

Carpenter, S. R., and W. A. Brock. 2006. Rising variance: a leading indicator of ecological transition. Ecology Letters 9:311–318.

Catford, J. A., R. J. Naiman, L. E. Chambers, J. Roberts, M. Douglas, and P. Davies. 2013. Predicting novel riparian ecosystems in a changing climate. Ecosystems 16:382–400.

Chandler, R. B., E. Muths, B. H. Sigafus, C. R. Schwalbe, C. J. Jarchow, and B. R. Hossack. 2015. Spatial occupancy models for predicting metapopulation dynamics and viability following reintroduction. Journal of Applied Ecology 52:1325–1333.

Chapman, B. B., C. Brönmark, J. Å. Nilsson, and L. A. Hansson. 2011. The ecology and evolution of partial migration. Oikos 120:1764–1775.

Cochran-Biederman, J. L., K. E. Wyman, W. E. French, and G. L. Loppnow. 2015. Identifying Correlates of success and failure of native freshwater fish reintroductions. Conservation Biology 29:175–186.

Conroy, M. J., and J. T. Peterson. 2013. Decision Making in Natural Resource Management: A Structured, Adaptive Approach. John Wiley and Sons, Hoboken, NJ.

Creech, T. G., C. W. Epps, R. J. Monello, and J. D. Wehausen. 2014. Using network theory to prioritize management in a desert bighorn sheep metapopulation. Landscape Ecology 29:605–619.

Crees, J. J., and S. T. Turvey. 2015. What constitutes a "native" species? Insights from the Quaternary faunal record. Biological Conservation 186:143–148.

Crooks, K. R., and M. Sanjayan. 2006. Connectivity Conservation. Conservation Biology Book Series, Cambridge University Press, Cambridge, UK.

Cushman, S. A., J. S. Evans, and K. McGarigal. 2010. Landscape ecology: past, present, and future. Pp.65–82. In S. A. Cushman and F. Huettmann, eds., Spatial Complexity, Informatics, and Wildlife Conservation. Springer, New York.

D'Elia, J., S. M. Haig, M. Johnson, B. G. Marcot, and R. Young. 2015. Activity-specific ecological niche models for planning reintroductions of California condors (*Gymnogyps californianus*). Biological Conservation 184:90–99.

Denny, E. G., K. L. Gerst, A. J. Miller-Rushing, G. L. Tierney, T. M. Crimmins, C. A. E. Enquist, P. Guertin et al. 2014. Standardized phenology monitoring methods to track plant and animal activity for science and resource management applications. International Journal of Biometeorology 58:591–601.

Driscoll, D. A., S. C. Banks, P. S. Barton, D. B. Lindenmayer, and A. L. Smith. 2013. Conceptual domain of the matrix in fragmented landscapes. Trends in Ecology and Evolution 28:605–613.

Dunham, J. B, K. Gallo, D. Shively, C. Allen, and B. Goehring. 2011. Assessing the feasibility of native fish reintroductions: a framework applied to threatened bull trout. North American Journal of Fisheries Management 31:106–115.

Dunham, J. B., B. E. Rieman, and J. T. Peterson. 2002. Patch-based models to predict species occurrence: lessons from salmonid fishes in streams. Pp.327–334. In J. M. Scott, P. Heglund, and M. L. Morrison, eds., Predicting Species Occurrences: Issues of Accuracy and Scale. Island Press, Covelo, CA.

Dunning, J. B., B. J. Danielson, and H. R. Pulliam. 1992. Ecological processes that affect populations in complex landscapes. Oikos 65:169–175.

Ehlers, E., and T. Krafft, eds. 2006. Earth System Science in the Anthropocene. Springer, New York.

Epps, C. W., J. D. Wehausen, V. C. Bleich, S. G. Torres, and J. S. Brashares. 2007. Optimizing dispersal and corridor models using landscape genetics. Journal of Applied Ecology 44:714–724.

Ewen, J. G., D. P. Armstrong, K. A. Parker, and P. J. Seddon, eds. 2012. Reintroduction Biology: Integrating Science and Management. Wiley-Blackwell Publishers, Hoboken, NJ.

Fausch, K. D., B. E. Rieman, J. B. Dunham, M. K. Young, and D. P. Peterson. 2009. The invasion versus isolation dilemma: tradeoffs in managing native salmonids with barriers to upstream movement. Conservation Biology 23:859–870.

Fenger, M., J. B. Buchanan, T. J. Cade, E. D. Forsman, S. M Haig, K. Martin, and W. A. Rapley. 2007. Northern Spotted Owl Population Enhancement and Recovery in British Columbia: Proposed Five-Year Action Plan. Report prepared for the Government of British Columbia. Available online at: http://www.env.gov.bc.ca/wld /speciesconservation/so/.

Ficetola, G. F., L. Marziali, B. Rossaro, F. De Bernardi, and E. Padoa-Schioppa. 2011. Landscape–stream interactions and habitat conservation for amphibians. Ecological Applications 21:1272–1282.

Fischer, J., and D. B. Lindenmayer. 2000. An assessment of the published results of animal relocations. Biological Conservation 96:1–11.

Forman, R. T. T., and M. Godron. 1986. Landscape Ecology. Wiley-Blackwell, Hoboken, NJ.

Gaines, S. D., and M. W. Denny. 1993. The largest, smallest, highest, lowest, longest, and shortest: extremes in ecology. Ecology 74:1677–1692.

Germano, J. M., and P. J. Bishop. 2009. Suitability of amphibians and reptiles for translocation. Conservation Biology 23:7–15.

Germano, J. M., K. J. Field, R. A. Griffiths, S. Clulow, J. Foster, G. Harding, and R. R. Swaisgood. 2015. Mitigation-driven translocations: are we moving wildlife in the right direction? Frontiers in Ecology and the Environment 13:100–105.

Gill, B. J. 1999. The Kiwi and Other Flightless Birds. David Bateman Limited, Auckland, New Zealand.

Gilpin, M. E., and M. E. Soulé. 1986. Minimum viable populations: processes of species extinction. Pp.19–34. In M. E. Soulé, ed., Conservation Biology: The Science of Scarcity and Diversity. Sinauer Associates Inc., Sunderland, MA.

Grady, K. C., S. M. Ferrier, T. E. Kolb, S. C. Hart, G. J. Allan, and T. G. Whitham. 2011. Genetic variation in productivity of foundation riparian species at the edge of their distribution: implications for restoration and assisted migration in a warming climate. Global Change Biology 17:3724–3735.

Graves, T., R. B. Chandler, J. A. Royle, P. Beier, and K. C. Kendall. 2014. Estimating landscape resistance to dispersal. Landscape Ecology 29:1201–1211.

Griffith, B., J. M. Scott, J. W. Carpenter and C. Reed. 1989. Reintroduction as a species conservation tool: status and strategy. Science 245:477–480.

Groves, C. R., D. B. Jensen, L. L. Valutis, K. H. Redford, M. L. Shaffer, J. M. Scott, J. V. Baumgartner, J. V. Higgins, M. W. Beck, and M. G. Anderson. 2002. Planning for biodiversity conservation: putting conservation science into practice. BioScience 52:499–512.

Finkelstein, M. E., D. F. Doak, D. George, J. Burnett, J., Brandt, M. Church, J. Grantham, and D. Smith. 2012. Lead poisoning and the deceptive recovery of the critically endangered California condor. Proceedings of the National Academy of Sciences of the United States of America 109(28):11449–11454.

Hagerman, S. M., and T. Satterfield. 2014. Agreed but not preferred: expert views on taboo options for biodiversity conservation, given climate change. Ecological Applications 24:548–559.

Hanski, I. A. 1999. Metapopulation Ecology. Oxford University Press, Oxford, UK.

Hendrickson, D. A., and J. E. Brooks. 1991. Transplanting short-lived fishes in North American deserts: review, assessment, and recommendations. Pp.283–298. In W. L. Minckley and J. E. Deacon, eds., Battle against Extinction: Native Fish Management in the American West. University of Arizona Press, Tucson.

Hobbs, R. J., E. Higgs, and J. A. Harris. 2009. Novel ecosystems: implications for conservation and restoration. Trends in Ecology and Evolution 24:599–605.

IUCN/SSC (International Union for Conservation of Nature/Species Survival Commission). 2013. Guidelines for Reintroductions and Other Conservation Translocations. Version 1.0. IUCN, Gland, Switzerland.

Jachowski, D. S., R. A. Gitzen, M. B. Grenier, B. Holmes, and J. J. Millspaugh. 2011. The importance of thinking big: large-scale prey conservation drives black-footed ferret reintroduction success. Biological Conservation 144:1560–1566.

Kellert, S. R. 1997. The Value of Life: Biological Diversity and Human Society. Island Press, Covelo, CA.

King, A. W. 1997. Hierarchy theory: a guide to system structure for wildlife biologists. Pp.185–212. In J. A. Bissonette, ed., Wildlife and Landscape Ecology. Springer, New York.

Kotliar, N. B., and J. A. Wiens. 1990. Multiple scales of patchiness and patch structure: a hierarchical framework for the study of heterogeneity. Oikos 55:253–260.

Landres, P. B., P. Morgan, and F. J. Swanson. 1999. Overview of the use of natural variability concepts in managing ecological systems. Ecological Applications 9:1179–1188.

Lawler, J. J., and J. D. Olden. 2011. Reframing the debate over assisted colonization.

Frontiers in Ecology and the Environment 9:569–574.

Le Gouar, P., J.-B. Mihoub, and F. Sarrazin. 2012. Dispersal and habitat selection: behavioral and spatial constraints for animal translocations. Pp.149–169. In J. G. Ewen, D. P. Armstrong, K. A. Parker, and P. J. Seddon, eds., Reintroduction Biology: Integrating Science and Management. John Wiley and Sons, Sussex, UK.

Lytle, D. A., and N. L. Poff. 2004. Adaptation to natural flow regimes. Trends in Ecology and Evolution 19:94–100.

Manel, S., M. K. Schwartz, G. Luikart, and P. Taberlet. 2003. Landscape genetics: combining landscape ecology and population genetics. Trends in Ecology and Evolution 18:189–197.

Marcot, B. G., C. S. Allen, S. Morey, D. Shively, and R. White. 2012a. An expert panel approach to assessing potential effects of bull trout reintroduction on federally listed salmonids in the Clackamas River, Oregon. North American Journal of Fisheries Management 32:450–465.

Marcot, B. G., R. E. Gullison, and J. R. Barborak. 2001. Protecting habitat elements and natural areas in the managed forest matrix. Pp.523–558. In R. A. Fimbel, A. Grajal, and J. G. Robinson, eds., The Cutting Edge: Conserving Wildlife in Logged Tropical Forests. Columbia University Press, New York, USA.

Marcot, B. G., M. G. Raphael, N. H. Schumaker, and B. Galleher. 2013. How big and how close? Habitat patch size and spacing to conserve a threatened species. Natural Resource Modeling 26:194–214.

Marcot, B. G., M. P. Thompson, M. C. Runge, F. R. Thompson, S. McNulty, D. Cleaves, M. Tomosy, L. A. Fisher, and A. Bliss. 2012b. Recent advances in applying decision science to managing national forests. Forest Ecology and Management 285:123–132.

Millar, C. I., N. L. Stephenson, and S. L. Stephens. 2007. Climate change and forests of the future: managing in the face of uncertainty. Ecological Applications 17:2145–2151.

Minteer, B. A., and J. P. Collins. 2010. Move it or lose it? The ecological ethics of relocating species under climate change. Ecological Applications 20:1801–1804.

Muhlfeld, C. C., V. D'Angelo, S. T. Kalinowski, E. L. Landguth, C. C. Downs, J. Tohtz, and J. L. Kershner. 2012. A fine-scale assessment of using barriers to conserve native stream salmonids: a case study in Akokala creek, Glacier National Park, USA. The Open Fish Science Journal 5:9–20.

Nash, K. L., C. R. Allen, D. G. Angeler, C. Bar-ichievy, T. Eason, A. S. Garmestani, N. A. J. Graham et al. 2014. Discontinuities, cross-scale patterns and the organization of ecosystems. Ecology 95:654–667.

Nathan, R., W. M. Getz, E. Revilla, M. Holyoak, R. Kadmon, D. Saltz, and P. E. Smouse. 2008. A movement ecology paradigm for unifying organismal movement research. Proceedings of the National Academy of Sciences of the United States of America 105:19052–19059.

Osborne, P. E., and P. J. Seddon. 2012. Selecting suitable habitats for reintroductions: variation, change and the role of species distribution modelling. Pp.90–117. In J. G. Ewen, D. P. Armstrong, K. A. Parker, and P. J. Seddon, eds., Reintroduction Biology: Integrating Science and Management. John Wiley and Sons, Sussex, UK.

Painter, L. E., R. L. Beschta, E. J. Larsen, and W. J. Ripple. 2015. Recovering aspen follow changing elk dynamics in Yellowstone: evidence of a trophic cascade? Ecology 96:252–263.

Parmesan, C., and G. Yohe. 2003. A globally coherent fingerprint of climate change impacts across natural systems. Nature 421(6918): 37–42.

Patten, M. A., and J. F. Kelly. 2010. Habitat selection and the perceptual trap. Ecological Applications 20:2148–2156.

Pe'er, G., M. A. Tsianou, K. W. Franz, G. Y, Matsinos, A. D. Mazaris, D. Storch, L. Kopsova et al. 2014. Toward better application of minimum area requirements in conservation planning. Biological Conservation 170:92–102.

Pérez, I., J. D. Anadón, M. Díaz, G. G. Nicola, J. L. Tella, and A. Giménez. 2012. What is wrong with current translocations? A review and a decision-making proposal. Frontiers in Ecology and the Environment 10:494–501.

Peters, R. H. 1986. The Ecological Implications of Body Size. Cambridge University Press, Cambridge, UK.

Peters, R. L. 1988. The effect of global climatic change on natural communities. Pp.450–461. In E. O. Wilson and F. M. Peter, eds., Biodiversity. National Academy Press, Washington DC.

Peterson, D. P., B. E. Rieman, J. B. Dunham, K. D. Fausch, and M. K. Young. 2008. Analysis of trade-offs between threats of invasion by nonnative brook trout (*Salvelinus fontinalis*) and intentional isolation for native westslope cutthroat trout (*Oncorhynchus clarkii lewisi*). Canadian Journal of Fisheries and Aquatic Sciences 65:557–573.

Pickett, S. T. A., and J. N. Thompson. 1978. Patch dynamics and the design of nature reserves. Biological Conservation 13:27–37.

Pickett, S. T. A., and P. S. White. 1985. The Ecology of Natural Disturbance and Patch Dynamics. Academic Press, New York.

Pillai, A., and D. Heptinstall. 2013. Twenty years of the Habitats Directive: a case study on species reintroduction, protection and management. Environmental Law Review 15:27–46.

Pollock, M. M., T. J. Beechie, J. M. Wheaton, C. E. Jordan, N. Bouwes, N. Weber, and C. Volk. 2014. Using beaver dams to restore incised stream ecosystems. BioScience 64:279–290.

Pope, S. E., L. Fahrig, and H. Merriam. 2000. Landscape complementation and metapopulation effects on leopard frog populations. Ecology 81:2498–2508.

Pulliam, H. R. 1988. Sources, sinks, and population regulation. American Naturalist 132:652–661.

Pulliam, H. R., and B. J. Danielson. 1991. Sources, sinks, and habitat selection: a landscape perspective on population dynamics. American Naturalist 137:S50–S66.

Rahel, F. J. 2013. Intentional fragmentation as a management strategy in aquatic systems. BioScience 63:362–372.

Rantanen, E. M., F. Buner, P. Riordan, N. Sotherton, and D. W. Macdonald. 2010. Habitat preferences and survival in wildlife reintroductions: an ecological trap in reintroduced grey partridges. Journal of Applied Ecology 47:1357–1364.

Reading, R. P., T. W. Clark, and S. R. Kellert. 2002. Towards an endangered species reintroduction paradigm. Endangered Species Update 19:142–146.

Reding, D. M., S. A. Cushman, T. E. Gosselink, and W. R. Clark. 2013. Linking movement behavior and fine-scale genetic structure to model landscape connectivity for bobcats (*Lynx rufus*). Landscape Ecology 28:471–486.

Reed, J. M., and A. P. Dobson. 1993. Behavioural constraints and conservation biology: conspecific attraction and recruitment. Trends in Ecology and Evolution 8:253–256.

Ripple, W. J., J. A. Estes, R. L. Beschta, C. C. Wilmers, E. G. Ritchie, M. Hebblewhite, J. Berger et al. 2014. Status and ecological effects of the world's largest carnivores. Science 343(6167). doi:10.1126/science.1241484.

Roe, J. H., M. R. Frank, and B. A. Kingsbury. 2015. Experimental evaluation of captive-rearing practices to improve success of snake reintroductions. Herpetological Conservation and Biology 10:711–722.

Rosenberg, D. K., and K. S. McKelvey. 1999. Estimation of habitat selection for central-place foraging animals. Journal of Wildlife Management 63:1028–1038.

Rudnick, D. A., S. J. Ryan, P. Beier, S. A. Cushman, F. Dieffenbach, C. W. Epps, L. R. Gerber et al. 2012. The role of landscape connectivity in planning and implementing conservation and restoration priorities. Issues in Ecology 16:1–20.

Scheerer, P. D. 2002. Implications of floodplain isolation and connectivity on the conservation of an endangered minnow, Oregon chub, in the Willamette River, Oregon. Transactions of the American Fisheries Society 131:1070–1080.

Scheerer, P. D., and P. J. McDonald. 2003. Age, growth, and timing of spawning of an endangered minnow, the Oregon chub (*Oregonichthys crameri*), in the Willamette Basin, Oregon. Northwestern Naturalist 84:68–79.

Schlaepfer, M. A., M. C. Runge, and P. W. Sherman. 2002. Ecological and evolutionary traps. Trends in Ecology and Evolution 17:474–480.

Schultz, C. B., and E. F. Crone. 2005. Patch size and connectivity thresholds for butterfly habitat restoration. Conservation Biology 19:887–896.

Schultz, C. B., and K. M. Dlugosch. 1999. Nectar and hostplant scarcity limit populations of an endangered Oregon butterfly. Oecologia 119:231–238.

Scott, J. M., P. J. Heglund, M. Morrison, M. Raphael, J. Haufler, and B. Wall, eds. 2002. Predicting Species Occurrences: Issues of Accuracy and Scale. Island Press. Covelo, CA.

Seddon, P. J. 2010. From reintroduction to assisted colonization: moving along the conservation translocation spectrum. Restoration Ecology 18:796–802.

Seddon, P. J. 2013. The new IUCN guidelines highlight the importance of habitat quality to reintroduction success: reply to White et al. Biological Conservation 164:177.

Seddon, P. J., D. P. Armstrong, and R. F. Maloney. 2007. Developing the science of reintroduction biology. Conservation Biology 21:303–312.

Smith, D. W., and E. E. Bangs. 2009. Reintroduction of wolves to Yellowstone National Park: history, values and ecosystem restoration. Pp. 92–125. In: M. W. Hayward and M. Somers, eds., Reintroduction of Top-Order Predators. Wiley-Blackwell, Hoboken, NJ.

Smith, D. W., E. E. Bangs, J. K. Oakleaf, C. Mack, J. Fontaine, D. Boyd, M. Jimenez et al. 2010. Survival of colonizing wolves in the northern Rocky Mountains of the United States, 1982–2004. Journal of Wildlife Management 74:620–634.

Soulé, M. E., J. A. Estes, B. Miller, and D. L. Honnold. 2005. Strongly interacting species: conservation policy, management, and ethics. BioScience 55:168–176.

Sousa, W. P. 1984. The role of disturbance in natural communities. Annual Review of Ecology and Systematics 15:353–391.

Suvorov, P., and J. Svobodová. 2012. The occurrence of ecological traps in bird populations: is our knowledge sufficient? A review. Journal of Landscape Ecology 5:36–56.

Taylor, P. D., L. Fahrig, K. Henein, and G. Merriam. 1993. Connectivity is a vital element of landscape structure. Oikos 68:571–573.

Treves, A., L. Naughton-Treves, and V. Shelley. 2013. Longitudinal analysis of attitudes toward wolves. Conservation Biology 27:315–323.

Treves, A., R. B. Wallace, and S. White. 2009. Participatory planning of interventions to mitigate human–wildlife conflicts. Conservation Biology 23:1577–1587.

Turner, M. G. 2005. Landscape ecology in North America: past, present, and future. Ecology 86:1967–1974.

Turner, M. G. 2010. Disturbance and landscape dynamics in a changing world. Ecology 91:2833–2849.

USFWS (US Fish and Wildlife Service). 1998. Oregon chub (*Oregonichthys crameri*): Recovery Plan. US Fish and Wildlife Service, Oregon State Office, Portland.

van Heezik, Y., R. F. Maloney, and P. J. Seddon. 2009. Movements of translocated captive-bred and released critically endangered kaki (black stilts) *Himantopus novaezelandiae* and the value of long-term post-release monitoring. Oryx 43:639–647.

van Strien, M. J., D. Keller, R. Holderegger, J. Ghazoul, F. Kienast, and J. Bolliger. 2014. Landscape genetics as a tool for conservation planning: predicting the effects of landscape change on gene flow. Ecological Applications 24:327–339.

Vandergast, A. G., E. A. Lewallen, J. Deas, A. J. Bohonak, D. B. Weissman, and R. N. Fisher. 2009. Loss of genetic connectivity and diversity in urban microreserves in a southern California endemic Jerusalem cricket (Orthoptera: Stenopelmatidae: *Stenopelmatus* n. sp. "santa monica"). Journal of Insect Conservation 13:329–345.

Verboom, J., P. Schippers, A. Cormont, M. Sterk, C. C. Vos, and P. F. Opdam. 2010. Population dynamics under increasing environmental variability: implications of climate change for ecological network design criteria. Landscape Ecology 25:1289–1298.

Westrum, R. 1994. An organizational perspective: designing recovery teams from the inside out. Pp.327–349. In T. W. Clark, R. R. Reading, and A. L. Clarke, eds., Endangered Species Recovery: Finding the Lessons, Improving the Process. Island Press, Washington, DC.

Whittaker, R. J., and J. M. Fernández-Palacios. 2007. Island Biogeography: Ecology, Evolution, and Conservation. Oxford University Press. Oxford, UK.

Wiens, J. A., and B. T. Milne. 1989. Scaling of "landscapes" in landscape ecology, or, landscape ecology from a beetle's perspective. Landscape Ecology 3:87–96.

Wilcove, D. S., M. J. Bean, B. Long, W. J. Snape, B. M. Beehler, and J. Eisenberg. 2004. The private side of conservation. Frontiers in Ecology and the Environment 2:326–331.

Wilkerson, M. L. 2013. Invasive plants in conservation linkages: a conceptual model that addresses an underappreciated conservation issue. Ecography 36:1319–1330.

Williams, B. K., and E. D. Brown 2012. Adaptive Management: The U. S. Department of Interior Applications Guide. US Department of Interior, Washington, DC.

Wolf, C. M., B. Griffith, C. Reed, and S. A. Temple. 1996. Avian and mammalian translocations: update and reanalysis of 1987 survey data. Conservation Biology 10:1142–1154.

Wu, J. 2009. Ecological dynamics in fragmented landscapes. Pp.438–444. In S. A. Levin, ed., Princeton Guide to Ecology. Princeton University Press, Princeton, NJ.

Zuckerberg, B., and W.F., Porter. 2010. Thresholds in the long-term responses of breeding birds to forest cover and fragmentation. Biological Conservation 143:952–962.

SIX

Setting Objectives and Defining the Success of Reintroductions

Alienor L. M. Chauvenet, Stefano Canessa, and John G. Ewen

THE LACK OF A CLEAR DEFINITION of what constitutes success seems to be a major concern in reintroduction biology (Robert et al. 2015), and it is a subject widely discussed in both animal (Seddon 1999, Fischer and Lindenmayer 2000) and plant reintroductions (Menges 2008). Reported definitions of reintroduction success in the literature vary widely (see Chauvenet et al. 2013, for a review), from vaguely quantitative assessments (e.g. "establishing a self-sustaining population"; Griffith et al. 1989, Dickens et al. 2010) to taxon-specific definitions (e.g., specific number of individuals attempting to breed or successfully fledgling at 2 years; Reynolds et al. 2008). Robert et al. (2015) recently suggested that reintroduction success could be measured quantitatively and systematically using the International Union for Conservation of Nature (IUCN) criteria for threatened species, but they acknowledge that there is no general agreement on when to call a project a success.

This diversity of reintroduction success measures could be reconciled by using a decision-making framework. Success criteria for any given reintroduction program should be chosen primarily to reflect its specific objectives, defined at the beginning (McCarthy et al. 2012, Nichols and Armstrong 2012). Objectives should be thus carefully defined to correspond to the aspirations and preferences that motivate the reintroduction effort (Nichols and Armstrong 2012) as most reintroduction programs will aim to achieve multiple but different ecological, social, and economic objectives (McCarthy et al. 2012, Converse et al. 2013). Decisions will be easier and more transparent, and results of all projects are comparable and readily assessed, if objectives are precisely stated and measureable quantities.

In this chapter, we investigate the objectives most commonly published by reintroduction programs and whether these tend to be clearly defined using quantifiable indicators of success. We then discuss how objectives can be defined and structured within a systematic and transparent decision-making process. We argue that, to set success criteria, reintroduction practitioners should frame reintroductions as decision problems and use formal approaches of

STRUCTURED DECISION-MAKING	Theoretical framework and suit of methods for approaching decisions under uncertainty.
DECISION PROBLEM	A concise summary of the problem. To ensure the right problem is being solved, this summary should include answers to the following: Why does this decision need to be made? What deliverable is required from the decision process? What is its relationship to other decisions? Who is the decision maker? What is the timing?
VALUES	What we care about; should drive our decision-making. In a decision problem, values are expressed as objectives.
GOALS	Reflect the ultimate aims of translocations; expressed in terms of conservation benefit within a set time frame (IUCN/SSC 2013).
OBJECTIVES	Concise statements about what matters; must include the desired direction of change; how goals will be realized according to the IUCN guidelines (IUCN/SSC 2013).
FUNDAMENTAL OBJECTIVES	Things that matter regardless of how we achieve them. One way to arrive at these objectives is to repeatedly ask "why is this important?" If the answer is "just because it is" then it is probably reflecting something fundamentally important.
MEANS OBJECTIVES	Methods that can help us achieve the fundamental objectives. One way to arrive at these is to ask "how can we achieve our fundamental objectives?"
PROCESS OBJECTIVES	Relate to how decisions are made, for example, by adopting group consultations. These objectives influence the process of making a decision but not the final choice between two or more alternative courses of action within a decision problem.
STRATEGIC OBJECTIVES	Reflect the strategic priorities of the groups involved; for example, local governments may require the decisions to be made using public engagement. These objectives influence the process of making a decision but not the final choice between two or more alternative courses of action within a decision problem.
MEASURABLE ATTRIBUTES	Specific performance metrics that can be used to consistently estimate and report the anticipated consequences of a management alternative with respect to a particular objective (Gregory et al. 2012).
PERFORMANCE METRICS	Used as measurable attributes. They can either be natural (something that can be directly measured), constructed (using a sliding or relative scale), or a proxy (a natural attribute that is highly correlated with the objective but does not directly measure it).
TARGETS	In decision-making, a target is a desired level of performance toward an objective (Gregory et al. 2012). Care needs to be taken when using targets as fundamental objectives as they usually have implications for other objectives. They may be best expressed as alternative actions for better comparison across fundamental objectives.

decision theory to guide the entire reintroduction process.

REINTRODUCTION OBJECTIVES SO FAR

In the IUCN guidelines for reintroductions and other conservation translocations (IUCN/SSC 2013), there are clear recommendations as to how goals and objectives of reintroductions should be set. It is stated that objectives "should be clear and specific and ensure they address all identified or presumed current threats to the species" (IUCN/SSC 2013). Objectives are how the goals will be realized (Box 6.1), while goals are the ultimate aims of translocations and expressed in terms of conservation benefit within a set time frame. We searched the literature to see whether reintroduction projects always set clear goals and objectives.

Using Web of Knowledge, we looked for all articles containing the terms "reintroduction OR re-introduction" as well as "biodiversity AND conservation" (search conducted on October 31, 2014) and found 4,915 document matches. Out of these, we identified 86 articles or book chapters, spanning from 1986 to 2014, pertaining to a specific species on-the-ground reintroduction project and looked for mentions of objectives, goals, or aims within them. This yielded 17 peer-reviewed publications that specifically mentioned one objective or more. We found that "objective," "goal," or "aim" tended to be used interchangeably in the published literature, an issue that might contribute to the confusion about the different types of objectives. To add to these 17 peer-reviewed publications, we also browsed through the Reintroduction Specialist Group (RSG) books published in the last 5 years (Soorae 2010, 2011, 2013) for the recorded objectives of all invertebrate, fish, bird, and mammal species reintroductions (see Ewen et al. 2014a, for a review of reintroduction objectives for reptiles and amphibians); this yielded an additional 112 case studies. Within these 129 studies, we examined the focus of reported objectives to gauge how restoration "success" has been treated in the literature. Actions can be implemented to directly benefit species' viability or persistence (often a "fundamental" objective in the context of structured decision-making [SDM]) or to benefit other parts of a project (often a "means" objectives in the SDM context). A clear definition and distinction between fundamental and means objectives is provided in a section below. We classified reported objectives as *species-based*, when they were about the fate of the population or species (e.g., establish a viable population), or *technical*, when they were about conducting monitoring or raising public awareness (Table 6.1). The aim of this literature search and classification was to assess the proportion of species-based objectives that were directly quantifiable or had quantifiable indicators stated for them as required by the IUCN guidelines (IUCN/SSC 2013).

We were able to further classify objectives in three categories: "yes" when objectives were clear, quantitative, thus measurable (e.g., the aim was to establish at least 10 populations with >30 breeding adults; Jachowski et al. 2011); "no" when objectives were vague and nonquantifiable (e.g., establishing a free-ranging self-sustaining population without clear indicators of success; Huber et al. 2014); and "mixed" when projects had both quantitative and nonquantitative species-based objectives. We applied a strict categorization; in order to be in the "yes" category, the indicator of success had to be something measurable, and associated with a time frame. For example, if the objective was establishing a population, a clear indicator would be "the population exhibits stable or positive annual dynamics," while a vague statement would be "the released individuals survived."

Overall, our literature review showed that there are a number of common objectives reported for wildlife reintroductions (Table 6.1). By far, the most common species-based objective is to establish a new population; this is usually qualified using the terms "viable" and/or "self-sustaining." However, clear indicators of success were listed in only around 50% of the mentions of establishing a population as an objective. Technical objectives, which do not directly benefit species' persistence, were frequently reported; out of all the distinct objectives we recorded, nearly two-thirds are technical ones (Table 6.1). When looking at species-based objectives, the majority had no clear indicators associated with them (52%; Table 6.2) and only a third were in the "yes" category. Therefore, objectives focused on the species being reintroduced were less common than those focused on technical implementation, and most often were not quantifiable. This was true across literature on mammals, fish, and invertebrates, although to a varying extent. Only published literature on bird reintroductions

TABLE 6.1

List of Species-Based and Technical Objectives of Reintroductions, Found in a Total of 129
Cases in the Literature and the RSG Reintroduction Books (Soorae 2010, 2011, 2013)

Objective	Count
Species-Based	
Establish (self-sustaining) population	60
Reestablish or recolonize historical range	19
Maximize, preserve, or improve genetic diversity	14
Population supplementation	11
Increase numbers or expand range through new populations	8
Successful breeding	5
Persistence of population	4
Connectivity with other populations	4
Maximize population number	3
Increase numbers through management	3
Successful breeding over a given time frame	3
Promote individual welfare	3
Stop species decline or extinction	2
Survival of reintroduced animals	2
Persistence of population over a given time frame	2
Reduce conflict with other individuals, species, and humans	2
Maximize chance of successful establishment	1
Survival of reintroduced animals over a given time frame	1
Minimize extinction risk	1
Contribute to global species conservation	1
Increase species range or carrying capacity	1
Technical	
Monitor population and perform adaptive management	35
Create captive populations or genetic bank for future release	27
Study ecology of species and maximize knowledge	20
Generate local stakeholder support	18
Capture animals and perform reintroduction	18
Increase public or authorities awareness	16
Identify best release strategy	16
Habitat restoration	16
Identify best management or improve management strategies	13
Habitat management	13
Develop monitoring and reintroduction techniques for species	11
Contribute to ecological restoration or preservation	10

Remove or mitigate threats	9
Reestablish or maintain ecological function	7
Use species reintroduction as benchmark for future reintroduction efforts	6
Evaluate species reintroduction as a tool for conservation, ecological restoration, or as an umbrella species	5
Identify best release source	4
Minimize cost	3
Assess suitability of proposed reintroduction sites	3
Increase government support	2
Provide recreational benefits to visitors	2
Provide benefits to human population from reintroduced population	2
Disease control and management	2
Consider and account for impact of climate change	2
Satisfy success indicators	2
Manage wild and captive population as metapopulation	2
Promote the establishment of new National Park	1
Increase local livelihood	1
Maximize public relations benefits	1
Create a species recovery working group	1
Release individuals genetically similar to extinct population	1
Investigate sites from which the species has gone extinct to determine reason for extinction	1
Determine impact of species on environment	1

reported more quantifiable than nonquantifiable objectives. Finally, we also observed many instances of mismatch between objectives and indicators of success, where the indicators listed were not able to inform on the achievement of the objectives, reflecting the potential of a poor understanding of the distinction between objectives and indicators of success or how they are linked.

To help define clear objectives against which success can be assessed quantitatively, managers should consider the use of SDM. SDM is only slowly becoming recognized as a best practice for planning reintroductions globally, and we focus the rest of this chapter on how and why SDM is useful.

STRUCTURED DECISION-MAKING FOR REINTRODUCTION

Setting Objectives

The definition of SDM encompasses both the theoretical framework and the ensemble of methods for approaching decisions under uncertainty (Gregory et al. 2012). A decision consists of several components, which SDM first analyses separately and then reintegrates in an iterative process: (1) carefully defining the decision problem, (2) formulating clear and measurable objectives, (3) identifying the alternative actions available, (4) predicting their expected outcomes, and (5) solving the decision problem

TABLE 6.2

Number (and Percentage) of Taxa-Specific Reintroduction Cases Where Objectives Were Classified as Quantifiable (Yes),
Nonquantifiable (No), or Were a Mix

	Invertebrate (%)	Fish (%)	Bird (%)	Mammal (%)	Total (%)
Quantifiable	3 (18.8)	4 (16.0)	18 (52.9)	18 (33.3)	43 (33.3)
Nonquantifiable	13 (81.2)	15 (60.0)	13 (38.3)	26 (48.2)	67 (52.0)
Mixed	–	6 (24.0)	3 (8.8)	10 (18.5)	19 (14.7)
Total	16	25	34	54	129

by addressing trade-offs (this step is also called optimization; see Converse and Armstrong, Chapter 7, this volume). These methods have been successfully employed in several branches of environmental resource management (Ralls and Starfield 1995, McDaniels et al. 1999, Failing et al. 2007, Irwin et al. 2008, Martin et al. 2009, Bunnefeld et al. 2011, Martin et al. 2011, Gannon et al. 2013, McGowan et al. 2015), and their use has long been advocated in reintroduction biology (Maguire 1986, Maguire et al. 1987, 1988, 1990). However, their application in real-world reintroduction programmes has only recently gained momentum (see, e.g., Collazo et al. 2013, Converse et al. 2013, Gedir et al. 2013, Ewen et al. 2014b, Servanty et al. 2014).

Converse and Armstrong (Chapter 7, this volume) describe each of the SDM steps in greater detail. Throughout the remainder of this chapter, we will focus on step (2), exploring how reintroduction objectives can be set and negotiated within the SDM framework, and how doing so can assist in making rational management decisions. This step arises after a group of expert stakeholders has already been formed. An example of the use of SDM for a reintroduced species is given in Box 6.2. This brief case study is sufficient to highlight the importance of objectives in the SDM context: objectives are the things that matter. In this sense, they represent the reason for making a decision: if no objectives exist, we would have no preferences about outcomes, and therefore it would not matter what action we choose. SDM explicitly recognizes that decisions are value

driven, and that defining those values as explicit objectives is an important step in any decision process. Science can inform decisions and assist in searching for optimal solutions, but only when objectives are clearly formulated to reflect the driving values behind a decision. So what does a clear formulation of objectives entail?

Structuring Objectives

According to Clemen (1996) and Gregory et al. (2012), a good set of objectives should be the following:

1. *Complete*: objectives should capture all the relevant values of decision makers and stakeholders.

2. *Concise*: the number of objectives chosen must be sufficient to provide a clear representation of what is actually important while ensuring they remain tractable and avoid redundancy.

3. *Sensitive*: actions under consideration must differ in their ability to influence different objectives. If different actions all result in the same outcome for one objective, then that objective has no relevance for the decision problem at hand.

4. *Understandable*: as far as possible, they should be understandable to everyone by avoiding ambiguous terms and jargon. If technical terms or jargon must be used, then these need to be clearly defined.

BOX 6.2 · A Case Study of Supplementary Feeding Regime for the Hihi on Kapiti Island, New Zealand

6a Female hihi drinking out of supplementary food dispenser (credit: P. Brekke).

The hihi (*Notiomystis cincta*) is an endangered endemic bird from New Zealand. Through almost two decades of recovery effort, the species has been reestablished to several managed populations offshore and on the mainland of North Island. For the population on Kapiti Island, Ewen et al. (2014b) used SDM to identify a supplementary feeding regime (figure 6a). The decision process entailed the following steps:

1. *Problem definition*. The hihi recovery team needed to identify whether and how to provide sugar water supplement to the reintroduced hihi population on Kapiti Island.

2. *Objectives*. Through email interaction prior to the meeting, the group identified three objectives: maximizing the number of hihi on the island, minimizing the cost of management, and minimizing the risk of extinction of the species on the island. Following the SDM process, each of these objectives was used to define a column of a consequence table (Table 6.3). Each was given a measurable attribute to be able to quantify it.

3. *Alternative actions*. The group then defined six candidate actions that reflected a range of management scenarios, ranging from ad libitum feeding to removal of birds from the island. These actions were used to define the rows of the consequence table (Table 6.3).

4. *Consequences*. During a meeting, group members used their expert judgment to fill in the consequence table. Each cell indicated the expected outcome of a given action in relation to a given objective: for example, how many birds we would expect as a result of the action "Maintain current feeding regime," if it is applied.

5. *Trade-offs*. The group then defined scores reflecting the relative importance of different objectives, and used these to interpret the consequence table, evaluating trade-offs and selecting a preferred strategy.

5. *Independent*: as far as practicable, it should be possible to evaluate the importance of each objective independently of others.

The first step toward meeting these criteria is a clear separation of *fundamental* and *means* objectives, for example, by building an "objective hierarchy" (Gregory et al. 2012; see Box 6.1). Fundamental objectives represent the core values that a decision maker wants to achieve, the things that really matter; means objectives are important only because they help achieve some other objective. Specifying fundamental objectives will help in setting the criteria for evaluating alternatives; a clear network of means objectives can assist in formulating alternatives, highlighting the key aspects that should be acted upon, and defining measurable attributes that reflect the fundamental objectives. Objectives can be arranged along the fundamental-means spectrum by the repeated use of two simple guiding questions: "why is that important?" and "how can we achieve that?" By asking these questions, we will move toward fundamental and means objectives, respectively (Clemen 1996).

We can better understand this distinction by considering the notion of "insurance populations" in reintroduction plans. Captive populations of threatened species are often considered insurance against species extinctions (Jones et al. 2007, Garnett et al. 2011, Zippel et al. 2011). What exactly this insurance entails, however, may change depending on how we view our fundamental and means objectives. In a simplified example, let us imagine three stakeholders involved in recovery planning for the conservation of a threatened species: one of their stated objectives is to establish and maintain a viable captive population. However, there are subtle differences in their core values that can influence the way in which this objective will be treated (figure 6.1).

- **Stakeholder A** has a specific fundamental objective: maximize the persistence of the species in the wild. She attaches a funda-

mental value to having a species live in its natural environment. *How could this be achieved?* It could be achieved either by maintaining the extant populations of the species in situ or by establishing and maintaining a captive population that will provide individuals for a reintroduction. In this sense, when she states the objective of establishing and maintaining a captive population, she is considering a means objective: it is only important because it might allow reintroduction, which in turn might increase the probability of the species persisting in the wild. If population modeling were to show that in situ management of the extant populations is expected to ensure higher persistence without involving captivity, then maintaining a captive population might no longer be of interest to Stakeholder A.

- **Stakeholder B** initially shares the same objective as Stakeholder A, giving importance only to persistence of the species in the wild. However, when modeling shows that in situ management is indeed expected to provide better persistence, he feels uneasy about abandoning the captive population. After all, having the ex situ population in a zoo also increases the exposure of visitors to the recovery program. But *why* is this exposure important? Well, it is just important because the zoo has a fundamental objective of community outreach: they care about people knowing about the species. Again, the captive population is just a means for this fundamental objective. If an action that did not involve a captive population was expected to provide better outcomes in terms of community engagement, there would be no need to maintain that captive population. However, without expressing this additional fundamental objective, the decision-making process would not have captured the full range of preferences of Stakeholder B.

- **Stakeholder C** also fundamentally values the persistence of the species in the wild; he also

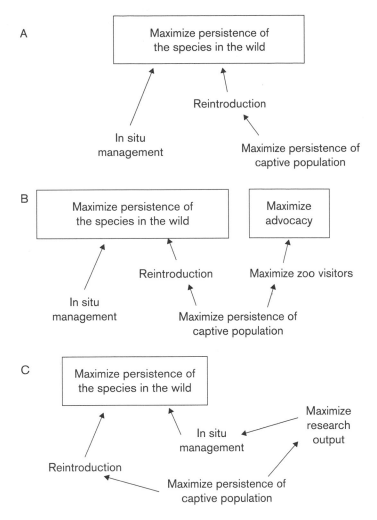

FIGURE 6.1. Different definitions of fundamental and means objectives for the problem of establishing a captive population. Each box reflects the values of one of three managers (A, B, and C; see text). Fundamental objectives are indicated by boxes; arrows indicate the link between means and fundamental objectives.

feels uneasy about abandoning the captive population. However, community outreach is not important for this stakeholder; what is important is the research that could be carried out on captive individuals. Establishing a captive population would allow laboratory experiments to resolve an important uncertainty, such as the causes of decline. Reducing uncertainty could improve development and implementation of management actions: for example, isolating a disease-causing pathogen and developing a treatment for it. So, establishing a captive population is once again identified as a means objective, linked to another objective of conducting research, which in turn is a means for the fundamental objective of persistence. And again, this implies that both establishing a captive population and conducting research have no fundamental value, but are only important in terms of the benefit they will provide to species persistence.

Although the differences in the way objectives are defined and structured in this example may appear marginal, they can have important consequences for decision-making. For example, if establishing a captive population were listed as a fundamental objective, the recovery team might be compelled to allocate resources to it, regardless of the actual benefit for the wild population, and potentially subtracting resources from more effective actions that do not involve captivity. If stakeholder C stated "conducting research" as a fundamental objective, again the recovery group might feel committed to maintaining the captive population, without explicitly evaluating how the results of research could improve persistence of the wild population. If additional objectives were not recognized, frustrating arguments about alternative actions might arise. In such scenarios, an objective hierarchy may help stakeholders see how their preferred objective or action fits into the bigger picture of the decision analysis. For example, as mentioned above, if the commitment of Stakeholder B to community outreach were not explicitly stated, then his resistance to abandoning the captive population would not be understood by other stakeholders, potentially leading to conflicts and inefficient final decisions. In general, a clear objective hierarchy can insure against the risk of becoming fixated on specific means objectives or management actions, maintaining instead a clear view of the bigger picture of the decision problem.

Finally, "fundamental" and "means" are not the only types of objectives we may come across. Two others are frequently encountered and should be recognized, *process* and *strategic* objectives (Box 6.1). Process objectives relate to how decisions are to be made, for example, by adopting group consultations or by producing transparent budget calculations. Strategic objectives reflect more the strategic priorities of the groups involved. Local governments may, for example, require that the process maximize public engagement. While process and strategic objectives influence the decision process, they do not directly affect the specific decision between two or more alternative courses of action, so we do not consider them further here (but see Gregory et al. 2012, for more detail).

Measuring Objectives

We have stated above that fundamental objectives are the very reason why decisions are necessary in the first place: they represent outcomes that we prefer or not. *Vice versa*, to make a decision means to compare possible outcomes in terms of the fundamental objectives or, where this is not possible, of the corresponding means objectives (such as a combination of vital rates reflecting a fundamental objective of population viability). This is why SDM requires objectives to be assigned unambiguous measurable attributes, which represent the link between objectives and alternative actions. Keeney and Gregory (2005) recommend the choice of measurable attributes that are *unambiguous, comprehensive, direct, understandable,* and *operational* (i.e., consequences and trade-offs can be described in practice). If decision makers care about something, it must be measured in the decision context to allow comparison of alternative management actions. Measurable attributes represent *performance* metrics for their respective objectives. In Box 6.2, in the example of decisions around supplementary feeding for New Zealand hihi (*Notiomystis cincta*), the number of individuals was directly measurable and relevant to decision makers, whereas management cost was more easily arranged along a constructed scale, in this case from 0 (best) to 100 (worst).

In some cases, it may be useful to formulate objectives as targets, where a specific performance level is stated. For example, a recovery plan may state the objective of achieving 95% probability of persistence by a species (see the "Reintroduction Objectives So Far" section). This is often the case for cost objectives, where a threshold or constraint can be set to "meet available budget" (e.g., Canessa et al. 2014). Although specifying objectives as targets can be useful, it is necessary to carefully consider the assumptions implied. By setting a specific performance level, an a priori assumption is made about the desirability of different levels: in the most extreme case, anything worse than the specified level is unacceptable, and anything better is unimportant. This may not be entirely realistic; in the example above, are persistence probabilities of 90% and 50% equally considered as failures? Would it be more realistic to formulate this objective as "maximize persistence"? As for means objectives, targets and threshold-type objectives can be useful but should not be confused with fundamental objectives. In some cases, specific targets might also represent process objectives. For example, "full compliance with the Species Protection Act" is an objective that can intuitively take on two values, yes or no (or 0/1 or similar binary metrics), and further complication may be unnecessary.

Addressing Multiple Objectives

As the examples in the previous section began to highlight, the majority of reintroduction programs will have more than one fundamental objective (Converse et al. 2013). In general, the underlying SDM approach does not differ for single- and multi-objective problems. Each action is judged on the basis of its expected outcomes in terms of each fundamental objective: this requires again to identify and to structure fundamental and means objectives, and to choose adequate measurable attributes for each one.

The main additional source of complexity in multi-objective problems is the need to identify and resolve trade-offs. For example, in the SDM evaluation of supplementary feeding for hihi (Box 6.2), the approach of ad libitum feeding provided the best result in terms of bird persistence and the worst in terms of management costs. Within SDM, a number of approaches can be used for solving such trade-offs. In one recent example, Canessa et al. (2014) used population models to identify the optimal

release rates of captive-bred eggs and subadult southern corroboree frogs (*Pseudophryne corroboree*) from captivity that maximized the size of the captive and reintroduced populations while meeting cost constraints. In that example, population persistence and costs were considered equally important: however, within a set of fundamental objectives, some may be more important than others. For example, the manager of a reintroduction program might be concerned about costs, but still consider the persistence of the focal species as the most important objective.

An intuitive method to account for such differences in the relative importance of objectives is the simple multi-attribute rating technique (SMART; Keeney and Raiffa 1993). The SMART uses an additive model to calculate the aggregate outcome of every action under evaluation:

$$EV_{Total(i)} = \sum_{j=1}^{J} EV(j)_j w_j, \qquad (1)$$

where $EV_{Total(i)}$ is the aggregate expected value of action i, $EV(j)$ is the expected outcome of action i relative to objective j, and w_j is a score between 0 and 1 reflecting the importance of objective j relative to the whole set of objectives (Keeney and Raiffa 1993).

The SMART approach was used in the case study presented in Box 6.2 (Ewen et al. 2014b). In this case, recovery group members provided consensus estimates of the predicted outcome of each action for each objective (Table 6.3). The recovery group members then expressed their individual preferences for each objective as relative weights. Weights (normalized to sum to 1) reflect the relative importance of objectives: for example, an objective weighted 0.5 is considered twice as important as one weighted 0.25. Objectives of the same importance are given equal weights. In this decision, the group agreed to average individual preference scores (Table 6.3); this process, while not easy, can be simplified by using appropriate expert elicitation technique (see, e.g., Gregory and Wellman 2001). The rescaled predicted

TABLE 6.3A

Weighting of Hihi Conservation by Individual Experts (the Higher the Score, the More Preferred the Objective)
and Their Averaged Scores, Which Were Then Normalized

Individual Objective Weights from Experts

Objective	Ind. 1	Ind. 2	Ind. 3	Ind. 4	Ind. 5	Ind. 6	Ind. 7	Total Score	Rank	Weight
Maximize number of hihi	100	100	100	100	100	100	80	97.14	1	0.45
Minimize cost of management	80	60	75	50	60	90	60	67.86	2	0.31
Minimize extinction risk	20	50	60	65	20	40	100	50.71	3	0.24

SOURCE: Modified from Ewen et al. (2014b).

outcomes of each action under each objective were then multiplied by the weights for the respective objectives (equation 1). In this case, the aggregate SMART score suggested two actions would provide similarly high expected outcomes and either may be a valid approach. The recovery group chose the option that was easier to implement (redistributing outlying feeder stations into the core area and providing more food) and this has now been implemented and the population monitored (Table 6.3B).

Value and Utility Functions

The objective weighting used in SMART recognizes that decision makers may value multiple objectives differently, considering some more important than others. Moreover, even for the same objective, the importance attributed to expected outcomes might be a linear function of the predicted values or be subject to increasing or diminishing returns. This variation can be captured using a value function, a transformation of the raw outcomes to reflect their perceived importance (Clemen 1996). For example, a logarithmic function could be used to describe outcomes that have diminishing returns.

Most importantly, trade-offs are likely to be influenced by value functions. Reintroductions, in particular, have been criticized for their high resource requirements, but at the same time

are often considered the only effective actions for endangered or critically endangered species. In a recent study, Canessa et al. (2016a) assessed the choice of ex situ-only versus in situ–only management for the critically endangered spotted tree frog (*Litoria spenceri*) under different value functions of decision makers. Actions focusing exclusively on in situ treatment of wild populations were cheaper, but they were only expected to provide results up to a certain level of persistence (approximately 60%). Actions that involved ex situ management were more expensive, but they were the only ones that could increase the short-term probability of persistence beyond 60%. As a result, managers preferring diminishing returns for persistence were more likely to choose in situ management only; those preferring increasing returns (i.e., who gave little value to improvements in the lower range of persistence) would choose ex situ management as their optimal action.

Moreover, the outcomes of conservation actions are generally uncertain; rather than point values, expectations will more likely encompass a range of possible outcomes, some of which may be less desirable than others. Making a decision on the basis of such uncertain expectations may entail a risk, and attitude to risk will be a determining factor in the decisions being made. For example, the harvest of remnant populations of endangered species

TABLE 6.3B

| | All Considered Feeding Alternatives and Consequences | | | | | |
Objective	Redistribute Feeders, Same Amount of Food	Redistribute Feeders, Increase Food	Increase Food to Existing Feeders	Maintain Current Feeding Regime	Redistribute Feeders, Dynamic Feeding[b]	Remove Hihi
Maximize number of hihi	80	160	180	80	180	0
Minimize cost of management[c]	60	65	100	75	70	0
Minimize extinction risk[d]	0	0	0	0	0	1
	Dominated Alternatives Discarded and Consequences Normalized					
Maximize number of hihi	0.44	0.89	–	–	1.00	0.00
Minimize cost of management	0.14	0.07	–	–	0.00	1.00
Minimize extinction risk	1.00	1.00	–	–	1.00	0.00
	Weighted Normalized Scores for Each Alternative Reflecting Relative Preferences for Different Objectives					
Maximize number of hihi	0.20	0.40	–	–	0.45	0.00
Minimize cost of management	0.05	0.02	–	–	0.00	0.32
Minimize extinction risk	0.24	0.24	–	–	0.24	0.00
Totals	0.49	0.66		–	0.69	0.32

SOURCE: Modified from Ewen et al. (2014b).

a. Numbers reported are the estimated consequence that each alternative has for each objective.

b. Providing as much food as required by the birds by responding to demand.

c. Cost of management is a constructed scale of hours of effort required by managers (0 if no feeding and 100 if ad libitum).

d. Probability of population extinction in 20 years.

to provide individuals for translocation can be seen as a high-risk activity. Harvesting may have a negative effect on the source, which may or may not be offset by the success of the translocation: both outcomes are likely to be uncertain. How does risk attitude influence our decision-making? Consider the following simplified scenario: the only extant population of an endangered species is facing a probability of extinction of 0.5 if no action is taken. It is possible to harvest some of its individuals and translocate them to a new location. There is a 0.4 chance that the new population establishes and persists, increasing the probability of persistence of the species to 0.65; however, if the translocation fails, the harvest will decrease the persistence of the extant population to 0.4. It may also be possible to move the entire population to a site that is believed to be threat free. If this action is successful, the persistence of the species is guaranteed; if it fails, it will go extinct, and there is a 0.5 chance either way. This problem, including recognized uncertainties, can be visualized using a decision tree (figure 6.2).

Simple techniques such as the SMART would approach this problem by comparing the expected outcome of actions. Expected outcomes are usually calculated as the mean of a

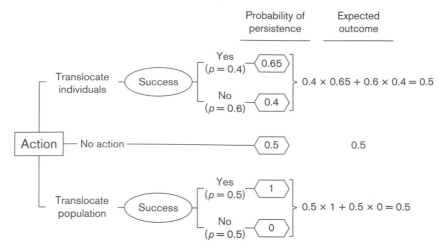

FIGURE 6.2. Decision tree describing a potential risk scenario facing reintroduction planners. Here we present a simple scenario where managers can make one of three decisions for a threatened population: do nothing, translocate part of the population, or translocate the entire population. The decision tree can help weigh up the costs and benefits of each action; the population has a 50% chance of going extinct if no action is taken, but each translocation strategy has a chance of being successful (<100%) and an associated cost to the population's probability of persistence.

distribution or (as in this case) as the average of all potential outcomes weighted by their respective probability of occurring. In this case, the expected outcome of the three actions is the same (a persistence of 0.5). However, different managers might consider the direct movement of the whole population as riskier than the limited translocation, and reflect this *risk-averse* attitude by discarding the former option. On the other hand, some may consider the possibility of securing the species desirable enough to justify the risk: these *risk-seeking* managers may be willing to pursue this action.

As with value functions, SDM provides a platform to apply a range of techniques for identifying optimal decisions in the face of risk. Risk attitudes can be evaluated through the use of certainty equivalents or risk premiums (see Haight 1995, for an example of risk premiums applied to extinction risks and management costs), of utility functions (Pratt 1964) or stochastic dominance (McCarthy 2014). For example, Canessa et al. (2016b) used stochastic dominance to reassess the *L. spenceri* decision problem, identifying the optimal decision for managers with different

risk attitudes. Reintroduction biology mostly focuses on highly threatened species, with small and declining populations and a common need to make urgent decisions under uncertainty. Further research on values and risk in reintroduction objectives can improve their evaluation in the broader context of recovery plans. We recognize that asking such questions increases the complexity of decision processes in comparison to simply applying intuition and commonly used methods, particularly where it may unearth emotionally charged debates. However, we do believe that iterative revisions of problem definitions and objectives should not be seen as a waste of time; rather, they indicate that complex problems such as reintroductions are being taken seriously, and provide the ideal framework for science to play its role.

SUMMARY

In this chapter, we have highlighted the apparent contrast between how reintroduction objectives are set by managers, or reported to be set, and the state-of-the-art approach for setting these objectives within a larger decision context. Using case studies, we have demonstrated how SDM can bridge the gap between theory and practice for setting objectives and measuring the success of reintroduction projects. We believe that more work is required on the importance of value functions when making decisions, i.e., how value depends on the return on investment, and context. Moreover, future work should focus on how to measure and incorporate acceptable levels of risks associated with uncertainty.

MANAGEMENT RECOMMENDATIONS

- In order to measure the success of reintroduction projects, including how efficiently limited funding was spent, it is important for the reintroduction to have clear objectives set from the get-go.

- Setting objectives is a notoriously difficult task. We reviewed the published literature for reintroduction case studies and found that objectives reported are often vague and without clear indicators of success.

- While objectives with a quantifiable target are an important part of reintroduction projects, objectives that are too narrow may be too limiting and fail to promote success. In order to avoid this problem, managers can use SDM as a tool to set objectives.

- We provide an overview of the process involved in SDM for reintroductions, including case studies and lessons learnt. Because SDM is a systematic and transparent process, it is easy to use and allows decisions to be justified. Moreover, it helps clarify the required actions to reach project objectives on the short term, avoiding long-term and vague goals for which there are no clear paths to success.

- To further develop the field of SDM for reintroductions, we suggest that value functions should be more readily incorporated in SDM, i.e., how values change over time and depending on the context. Second,

because of the uncertainty associated with decisions, the level of risk deemed acceptable for reintroduction projects should always be explicitly defined.

LITERATURE CITED

Bunnefeld, N., E. Hoshino, and E. J. Milner-Gulland. 2011. Management strategy evaluation: a powerful tool for conservation? Trends in Ecology and Evolution 26:441–447.

Canessa, S., S. J. Converse, M. West, N. Clemann, G. Gillespie, M. McFadden, A. J. Silla, K. M. Parris, and M. A. McCarthy. 2016a. Planning for ex-situ conservation in the face of uncertainty. Conservation Biology 30:599–609.

Canessa, S., J. G. Ewen, M. West, M. A. McCarthy, and W. T. V. 2016b. Stochastic dominance to account for uncertainty and risk in conservation decisions. Conservation Letters. doi:10.1111/conl.12218.

Canessa, S., D. Hunter, M. McFadden, G. Marantelli, and M. A. McCarthy. 2014. Optimal release strategies for cost-effective reintroductions. Journal of Applied Ecology 51:1107–1115.

Chauvenet, A. L. M., J. G. Ewen, D. P. Armstrong, T. M. Blackburn, and N. Pettorelli. 2013. Maximizing the success of assisted colonizations. Animal Conservation 16:161–169.

Clemen, R. T. 1996. Making Hard Decisions: An Introduction to Decision Analysis. Duxbury Press, Belmont, CA.

Collazo, J. A., P. L. Fackler, K. Pacifici, T. H. White, I. Llerandi-Roman, and S. J. Dinsmore. 2013. Optimal allocation of captive-reared Puerto Rican parrots: decisions when divergent dynamics characterize managed populations. Journal of Wildlife Management 77:1124–1134.

Converse, S. J., C. T. Moore, M. J. Folk, and M. C. Runge. 2013. A matter of tradeoffs: reintroduction as a multiple objective decision. Journal of Wildlife Management 77:1145–1156.

Dickens, M. J., D. J. Delehanty, and L. Michael Romero. 2010. Stress: an inevitable component of animal translocation. Biological Conservation 143:1329–1341.

Ewen, J. G., P. Soorae, and S. Canessa. 2014a. Reintroduction objectives, decisions and outcomes: global perspectives from the herpetofauna. Animal Conservation 17:74–81.

Ewen, J. G., L. Walker, S. Canessa, and J. J. Groombridge. 2014b. Improving supplementary feeding in species conservation. Conservation Biology 29:341–349.

Failing, L., R. Gregory, and M. Harstone. 2007. Integrating science and local knowledge in environmental risk management: a decision-focused approach. Ecological Economics 64:47–60.

Fischer, J., and D. Lindenmayer. 2000. An assessment of the published results of animal relocations. Biological Conservation 96:1–11.

Gannon, J. J., T. L. Shaffer, and C. T. Moore. 2013. Native Prairie Adaptive Management: A Multi Region Adaptive Approach to Invasive Plant Management on Fish and Wildlife Service Owned Native Prairies. Report 2013-1279. US Geological Survey, Reston, VA.

Garnett, S., J. Szabo, and G. Dutson. 2011. Action Plan for Australian Birds 2010. CSIRO Publishing, Collingwood, VIC.

Gedir, J. V., J. M. Thorne, K. Brider, and D. P. Armstrong. 2013. Using prior data to improve models for reintroduced populations: a case study with North Island saddlebacks. Journal of Wildlife Management 77:1114–1123.

Gregory, R., L. Failing, M. Harstone, G. Long, T. McDaniels, and D. Ohlson. 2012. Structured Decision Making: A Practical Guide to Environmental Management Choices. John Wiley & Sons, Chichester, UK.

Gregory, R., and K. Wellman. 2001. Bringing stakeholder values into environmental policy choices: a community-based estuary case study. Ecological Economics 39:37–52.

Griffith, B., J. M. Scott, J. W. Carpenter, and C. Reed. 1989. Translocation as a species conservation tool: status and strategy. Science 245:477–480.

Haight, R. G. 1995. Comparing extinction risk and economic cost in wildlife conservation planning. Ecological Applications 5:767–775.

Huber, P. R., S. E. Greco, N. H. Schumaker, and J. Hobbs. 2014. A priori assessment of reintroduction strategies for a native ungulate: using HexSim to guide release site selection. Landscape Ecology 29:689–701.

Irwin, B. J., M. J. Wilberg, J. R. Bence, and M. L. Jones. 2008. Evaluating alternative harvest policies for yellow perch in southern Lake Michigan. Fisheries Research 94:267–281.

IUCN/SSC (International Union for Conservation of Nature Species Survival Commission). 2013. Guidelines for Reintroductions and Other Conservation Translocations. Version 1.0. IUCN Species Survival Commission, Gland, Switzerland.

Jachowski, D. S., R. A. Gitzen, M. B. Grenier, B. Holmes, and J. J. Millspaugh. 2011. The importance of thinking big: large-scale prey conservation drives black-footed ferret reintroduction success. Biological Conservation 144:1560–1566.

Jones, M. E., P. J. Jarman, C. M. Lees, H. Hesterman, R. K. Hamede, N. J. Mooney, D. Mann, C. E. Pukk, J. Bergfeld, and H. McCallum. 2007. Conservation management of Tasmanian devils in the context of an emerging, extinction-threatening disease: devil facial tumor disease. EcoHealth 4:326–337.

Keeney, R. L., and R. S. Gregory. 2005. Selecting attributes to measure the achievement of objectives. Operations Research 53:1–11.

Keeney, R. L., and H. Raiffa. 1993. Decisions with Multiple Objectives: Preferences and Value Trade-Offs. Cambridge University Press, Cambridge, UK.

Maguire, L. A. 1986. Using decision analysis to manage endangered species populations. Journal of Environmental Management 22:345–360.

Maguire, L. A., T. W. Clark, R. Crete, J. Cada, C. Groves, M. L. Shaffer, and U. S. Seal. 1988. Black-footed ferret recovery in Montana: a decision analysis. Wildlife Society Bulletin 16:111–120.

Maguire, L. A., R. Lacy, R. J. Begg, and T. Clark. 1990. An Analysis of Alternative Strategies for Recovering the Eastern Barred Bandicoot in Victoria. Chicago Zoological Society, Brookfield, IL.

Maguire, L. A., U. S. Seal, and P. F. Brussard. 1987. Managing critically endangered species: the Sumatran rhino as a case study. Pp.141–158. In M. Soulé, ed., Viable Populations for Conservation. Cambridge University Press, Cambridge, UK.

Martin, J., P. L. Fackler, J. D. Nichols, M. C. Runge, C. L. McIntyre, B. L. Lubow, M. C. McCluskie, and J. A. Schmutz. 2011. An adaptive-management framework for optimal control of hiking near golden eagle nests in Denali National Park. Conservation Biology 25:316–323.

Martin, J., M. C. Runge, J. D. Nichols, B. C. Lubow, and W. L. Kendall. 2009. Structured decision making as a conceptual framework to identify thresholds for conservation and management. Ecological Applications 19:1079–1090.

McCarthy, M. A. 2014. Contending with uncertainty in conservation management decisions. Annals of the New York Academy of Sciences 1322:77–91.

McCarthy, M. A., D. P. Armstrong, and M. C. Runge. 2012. Adaptive management of reintroduction. Pp.256. In J. G. Ewen, D. P. Armstrong, and K. A. Parker, eds., Reintroduction Biology: Integrating

Science and Management. Wiley-Blackwell, Oxford, UK.

McDaniels, T. L., R. S. Gregory, and D. Fields. 1999. Democratizing risk management: successful public involvement in local water management decisions. Risk Analysis 19:497–510.

McGowan, C., J. Lyons, and D. Smith. 2015. Developing objectives with multiple stakeholders: adaptive management of horseshoe crabs and red knots in the Delaware Bay. Environmental Management 55:972–982.

Menges, E. S. 2008. Turner review no. 16. Restoration demography and genetics of plants: when is a translocation successful? Australian Journal of Botany 56:187–196.

Nichols, J. D., and D. P. Armstrong. 2012. Monitoring for reintroductions. Pp.223–255. In J. G. Ewen, D. P. Armstrong, and K. A. Parker, eds., Reintroduction Biology: Integrating Science and Management. Wiley-Blackwell, Oxford, UK.

Pratt, J. W. 1964. Risk aversion in the small and in the large. Econometrica 32:122–136.

Ralls, K., and A. M. Starfield. 1995. Choosing a management strategy: two structured decision-making methods for evaluating the predictions of stochastic simulation models. Conservation Biology 9:175–181.

Reynolds, M. H., N. E. Seavy, M. S. Vekasy, J. L. Klavitter, and L. P. Laniawe. 2008. Translocation and early post-release demography of endangered Laysan teal. Animal Conservation 11:160–168.

Robert, A., B. Colas, I. Guigon, C. Kerbiriou, J.-B. Mihoub, M. Saint-Jalme, and F. Sarrazin. 2015. Defining reintroduction success using IUCN criteria for threatened species: a demographic assessment. Animal Conservation 18: 397–406.

Seddon, P. J. 1999. Persistence without intervention: assessing success in wildlife reintroductions. Trends in Ecology and Evolution 14:503.

Servanty, S., S. J. Converse, and L. L. Bailey. 2014. Demography of a reintroduced population: moving toward management models for an endangered species, the whooping crane. Ecological Applications 24:927–937.

Soorae, P. S. 2010. Global Re-Introduction Perspectives: Additional Case Studies from Around the Globe. IUCN/SSC Re-introduction Specialist Group, Gland, Switzerland/Environment Agency-Abu Dhabi, Abu Dhabi, UAE.

Soorae, P. S. 2011. Global Re-Introduction Perspectives, 2011: More Case Studies from Around the Globe. IUCN/SSC Re-introduction Specialist Group & Environment Agency-Abu Dhabi, Gland, Switzerland and Abu Dhabi, UAE.

Soorae, P. S. 2013. Global Re-Introduction Perspectives: 2013. Further Case-Studies from Around the Globe. IUCN/SSC Re-Introduction Specialist Group, Gland, Switzerland/Environment Agency-Abu Dhabi, Abu Dhabi, UAE.

Zippel, K., K. Johnson, R. Gagliardo, R. Gibson, M. McFadden, R. Browne, C. Martinez, and E. Townsend. 2011. The Amphibian Ark: a global community for ex situ conservation of amphibians. Herpetological Conservation and Biology 6:340–352.

SEVEN

Demographic Modeling for Reintroduction Decision-Making

Sarah J. Converse and Doug P. Armstrong

THE PRIMARY OBJECTIVE OF MOST reintroductions is to establish a self-sustaining population with a high probability of persistence (e.g., Scott and Carpenter 1987, Griffith et al. 1989, Fischer and Lindenmayer 2000). A self-sustaining population, broadly, is one that is stable or growing in size, without additional releases. Population growth and persistence, and therefore reintroduction success, are determined by demography: specifically, the BIDE factors (births, immigration, deaths, and emigration). For a population to be self-sustaining, additions (births and immigration) need to equal or exceed losses (deaths and emigration) over the time period of interest.

It is well known that reintroductions have had a poor success rate in the past (e.g., Lyles and May 1987, Scott and Carpenter 1987, Griffith et al. 1989), leading to the creation of the IUCN Reintroduction Specialist Group (IUCN 1987). This has also inspired a major push to strengthen research in reintroduction biology (Armstrong and Seddon 2008, Ewen et al. 2012), and in recent years to incorporate decision analysis into reintroduction management

(Rout et al. 2009, McCarthy et al. 2012, Nichols and Armstrong 2012, Converse et al. 2013a). Decision analysis is the systematic analysis of decisions for the purpose of identifying optimal choices, using the theory and methods of decision science (Keeney 1992, Possingham et al. 2001, Gregory et al. 2012, Converse et al. 2013a). As is evident from published reintroduction guidelines (e.g., IUCN/SSC 2013), managers undertaking reintroductions face many decisions. These include whether to proceed with a proposed reintroduction, how many founder animals to source and from where, the sex and age composition of the founder group, the release site, the release method, whether and how to supplement the founder group after initial release, and how to manage habitat at the reintroduction site. For difficult decisions—those that are complicated by obscured or competing objectives, complex or unidentified alternatives, and/or substantial uncertainty—decision analysis offers a pathway through the challenges of identifying the best choices.

A major component of reintroduction research is development of methods for modeling reintroduced populations (Armstrong and Reynolds 2012). However, it is often not made explicit how the models will be used to make management decisions. We suggest that the best approach is to begin by recognizing that reintroduction management is the process of making decisions under uncertainty with the objective of establishing a viable population, and then begin an iterative process of predicting (i.e., modeling) the impacts of management actions on demographic outcomes.

In this chapter, we consider the construction and use of population models to support reintroduction decision-making. We begin by reviewing the decision-analytic process, also known as structured decision-making. The material on structured decision-making builds on the chapter by Chauvenet et al. (Chapter 6, this volume) who focus their attention on the objective setting step of structured decision-making. In a decision-analytic setting, modeling cannot commence before objective setting, because the purpose of models is to provide predictions about how different management alternatives will influence the attainment of management objectives. After the introductory section on structured decision-making, we turn our attention to the construction and use of population models. We discuss the steps in modeling, including structuring and planning models, obtaining parameter estimates from data, filling information gaps using expert judgment, constructing models, and using models to evaluate action alternatives. We close the chapter with a discussion of research and development needs that we believe will advance the use of population models to support the effective management of reintroduced populations.

REINTRODUCTION AS A DECISION-MAKING PROCESS

Decision analysis, or structured decision-making (figure 7.1), is designed to produce choices most likely to lead to favorable outcomes, given the set of options available. A feature of decision analysis is that the logic underlying decisions is transparent, providing greater clarity and accountability. Transparent decisions are also more likely to be documented, facilitating learning about reintroduction biology (Fischer and Lindenmayer 2000, Armstrong and Seddon 2008).

Decision analysis draws on a varied literature, from psychology, engineering, economics, and related fields (Keeney 1992, Gregory et al. 2012, Converse et al. 2013a). The process requires identifying and analyzing the decision components and reintegrating those components to reach a decision. Basic components of all decisions, upon which we elaborate below, include (1) a clear articulation of the decision to be made, (2) a statement of the management objective(s), (3) the set of management alternatives under consideration, (4) one or more models to predict outcomes under each alternative, and (5) a method for solving the decision problem (often called optimization, or trade-offs analysis). These components are the same, in their essentials, as the steps laid out by Chauvenet et al. (Chapter 6, this volume).

As a first step, it is important to spend time developing a clear articulation of the decision. The potential for a lack of clarity in the nature of the decision itself increases quickly as the number of people involved in the decision-making process increases. In defining the decision, it is important to be clear about who has the power to make the decision. In reintroductions, there are often multiple decision makers: the manager of the area to receive the reintroduced population, the manager of the wild or captive source population, the agency with authority for managing the species overall, and the entity paying for the effort. Other aspects of the decision articulation step include the timing of the decision, legal and regulatory considerations, and the scale of the decision. The definition of the problem will influence all subsequent components of the decision. For example, the decision by a refuge manager about how to manage habitat for a species to be reintroduced is an entirely different decision from that made by a species recovery team about what reintroduc-

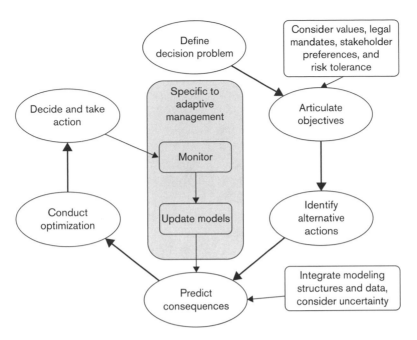

FIGURE 7.1. A simplified diagram representing the structured decision-making process. Major steps in structured decision-making include articulation of the decision problem, identification of objectives, identification of a feasible set of alternative actions, predicting consequences of alternative actions via models, and optimization to identify the best possible action to take. A special case of structured decision-making, adaptive management, involves an iterative (e.g., annual) loop, which involves monitoring the outcome of decisions, updating models, and using updated models to inform decisions at the next time step. Reproduced from Converse et al. (2013a).

tion site to choose. Both decisions nominally involve a reintroduced population and its habitat, but the decisions will vary in objectives (refuge-specific outcomes vs. species-specific outcomes), alternatives (different habitat management actions on the refuge vs. different release sites), and in the predictive models necessary to evaluate alternatives. As an aside, we note that these two decisions are what are known as linked decisions. In this case, the decision of the refuge manager will certainly be influenced by whether the recovery team chooses that refuge for a reintroduction, and the recovery team will want to anticipate how the area will be managed when making their decision. Linked decisions are solved most efficiently if considered in tandem (see Canessa et al. (2016) for an in-depth treatment of linked decision problems).

Setting thoughtful and operationally defined objectives has, unfortunately, been a frequently missed step in reintroductions (Fischer and Lindenmayer 2000; Chauvenet et al., Chapter 6, this volume). In many cases, this has led to reintroduction efforts that place inordinate focus on means objectives rather than on fundamental objectives. For example, efforts are sometimes overly focused on survival over the first months after release, rather than on the longer-term objective of having a self-sustaining population (Lintermans et al. 2015). A fundamental objective is defined as "the broadest objective that will be directly influenced by the alternatives and within the control of the decision-maker" (Gregory et al. 2012, 74). In other words, fundamental objectives are the most basic outcomes desired from a decision. For a reintroduction effort, establishment of a self-sustaining population is likely to be fundamental. Good short-term survival of released animals is a means: important but insufficient for meeting the fundamental objective. By putting adequate attention toward identifying the fundamental objective(s), defining them operationally, and referring back to them in all decision processes, managers planning reintroductions can avoid the situation of solving the wrong problem, and, instead, focus efforts on actions that will lead to attainment of a self-sustaining population.

Decision analysis involves identifying and analyzing a set of alternative actions rather than just evaluating whether one action, often the status quo, is acceptable (Gregory et al. 2012). There is a strong focus on creating a broad array of alternatives. Use of the best brainstorming techniques (e.g., Brown and Paulus 2002) and emphasis on idea generation, rather than idea evaluation, in the brainstorming step, can unlock the creative ideas that people may feel inhibited to put forward because they fall outside the mainstream, or because they are not the way that "we've always done it." Examples wherein creative solutions have been implemented, such as ultralight-led migration to reintroduce migratory bird species (Urbanek et al. 2005, Mueller et al. 2013), can be inspiring for a group in the process of generating ideas.

Once the objectives and alternative management actions are identified, the role of demographic models becomes clear. Models are used to make predictions about the impacts of alternative actions in terms of the objectives. In situations where objectives involve demographic outcomes, as is invariably the case with reintroductions, demographic models will be needed. These models will typically include expected impacts of alternative management actions on births, deaths, and movements. These effects can then be translated, through the model, into metrics such as probability of persistence. Formalized demographic models integrate complex information that cannot be integrated informally, and help managers and scientists working in teams to share information in a transparent way.

Uncertainty in reintroductions is ubiquitous. Uncertainty should be fully integrated into models, such that predictions reflect the uncertainty, and decision makers can understand not just the most likely outcome, but also the range of possible outcomes that may be realized under a given action. In the case where management actions are repeated over time, an adaptive management approach allows uncertainty about system function to be reduced through monitoring, when that uncertainty can be articulated as different models of system function or as different values of model parameters. Others have dealt with adaptive management of reintroductions at length (e.g., McCarthy et al. 2012, Runge 2013). The key components include all those previously mentioned (problem statement, objectives, and alternatives) plus model(s) reflecting the uncertainty: either in the form of a single model with uncertainty in one or more parameter values, or a set of competing models with different structural forms. Finally, adaptive management requires a monitoring program designed to reduce the uncertainty. Adaptive management can proceed via either a passive approach, in which management actions are not specifically selected to reduce uncertainty or an active approach, in which management actions that are apparently nonoptimal in the short term may be selected because of the expected long-term benefits resulting from a reduction in uncertainty (so-called dual control; Williams 2001).

The final component of a decision framework is optimization (sometimes known as trade-offs; sensu Hammond et al. 1999). Optimization is the process of identifying the optimal management action from those considered. A detailed accounting of approaches for optimization is beyond the scope of this chapter. Instead, we mention a few that are particularly relevant. First, in the context of decisions involving complex demographic models, stochastic simulation approaches are often useful. Stochastic simulation allows for prediction of demographic responses to management actions under uncertainty. In many cases, however, evaluation of alternatives in terms of demographic outcomes alone will be insufficient. In addition to desired demographic outcomes, objectives may also include costs, impacts on source populations, or other social and ecological considerations. When additional objectives exist, multi-criteria tools are often useful (Keeney and Raiffa 1976). These tools are designed to handle the fact that different objectives may have differential importance to the decision maker, and analyses must apply weights to the various objectives. Simulated outcomes from demographic models as described above can be nested within a multi-criteria decision framework (Converse et al. 2013b), and the solution to the problem can be based on approaches such as the Simple Multi-Attribute Rating Technique (Edwards 1977, Goodwin and Wright 2004). Finally, in the context of adaptive management problems, dynamic optimization approaches, such as stochastic dynamic programming (Walters 1986, Williams 2011), are valuable. These approaches allow the dynamic aspects of management, and the value of learning about system function, to be anticipated formally in the decision-making process. It is important to recognize, however, that these approaches generally require rather simple predictive models to be computationally tractable. Thus, a trade-off must be made between the value of considering dynamic aspects of the decision and anticipating the value of learning, versus construction of more detailed predictive models. Experience will be needed in making this trade-off. When fully realized dynamic optimization approaches are being considered, the development team must include substantial technical expertise in implementing such approaches.

BUILDING PREDICTIVE MODELS OF POPULATIONS

Once the decision context is clear, and the objectives and alternatives have been identified, the next step is construction of demographic models. We recognize a set of basic steps in building demographic decision models. These include structuring and planning the model, demographic estimation using available data, filling information gaps using expert judgment, construction of the population model, and, finally, use of the model for evaluating the effects of action alternatives. Armstrong and Reynolds (2012) laid out similar, though more detailed, steps for building models of reintroduced populations.

Progress through these steps will not proceed forward only; one will often revisit earlier steps while working through later steps, as the need for modification and revision becomes apparent. This is closely related to the concept of rapid prototyping, which we advocate for model development (Nicolson et al. 2002). Rapid prototyping is the process of constructing models through iterative, rapidly completed and successive prototypes of a model. The value of rapid prototyping is that it is often impossible to understand, at the outset of a modeling effort, exactly what the final product could or should look like. For example, we might begin by conceiving of a model for a particular species that does not include the process of animals dispersing away from the release site. However, upon analysis of available data, it could become clear that dispersal is an important process in the population. Therefore, rather than spending a lot of time planning a very detailed model structure, only to have that structure crumble at the next step in the model-building process, it is sensible to proceed relatively quickly through the stages of model development. In our example, we would realize at the second step that the work completed at the first step required revision. Therefore, we might quickly go back, develop a modified model structure that does accommodate dispersal, and then move forward to the next step. We should complete the first round of model development with the simplest reasonable model of the population, and then ask the question: does the model contain the elements that are likely to be important in the context of the decision-making process? Only if the answer to this question is "no" should additional detail be added.

Structuring and Planning the Model

Quantitative models almost always begin as conceptual versions of themselves. In fact, models are ubiquitous in reintroduction management, but, because most models are expressed through words rather than through formal equations, they are not recognized as such (Armstrong and Reynolds 2012). However, such informal models are as much predictions about how the world works as are quantitative models. Often, the first step in building a model is to work within a team including managers, stakeholders, and modelers to turn informal models into shared, conceptual models. Rapid prototyping provides a low-risk method for getting started in a collaborative model-building process. If the first prototype is not quite right, it can be discarded or modified without much regret because little time has been invested.

At the most basic level, models used in decision-making must link management actions, represented by particular model inputs, with objectives, represented by particular model outputs. For example, if a decision concerns the question of how many animals to release to achieve a given probability of population persistence, release strategies involving different numbers of individuals would be represented by different model inputs, and the output of concern would be some appropriate measure of population persistence. Models of reintroduced populations will generally fit under the class known as population viability analysis (PVA) models (Morris and Doak 2002). More specifically, these models allow for predicting the impacts of alternative actions in terms of viability, resulting in what has been referred to as a PVA-based management model (Servanty et al. 2014).

In the structuring phase, a model-building team will work to determine the basic structural elements of the model. Armstrong and

Reynolds (2012) provided a useful set of questions about model structure. We paraphrase and expand on their list here:

1. Should the model include females only or both sexes?
2. What age and/or life stages should be defined?
3. Should positive (e.g., Allee effects) and/or negative density-dependence be included?
4. Should the model be individual based?
5. Should the model include dispersal in addition to survival and reproduction?
6. Is there potential metapopulation structure that should be accounted for?
7. Should the model be spatially explicit?
8. How should uncertainty in parameter values be handled?
9. Should uncertainty be represented through competing models?
10. To what extent should temporal variation and/or catastrophes be modeled?
11. Should individual variation be included, and if so, how?
12. Should demographic stochasticity—random variation in demographic outcomes—be included?
13. Should genetic effects be included?
14. How will the effects of the management actions of interest be modeled?

When determining the structure of a model, one must consider the unique aspects of reintroduced populations. First, reintroduced populations may, at least initially, be influenced by post-release effects. Post-release effects are changes, usually declines, in survival and/or breeding in the period just after release, due to stress and/or lack of familiarity with new habitat (Sarrazin and Legendre 2000, Le Gouar et al. 2008, Tavecchia et al. 2009, Dickens et al. 2010; see Box 6.1 of Armstrong and Reynolds 2012, for a detailed review). The temporal structure of the model must capture the length

BOX 7.1 · The Importance of Accounting for Post-release Effects When Modeling Reintroduced Populations

In March 2013, the hihi, a rare New Zealand forest bird, was reintroduced to Bushy Park, a predator-fenced conservation reserve. The population declined, and after 2 years was still lower than the initial number of birds released. However, the dynamics appeared to be strongly affected by poor survival of females in the months after release. Panfylova (2015) modeled the survival of the population, considering a range of candidate models reflecting different hypotheses about the effects of sex, age, and post-release acclimation. Based on expert knowledge of the species, she considered two possible time frames for the acclimation period: 1 month and 6 months. She then used model averaging to obtain survival estimates reflecting the relative support for the survival models, and incorporated these into a stochastic population model. The survival modeling gave strongest support for models that included a 6-month acclimation effect on female survival. Consequently, the population projections indicated a low probability of the population declining over the next 10 years despite the decline over the first 2 years. These projections led to a decision to continue the existing habitat management regime.

and magnitude of release effects, thus accounting for early higher mortality or lower breeding, but allowing for a termination of those effects after some period of time (Box 7.1). A special case of release effects are those effects that extend throughout the life of the released animal, due to differences between the situation the animal was reared in and the release situation. An obvious example is a captive-rearing effect, in which animals released from captivity have poorer demographic performance than do wild-reared animals (Buner and Schaub 2008, Heath et al. 2008, Roche et al. 2008, Moore et al. 2012). This can be particularly important for animals where learning has a major influence on behavior (Moehrenschlager and Lloyd, Chapter 11, this volume). Second, Allee effects may need to be considered due to the small initial size of reintroduced populations (Deredec and Courchamp 2007, Armstrong and Wittmer 2011; Biebach et al., Chapter 8, this volume). Small population sizes may disrupt social functioning, for example, or increase vulnerability to predators (Jachowski et al. Chapter 9, this volume). Third, it is important to include demographic stochasticity, as chance outcomes in demographic events can pose a serious risk to the viability of small populations, and these risks may affect man-

agement decisions (e.g., Armstrong et al. 2007). Finally, the initial age and sex structures of reintroduced populations may be different than the stable structures expected; thus, the initial population growth (or decline) may be a misleading indicator of future growth. For example, a founder group with a high proportion of breeding-age females might show initial population growth even though the population will decline in the long term. Incorporating sex and age specificity in vital rates can solve this problem, either by calculating the population's finite rate of increase using matrix algebra (Caswell 2001; see Box 6.4 in Armstrong and Reynolds 2012) or by simulating into the future to allow for stabilization of dynamics.

Managers will often be concerned about the source population in addition to the reintroduced population. Models of source populations should allow for evaluation of removal, or harvest, actions. In the case of captive sources, models will often include both demographic and genetic considerations in tandem, as captive-population managers typically are concerned about the conservation of genetic diversity (Ballou and Lacy 1995, Converse et al. 2012, Jamieson and Lacy 2012).

A first-cut distinction can be made between models that are individual-based (IBMs) and those that are not. IBMs have the ability to include more complex dynamics, such as complex social structure, unbalanced sex ratio, genetic effects, and movements (Lacy 2000a, Chauvenet et al. 2015). Lacy (2000a) argued that, as populations get smaller and the importance of small population processes (sensu Caughley 1994) increases—including demographic, environmental, and genetic stochasticity—the value of IBMs increase. However, IBMs are often substantially more complex to implement. Customized IBMs have been used infrequently in PVA for reintroductions, though there are exceptions (e.g., Gusset et al. 2009). However, the most widely used generic PVA software package, VORTEX, is an IBM (Lacy 2000b). While a software program like VORTEX can be helpful by allowing the inclusion of a number of processes requiring more sophisticated coding, we believe that managers and biologists engaged in reintroduction activities can benefit, at least initially, from building a model of their population that is transparent and understandable for them, and then add complexity to the model only as needed.

Perhaps the most transparent approach to building population models is the Leslie matrix (Caswell 2001). A matrix model consists of a matrix that contains all of the rates—survival and breeding—that are needed to calculate change in population size over some specified time frame. Matrix models lend themselves to tractable and transparent implementation in spreadsheet software (White 2000). More sophisticated versions of matrix models can be implemented in software such as MATLAB or the popbio package (Stubben and Milligan 2007) for R (R Development Core Team 2012) or through the functionality of an add-in such as PopTools (Hood 2010) for Excel® spreadsheets (any use of trade, firm, or product names is for descriptive purposes only and does not imply endorsement by the US government). Analytical solutions of matrix models produce predictions of asymptotic growth rate given a matrix of vital rates. Because we often may be concerned about initial, or transient, dynamics

in a reintroduced population, projection-based evaluation of matrix population models may be more useful (Caswell 2001), wherein age- or stage-specific abundance is calculated through iterative multiplication of the matrix by the previous age- or stage-specific abundance vector. These projections can also integrate demographic stochasticity (via Monte Carlo simulation of survival and breeding outcomes), process variation (via, e.g., random annual variation in survival and/or breeding rates), and parametric uncertainty. Structural uncertainty can also be represented through competing models (Box 7.2).

Demographic Estimation Using Available Data

Demographic estimation is the process of using data to estimate parameters describing demographic processes such as survival and reproduction. The use of statistical models for estimation of demographic parameters is almost always a necessary step in building population models. Statistical models rely on data collected in the past, so projections from these models are based on the assumption that future conditions will be similar to conditions present when data were collected. If this cannot be assumed, expert judgment may be necessary to generate parameter estimates (see below).

If a given decision concerns ongoing management of an established population, there should be data available directly from the population of interest for estimation of at least some of the demographic parameters. These data should be exploited to the greatest extent possible. However, we advocate building models of reintroduced populations as early as possible in the process of managing a reintroduced population, even, ideally, before the reintroduction is initiated. In these cases, relevant data will not be available from the population of interest. However, there are often other sources of information on demographic parameters, including information from the source population, from other populations of the species, or from closely related

BOX 7.2 · Demographic Simulation Models for Evaluating Whooping Crane (*Grus americana*) Reintroduction Strategies under Uncertainty

Moore et al. (2012) built a series of PVA models to evaluate the influence of different reintroduction strategies on persistence of a reintroduced population of nonmigratory whooping cranes in Florida, USA (figure 7a). Captive-bred young were released to establish the population between 1993 and 2004, but poor breeding and survival resulted in the decision to stop releases. The question to be addressed, based on data through mid-2007, was whether the releases should be restarted, and if so, what was the best strategy for doing so. The population persistence objective was evaluated based on whether the population was expected to be extant and either stable or growing 100 years after the last release (no density dependence was included).

The authors evaluated 29 alternative release strategies. Strategies varied by the number of birds released per year, the numbers of years for releases to continue, whether releases should commence immediately or after a 10-year delay, and whether releases should occur every year or every other year. In addition, the alternative of no further releases was evaluated. Aspects of these strategies were developed to accommodate other objectives. For example, a 10-year delay would have allowed another ongoing reintroduction program to receive all birds available for release and thus, it was hoped, successfully establish a population in that period.

An important source of uncertainty for the whooping crane program was the demographic performance of second- and later-generation (wild-hatched) offspring of captive-bred population founders. Differences between captive-reared and wild-hatched individuals have been documented in translocations of other bird species (Buner and Schaub 2008, Heath et al. 2008, Roche et al. 2008). However, because breeding onset is delayed in whooping cranes and breeding success had been poor, there were few data to assess whether better performance could be expected in wild-hatched birds; only a few wild-hatched birds were extant in the population.

Therefore, three alternative models were constructed. The first model assumed that future wild-hatched offspring would have vital rates (survival and reproduction) similar to captive-reared released birds (Baseline model). The second

model assumed that wild-hatched offspring would have vital rates like birds from the sole wild (non-reintroduced) population (Wild model). These rates were higher than those observed in the captive-reared birds in the reintroduced population. The third model hypothesized an Allee effect operating in the population, such that the breeding rate of wild-hatched birds would be equal to that of the birds in the wild population only if the population was above some threshold population size (Allee model), and otherwise would be equal to that of the captive-reared population. Various different threshold levels were considered, constituting eight different Allee submodels.

Demographic estimation was accomplished through a hierarchical Bayesian estimation model, with age structure and breeding stage structure in the population. The hierarchical structure allowed for the inclusion of individual, cohort, and temporal random effects to account for process variation. The population was then simulated forward under these rates by selecting parameter values from the Markov chains resulting from the fitting procedure, and then simulating temporal variation and adding birds to the population (with their own individual- and cohort-specific random effects) according to the release alternative being simulated.

The optimal decision was model-dependent. Under the Baseline model, no amount of additional releases could overcome the poor expected performance of the wild-hatched descendants of captive-reared birds, so all the strategies had a high and nearly equal probability of failure. Under the Wild model, the improved performance expected from wild-hatched birds would nearly guarantee success given the current population, so all strategies had a high and nearly equal probability of success. In both cases, managers would be expected to prefer no additional releases, due to the costs of release (see below). However, under the Allee model, the expected outcome was highly sensitive to the number of birds released: the more birds released, the more quickly and completely the population threshold would be exceeded, and the more likely the predicted increase in performance of wild-hatched birds would be realized.

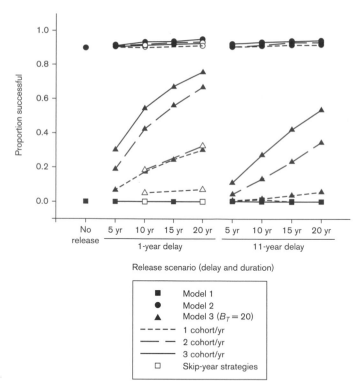

7a Population outcomes (defined as the proportion of stochastic simulations that produce a growing population 100 years after releases terminate) are displayed for three different models (Model 1: Baseline; Model 2: Wild; Model 3: Allee) of demographic performance of wild-hatched whooping cranes in a reintroduced population, as a function of release strategy (numbers of birds released, numbers of years released, whether releases should start immediately, and whether releases should be annual or every other year). In the case of Models 1 and 2, the preferred action is no further releases (because the population is either doomed to failure—Model 1—or nearly guaranteed of success—Model 2). Under Model 3, the Allee effect model, the outcome is highly sensitive to the decision made. Without clear data that would result in preference of one model over another, the management decision can be most reasonably based on a model-weighted prediction, where model credibility measures serve as weights, and are derived from, for example, an expert elicitation process.

SOURCE: Reproduced from Moore et al. (2012).

Only very limited empirical information was available to distinguish among these models, and then only between the Baseline and Wild models, both of which suggested the same action: no further releases, either because the population was nearly doomed to failure (for the Baseline model) or because the population was nearly guaranteed success (for the Wild model). There was no empirical information to argue for or against the credibility of the Allee model.

In a follow-up paper, Converse et al. (2013b) used an expert elicitation process to assign credibility weights to the models described by Moore et al. (2012). They then considered the model-averaged predictions within a larger structured decision-making framework in which model-averaged pre-

dictions of population persistence were considered alongside other objectives, including the desire of managers to minimize costs and maximize the number of captive-reared birds available for another ongoing reintroduction effort. The expert elicitation process suggested that there was relatively little belief in the Wild model, and relatively strong belief in the Baseline model. This resulted in a predicted probability of success of around 41% with the most aggressive release strategy. In the end, the decision was made to finally terminate all releases to the nonmigratory whooping crane population in Florida. As of October 2014, fewer than 14 birds were believed to be alive in this population (W. Harrell, US Fish and Wildlife Service, and M. Bidwell, Canadian Wildlife Service, unpublished report).

species. Parlato and Armstrong (2012), for example, used information from multiple reintroduced populations of the North Island robin (*Petroica longipes*) to estimate survival and fecundity rates and make predictions about expected growth rates in novel reintroduced populations.

Demographic events are typically modeled as outcomes of binomial (e.g., survival, breeding probability) or Poisson (e.g., number of offspring) processes. We are concerned with estimating the parameters underlying these stochastic processes, as well as estimating structural effects in those parameters, e.g., whether the parameter governing survival is different for males and females of a species. In some cases, then, logistic or Poisson regression are useful tools for modeling demographic processes. For example, Armstrong et al. (2002) used Poisson regression to model the number of offspring fledged per female in intensively monitored reintroduced forest birds in New Zealand. The authors were able to document effects of forest patch, bird age, population density, and several other factors on fecundity. The resulting parameter estimates were used in the construction of population models to guide management of those populations.

In many cases, however, basic logistic or Poisson models are inadequate because we must account for the fact that the detection probability of animals is <1. In those cases, mark-recapture models are required to estimate demographic parameters. Mark-recapture models most relevant for survival analysis in reintroduced populations are variations of the Cormack-Jolly-Seber (Cormack 1964, Jolly 1965, Seber 1965) and multistate (Hestbeck et al. 1991, Brownie et al. 1993) models (see also Lebreton et al. 1992, 2009, Williams et al. 2002, Kéry and Schaub 2012). For example, Servanty et al. (2014) used a multistate model to estimate survival and breeding state transitions (probability of becoming paired, probability of nesting) for a reintroduced population of whooping cranes (*Grus americana*) and considered the effects of release type, age, sex, season, and year on these parameters. Again, the pur-

pose of the analysis was to develop parameters for use in building a population model to guide decisions about the reintroduced population.

In both of the examples noted above (New Zealand forest birds and whooping cranes; Armstrong et al. 2002, Servanty et al. 2014), the investigators had to decide how to structure their estimation models. That is, they had to decide whether to estimate demographic rates as a function of factors such as age, sex, or patch type. In making these decisions, there should be substantial feedback between the model structuring and planning step and the estimation step. For example, if managers, biologists, and modelers believe that temporal variability is substantial, it would be sensible to construct estimation models with temporal effects (as described by Servanty et al. 2014). In this case, the desired population model structure informs the estimation model. Alternatively, the data themselves can help us determine how to structure our models, via model selection. Armstrong et al. (2002) used information-theoretic model selection methods (Burnham and Anderson 2002) to select models most appropriate for developing their parameter estimates. This approach allowed them to identify the importance of factors such as supplemental feeding on fecundity in stitchbirds (hihi; *Notiomystis cincta*). Analogous model selection tools are available for Bayesian analysis as well (Link and Barker 2006, Hooten and Hobbs 2015). Again, this emphasizes the recursive nature of the model-building process: lessons learned at the estimation stage can influence the model structuring phase.

Filling Information Gaps Using Expert Judgment

In many cases, when developing management models, available data will not reasonably reflect conditions that may be hypothesized to exist in the future. For example, if a novel release method is to be used, there may be no relevant data to predict the effect on vital rates. In these cases, structured elicitation of expert judgment can be used.

Martin et al. (2011) laid out the steps of an expert elicitation process, including (1) determining how elicited information will be used, (2) determining the information to be elicited, (3) designing the elicitation process, (4) performing the elicitation, and (5) translating the information into the form required for a model. These steps emphasize that the elicitation of expert judgment is not just asking people to provide informal "guesses." Instead, it is a highly structured process designed to produce the most accurate information possible. Expert elicitation methodology is an area of active scientific research and development (e.g., Meyer and Booker 1990, Ayyub 2001, Burgman 2004, 2005, Kuhnert et al. 2010, Speirs-Bridge et al. 2010, McBride et al. 2012). Evidence indicates that a version of the Delphi method—in which experts provide individual judgments, discuss their results collectively, and then provide final individual judgments (MacMillan and Marshall 2006, Martin et al. 2011)—tends to produce more accurate judgments than will the best-regarded expert alone (Burgman et al. 2011).

There are still relatively few examples of the use of formally elicited expert judgment to inform the modeling and management of reintroduced populations. However, the examples available indicate the broad utility of elicitation, and their recency suggests that expert elicitation may be increasingly used in the modeling of reintroduced populations (see also Canessa et al. 2014; Case Study 3). Runge et al. (2011) used expert elicitation to develop judgments of the effects of various management actions on demographic rates (e.g., fledging rate, survival rate) in reintroduced whooping cranes. In this case, the judgments were elicited conditional on a set of hypotheses about the cause of reproductive failure in this population, and the resulting judgments were used to identify priorities for learning (based on value of information; see discussion in the "Evaluating the Effects of Action Alternatives" section). Canessa et al. (2016) elicited judgments relevant to selecting optimal management actions for the

spotted tree frog (*Litoria spenceri*), an Australian species that has been reintroduced via releases of captive-reared individuals. The authors worked with experts to identify 32 management strategies for the species, and then elicited from experts their judgments on the probability of persistence and cost of each strategy. These judgments were used to parameterize a decision tree and examine the cost-effectiveness of various strategies. Converse et al. (2013b) used expert elicitation not to parameterize population models, but rather to assign credibility weights to a set of population models that represented competing hypotheses about population function. The credibility weights were used to produce weighted-average predictions of the effects of various management actions on population persistence.

Predictions based on expert judgment will inevitably be characterized by uncertainty if the elicitation accurately reflects expert knowledge. If data are so sparse that expert judgment is needed, managers should carefully consider whether there will be an opportunity to revisit the decision in the future. In that case, there is potential for monitoring the effects of initial management actions, updating models, and revising management in the future, i.e., adaptive management. Bayesian approaches in particular are useful for integrating new data with expert-generated priors via Bayesian updating (Dorazio and Johnson 2003).

Construction of the Population Model

Although we have placed this step toward the end, we believe it is useful to construct simplified quantitative versions of the model throughout the process, even if placeholder guesses for demographic parameters are used. This is in line with the rapid prototyping approach that we advocate. The process of model building will allow managers and biologists to gain confidence in working with the model, anticipate technical challenges in model building, and allow the model-building process to inform the process of obtaining parameter estimates either through data-based estimation or elicitation of expert judgment.

Consistent with the rapid prototyping approach, model complexity is best added in stages (Armstrong and Reynolds 2012). For example, one can begin by integrating parametric uncertainty into the model, then demographic stochasticity, and so on. The resulting model should be no more complex than is needed to do the job. Judging how much complexity is enough is partly informed by the statistical model selection procedures applied to the data (Burnham and Anderson 2002, Link and Barker 2006, Hooten and Hobbs 2015) but also requires population modeling experience and familiarity with the system of interest. Model building can usefully be accomplished as a collaborative process between managers, biologists, and population modelers. This allows the experience with model building and the familiarity with the particulars of a given system to be brought together in a team. This is further facilitated if individual team members can bridge these different areas of expertise.

Evaluating the Effects of Action Alternatives

The final step is to run analyses, often stochastic simulations, to assess the expected outcomes of different management actions in terms of management objectives. Once the model has been developed, the full complement of relevant uncertainty should be integrated into the predicted outcomes. However, it can also be useful to conduct sensitivity analysis to understand where the most important uncertainties lie. The sensitivity that is relevant is the sensitivity of the decision, i.e., if the optimal decision changes over the confidence limits of a parameter, or over different structural models, this is uncertainty that is of management relevance. Identifying such uncertainty will focus future research. Value of information constitutes a set of tools that are designed for evaluating the improvement that could be realized, in terms of management outcomes, if uncertainty were resolved or reduced (Runge et al. 2011, Williams et al. 2011, Canessa et al. 2015).

Presentation of model results to the decision makers is an important aspect of modeling. Decision makers need an understanding of model results in order to use them for decision-making, and should not be encouraged to believe the model results without question. A careful and measured explanation of the assumptions, how assumptions might be violated, and how violation of assumptions would affect the model performance is important for communicating a basic understanding of the model. No model will produce fully accurate predictions; all models are wrong. The question is whether the model is useful for selecting management actions that contribute to the attainment of management objectives.

Some authors have suggested that the absolute value of predictions from PVA models should not be relied upon, but instead the relative predictions should be the focus, i.e., whether one action performs better than another (Beissinger and Westphal 1998, McCarthy et al. 2003), and these authors present evidence that PVA models often fall short in making reliable predictions. However, we believe that it is important to continue to strive toward producing models that generate credible absolute predictions, and we emphasize that the above papers were written before the advent of modern methods for fully incorporating uncertainty into projections. Full incorporation of uncertainty will ensure that the range of model outcomes is more likely to include the true outcome. While PVA models have long incorporated demographic and environmental stochasticity (Lacy 1993), until recently most models failed to account for uncertainty associated with model structure and parameter estimation (Table 6.4 in Armstrong and Reynolds 2012). This uncertainty is most effectively incorporated using integrated models, where estimation uncertainty is automatically propagated into population projections (Besbeas et al. 2003, Brooks et al. 2004, Barker et al. 2009). However, it is also possible to fully incorporate uncertainty using methods as simple as spreadsheet models (Table 6.3 in Armstrong and Reynolds 2012).

The problem with relying on relative predictions is that they do not facilitate trade-offs among objectives. For example, a modeling exercise might suggest that a reintroduced population under management action A has X% probability of becoming self-sustaining, whereas the population under management action B has (X − 10)% probability. Apparently, then, management action A should be preferred. However, if X = 15, the manager may not wish to go forward with the reintroduction at all, because the costs incurred are not worth the small chance of success, whereas if X = 90, the same manager may feel differently. This underscores the importance of absolute versus relative predictions (Armstrong and Reynolds 2012, Moore et al. 2012) and also that we must continue to strive to improve the predictive capability of demographic models while also fully representing uncertainty in the models.

The last step in any decision-analytic process will be the solution step: sometimes known as optimization or trade-off analysis. This is the point at which the preferred action, of all those considered, is identified. The preferred action will be a function of not just demographic outcomes, but also outcomes for all management objectives (Box 7.3). For example, in almost all cases, a manager will need to consider costs. It is possible that a relatively tractable solution can be identified by simply taking the management action that predicts the best demographic performance and can be done for less than some fixed budget. In other cases, however, there may be a desire to minimize costs, because costs on one project restrict opportunities to conserve the target species or another species using some other management action. Or there may be a trade-off between probability of success of the reintroduction and the risk to the source population or the ecosystem, such as the risk of novel disease introduction (Ewen et al. 2015a). In all these cases, it will be necessary to grapple with trade-offs

explicitly. There is a growing body of literature on how to approach multiple-objective trade-offs in natural resource management (e.g., Gregory et al. 2012), with some examples involving reintroductions (Converse et al. 2013b, Canessa et al. 2014, Ewen et al. 2015b).

If the management decision of interest can reasonably be revisited through time (e.g., the decision about the sex ratio and age ratio of animals to release in any given year), there is the opportunity to reduce key uncertainties through monitoring, and thus improve decision-making. In adaptive management frameworks, uncertainty that impedes management decision-making can be expressed either as competing model structures representing different hypotheses about how the system works or as a single model including parametric uncertainty (McCarthy et al. 2012). Competing models may correspond to beliefs presented by different experts, for example, and initially the weights (degree of credibility) assigned to the models may be equal. Model-averaged predictions can be used, and decisions made based on those model-averaged predictions. The relative credibility of the models can be updated as monitoring data become available. As relative belief in one hypothesis grows over time, this will shift the weight of model-averaged predictions, such that management actions chosen are increasingly those that are optimal under the more credible model.

FUTURE RESEARCH AND DEVELOPMENT

Research and development in reintroduction biology and management have accelerated in recent decades with the formation of the IUCN Reintroduction Specialist Group (IUCN 1987) and greater attention on the shortcomings of reintroduction programs overall (Lyles and May 1987, Griffith et al. 1989). Perhaps the most important development that can result is greater attention on planning reintroduction efforts, where the focus of planning is placed squarely on determining the optimal methods for achieving the fundamental objectives. We

7b The southern corroboree frog (*Pseudophryne corroboree*)—endemic to southeastern Australia—is listed as critically endangered on the IUCN Red List. The species is bred in captivity, with captive-bred individuals used to support reintroduction and research programs. Canessa et al. (2014) developed optimal release strategies with the objectives of maximizing the size of reintroduced populations, maintaining the captive population, and minimizing costs. Credit: Michael McFadden, Taronga Zoo, Sydney, Australia.

Negative effects on source populations are a common concern in reintroduction programs, and population models will facilitate the evaluation of source population harvest strategies. The theory underlying harvest management can be useful in determining the size and age or sex structure of animals that can be released while maintaining the source population. At the same time, the age structure in a release cohort will influence the age structure in the remaining captive population, and this can affect costs. For example, if animals are released as juveniles, the number of years they must be housed in captivity, and so the cost, may decline. However, in at least some cases, release of juveniles is less likely to lead to a successful reintroduction, because of lower survival in the release cohort. To examine such trade-offs between costs and reintroduction outcomes, Canessa et al. (2014) considered optimal release strategies for an endangered amphibian, the southern corroboree frog (*Pseudophryne corroboree*; figure 7b).

The authors constructed an age-structured matrix model for both the captive and the reintroduced populations of this Australian species using published information and expert judgment; the models varied only in their parameter values. In one scenario, the population was assumed to be at a stable age distribution. In this scenario, the maximum proportion of either eggs or subadults, or an equal proportion of eggs and subadults, which could be harvested from the captive population indefinitely (ensuring maintenance of the captive population), was calculated. The calculation was constrained such that the captive population had to maintain a growth rate ≥1 (based on an analytical evaluation of the matrix model). These releases were added to a reintroduced population and its trajectory was calculated. The cost of maintaining the captive population was also assessed; different release strategies would result in different costs (e.g., retaining animals until they were subadults would incur greater costs).

The authors predicted that, under maximum releases of eggs only and assuming a stable age distribution in the captive population, the size of the reintroduced population would stabilize at 41% the size of the captive population. In other words, for each individual maintained in captivity, an eggs-only strategy would result in 0.41 individuals in the released population. In this scenario, the annual cost of the captive population would be AU$51.10 per individual. For releases of subadults only, the reintroduced population would be 249% the size of the captive population, and the captive population would cost AU$57.10 per individual. The joint eggs/subadults release strategy resulted in approximately an equal cost per captive individual as the eggs-only strategy, but resulted in a reintroduced population 102% the size of the captive population. The reintroduced population could not attain a positive growth rate given the vital rates, and so would be dependent on continued releases. The optimal strategy depends on the sensitivity of the decision maker to both persistence and cost. For example, in this case the authors judged the overall cost difference between the subadults-only and the mixed eggs/subadults strategies to be marginal, and so the subadults-only strategy was judged to be optimal.

In another, arguably more realistic scenario, the captive population was not assumed to be at a stable age distribution. Instead, the true starting population size and structure were assumed, and the population was simulated over a 10-year period. The optimization routine sought the number of individuals to be harvested for release each year—again, under eggs-only, subadults-only, or mixed eggs/subadults strategies—to maximize the average reintroduced population size, while maintaining the number of captive breeding adults, and staying within an annual budget of AU$250,000 for maintenance of the captive population. In this scenario, subadults-only releases could not simultaneously meet the constraints of staying within budget and maintaining the number of captive breeding adults. Both an eggs-only and a mixed eggs/subadults release strategy allowed the constraints to be met, and of these, the mixed release strategy led to higher population sizes in the reintroduced population.

The work of Canessa et al. (2014) constitutes a multiple-objective evaluation of a reintroduction effort which makes use of two separate population models: one for the source and one for the reintroduced population. The analysis demonstrates the sensitivity of the result to decision-framing and model construction; the more realistic scenario, with a nonstable age distribution in the source population, resulted in a management action that struck a balance between costs, maintaining a captive population, and the outcome for the reintroduced population.

argue here that the best approach is through the application of tools developed specifically to improve decision-making. We suggest that managers and biologists engaged in planning reintroduction efforts begin early—before the first animals are released—to anticipate the decisions that will need to be made, to formalize predictions of how reintroduced populations will respond to management actions via population models, to make use of data from other populations or other species to parameterize those models (and to make use of expert judgment when relevant data cannot be obtained), and, finally, to put in place monitoring programs to evaluate population outcomes empirically and gather further data to improve model predictions. There are a number of other areas where we hope to see further developments that could improve the prediction of population responses to management.

One area of future growth should be the use of more flexible modeling techniques, particularly Bayesian hierarchical models using, for example, relatively approachable software such as WinBUGS (Lunn et al. 2000, Kéry and Schaub 2012). These methods allow modelers to jointly estimate all demographic parameters for a given population model (Barker et al. 2009, Moore et al. 2012); integrate multiple data sources from a given population, such as abundance data and demographic rate data, via integrated population models (Besbeas et al. 2002, 2003, Brooks et al. 2004); or integrate data from multiple reintroduction programs for single or multiple species (Holland et al. 2009, McCarthy et al. 2012, Parlato and Armstrong 2012). Bayesian inference, and Markov Chain Monte Carlo methods specifically, allows for more complete characterization of uncertainty (Wade 2002, Servanty et al. 2014), and also facilitates joint estimation and projection in a single analysis.

More complete integration of genetic and demographic considerations is needed in PVA (Lande 1988, Pierson et al. 2015). The impacts of genetic effects on small populations, including inbreeding depression and loss of adaptive potential, may be more serious than is often appreciated (Lacy 2000a). Modern genomics offer great promise for improving our understanding of how genetic factors may influence fitness and resulting long-term demographic outcomes. Such tools can help to address questions about the impact of genetic factors on reintroduction outcomes (Lacy 2000a, Kirchner et al. 2006, Robert et al. 2007, Converse et al. 2012). Another possibility is modeling the impact of artificial or natural selection for particular genetic traits, which may hold some promise for species facing serious threats such as the pathogenic amphibian chytrid fungus, *Batrachochytrium dendrobatidis* (Retallick et al. 2004, Venesky et al. 2012).

Adaptive management is perhaps the most underused tool in reintroduction management. Many others have written about the challenges to implementation of adaptive management, and some of these challenges are certainly technical in nature (McLain and Lee 1996, Allen and Gunderson 2011). However, we advocate that reintroduction programs should start with adaptive management at the planning stage. The key aspects of adaptive management, in addition to those common to all decision-analytic processes, are model(s) that express uncertainty about system function, and a monitoring program that allows knowledge to be updated over time (Williams et al. 2002, Runge 2011, McCarthy et al. 2012). It is important to note that competing system models must result in different preferred actions for adaptive management to be useful: if the same management action is optimal under all models of system behavior or parameter values, there is no management-relevant uncertainty and adaptive management will not be useful (Runge et al. 2011). Once competing models are articulated,

a monitoring plan is needed. The monitoring plan should focus on collection of data that will reduce uncertainty (Nichols and Williams 2006, Nichols and Armstrong 2012). For example, if competing models differ in predictions about how breeding success will respond to different habitat types, then breeding success should be a major focus of monitoring. There are two major types of adaptive management: passive and active (McCarthy et al. 2012). In passive adaptive management, the preferred management action under the model-averaged prediction is taken, and monitoring data are used to determine whether the observed outcomes under that action are in accord with outcomes predicted from the models. At the point when the observed outcomes are poorer than the predicted outcomes under some other management action, the new management action is chosen. In active adaptive management, by contrast, management actions that are predicted to be nonoptimal in the short term may be implemented because they are expected to result in accelerated learning. Active adaptive management in particular requires more sophisticated technical tools for dynamic optimization, to manage the trade-off between learning and achieving objectives (McCarthy and Possingham 2007). The development of more approachable and tractable tools for implementation of adaptive management (e.g., Nicol and Chadès 2011) is a major research need.

SUMMARY

The primary objective of a reintroduction is typically the establishment of a population with a high probability of persistence. Therefore, demography will be a major focus of planning and evaluating reintroduction efforts. Reintroductions can usefully be viewed as a series of challenging decisions, made under uncertainty. This view suggests the use of a decision-analytic approach to reintroduction management. Decision analysis is the process of deconstructing and analyzing decisions to promote the transparent selection of optimal choices. Decision analysis is particularly useful in reintroduction decisions, which are almost always complicated by uncertainty and challenging trade-offs.

The first steps in decision-making are articulating the decision problem and identifying the objectives. Next, the set of alternative management actions must be identified—these may include, for example, different release sites, release methods, sex/age structures in the release cohorts, or post-release habitat management strategies. Demographic models provide the predictive link between the alternative management actions under consideration and the population objectives, such that managers can assess how different alternatives are expected to perform in terms of desired demographic outcomes. The model predictions are then used to inform the process of selecting the optimal actions.

Demographic models will be of varying complexity depending on the population, the needs of decision makers, and the available data and modeling resources. While conceptual models may be a starting point, quantitative models will be most useful for guiding decisions and integrating monitoring data. We recognize five steps in building demographic models: structuring and planning the model, demographic estimation using available data, filling information gaps using expert judgment, construction of the population model, and use of the model for evaluating the effects of alternative actions. Progress through these steps will not be entirely linear; in particular, while final model construction will not be completed until relevant parameter estimates are available, the first steps in construction of the model cannot begin too early. We suggest a rapid prototyping approach to model development, wherein modeling progresses through iterative, rapidly completed prototypes.

Ultimately, predictions of the impacts of alternative management actions on demographic outcomes will be only part of the picture. It is almost always the case that managers have to consider other objectives, such as budget, impacts on source populations, or other ecological or social effects. Evaluation of trade-offs, or optimization, is the final step in a decision-analytic process. Transparent consideration of trade-offs is aided by a variety of decision-analytic tools, including tools designed to deal with multi-criteria decision problems.

Demographic models for reintroduction management need to be designed to accommodate the unique characteristics of reintroduced populations. Particular concerns include post-release effects, Allee effects, and transitory dynamics caused by the initial age and sex composition of the release group. In addition, there will often be the need to build models predicting effects of removals on the source population—whether wild or captive—to identify what removals are sustainable. For all populations of interest, major sources of uncertainty should be included, which will allow for the identification of action alternatives that are optimal given both what is known and, importantly, what is unknown. In cases where uncertainty impedes the choice of management action, and management actions are repeated over time, there is the potential for implementing adaptive management. More thoughtful articulation of key uncertainties, done earlier in and throughout the process of reintroduction management, would lead to greater progress in both reintroduction biology and reintroduction management.

MANAGEMENT RECOMMENDATIONS

- Demographic considerations in reintroductions are paramount, as the overall success of a reintroduction program will ultimately be a function of demographic processes. A typical primary objective of a reintroduction effort is the establishment of a self-sustaining population: one in which losses (mortality and emigration) are exceeded by gains (reproduction and immigration) over some acceptably long time period.

- Population-level outcomes are the product of complex processes at lower levels—including survival, reproduction, and movement— which are themselves a function of many factors, such as habitat, release cohort source, release method, genetic factors, and others. The complexity of demographic outcomes requires quantitative population models to develop predictions of how management of the reintroduction may influence outcomes.

- Decision analysis provides an approach and a useful set of tools for solving challenging decision problems, such as choosing among management alternatives for a reintroduction effort. In decision analysis, decisions are deconstructed into their component parts to facilitate analysis.

- Making a decision about the management of a reintroduction will be facilitated by recognizing and appropriately analyzing the component parts. The first three of these components must be established before modeling or prediction can commence: a clear statement of the decision problem, the management objective(s) to be met, and the available management alternatives that may be considered. Additional steps are predicting (modeling) the impact of management alternatives on management objectives, and optimization, or solving the problem.

- Once the decision problem, objectives, and alternative actions are specified, the construction of demographic models can get underway. The goal of demographic models, in the context of reintroduction management, is to allow for prediction of the impacts of different management actions on demographic outcomes, such as persistence probability.

- Model building should progress through a number of steps. A simple accounting of these steps includes the following: structuring and planning the model, demographic estimation using available data, filling information gaps using expert judgment, construction of the population model, and,

finally, use of models for evaluating the effects of action alternatives.

- It is important to recognize that models are ubiquitous in management: any prediction about the way the world works, whether conceptual, verbal, or quantitative, is a model. Structuring a model is the process of beginning to create quantitative models from conceptual or verbal models. That will allow for a common understanding of system function to develop among decision makers and those aiding them, and will ultimately facilitate the inclusion of data into predictions. The major elements of the model— including age or stage structure, density dependence, uncertainty and stochasticity, and spatial or metapopulation structure— must be identified in the structuring and planning step.

- Demographic estimation will proceed after initial model structuring. There are a wide variety of approaches to demographic estimation, and the approach appropriate in a given case will be a function of the data available and the model that is envisaged. It is often useful to conceive of data that are appropriate for a given estimation task in a broader context: for example, data from other populations of the species of interest or even from closely related taxa can be used to develop demographic estimates if these are the best data available.

- In some cases, appropriate data are not available, particularly when novel management actions are being contemplated. In these cases, elicitation of expert judgment can be a powerful tool for making predictions.

- Constructing the population model is a process that can be made easier by breaking it into discrete steps, where complexity is added progressively and only as needed. The completed population model—or models in the case where structural uncertainty is represented by competing models of the system—can then be used to make predictions about how the population may respond to alternative management actions.

- Solving the decision problem may require consideration of additional objectives. For example, demographic outcomes may be traded off against costs, impacts on source populations, or other social or ecological values that may be impacted by the reintroduction. In these cases, predictions should be made for each objective as a function of the management alternatives under consideration, and multi-criteria trade-off tools can be used to evaluate trade-offs.

- Adaptive management is one of the most powerful and most underused tools in reintroduction management. Adaptive management requires the opportunity to revisit ongoing management actions periodically, the articulation of competing hypotheses of system behavior via models that make distinct predictions about what actions are optimal, and a monitoring plan for reducing uncertainty about which competing model is preferred. More upfront thought about hypotheses regarding how reintroduced populations may respond to management would greatly accelerate the progress in both management outcomes and basic reintroduction biology.

LITERATURE CITED

Allen, C. R., and L. H. Gunderson. 2011. Pathology and failure in the design and implementation of adaptive management. Journal of Environmental Management 92:1379–1384.

Armstrong, D. P., I. Castro, and R. Griffiths. 2007. Using adaptive management to determine requirements of re-introduced populations: the case of the New Zealand hihi. Journal of Applied Ecology 44:953–962.

Armstrong, D. P., R. S. Davidson, W. J. Dimond, J. K. Perrott, I. Castro, J. G. Ewen, R. Griffiths, and J. Taylor. 2002. Population dynamics of reintroduced forest birds on New Zealand islands. Journal of Biogeography 29:609–621.

Armstrong, D. P., and M. H. Reynolds. 2012. Modelling reintroduced populations: the state of

the art and future directions. Pp.165–222. In J. G. Ewen, D. P. Armstrong, K. A. Parker, and P. J. Seddon, eds., Reintroduction Biology: Integrating Science and Management. Wiley-Blackwell, Oxford, UK.

Armstrong, D. P., and P. J. Seddon. 2008. Directions in reintroduction biology. Trends in Ecology and Evolution 23:20–25.

Armstrong, D. P., and H. U. Wittmer. 2011. Incorporating Allee effects into reintroduction strategies. Ecological Research 26:687–695.

Ayyub, B. M. 2001. Elicitation of Expert Opinions for Uncertainty and Risks. CRC Press, Boca Raton, FL.

Ballou, J. D., and R. C. Lacy. 1995. Identifying genetically important individuals for management of genetic variation in pedigreed populations. Pp.76–111. In J. D. Ballou, M. Gilpin, and T. J. Foose, eds., Population Management for Survival and Recovery. Columbia University Press, New York.

Barker, R. J., M. R. Schofield, D. P. Armstrong, and R. S. Davidson. 2009. Bayesian hierarchical models for inference about population growth. Pp.3–17. In D. L. Thomson, E. G. Cooch, and M. C. Conroy, eds., Modeling Demographic Processes in Marked Populations. Springer, New York.

Beissinger, S. R., and M. I. Westphal. 1998. On the use of demographic models of population viability analysis in endangered species management. Journal of Wildlife Management 62:821–841.

Besbeas, P., S. N. Freeman, B. J. T. Morgan, and E. A. Catchpole. 2002. Integrating mark-recapture-recovery and census data to estimate animal abundance and demographic parameters. Biometrics 58:540–547.

Besbeas, P., J.-D. Lebreton, and B. J. T. Morgan. 2003. The efficient integration of abundance and demographic data. Applied Statistics 52:95–102.

Brooks, S., R. King, and B. Morgan. 2004. A Bayesian approach to combining animal abundance and demographic data. Animal Biodiversity and Conservation 27:515–529.

Brown, V. R., and P. B. Paulus. 2002. Making group brainstorming more effective: recommendations from an associative memory perspective. Current Directions in Psychological Science 11:208–212.

Brownie, C., J. E. Hines, J. D. Nichols, K. H. Pollock, and J. Hestbeck. 1993. Capture-recapture studies for multiple strata including non-Markovian transitions. Biometrics 49:1173–1187.

Buner, F., and M. Schaub. 2008. How do different releasing techniques affect the survival of reintroduced grey partridges *Perdix perdix*. Wildlife Biology 14:26–35.

Burgman, M. 2004. Expert frailties in conservation risk assessment and listing decisions. Pp.20–29. In P. Hutchings, D. Lunney, and C. Dickman, eds., Threatened Species Legislation: Is It Just an Act? Royal Zoological Society of New South Wales, Mosman, NSW.

Burgman, M. 2005. Risks and Decisions for Conservation and Environmental Management. Cambridge University Press, Cambridge, UK.

Burgman, M.A., M.F. McBride, R. Ashton, A. Speirs-Bridge, L. Flander, B. Wintle, F. Fidler, L. Rumpff, and C. Twardy. 2011. Expert status and performance. PLoS ONE 6(7):e22998. doi:10.1371/journal.pone.0022998.

Burnham, K.P., and D.R. Anderson. 2002. Model Selection and Multimodel Inference: a Practical Information-Theoretic Approach. 2nd ed. Springer-Verlag, New York.

Canessa, S., S.J. Converse, M. West, N. Clemann, G. Gillespie, M. McFadden, A.J. Silla, K.M. Parris, and M.A. McCarthy. 2016. Planning for ex-situ conservation in the face of uncertainty. Conservation Biology. 30:599–609.

Canessa, S., G. Guillera-Arroita, J.J. Lahoz-Monfort, D.M. Southwell, D.P. Armstrong, I. Chadès, R.C. Lacy, and S.J. Converse. 2015. When do we need more data? A primer on calculating the value of information for applied ecologists. Methods in Ecology and Evolution 6:1219–1228.

Canessa, S., D. Hunter, M. McFadden, G. Marantelli, and M.A. McCarthy. 2014. Optimal release strategies for cost-effective reintroductions. Journal of Applied Ecology 51:1107–1115.

Caswell, H. 2001. Matrix Population Models: Construction, Analysis, and Interpretation. 2nd ed. Sinauer Associates, Sunderland, MA.

Caughley, G. 1994. Directions in conservation biology. Journal of Animal Ecology 63:215–244.

Chauvenet, A.L.M., E.H. Parlato, J.V. Gedir, and D.P. Armstrong. 2015. Advances in modelling projections for reintroduced populations. Pp.91–104. In D.P. Armstrong, M.W. Hayward, D. Moro, and P.J. Seddon, eds., Advances in Reintroduction Biology of Australian and New Zealand Fauna. CSIRO Press, Melbourne.

Converse, S.J., C.T. Moore, and D.P. Armstrong. 2013a. Demographics of reintroduced populations: estimation, modeling, and decision analysis. Journal of Wildlife Management 77:1081–1093.

Converse, S.J., C.T. Moore, M.J. Folk, and M.C. Runge. 2013b. A matter of tradeoffs: reintroduc-tion as a multiple objective decision. Journal of Wildlife Management 77:1145–1156.

Converse, S.J., J.A. Royle, and R.P. Urbanek. 2012. Bayesian analysis of multi-state data with individual covariates for estimating genetic effects on demography. Journal of Ornithology 152:S561–S572.

Cormack, R. 1964. Estimates of survival from the sighting of marked animals. Biometrika 51:429–438.

Deredec, A., and F. Courchamp. 2007. Importance of the Allee effect for reintroductions. Ecoscience 4:440–451.

Dickens, M.J., D.J. Delehanty, and L.M. Romero. 2010. Stress: an inevitable component of animal translocation. Biological Conservation 143:1329–1341.

Dorazio, R.M., and F.A. Johnson. 2003. Bayesian inference and decision theory: a framework for decision making in natural resource manage-ment. Ecological Applications 13:556–563.

Edwards, W. 1977. How to use multiattribute utility measurement for social decisionmaking. IEEE Transactions on Systems, Man and Cybernetics 7:326–340.

Ewen, J.G., D.P. Armstrong, K.A. Parker, and P.J. Seddon. 2012. Reintroduction Biology: Integrat-ing Science and Management. Wiley-Blackwell, Oxford, UK.

Ewen, J.G., A.W. Sainsbury, B. Jackson, and S. Canessa. 2015a. Disease risk management in reintroduction. Pp.43–57. In D.P. Armstrong, M.W. Hayward, D. Moro, and P.J. Seddon, eds., Advances in Reintroduction Biology of Australian and New Zealand Fauna. CSIRO Publishing, Collingwood, VIC.

Ewen, J.G., L. Walker, S. Canessa, and J.J. Groom-bridge. 2015b. Improving supplementary feeding in species conservation. Conservation Biology 29:341–349.

Fischer, J., and D.B. Lindenmayer. 2000. An assessment of the published results of animal relocations. Biological Conservation 96:1–11.

Goodwin, P., and G. Wright. 2004. Decision Analysis for Management Judgement. Wiley, London, UK.

Gregory, R., L. Failing, M. Harstone, G. Long, T. McDaniels, and D. Ohlson. 2012. Structured Decision Making: A Practical Guide to Environ-mental Management Choices. Wiley-Blackwell, Oxford, UK.

Griffith, B., J.M. Scott, J.W. Carpenter, and C. Reed. 1989. Translocation as a species conservation tool: status and strategy. Science 245:477–480.

Gusset, M., O. Jakoby, M. S. Müller, M. J. Somers, R. Slotow, and V. Grimm. 2009. Dogs on the catwalk: modelling re-introduction and translocation of endangered wild dogs in South Africa. Biological Conservation 142:2774–2781.

Hammond, J. S., R. L. Keeney, and H. Raiffa. 1999. Smart Choices: a Practical Guide to Making Better Life Decisions. Broadway Books, New York.

Heath, S. R., E. L. Kershner, D. M. Cooper, S. Lynn, J. M. Turner, N. Warnock, S. Farabaugh, K. Brock, and D. K. Garcelon. 2008. Rodent control and food supplementation increase productivity of endangered San Clemente loggerhead shrikes (*Lanius ludovicianus mearnsi*). Biological Conservation 141:2506–2515.

Hestbeck, J. B., J. D. Nichols, and R. A. Malecki. 1991. Estimates of movement and site fidelity using mark-resight data of wintering Canada Geese. Ecology 72:523–533.

Holland, E. P., J. F. Burrow, C. Dythama, and J. N. Aegerter. 2009. Modelling with uncertainty: introducing a probabilistic framework to predict animal population dynamics. Ecological Modelling 220:1203–1217.

Hood, G. M. 2010. PopTools, version 3.2.3. Available online at: http://poptools.org.

Hooten, M. B., and N. T. Hobbs. 2015. A guide to Bayesian model selection for ecologists. Ecological Monographs 85:3–28.

IUCN (International Union for Conservation of Nature). 1987. IUCN Position Statement on the Translocation of Living Organisms: Introductions, Re-introductions, and Re-stocking. Prepared by the Species Survival Commission in Collaboration with the Commission on Ecology and the Commission on Environmental Policy, Law and Administration. Available online at: http://www.iucnsscrsg.org/.

IUCN. 2011. *Pseudophryne corroboree*. IUCN Red List of Threatened Species, version 2011.2.

IUCN/SSC. 2013. Guidelines for Reintroductions and Other Conservation Translocations. Version 1. IUCN Species Survival Commission, Gland, Switzerland.

Jamieson, I. G., and R. C. Lacy. 2012. Managing genetic issues in reintroduction biology. Pp.441–475. In J. G. Ewen, D. P. Armstrong, K. A. Parker, and P. J. Seddon, eds., Reintroduction Biology: Integrating Science and Management. Wiley-Blackwell, Oxford, UK.

Jolly, G. 1965. Explicit estimates from capture-recapture data with both death and immigration-stochastic model. Biometrika 52:225–247.

Keeney, R. L. 1992. Value-Focused Thinking: a Path to Creative Decisionmaking. Harvard University Press, Cambridge, MA.

Keeney, R. L., and H. Raiffa. 1976. Decisions with Multiple Objectives: Preferences and Value Tradeoffs. John Wiley & Sons, New York.

Kéry, M., and M. Schaub. 2012. Bayesian Population Analysis Using WinBUGS: A Hierarchical Perspective. Academic Press, Waltham, MA.

Kirchner, F., A. Robert, and B. Colas. 2006. Modelling the dynamics of introduced populations in the narrow-endemic *Centaurea corymbosa*: a demo-genetic integration. Journal of Applied Ecology 43:1011–1021.

Kuhnert, P. M., T. G. Martin, and S. P. Griffiths. 2010. A guide to eliciting and using expert knowledge in Bayesian ecological models. Ecology Letters 13:900–914.

Lacy, R. C. 1993. VORTEX: a computer simulation model for population viability analysis. Wildlife Research 20:45–65.

Lacy, R. C. 2000a. Considering threats to the viability of small populations using individual-based models. Ecological Bulletins 48:39–51.

Lacy, R. C. 2000b. Structure of the VORTEX simulation model for population viability analysis. Ecological Bulletins 48:191–203.

Lande, R. 1988. Genetics and demography in biological conservation. Science 241: 1455–1460.

Le Gouar, P., A. Robert, J.-P. Choisy, S. Henriquet, P. Lecuyer, C. Tessier, and F. Sarrazin. 2008. Roles of survival and dispersal in reintroduction success of griffon vulture (*Gyps fulvus*). Ecological Applications 18:859–872.

Lebreton, J.-D., K. P. Burnham, J. Clobert, and D. R. Anderson. 1992. Modeling survival and testing biological hypotheses using marked animals: a unified approach with case studies. Ecological Monographs 62:67–118.

Lebreton, J.-D., J. D. Nichols, R. J. Barker, R. Pradel, and J. A. Spendelow. 2009. Modeling individual animal histories with multistate capture-recapture models. Pp.87–173. In H. Caswell, ed., Advances in Ecological Research, Vol. 41. Academic Press, Burlington, MA.

Link, W. A., and R. J. Barker. 2006. Model weights and the foundations of multimodel inference. Ecology 87:2626–2635.

Lintermans, M., J. P. Lyon, M. P. Hammer, I. Ellis, and B. C. Ebner. 2015. Underwater, out of sight: lessons from threatened freshwater fish translocations in Australia. Pp.237–253. In D. P. Armstrong, M. W. Hayward, D. Moro, and P. J. Seddon, eds., Advances in Reintroduction

Biology of Australian and New Zealand Fauna. CSIRO Publishing, Collingwood, VIC.

Lunn, D. J., A. Thomas, N. Best, and D. Spiegelhalter. 2000. WinBUGS – a Bayesian modeling framework: concepts, structure, and extensibility. Statistics and Computing 10:325–337.

Lyles, A. M., and R. M. May. 1987. Problems in leaving the ark. Nature 326:245–246.

MacMillan, D. C., and K. Marshall. 2006. The Delphi process-an expert-based approach to ecological modelling in data-poor environments. Animal Conservation 9:11–19.

Martin, T. G., M. A. Burgman, F. Fidler, P. M. Kuhnert, S. Low-Choy, M. F. McBride, and K. Mengersen. 2011. Eliciting expert knowledge in conservation science. Conservation Biology 26:29–38.

McBride, M. F., F. Fidler, and M. A. Burgman. 2012. Evaluating the accuracy and calibration of expert predictions under uncertainty: predicting the outcomes of ecological research. Diversity and Distributions 2012:782–794.

McCarthy, M. A., S. J. Andelman, and H. P. Possingham. 2003. Reliability of relative predictions in population viability analysis. Conservation Biology 17:982–989.

McCarthy, M. A., D. P. Armstrong, and M. C. Runge. 2012. Adaptive management of reintroduction. Pp.256–289. In J. G. Ewen, D. P. Armstrong, K. A. Parker, and P. J. Seddon, eds., Reintroduction Biology: Integrating Science and Management. Wiley-Blackwell, Oxford, UK.

McCarthy, M. A., and H. P. Possingham. 2007. Active adaptive management for conservation. Conservation Biology 21:956–963.

McLain, R. J., and R. G. Lee. 1996. Adaptive management: promises and pitfalls. Environmental Management 20:437–448.

Meyer, M. A., and J. M. Booker. 1990. Eliciting and Analyzing Expert Judgment: A Practical Guide. Office of Nuclear Regulatory Research, Division of Systems Research, US Nuclear Regulatory Commission, Washington, DC.

Moore, C. T., S. J. Converse, M. J. Folk, M. C. Runge, and S. A. Nesbitt. 2012. Evaluating release alternatives for a long-lived bird species under uncertainty about long-term demographic rates. Journal of Ornithology 152:S339–S353.

Morris, W. F., and D. F. Doak. 2002. Quantitative Conservation Biology. Sinauer Associates, Sunderland, MA.

Mueller, T., R. B. O'Hara, S. J. Converse, R. P. Urbanek, and W. F. Fagan. 2013. Social learning of migratory performance. Science 341:999–1002.

Nichols, J. D., and D. P. Armstrong. 2012. Monitoring for reintroductions. Pp.223–255. In J. G. Ewen, D. P. Armstrong, K. A. Parker, and P. J. Seddon, eds., Reintroduction Biology: Integrating Science and Management. Wiley-Blackwell, Oxford, UK.

Nichols, J. D., and B. K. Williams. 2006. Monitoring for conservation. Trends in Ecology and Evolution 21:668–673.

Nicol, S., and I. Chadès. 2011. Beyond stochastic dynamic programming: a heuristic sampling method for optimizing conservation decisions in very large state spaces. Methods in Ecology and Evolution 2:221–228.

Nicolson, C. R., A. M. Starfield, G. P. Kofinas, and J. A. Kruse. 2002. Ten heuristics for interdisciplinary modeling projects. Ecosystems 5:376–384.

Panfylova, J. 2015. Applying structured decision making to management of the reintroduced hihi population in Bushy Park. MSc thesis, Massey University, New Zealand.

Parlato, E. H., and D. P. Armstrong. 2012. An integrated approach for predicting fates of reintroductions with demographic data from multiple populations. Conservation Biology 26:97–106.

Pierson, J. C., S. R. Beissinger, J. G. Bragg, D. J. Coates, J. G. B. Oostermeijer, P. Sunnucks, N. H. Schumaker, M. V. Trotter, and A. G. Young. 2015. Incorporating evolutionary processes into population viability models. Conservation Biology 29:755–764.

Possingham, H. P., S. J. Andelman, B. R. Noon, S. Trombulak, and H. R. Pulliam. 2001. Making smart conservation decisions. Pp.225–244. In M. Soulé, and G. Orians, eds. Conservation Biology: Research Priorities for the Next Decade. Island Press, Washington, DC.

R Development Core Team. 2012. R: A Language and Environment for Statistical Computing. R Foundation for Statistical Computing, Vienna, Austria. Available online at: http://www.R-project.org.

Retallick, R. W. R., H. McCallum, and R. Speare. 2004. Endemic infection of the amphibian chytrid fungus in a frog community post-decline. PLoS Biology 2:e351.

Robert, A., D. Couvet, and F. Sarrazin. 2007. Integration of demographic and genetics in population restorations. Ecoscience 14:463–471.

Roche, E. A., F. J. Cuthbert, and T. W. Arnold. 2008. Relative fitness of wild and captive-reared piping plovers: does egg salvage contribute to recovery of the endangered Great Lakes population? Biological Conservation 141:3079–3088.

Rout, T. M., C. E. Hauser, and H. P. Possingham. 2009. Optimal adaptive management for the translocation of a threatened species. Ecological Applications 19:515–526.

Runge, M. C. 2011. An introduction to adaptive management for threatened and endangered species. Journal of Fish and Wildlife Management 2:220–233.

Runge, M. C. 2013. Active adaptive management for reintroduction of an animal population. Journal of Wildlife Management 77:1135–1144.

Runge, M. C., S. J. Converse, and J. E. Lyons. 2011. Which uncertainty? Using expert elicitation and expected value of information to design an adaptive program. Biological Conservation 144:1214–1223.

Sarrazin, F., and S. Legendre. 2000. Demographic approach to releasing adults versus young in reintroductions. Conservation Biology 14:488–500.

Scott, J. M., and J. W. Carpenter. 1987. Release of captive-reared or translocated endangered birds: what do we need to know? Auk 104:544–545.

Seber, G. A. F. 1965. A note on the multiple-recapture census. Biometrika 52:249–259.

Servanty, S., S. J. Converse, and L. L. Bailey. 2014. Demography of a reintroduced population: moving toward management models for an endangered species, the whooping crane. Ecological Applications 24:927–937.

Speirs-Bridge, A., F. Fidler, M. McBride, L. Flander, G. Cumming, and M. Burgman. 2010. Reducing overconfidence in the interval judgments of experts. Risk Analysis 30:512–523.

Stubben, C. J., and B. G. Milligan. 2007. Estimating and analyzing demographic models using the popbio package in R. Journal of Statistical Software 22:1–23.

Tavecchia, G., C. Viedma, A. Martínez-Abraín, M.-A. Bartolomé, J. A. Gómez, and D. Oro. 2009. Maximizing re-introduction success: assessing the immediate cost of release in a threatened waterfowl. Biological Conservation 142:3005–3012.

Urbanek, R. P., L. E. A. Fondow, C. D. Satyshur, A. E. Lacy, S. E. Zimorski, and M. Wellington. 2005. First cohort of migratory whooping cranes reintroduced to eastern North America: the first year after release. Proceedings of the North American Crane Workshop 9:213–223.

Venesky, M. D., J. R. Mendelson, III, B. F. Sears, P. Stiling, and J. R. Rohr. 2012. Selecting for tolerance against pathogens and herbivores to enhance success of reintroduction and translocation. Conservation Biology 26:586–592.

Wade, P. R. 2002. Bayesian population viability analysis. Pp. 213–238. In S. R. Beissinger, and D. R. McCullough, eds., Population Viability Analysis. University of Chicago Press, Chicago, IL.

Walters, C. 1986. Adaptive Management of Renewable Resources. MacMillian, New York.

White, G. C. 2000. Modeling population dynamics. Pp. 85–107. In S. Demarais, and P. R. Krausman, eds., Ecology and Management of Large Mammals in North America. Prentice-Hall, Englewood Cliffs, NJ.

Williams, B. K. 2001. Uncertainty, learning, and the optimal management of wildlife. Environmental and Ecological Statistics 8:269–288.

Williams, B. K. 2011. Passive and active adaptive management: approaches and an example. Journal of Environmental Management 222:3429–3436.

Williams, B. K., M. J. Eaton, and D. R. Breininger. 2011. Adaptive resource management and the value of information. Ecological Modelling 222:3429–3436.

Williams, B. K., J. D. Nichols, and M. J. Conroy. 2002. Analysis and Management of Animal Populations. Academic Press, San Diego, CA.

PART THREE

Obstacles to Successful Reintroductions

EIGHT

Genetic Issues in Reintroduction

Iris Biebach, Deborah M. Leigh, Kasia Sluzek,
and Lukas F. Keller

REINTRODUCING PLANTS OR ANIMALS implies moving individuals. When individuals are moved, so are the parasites, pathogens, symbionts, and mutualists that live in and on them, and the genes that each individual carries. This movement of genes permanently changes the genetic composition of the source and the released populations in ways that may affect the success of the reintroductions (Frankham et al. 2010, Allendorf et al. 2013).

Scientists and managers involved in reintroductions, or other forms of translocations, have to make a number of decisions: which source populations to choose, how many individuals of what age and sex to release, where and when to release them, over how many years, with or without a captive breeding phase, what postrelease monitoring to carry out, and so on. Most of these decisions, directly or indirectly, will affect the genetic composition of the released populations, and these effects may last for hundreds of generations (Varvio et al. 1986). To maximize reintroduction success in the long term, it is therefore important that sci-

entists and managers are aware of the genetic consequences of the decisions they take.

The aim of this chapter is to highlight the most important genetic consequences of management actions during reintroductions. Appreciating and understanding these consequences is not possible without going into population genetics in a fair bit of detail, because reintroduced populations, by their very nature, are not in equilibrium. That is, the various genetic forces that affect them—mutation, migration, drift, and selection—have not yet reached a steady state. The genetics of reintroductions thus is the genetics of nonequilibrium populations (Maruyama and Fuerst 1985a, b). This is more complex than if we could assume steady state—but also more interesting for a research scientist.

We start this chapter with a review of the major genetic issues in reintroductions: genetic drift, loss of evolutionary potential, and inbreeding. We then discuss some special genetic concerns that arise when reintroductions have to rely on captive breeding programs, and when released individuals hybridize with other wild or domestic species. We present methods of estimating some of the most important variables, such as the effective population size, evolutionary adaptive potential, inbreeding, inbreeding depression, and the degree of hybridization. Specific recommendations for the planning and execution of a reintroduction are given in the penultimate section, divided into prerelease and post-release issues. Finally, we present an outlook on future research and end with management recommendations.

GENETIC ISSUES IN REINTRODUCTIONS

Reintroduced Populations Begin Small

Almost all reintroduced populations are small for a substantial length of time. To begin with, reintroduced populations often have a history of severe reductions in population size (so-called bottlenecks) prior to reintroduction. They are also small following reintroduction because founder numbers are nearly always limited by availability of individuals for release, funds, or political will. Small founder numbers are also typical of populations that are bred in captivity. Populations reintroduced from captivity thus experience founder events and periods of small population size at least twice.

How long reintroduced populations remain small depends on the intrinsic population growth rate of the species, post-release mortality, and other components of reintroduction success. Populations of species with typically low population growth rates (e.g., many large mammals) will remain small for longer following a bottleneck than populations of species with higher population growth rates, e.g., fish (Denney et al. 2002). Small population size has many important consequences for conservation and reintroduction biology, as noted decades ago when conservation biology emerged as a new field (e.g., Soulé 1986, Ballou et al. 1995). These consequences are often demographic (see Chapter 7), but small population size also raises genetic concerns. This is because random genetic fluctuations, known as genetic drift, become more pronounced the smaller the size of a population, with potentially deleterious long-term consequences. Most of the detrimental genetic consequences that may arise from small population size such as an increased frequency of deleterious mutations, reduced adaptive evolutionary potential, and inbreeding depression (Allendorf et al. 2013), are direct or indirect consequences of random genetic drift. Hence, in the following we first introduce genetic drift in some detail before moving on to its detrimental consequences in reintroduced populations.

Genetic Drift

If one follows allele frequencies in small populations, one can see that they change erratically from one generation to the next. After a number of generations, the allele frequencies are often very different from those at the outset. This random variation of allele frequencies from one generation to the next is called genetic drift. Genetic drift is ubiquitous because it stems from a very basic feature of the process of inheritance: whenever an offspring is produced, each parental allele at a particular locus has a probability of 0.5 to be passed on to that offspring. This creates random variation in how often a particular allele is passed on to the next generation and, in turn, random variation in allele frequencies between generations. This random variation is greater when fewer offspring are produced, as occurs when populations are small. Take, for example, the case of an individual that is heterozygous at a particular locus (A/a). Let us assume that it mates with an individual that is homozygous a/a and that they have two offspring during their lifetime. The probability that the first offspring is an a/a homozygote is 0.5. Hence, the probability that the A allele is *not* passed on to this offspring is 0.5. The probability that the A allele is also not passed on to the second offspring is also 0.5. Hence, the probability that neither offspring inherits a copy of the A allele is $0.5 \times 0.5 = 0.25$. Thus, in this scenario, there is a one-fourth chance that the *A* allele is lost from one generation to the next and that the offspring population is entirely homozygous (*a*/*a*). Conversely, the probability of the *A* allele being passed on to both offspring is also one-fourth. Hence, there is one-fourth probability that the *A* allele is underrepresented, and one-fourth probability that it is overrepresented in the next generation. This basic feature of Mendelian inheritance, sometimes called Mendelian lottery, is the basis for random genetic drift. Clearly, the probabilities we just calculated depend on the frequency of an allele in the population, the population size, and the number of offspring produced. The equations that make the relationship between allele frequency, population size, and number of offspring explicit are relatively complex, so we will not present them here. The interested reader is instead referred to Frankham et al. (2010) for details.

Even without going into the mathematical details of genetic drift, this short introduction allows us to reach a few conclusions that are important for reintroductions:

a. From one generation to the next, genetic drift leads to changes in allele frequency in random directions, with no tendency for allele frequencies to return to the original values.

b. These random changes are more pronounced the smaller a population is.

c. On average, the number of offspring produced is greater in growing populations than in stable populations. The reverse is true in shrinking populations. Hence, genetic drift is less pronounced in growing populations than in stable populations of the same size. Conversely, in shrinking populations genetic drift is more pronounced.

d. Ultimately, genetic drift leads to the loss of alleles and, hence, to a loss of genetic variation, with rare alleles being lost faster than common alleles. Already before an allele is completely lost, reduced genetic variation is evident because heterozygosity in the population declines. The decline in heterozygosity as a consequence of genetic drift is proportional to $\frac{1}{2N_e}$, where N_e is the effective population size (Box 8.1).

e. In independent populations, different alleles will increase or decrease in frequency through time as a result of genetic drift. Consequently, as time goes on, populations will become increasingly differentiated. Thus, genetic drift also results in genetic divergence between populations. This has important consequences for reintroductions: when several independent populations are reestablished from a single source, genetic drift will lead to genetic differentiation among the reintroduced populations, even though they are all descendants of the same source population. This effect has been observed in genetic analyses of reintroductions (Biebach and Keller 2009). In the case of the Alpine ibex (*Capra ibex*), these effects of drift are evident even a hundred years after successful reintroduction (Box 8.2).

Natural populations of mockingbirds on the Galápagos Islands illustrate some of these effects of genetic drift clearly (figure 8.1). First, the smaller the population size, the higher the random changes in allele frequencies over a period of 100 years, which corresponds to approximately 25 generations. Second, the smaller the population size, the smaller the proportion of rare alleles (alleles with a frequency <0.1) that exist in a population. Finally, although this cannot be seen directly from figure 8.1, heterozygosity and the number of alleles at a locus are lower in small populations of mockingbirds (Hoeck et al. 2010, Keller et al. 2012). Note, however, that across all islands in Galápagos, genetic diversity did not change over the 100 years, despite strong genetic drift within islands. This may at first seem counterintuitive, but it is actually an expected result for two reasons. First, as outlined in point (e)

To keep equations manageable, much of population genetics theory makes use of the concept of an idealized population. Such an ideal population has no migration, an equal sex ratio, random mating, Poisson distributed offspring numbers, and a constant size (Allendorf et al. 2013). Natural populations clearly deviate from these characteristics. To nevertheless allow the application of the theory to real-life populations, population geneticists invented a trick: the effective population size (N_e). N_e is the population size of an idealized population, which meets the prior criteria, that experiences equal variance in allele frequencies or equal rate of inbreeding in each generation as the population under study (Allendorf et al. 2013). In natural populations, N_e is generally smaller than the census population size, N_c, although exceptions exist where the opposite is true (Waples et al. 2013). This reduction in N_e compared to N_c depends on how the life history of a species leads to deviations

from the idealized population characteristics. Particularly important parameters that affect the N_e/N_c ratio are variance in reproductive success and varying population size (Frankham 1995; see also Box 8.2). Some species have high census population sizes, but relatively low N_e. For example, in Chinook salmon populations an N_e/N_c ratio below 0.2 is common (Shrimpton and Heath 2003). In species that live in social groups, variance in reproductive success may be high due to the suppression of reproduction in subordinate individuals. Consequently, the N_e/N_c ratio is small. For example in wild dogs (*Lycaon pictus*) N_e/N_c ratios below 0.1 were found (Marsden et al. 2012).

Note that there are several definitions of N_e and N_c (e.g., Crow and Denniston 1988, Luikart et al. 2010, Waples et al. 2013) that can differ substantially in reintroduced populations (Ewing et al. 2008). Details in the context of reintroductions can be found in Keller et al. (2012).

above, when populations drift independently, a particular allele at a locus is likely to become more frequent in some populations and less frequent in others. Although allele frequencies change substantially within populations, on average across all populations the frequencies remain roughly constant. Hence, while genetic drift leads to allele frequency changes and loss of genetic variation within each individual population, across a set of populations genetic drift does not reduce genetic variation in the same way. A second reason why across several populations one does not always expect a loss of genetic diversity is that in the presence of some gene flow among populations, an equilibrium develops between genetic drift and gene flow (Varvio et al. 1986). Genetic drift reduces genetic variation within populations and increases genetic differentiation among populations, while gene flow has the opposite effects. Hence, over time, an equilibrium is approached and the levels of genetic variation and differentiation across several populations remain roughly constant. The

number of generations required to attain equilibrium depends on various factors (number of populations, population sizes, rates of gene flow) but will often be in the hundreds or thousands of generations (Varvio et al. 1986). This equilibrium is evident in the Galápagos mockingbirds. Reintroduced populations, however, are rarely at equilibrium and will become increasingly differentiated over generations, as observed among reintroduced Alpine ibex populations (Box 8.2).

Founder Effects Are Periods of Particularly Strong Genetic Drift

In reintroductions, a limited number of founder individuals are released in a new location. Of those founders, not all will survive and reproduce. In fact, post-release mortality can be high. In crested coot (*Fulica cristata*), annual mortality and dispersal was about 80% the first year after release and 34% once an animal survived the first year (Tavecchia et al. 2009). In Alpine ibex, less than half of all released indi-

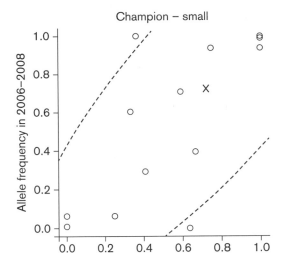

Champion – small

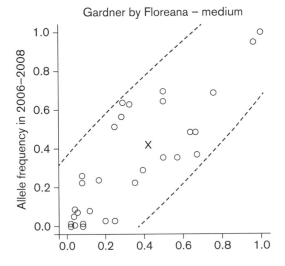

Gardner by Floreana – medium

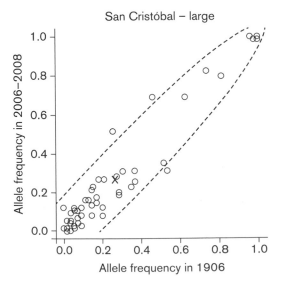

San Cristóbal – large

Allele frequency in 1906

FIGURE 8.1. Comparison of allele frequencies at 16 microsatellite loci in mockingbird populations (*Mimus trifasciatus* and *Mimus melanotis*) in Galápagos in 1906, and approximately 100 years later in the same populations. Open circles represent the frequencies of each allele, the x represents the average allele frequency in each population across all 16 loci, and the dashed lines represent the 95% normal probability contour ellipses. The smallest population (Champion Island) shows the greatest differences in allele frequencies (widest 95% contour ellipses) and the lowest proportion of rare alleles (alleles with frequencies <0.1), followed by the medium-sized population on Gardner Island by Floreana and the large population on Cristobal Island. Note that the average allele frequency across all loci did not change appreciably over time in any of the populations.

BOX 8.2 · Genetic Management of Reintroduced Alpine Ibex

Alpine ibex (*Capra ibex*) are a wild goat species endemic to the European Alps that nearly went extinct in the eighteenth century due to overhunting. This near extinction led to a strong bottleneck, with only one population of fewer than 100 individuals surviving in northern Italy, in what is now the Gran Paradiso National Park (Grodinsky and Stüwe 1987). Alpine ibex recovered from this first known bottleneck and by the beginning of the twentieth century the population numbered about 2,000 individuals (Maudet et al. 2002). Starting in 1906, a total of 88 young ibex were taken from the sole remaining population in Italy and raised in two zoos in Switzerland (Stüwe and Nievergelt 1991). Individuals produced from this captive breeding stock were used for later reintroductions in Switzerland and elsewhere.

In Switzerland, approximately 50 ibex populations were reintroduced between 1911 and 1984. Other countries also reintroduced Alpine ibex, and today an estimated 40,000 individuals inhabit the region, 14,000 of which can be found in Switzerland (Shackleton and IUCN/SSC Caprinae Specialist Group 1997). These reintroductions created up to four bottlenecks in addition to the one, which drove the species precipitously close to extinction (Biebach and Keller 2009). The first additional bottleneck occurred when the ibex kids were transferred to Swiss zoos. Reintroductions of zoo-bred individuals into the wild caused the second bottleneck. The third took place when three of these wild, but captive-founded, populations (Mont Pleureur, Albris, Brienzer Rothorn) served as the main sources for almost all subsequently founded populations. Some of these wild-founded populations served as further source populations, leading in some cases to an additional fourth bottleneck. These serial founder events with iterative periods of small population size resulted in strong genetic drift and inbreeding in Alpine ibex populations (Biebach and Keller 2009, 2010, 2012).

The combination of many replicated reintroductions and detailed demographic data over several decades makes the Alpine ibex an ideal study system to explore the genetic consequences of reintroductions. In the following we will highlight some of the ways in which the reintroductions affected the genetic composition of these populations and their subsequent development.

Genetic Contribution of Founders

In a reintroduction, not all founders survive long enough to contribute genetically to the next generations (Jule et al. 2008). Hence, the number of founders that contribute genetically to a reintroduced population may differ substantially from the number of released individuals. That was the case in Alpine ibex, even after accounting for the loss of genetic lineages through genetic drift after the reintroductions (Anderson and Slatkin 2007). Only about half of all the released individuals contributed genes to the next generations, probably due to heavy post-release mortality and lack of successful reproduction (Biebach and Keller 2012). However, the release of closely related founder individuals could also contribute to this pattern. Since the genetic composition of reintroduced populations will depend on the number of founders that contribute genetically to the next generations and not on the number of released founders, post-release mortality, lack of successful reproduction, and the relatedness of the founders need to be considered when planning reintroductions.

Genetic Drift

Alpine ibex populations exhibit the genetic structure typical of populations shaped by strong genetic drift: they differed substantially in their allele frequencies and showed significant genetic differentiation from each other, even though they all descended from the same original source population (Biebach and Keller 2009). The population differentiation reflected the reintroduction history rather than the biology of ibex (figure 8a): there are three main clusters representing the three captive-founded populations and their descendant populations. Thus, even 100 years after the beginning of the reintroductions, genetic drift caused by the reintroduction history was the main determinant of today's genetic structure. Decisions made during reintroductions thus affect the genetic makeup of reintroduced populations for a very long time.

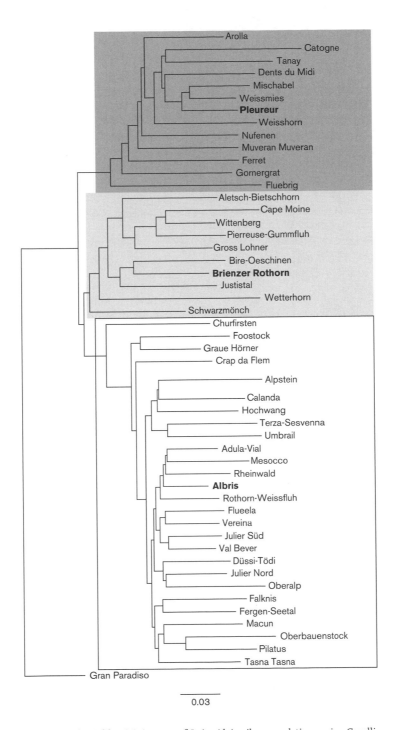

0.03

8a Unrooted neighbor-joining tree of Swiss Alpine ibex populations using Cavalli-Sforza's and Edwards' chord distance (D_c). The ancestral Gran Paradiso population was used as the outgroup or source population. The three captive-founded populations (in bold) and their descendant populations represent the three main clusters (dark gray, light gray, and white). For details, see Biebach and Keller (2009).

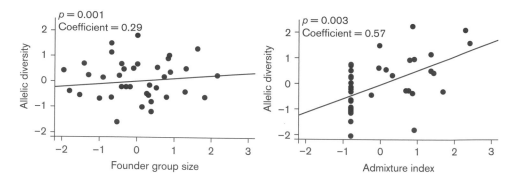

8b Effects of founder group size (left) and admixture of the founder group (right) on allelic diversity in 41 Swiss Alpine ibex populations. All variables are scaled to a mean of 0 and a standard deviation of 1; hence, the regression coefficients given in the top left corner of each panel can be compared directly. For details, see Biebach and Keller (2012).

Effective Population Size

The most straightforward way to quantify the strength of genetic drift is to estimate the effective population size, N_e. We used linkage disequilibrium at neutral loci (Waples and Do 2008) to estimate the effective population size of Alpine ibex populations at the present time (Biebach and Keller 2010). The average N_e of Swiss Alpine ibex populations was about 100, indicating that extant ibex populations are losing on average 0.5% of their expected heterozygosity per generation. As expected for N_e estimates from only one sampling period, confidence intervals were large. Interestingly, current population sizes were not a good predictor of N_e in Alpine ibex, because only 16% of the variance in N_e among populations was explained by current population sizes. This highlights the need for genetic studies in addition to careful demographic monitoring.

Loss of Genetic Variation

Alpine ibex populations generally exhibited low but differing levels of genetic variation. The observed levels of genetic variation were in part a consequence of the founder group composition. Both the founder group size and the admixture of the founder group affected genetic variation, with admixture, i.e., the contribution of the different source populations in Switzerland to the founder group, being twice as important as the founder group size (figure 8b). This highlights that releasing animals from different sources can be very important for the maintenance of genetic varia-tion, even in reintroductions where all individuals ultimately stem from a single ancestral population and where admixture is thus only between sub-populations created by the reintroduction management.

In addition, observed levels of genetic variation declined with increasing numbers of bottlenecks. In line with expectations, heterozygosity declined continuously across bottleneck number, but allelic diversity was lost at a greater rate in earlier bottle-necks owing to the preferential loss of rare alleles (Wright 1931, Maruyama and Fuerst 1985a, Allendorf 1986).

The combined Swiss ibex populations and the ancestral Gran Paradiso population exhibited very similar levels of genetic variation, indicating that the estimated 88 ibex that were brought from Gran Paradiso to Switzerland were sufficient to transfer most of the genetic variation. However, each Swiss population on its own exhibited lower genetic variation than the Gran Paradiso popula-tion, showing that none of the Swiss populations received the total genetic variation brought into the captive breeding programs in Switzerland. Instead, this variation was split up between rein-troduced ibex populations, with genetic drift resulting in further loss of genetic variation within each population.

Inbreeding

Substantial inbreeding accumulated in the Swiss Alpine ibex populations in the approximately 12 generations since the beginning of the reintroductions (Biebach and Keller 2010).

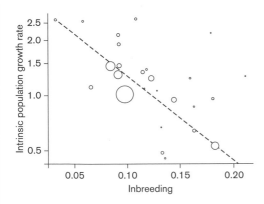

8c Effect of inbreeding (measured as population-specific F_{ST}) on intrinsic population growth rate (measured as the log-transformed population growth rate, ro). The sizes of the circles represent the weights of the data points in the regression analysis.

The average inbreeding coefficient was 0.11, nearly as high as that expected from one generation of half-sib mating. Inbreeding varied substantially among populations as a function of the size of the founder group, its admixture, and the harmonic mean population size over time (see equation 1).

Inbreeding Depression

Inbreeding depression was evident at both the individual and the population level. At the individual level, inbreeding measured as multi-locus heterozygosity at 37 microsatellite loci revealed inbreeding depression in several fitness-related traits (body mass, horn length, and parasite load) in adult males in the Gran Paradiso population (Brambilla et al. 2015). The population-level analysis also revealed inbreeding depression: more inbred populations grew more slowly than less inbred populations, when other factors, such as climatic variation, were accounted for (figure 8c; Bozzuto et al., forthcoming).

viduals survived long enough to contribute genes to the following generations (Box 8.2). Thus, from a genetic point of view, the bottlenecks in Alpine ibex associated with the reintroductions were twice as pronounced than expected from the number of released founders. Reintroductions therefore generally lead to periods of pronounced genetic drift in the newly founded populations, a consequence termed the "founder effect." As expected from the effects of genetic drift outlined above, founder effects cause drastic changes in allele frequencies, heterozygosity, and number of alleles at a locus. However, heterozygosity and the number of alleles react differently to the founder effect.

To see why, imagine a reasonably large population that has 10 alleles at a particular locus with a heterozygosity of 0.8. Two founders are taken from this population to create a new, reintroduced population. For simplicity, we assume that the population immediately grows to a large population size one generation after founding. This is admittedly an unrealistic scenario, but it illustrates the general principle.

More realistic scenarios can be found in Nei et al. (1975) and Denniston (1978). Since heterozygosity declines in small populations at a rate of $\frac{1}{2N_e}$ each generation, the reintroduced population will have lost $\frac{1}{2*2} = 25\%$ of its heterozygosity, retaining a heterozygosity of 0.6 at the hypothetical locus. The expected number of alleles surviving at this same locus is $m - \sum_j (1-p_j)^{2N}$, where m is the number of alleles before the founder event, p_j is the frequency of the jth allele prior to the founder event, and N is the number of founder individuals (Denniston 1978). Under this expectation, alleles at previously high frequencies have a high likelihood of being present within surviving individuals and a low probability of loss. Rare alleles, in contrast, are unlikely to be represented among the founders unless a large number of founders are used. In our example above with $m = 10$, we would expect on average 7.5 alleles to be lost in a founder event with two founders, given reasonable assumptions about the allele frequencies p_j. Thus, the relative loss of alleles through the founder effect (75%) is much more pronounced than the relative loss of heterozygosity (25%).

Reintroductions sometimes result in serial founder effects (Box 8.2). While heterozygosity is predicted to decline by $\frac{1}{2N_e}$ through each of these founder effects regardless of levels of heterozygosity, the loss of alleles will slow down. As many rare alleles are lost through the first founder effect, those alleles that remain are at higher frequency. Therefore, they are lost at a reduced rate through further founder effects (Biebach and Keller 2009). Highly endangered species that have gone through a global bottleneck before becoming the focus of active conservation efforts may have already lost most rare alleles. In such situations, further bottlenecking will lead to little additional loss of alleles (Box 8.2) (e.g., Taylor and Jamieson 2008, Brekke et al. 2011). Note that this does not imply that there is no further genetic drift or no further loss of genetic variation. Heterozygosity, and hence genetic variation, continues to be lost through genetic drift through each additional founder effect (Box 8.2).

Genetic Drift Can Increase the Frequency of Deleterious Alleles

All populations harbor deleterious alleles that arise continuously through mutation. In large populations, such deleterious alleles are normally found at low frequencies because selection operates against them (Whitlock and Bürger 2004). However, in small populations, selection is less effective, and genetic drift can even lead to an increase in the frequency of deleterious alleles. Specifically, when the detrimental effects of an allele are less than one over twice the effective population size (i.e., $s \leq \frac{1}{2N_e}$, where s is the selection coefficient and N_e the effective population size), the frequency of an allele is more strongly affected by genetic drift than by selection (Lynch 1996). When that is the case, harmful alleles can increase or decrease in frequency, depending on the vagaries of genetic drift. The majority of harmful alleles will be lost through genetic drift, but

some will become frequent, perhaps even fixed (Whitlock and Bürger 2004). The increased frequency of harmful alleles in small populations reduces the average fitness of these populations. For this reason, the phenomenon is also called "drift load" (Willi et al. 2013). The probability that harmful alleles drift to high frequencies increases with the length of time a population is small. Populations that have been small for a long period due to habitat fragmentation or repeated bottlenecks may be particularly vulnerable to fitness loss through this process. If the fitness loss is large enough, a population may not be able to sustain itself and will decline to extinction (Whitlock and Bürger 2004).

Unfortunately, the effects of the increased frequency of harmful alleles are difficult to detect empirically in reintroduced populations. For one, it takes tens or hundreds of generations for the effects of drift load to become apparent (Whitlock and Bürger 2004), so most reintroductions are not yet old enough to show such effects. Second, it is difficult to conclusively show that fixation of deleterious alleles is the cause of an observed fitness decline, because crossings among populations are required to demonstrate this, something that is not possible in most populations of conservation concern. However, Luquet et al. (2011, 2013) have recently proposed a method that yields estimates of the relative magnitude of drift load in natural populations. Applications of this methodology to reintroduced populations are still lacking. Hence, at present there is no empirical study of the effects of drift load in reintroduced populations. However, given the theory and the empirical results from laboratory crosses (Coutellec and Caquet 2011, Willi et al. 2013), it seems prudent to assume that reintroduced populations could also suffer from increased frequencies of harmful alleles given their often prolonged small size.

Loss of Genetic Variation Reduces Evolutionary Potential

No environment is constant: predators and competitors come and go, novel diseases emerge, habitat conditions change. To some degree, organisms can cope with such changes by being phenotypically plastic, that is, by adjusting their physiology, morphology, or behavior to the environmental conditions they experience. For example, great tits (*Parus major*) have managed so far to adjust their breeding phenology to warmer spring temperatures by laying their eggs earlier (Vedder et al. 2013). However, not all species show such phenotypic plasticity. The breeding phenology of roe deer (*Capreolus capreolus*), for example, has not tracked climate change over the past 27 years and roe deer hence suffer reduced fitness (Plard et al. 2014). In the absence of phenotypic plasticity, adaptation to changing environmental conditions requires adaptive evolutionary change (see part III in Carroll and Fox 2008). Hence, one major goal of reintroductions is to provide newly established populations with the necessary raw material for adaptive evolutionary change in response to selective challenges in their environment (e.g., Lynch 1996). This ability for adaptive evolutionary change is called evolutionary potential (e.g. Frankham et al. 2010) or adaptive potential.

The raw material for evolutionary potential is genetic variation (Frankham et al. 2010, 42) with higher variation increasing evolutionary potential. Genetic variation is important for evolutionary potential in two ways. First, the degree to which traits that are affected by many genes can adapt in response to selection is a function of the heterozygosity at the underlying genetic loci. This dependence on heterozygosity arises from the fact that the additive genetic variance (V_A), that is, the genetic variance upon which selection can act, is proportional to the expected heterozygosity (see Falconer and Mackay 1996, Frankham et al. 2010, for a technical explanation). Thus, when reintroduced populations go through bottlenecks and lose heterozygosity, they also lose V_A and hence evolutionary potential (Franklin 1980).

A second way in which genetic variation is important for evolutionary potential is through allelic diversity, the number of alleles present at a locus. Allelic diversity is important because it affects the long-term adaptive responses of a population (Allendorf 1986). It does so through rare alleles, which are found more frequently in large populations (see figure 8.1 for an empirical example). Rare alleles do not contribute substantially to V_A and, hence, immediate response to selection; however, if favorable, they may be important for the long-term adaptive response of a population (Robertson 1960, James 1970, Frankham 1980, Allendorf 1986).

A recent simulation study found that measures based on heterozygosity and V_A predicted short-term response to selection best, while allelic diversity was the best predictor of the total long-term adaptive response to selection (Caballero and Garcia-Dorado 2013). The reason that allelic diversity captures the long-term responses to selection better than V_A and heterozygosity is that long-term response to selection depends on the input of new mutations. The number of new mutations is better captured by allelic diversity than heterozygosity. The explanation for this is somewhat technical, and can be found in Caballero and Garcia-Dorado (2013). The important conclusion is that both heterozygosity and allelic diversity must be considered to conserve a population's evolutionary potential in the long run. While no data exist from reintroduced populations, experiments with bottlenecked *Drosophila* support this view: changes in heterozygosity alone were not sufficient to predict adaptive responses after extreme bottlenecks, suggesting an important role for allelic diversity (Swindell and Bouzat 2005).

Ensuring sufficient evolutionary potential in the long run must therefore be an important goal of every reintroduction. Interestingly, ensuring high heterozygosity and allelic diversity may also have immediate payoffs for reintroductions. A meta-analysis of several plant and animal species demonstrated that in 17 out of 18 cases more diverse populations had a greater chance of successful establishment in the first few generations after founding, and this effect was most pronounced among populations established in the wild (Forsman 2014).

Thus, maintaining high allelic diversity and high heterozygosity may have both short- and long-term benefits.

Reintroductions have attempted to maximize the adaptive potential of new populations through a number of strategies, such as admixing founders from different populations (e.g., *Notiomystis cincta*, Ewen et al. 2011). Admixed populations frequently display greater genetic variation than populations from a single source. In Alpine ibex, the positive effect of admixture was pronounced: it had two to three times the impact of the number of released individuals on metrics of genetic variation (Biebach and Keller 2012).

Inbreeding

Because reintroduced populations are typically small to begin with, and in small populations animals often will have no other choice than to mate with relatives, inbreeding is another important genetic issue that needs to be considered when planning and executing reintroductions. Inbreeding has received a lot of attention in conservation biology over the past few decades. We will give a brief overview here, and a more detailed account of inbreeding can be found in Keller et al. (2012).

Inbreeding is quantified using the inbreeding coefficient (commonly denoted by F or f), which ranges from 0 to 1. For example, offspring of a full-sib mating or a parent-offspring mating have $F = 0.25$ and offspring of a cousin mating have $F = 0.0625$. The inbreeding coefficient estimates the proportion of loci across the genome of a diploid organism at which the two alleles are identical by descent (IBD; i.e., they are copies of the same ancestral DNA sequence) as a result of the recent common ancestry of its parents (Charlesworth and Willis 2009). It is impossible to trace all ancestors all the way back and thus inbreeding is always a measure relative to something and not a measure on an absolute scale (see "Techniques for Measuring Important Genetic Variables in Reintroductions" later in this chapter, and

Keller and Waller 2002). It is crucial to be aware of this relativity to avoid confusing different meanings of inbreeding. For more details about the different meanings of the term inbreeding, see Keller et al. (2012).

In reintroduced populations that have undergone periods of small population size, inbreeding will be unavoidable simply due to the small pool of breeding individuals. Inbreeding will arise in reintroduced populations even in the presence of inbreeding avoidance mechanisms in the species' behavioral repertoire. Even if such avoidance behaviors exist, inbreeding will still occur relative to the original founder population. Such inbreeding due to restricted population size is also known as the "inbreeding effect of small populations" (Allendorf et al. 2013). Under random mating, inbreeding is expected to increase per generation by $\frac{1}{2N_e}$ (N_e is the effective population size, Box 8.1). As a consequence, inbreeding will accumulate over multiple generations according to

$$F_t \approx 1 - e^{\frac{-t}{2N_e}}, \qquad (1)$$

where t is the number of generations and N_e is the harmonic mean population size over different years (Crow and Kimura 1970). A reintroduced population of Mauritius kestrel (*Falco punctatus*) exemplifies well this accumulation of inbreeding over time. As the population grew from 12 to 154 birds, the inbreeding coefficient increased from $F = 0.02$ to $F = 0.173$ (Ewing et al. 2008). This was equivalent to a 2.6% increase in inbreeding per generation. Equation (1) could be taken to mean that any population would get completely inbred with time. However, this would only be the case in the absence of immigration and mutation, which reduce inbreeding in the short and long run (Höglund 2009).

Inbreeding Reduces Fitness

In most species, including in species that inbreed regularly, inbreeding has harmful effects on fertility, survival, and other fitness-related traits, a phenomenon summarily referred to as inbreeding depression (Keller and Waller 2002, Kristensen and Sørensen 2005, Charlesworth and Willis 2009). The primary reason for inbreeding depression is that numerous mildly deleterious recessive alleles become expressed in homozygous inbred individuals (Charlesworth and Willis 2009). The extent of inbreeding depression varies greatly, especially among populations of small size (e.g., Ralls et al. 1979), because there is a large stochastic element in which deleterious alleles are present in a population. Thus, variation in inbreeding depression is as much a "rule" as its presence on average (Keller et al. 2012).

A long list of fitness-related traits have been demonstrated to be negatively affected by inbreeding, including morphological traits, which can serve as an assessment of general condition (see Box 8.2), physiological traits, and life-history traits directly linked to fitness such as fecundity and survival (Crnokrak and Roff 1999). Unsurprisingly, the direct components of fitness are more susceptible to inbreeding depression than proxies such as body mass (Forstmeier et al. 2012). Inbreeding depression is also known to be environmentally dependent, with its effects strongest at times of high environmental stress such as extreme climatic events (Keller et al. 1994, Armbruster and Reed 2005, Fox and Reed 2011).

Inbreeding depression is by no means a syndrome to which reintroduced populations are immune. In a population of North Island robins (*Petroica longipes*), for instance, translocated to a New Zealand island, juvenile survival plummeted from 31% for non-inbred offspring (with F [pedigree inbreeding coefficient] of 0) to 11% for offspring with $F = 0.25$ (Jamieson et al. 2007). Likewise, in the Mexican wolf (*Canis lupus baileyi*), where all living individuals stem from three lineages founded by only seven individuals, inbred males of the McBride lineage had an extremely low mating success, to the point of being virtually infertile. Several other traits in fathers, mothers, and pups were also affected by inbreeding (Fredrickson et al. 2007).

When a considerable proportion of individuals in a population suffer from inbreeding depression, this can have effects on population dynamics, although this is not always the case (Keller et al. 2007). Most worrying in the conservation and reintroduction biology context is that inbreeding can decrease the population growth rate (Box 8.2) and increase the likelihood that a population will go extinct. Simulations suggest that the median time to extinction decreases by 37% in populations with an average amount of inbreeding depression (O'Grady et al. 2006). Increased extinction as a consequence of inbreeding depression has been demonstrated empirically in the Glanville fritillary butterfly (*Melitaea cinxia*, Saccheri et al. 1998). Populations that show increased population growth after the introduction of new genetic material (e.g., Westemeier et al. 1998, Hogg et al. 2006) also indicate a previous effect of inbreeding on population dynamics, which is alleviated by the influx of new genetic material. For example, in the South Island robin (*P. australis*) introduction of new genetic material even from an inbred population was sufficient to more than double juvenile survival and to increase recruitment into the focal breeding populations from 59% to 95% (Heber et al. 2012). Thus, inbreeding can have substantial consequences for individual fitness and for population dynamics, illustrating the importance of minimizing inbreeding in reintroduction programs.

As noted earlier, variation in inbreeding depression is very common, and hence not all species, population, or traits show adverse effects of inbreeding (Keller and Waller 2002, Keller et al. 2012). However, it is often difficult to determine whether an apparent lack of inbreeding depression in a particular population is due to a lack of data and statistical power, or whether it represents a true absence of deleterious alleles. Given that inbreeding depression is very common, even in species that habitually inbreed (e.g., Ross-Gillespie et al.

2007) and given the serious potential consequences of inbreeding for reintroduced population, it seems wise to plan and execute reintroductions in ways that minimize inbreeding (Allendorf et al. 2013).

Purging Is Unlikely to Reduce Genetic Load Effectively in Reintroduced Populations

By increasing homozygosity, inbreeding exposes the so-called genetic load of deleterious recessive alleles to natural selection and, hence, may lead to a reduction in the frequency of deleterious alleles (Allendorf et al. 2013). This process, known as genome purging, may appear attractive from a reintroduction point of view because it suggests that inbreeding problems may eventually disappear in reintroduced populations. However, a number of studies have made it clear that the right cocktail of demographic conditions is required for this process to work: population size needs to decline slowly enough during a bottleneck so that inbreeding accumulates gradually, alleles need to be deleterious enough so that selection can act on them ($s \geq \frac{1}{2N_e}$), the population must be sufficiently isolated for these alleles not to be reintroduced with gene flow from neighboring populations, and inbreeding should arise from nonrandom mating rather than random mating (see Keller et al. 2012, for more details and a summary of the literature). In reintroductions, these conditions seem to be met rarely so that conclusive empirical evidence supporting the occurrence of purging in reintroductions remains absent. Although purging undoubtedly can be effective under the right circumstances, these circumstances are rare and purging is thus far too risky a management strategy to be employed systematically (Crnokrak and Roff 1999, Frankham et al. 2001, Keller et al. 2012).

Captive Breeding

Wild source populations are not always available for reintroductions. In those cases, captive-bred individuals are reintroduced into the wild.

Captive breeding is required for a number of diverse reasons (Allendorf et al. 2013). For instance, individuals from a wild source population may not be available in sufficient numbers to found a new population (Box 8.2), or declining wild populations may need supplementation with individuals from captivity, as is often the case in overexploited species of commercial value (Box 8.3).

The genetic issues that arise during a captive breeding program are largely the same as we have discussed so far. Captive populations experience the consequences of genetic drift and inbreeding like wild populations, including loss of genetic variation, increased frequency of harmful alleles, increased inbreeding, and inbreeding depression (Frankham 2008). These effects are well documented in captive populations, e.g., in captive-bred populations of the Mexican wolf (Fredrickson et al. 2007, Hedrick and Fredrickson 2008; see also Box 8.3).

However, one genetic process differs in important ways between captive and wild populations: selection. Selective pressures differ between captive and natural environments, leading to relaxed natural selection against deleterious alleles in captivity and to adaptation to captivity (Allendorf et al. 2013, 405–410). These effects can be detrimental to reintroduction success (Frankham 2008), and may be among the reasons for the lower success rates of reintroductions using individuals raised in captivity rather than in the wild (13% vs. 31% [Fischer and Lindenmayer 2000]; 28% vs. 75% [Griffith et al. 1989]; 50% vs. 71% [Wolf et al. 1996]). In the following, we briefly discuss the genetic issues arising from the changed selective pressures in captivity.

Adaptation to Captivity Is a Serious Problem, Which Increases with Duration in Captivity and N_e

Selection pressures in captivity differ from those in the natural habitat, leading to genetic adaptation to the captive environment, a process that has been documented in many

BOX 8.3 · Genetic Management of Anadromous Steelhead Trout

Steelhead trout (*Oncorhynchus mykiss*) are a migratory life-history form of the resident rainbow trout. Steelhead trout hatch in freshwater streams, migrate to the sea, and return as adults 3–6 years later to their natal freshwater stream for spawning. The return year of fish is called the run year and the year after, the brood year (figure 8d). In the Hood River in Oregon, United States, two distinct steelhead populations are abundant, which migrate at different times of the year (summer run and winter run). Both populations breed in spring, but in different forks of the river (Kostow 2004). There is little or no hybridization between the winter- and summer-run populations, but interbreeding between the two life-history forms of *O. mykiss* (steelhead and rainbow trout) does occur in the Hood River (Araki et al. 2007a).

Hatchery Programs

Demographic supplementation of the steelhead trout populations in the Hood River started with a so-called traditional hatchery program. A collection of fish from other river populations were used to build a brood stock, with the brood stock for the winter and summer runs founded in 1941 and the 1950s, respectively (Araki et al. 2007a). Offspring from this brood stock, which was in captivity for multiple generations and originated from nonlocal populations, were released into the Hood River. This traditional hatchery program was phased out in the 1990s and replaced with a supplemental hatchery program that started in 1991. In the supplemental hatchery program, wild fish from local populations in the Hood River were caught each year anew and used as brood stock. Offspring of these wild-caught fish were reared in a hatchery in the Hood River and released as smolts after 1 year (Araki et al. 2007a). This hatchery program therefore involves the release of local fish, which remain in captivity only during the freshwater phase of their life cycle.

A dam in the Hood River acts as a complete barrier to all salmonids, where all returning adults are counted, trapped, and sampled for DNA analysis. Additionally, various data such as body size, run time, age, and gender are recorded (Araki et al. 2007a). After being handled, fish are either taken as brood stock, allowed to pass through the dam to continue on to the spawning grounds, or released downstream. Of the fish allowed to pass through the dam, numbers of hatchery-born fish are not allowed to exceed numbers of wild-born fish, to avoid a predominance of hatchery-born fish in the gene pool of the spawning fish (Araki et al. 2007a).

Fitness of Local versus Nonlocal Brood Stock

The reproductive success (RS, hereinafter) of released fish raised in the traditional hatchery (H_{trad}) was 62% lower than of fish raised in the supplemental hatchery (H_{supp}; Araki et al. 2007a). Fish experienced similar environments in both hatchery systems and thus it seems reasonable to conclude that most of this large difference in RS between fish from H_{trad} and H_{supp} is of genetic rather than environmental origin (Araki et al. 2007a). Released fish from H_{trad} could carry maladapted genes either because the brood stock originated from a population that was adapted to a different riverine environment or due to adaptation to the captive environment. It is likely that both mechanisms contributed to the lower RS of fish from H_{trad} than H_{supp}. These results exemplify the risk of releasing individuals that are not adapted to the reintroduction site.

Genetic Consequences for Fitness of Captive Rearing

To investigate the genetic effects of hatcheries on the fitness of fish, RS of fish with one parent reared in the wild and one parent reared in captivity was compared to RS of fish from two wild parents (figure 8d, study design 2) (Araki et al. 2007b). Both fish types were reared under standard conditions. Therefore, the only difference between the two fish types was that either none or half of the genome came from a captive-reared parent. Fish were released as juveniles and returned 2–5 years later for spawning, when RS was measured. RS of fish with one captive-reared parent was only 55% that of fish with two wild parents (figure 8e, panel A). These results together with data from four other salmonid hatchery stocks with different numbers of generations in captivity show a 37.5%

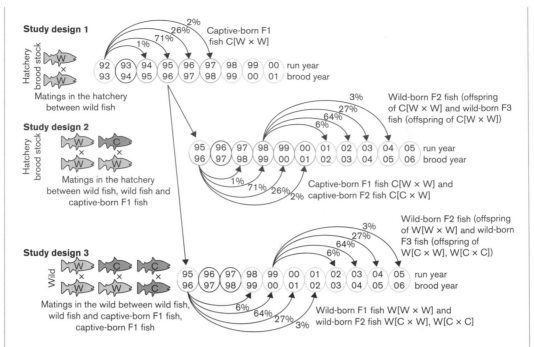

8d Study design used to investigate the genetic consequences of adaptation to captivity in steelhead trout. Light gray denotes wild-born fish and dark gray captive-born fish. Captive-born fish were born and reared in the hatchery and released as juveniles into the wild. Wild-born fish spent their entire lives in the wild. Numbers in circles show the run years in which steelhead trout parents returned (top) and their brood year (bottom). The proportion of adults returning in each year is indicated next to each arrow. Note that these proportions differ between wild-born (arrows above) and captive-born fish (arrows below). Study design 1 created captive-born fish from two wild parents (C[W × W]). Study design 2 quantified the genetic effects of being raised in captivity on reproductive success by crossing captive-born individuals with wild fish (C[C × W]) and comparing their reproductive success to fish with two wild-born parents (C[W × W]). Reproductive success was estimated as the number of returning offspring. Study design 3 extended the analysis of an additional generation to wild-born descendants of wild- and captive-born fish (W[W × W], W[C × W] and W[C × C]). For clarity, only a subset is shown of the years in which the studies were conducted.

SOURCE: After Araki et al. (2007b), Araki et al. (2009), and Christie et al. (2012a).

fitness decline per captive-reared generation (Araki et al. 2007b). Results of this study show strikingly that captive breeding can lead to substantial fitness reductions, even when fish are only born and reared in captivity and remain in the wild the rest of their lives. Such a substantial fitness decline within only one generation is plausible if domestication selection acts on multiple traits throughout the life cycle (Araki et al. 2008).

Effects of Captive Breeding on Wild Descendants

Further studies showed that the genetic effects of captive breeding are not erased by a full generation in the wild, but are still evident in the second generation after captive breeding (Araki et al. 2009). RS of wild-born fish with either two captive-reared parents, one captive-reared and one wild-born parent, or two wild-born parents was compared (figure 8d, study design 3). Although there was high variation between years, on average wild-born fish with two captive-bred parents had only about one-third of the fitness (37%) of wild-born fish from two wild parents (figure 8e, panel B). Reproductive fitness of wild-born fish with only one captive-reared parent was 87% of the fitness of wild-born fish from two wild parents. In total, this carryover effect of releasing captive-reared fish is estimated to reduce population fitness by 8% relative to a purely wild population of the same size (Araki et al. 2009). This highlights

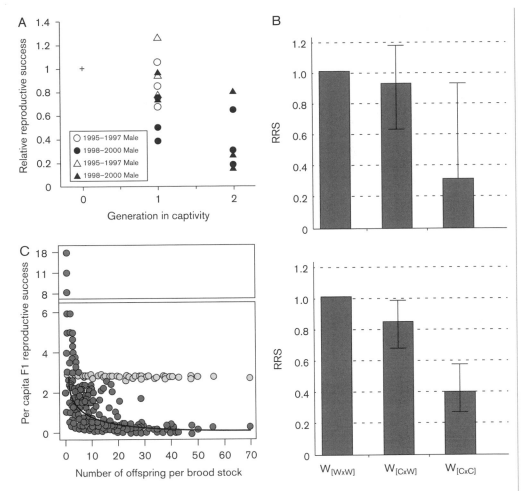

8e A: Relative reproductive success (RS) of captive-reared fish with one captive-born and one wild-born parent (C[C × W]) and with two wild-born parents (C[W × W]). RS was measured relative to wild fish (cross at generation 0). Relative RS of C[W × W] is plotted at generation 1 and relative RS of C[C × W] at generation 2 (after Araki et al. 2007b, reprinted with permission from AAAS).

B: Relative RS of wild-born male (above) and female (below) fish with either one W[C × W] or two W[C × C] captive-born parents relative to fish with only wild parents (W[W × W], relative RS = 1) (after Araki et al. 2009, by permission of the Royal Society).

C: Trade-off between performance in captivity (number of offspring per boodstock) and in the wild (per capita F1 reproductive success). Dark gray points show that wild fish with higher reproductive success in the captive brood stock produced F1 offspring with lower reproductive success in the wild. The solid line represents the fitted model. Light gray points show the expected reduction of per capita F1 reproduction success due to inbreeding depression. This reduction is minimal. Thus, the reduction in F1 performance in the wild is a consequence of adaptation to captivity and not inbreeding depression (after Christie et al. 2012a, with kind permission of PNAS).

that captive rearing can reduce fitness not just in the first, but also in subsequent generations.

Adaptation to Captivity and Inbreeding

Several mutually nonexclusive mechanisms could be the cause for the reduced fitness of hatchery-reared fish and their descendants. The two most plausible mechanisms in this study are inbreeding in offspring of hatchery fish and selection for the captive environment (Araki et al. 2008, Christie et al. 2012a). Both mechanisms were tested in steelhead trout in the Hood River.

In the supplemental hatchery program of steelhead trout, inbreeding arises when captive-reared fish return as adults to the spawning grounds and mate in the wild with half-sibs or full-sibs. Christie et al. (2014) estimated that inbreeding was responsible for 1–4% of the 15% fitness reduction in hatchery fish relative to wild fish.

Adaptation to captivity was studied using designs 1, 2, and 3 in figure 8d (Christie et al. 2012a). Captive-born F1 fish had nearly twice the lifetime RS of wild fish under identical captive conditions (figure 8d, study design 2), suggesting adaptation to captivity. Moreover, a trade-off between performance in captivity and in the wild was evident (figure 8d, study designs 1 and 3). Brood stock of wild fish that had higher fitness in the captive environment produced F1 offspring that performed worse in the wild (figure 8e, panel C). Thus, adaptation to captivity seems to be the primary explanation for the fitness reduction in captive-reared fish.

Reduced N_e Due to Supplemental Reintroduction

Supplementation programs harbor the risk of increasing the population size at the cost of the genetic viability of the population. This can occur because supplementation programs can reduce the effective population size of the supplemented population. In the hatchery program, between 40 and 80 fish were used each year as brood stock, yet the effective number of breeders ranged only from 17 to 37 with a harmonic mean of 25 (Christie et al. 2012b). In the wild, the effective number of breeders was 373. By releasing hatchery-raised fish with much lower genetic variation and higher relatedness, the supplementation program reduced the effective population size: The effective number of breeders of the entire population was only 41.6% of that of a population without the supplementation program, despite the supplementation program nearly doubling the population size (Christie et al. 2012b). These results demonstrate the trade-off between increasing a wild population's size through release of captive-bred individuals and preserving its genetic diversity. Given this trade-off, supplementation programs are best carried out only for a short period of time and when RS of released fish is low (Waples 2004).

taxa (see Box 8.3 for an example, Frankham 2008). Such genetic adaptation to captivity can be a serious problem when trait values that are favored in captivity differ from those in the wild, because the trait values favored in captivity are often disadvantageous once individuals are released (Frankham 2008, Box 8.3).

The rate of genetic adaptation is given by

$$\text{GA}_t \sim Sh^2 \, \Sigma \left(1 - \frac{1}{2N_e}\right)^{t-1}, \qquad (2)$$

where GA is the extent of adaptation to captivity, t the number of generations, Sh^2 the response to selection in the first generation (S = selection differential and h^2 = narrow sense heritability), and N_e the effective population size (Frankham 2008). Hence, the higher N_e, the number of generations in captivity, the selection differential, and the heritability, the greater the extent of genetic adaptation to captivity. Given the potentially serious consequences of adaptation to cap-

tivity, breeding programs need to reduce its impacts as much as possible. We will return to possible strategies later.

Reintroduced Species and Hybridization

Reintroduction success can also be affected by the presence of related species. For example, recovery and reintroduction efforts for the European mink (*Mustela lutreola*), in Europe are hindered by the presence of the invasive American mink (*Neovison vison*), which might outcompete, prey upon, and even mate with the native species, disrupting natural reproductive processes (Maran and Henttonen 1995, Põdra et al. 2013). Potential for hybridization also raises concerns about the genetic integrity of reintroduced species, an additional genetic issue to consider in reintroductions (Allendorf et al. 2013). Hybridization might occur between wild species but also between reintroduced and domesticated forms. For example, the European wild boar (*Sus scrofa*) has been the subject of considerable conservation effort, including translocations. Genes from domestic pigs are present in some populations of wild boars, suggesting that hybridization has occurred (Scandura et al. 2008, Koutsogiannouli et al. 2010). Managing potential hybridization is thus an additional genetic issue that needs consideration when planning and executing reintroductions.

TECHNIQUES FOR MEASURING IMPORTANT GENETIC VARIABLES IN REINTRODUCTIONS

Obtaining Genetic Data

To obtain information about genetic drift, adaptive potential, or inbreeding from molecular markers, suitable material containing DNA needs to be collected from individuals under study. While DNA can be extracted from many parts of an animal, quality and yield differ greatly among them. DNA extracted from teeth, hair, and feces has generally lower quality and yield than DNA from tissue and blood. Note that in mammals, red blood cells are not nucleated, and tissue therefore yields higher quantities of DNA than blood in mammalian species. The issue of DNA quantity has recently gained importance again, because genomic approaches tend to require larger amounts of DNA.

Choice of Genetic Marker

Once samples have been obtained, they need to be genotyped with a suitable set of genetic markers. Changes in sequencing technology are continually advancing the marker types and scale of variation accessible to studies on non-model organisms. In the last decade, conservation genetic studies have moved from genotyping microsatellites at tens of loci to quantifying variation at tens of thousands of single nucleotide polymorphisms (SNP) across the entire genome of endangered species (Miller et al.

2011, Perry et al. 2013). Marker choice has been covered elsewhere (Amato et al. 2009), so we will not repeat this information here.

Quantifying Genetic Drift by N_e

The most straightforward way to quantify the strength of genetic drift is to estimate the effective population size, N_e (Box 8.1). Typically, this is achieved by estimating the change of allele frequencies at a set of loci over several generations. More details about these temporal and other methods to estimate N_e can be found in Luikart et al. (2010). Generally, care should be taken when different N_e estimators are compared, because estimators may refer to different spatial and time scales (Luikart et al. 2010).

Estimating Evolutionary Potential

Adaptive evolutionary potential is a central concept in conservation genetics (see "LOSS OF Genetic Variation Reduces Evolutionary Potential" earlier in this chapter), but its estimation is difficult. Since evolutionary potential is proportional to the additive genetic variance (V_A) of traits under selection, V_A is a measure of evolutionary potential (Lynch 1996) that can be estimated using standard quantitative genetic protocols (Falconer and Mackay 1996). However, in practice this has rarely been done in a conservation context, and we are not aware of a single estimate from a reintroduced population. The reason is that quantitative genetic approaches require detailed phenotypic and pedigree or genotypic data, which are often not available in enough detail in reintroduced populations.

Efforts to estimate evolutionary potential in conservation biology have focused instead on the use of molecular markers. A handful of neutral molecular markers is often not enough to infer adaptive potential (Reed and Frankham 2001). This is changing with the recent increase in the number of markers available to conservation studies through next-generation sequencing techniques. A recent simulation study suggested that genetic diversity measures at 100+ neutral markers might accurately predict evolutionary potential in populations with variable demographic parameters (Caballero and Garcia-Dorado 2013). This finding suggests that we may be able to measure evolutionary potential through neutral markers in the near future.

An alternative approach is to identify regions of the genome that exhibit signs of selection, using, for example, SNP (Keller et al. 2013) or transcriptome data (Angeloni et al. 2011, Smith et al. 2013). However, loci that harbor adaptive genetic diversity in the current environment may not be the ones required for adaptation to a new environment (Allendorf et al. 2013). Thus, it is unclear at present how non-neutral genetic variation can be used to measure evolutionary potential.

Measuring Inbreeding and Inbreeding Depression

Inbreeding is typically measured with the inbreeding coefficient (F). The inbreeding coefficient estimates the probability of two randomly drawn gene copies being IBD in an individual as a consequence of recent ancestry. Depending on the type of data available, this can be accomplished using two complimentary approaches: pedigrees and/or molecular markers.

Quantifying Inbreeding with Pedigrees
If sufficiently detailed long-term records have been kept on the relatedness among individuals, one has the option to estimate the pedigree inbreeding coefficient (F), equivalent to Wright's F_{IT} when it is averaged over all individuals in the population (Wright 1965, Keller and Waller 2002). Once a pedigree is constructed, this measure can be obtained in one of several ways, including the path analysis approach proposed by Wright (1922, 1969; see also Höglund 2009, Allendorf et al. 2013). The usefulness of inbreeding coefficients estimated from a pedigree rests on its degree of completeness, accuracy, and depth. When computed

from a shallow pedigree (one that is only two or three generations deep), the inbreeding coefficients would only reflect the very recent inbreeding that occurred during those two or three generations. Any inbreeding in preceding generations would go unnoticed, because pedigree analysis assumes that the individuals founding a pedigree are outbred (i.e., $F = 0$) and unrelated. For instance, the lack of correlation between the pedigree and molecular marker inbreeding coefficients in a population of captive-bred, critically endangered Mohor gazelle (*Gazella dama mhorr*) was likely due to a serious violation of this assumption of outbred and unrelated founding individuals (Ruiz-Lopez et al. 2009). A study by Hammerly et al. (2013) also illustrates the risk of running a captive breeding program relying solely on inaccurate pedigrees to prevent breeding between relatives. Pedigrees of a captive population of the critically endangered Attwater's prairie chicken (*Tympanuchus cupido attwateri*) suggested that average inbreeding was fairly low ($F = 0.025$), while molecular data revealed much higher degrees of inbreeding ($F = 0.087$) and consequent inbreeding depression. Such inaccuracies in pedigrees can also be caused by extra-pair fertilizations creating false paternity assignments and, hence, error-prone estimates of inbreeding coefficients (Reid et al. 2014). Thus, if calculating F from pedigrees, it is advisable to infer unknown relationships with molecular markers (Ewing et al. 2008) and to check whether these estimates align with those obtained from molecular data (Pemberton 2008, Hammerly et al. 2013, Townsend and Jamieson 2013).

Quantifying Inbreeding with Molecular Markers
Pedigree inbreeding coefficients can only provide an average expectation for the proportion of the genome IBD. However, due to the inherent stochasticity of the segregation of alleles into gametes in meiosis and subsequent uniting of parental gametes during fertilization, the true level of IBD in an individual may deviate from this estimate. Consequently, inbreed-

ing coefficients estimated from molecular markers can come closer to the true IBD by directly assessing the realized genome-wide homozygosity, provided enough molecular markers are available.

Previously, when only a handful of microsatellite loci were used to infer the inbreeding coefficients of individuals, the multi-locus heterozygosity at this set of markers often did not correlate well with that across the genome (Aparicio et al. 2007, Szulkin et al. 2010). With the advent of modern sequencing technologies, it is now feasible to sequence larger portions of the genome of non-model organisms, enabling the estimation of inbreeding with unparalleled accuracy (Kirin et al. 2010, Keller et al. 2011).

Modern sequencing techniques such as restriction-site associated DNA sequencing (RADseq) (Baird et al. 2008) allow thousands of SNPs to be genotyped per individual, providing detailed measures of the realized genome-wide homozygosity. Another method for the estimation of F from molecular data involves looking at runs of homozygosity (ROH), long chromosomal segments (>1 Mb, rarely up to 100 Mb) that are IBD (Keller et al. 2011). The fraction of an individual's genome that is in such ROHs can be used to calculate the inbreeding coefficient. While this technique has so far mostly been used in humans and livestock (McQuillan et al. 2008, Purfield et al. 2012), its excellent performance in estimating inbreeding coefficients, particularly in small and isolated populations having recently experienced a sudden reduction in N_e, suggests that it will likely be incorporated into ecological studies on inbreeding in the near future (Kardos et al. 2015).

Although a smaller set of markers (e.g., tens of microsatellites) may not provide enough power to estimate individual inbreeding coefficients, they are suitable for estimating inbreeding at the population level (Biebach and Keller 2010, Keller et al. 2012, Box 8.2) by estimating F_{IT} through its components F_{IS} and F_{ST} of Wright's F-statistics (Wright 1965, Hartl and Clark 2007). These population genetics met-

rics describe how genetic variance is partitioned across different levels of organization (individuals, subpopulations, and the total population) and allow the estimation of the degree of inbreeding as a function of population structure and consequently of genetic structure.

The three F-statistics are related in the following way (Wright 1969, 295):

$$(1 - F_{IT}) = (1 - F_{IS})(1 - F_{ST}) \qquad (3)$$

F_{IT} quantifies a population's total inbreeding as a function of the combined effects of nonrandom mating within a subpopulation (F_{IS}) and inbreeding due to small population size (F_{ST}). In reintroduced populations of animals, most inbreeding tends to stem from the small population size (F_{ST}) rather than nonrandom mating (F_{IS}). In these cases, population-specific F_{ST} as defined by Vitalis et al. (2001) can be a very useful measure of inbreeding in reintroduced populations (see Keller et al. 2012, for more details). Note that this population-specific F_{ST} is distinct from the well-known pairwise F_{ST} (Weir and Cockerham 1984).

Inbreeding Depression: Fitness as a Function of Inbreeding Coefficient or Heterozygosity

Once inbreeding coefficients have been estimated for individuals in the study population, the magnitude of fitness depression associated with the observed degree of inbreeding can be quantified.

A way to do this is to estimate the inbreeding load as the number of lethal equivalents, a set of recessive alleles whose combined effect when homozygous would be equivalent to that of a lethal allele, i.e., their expression would render the zygote unviable (Keller and Waller 2002, Höglund 2009, Allendorf et al. 2013). This is normally achieved by obtaining the slope of the linear regression of log-survival (or the log of other fitness traits) on individual inbreeding coefficients (Morton et al. 1956, Keller and Waller 2002).

Another commonly implemented method for the detection of inbreeding depression involves testing for a negative correlation between multi-locus heterozygosity (a proxy of the inbreeding coefficient) and fitness measures. For example, Ruiz-Lopez et al. (2012) used such heterozygosity-fitness correlations (HFCs) to quantify inbreeding depression in sperm quality in a captive population of endangered Mohor gazelles. Quantifying inbreeding load from HFCs requires careful analysis and sufficient molecular data, and the reader is referred to this literature for further details (Chapman et al. 2009, Szulkin et al. 2010).

When marker data are available from several populations, inbreeding depression can be estimated across populations. Population structure creates variance in inbreeding coefficients among populations, and thus there may be more power to detect inbreeding depression among populations than among individuals within one population. Such analyses among populations can test for the effects of inbreeding on population dynamics (Box 8.2).

Detecting Hybridization

Detecting hybrids using genetic markers is difficult, particularly beyond the first generation of hybrids (F1), because hybridized individuals share a large part of their genome with one of the parental species (Nussberger et al. 2013). For example, a first-generation backcross (i.e., an offspring of a mating between a F1 and the wild species) has on average 75% of its genes in common with the wild parental species. Hence, detecting the presence of genes from other species in a population of conservation concern requires data from many loci, or from carefully selected loci. The most useful loci are those that are highly differentiated between the species, i.e., here the allele frequency differences between the species are high (high pairwise F_{ST}). Developing a panel of such loci is labor intensive but well worth the effort, because it enables the genetic detection and classification of hybrids and backcrosses without a reference set of "pure" samples of the parental species (Anderson and Thompson 2002). For instance, a panel of 48 SNPs with high pairwise F_{ST} values among parental species ($F_{ST} > 0.8$) allowed the reliable detection of up to third-generation hybrids between European wildcats (*Felis silvestris silvestris*) and domestic cats (*F. s. catus*) in Switzerland (Nussberger et al. 2013).

RECOMMENDATIONS FOR PLANNING AND CONDUCTING A REINTRODUCTION

Prerelease Management Decisions

The Number of Populations to Be Established

The number of populations a reintroduction should establish must be determined on a case-by-case basis. The decision depends on the existence of remnant wild populations, the environments these populations inhabit, and the degree of local adaption to these different environments. Generally, managers should aim for at least two populations in each environment. If genetically diverged lineages exist, two of each should be established (Allendorf et al. 2013, 415). These populations should be demographically independent to protect against demographic stochastic threats and extinction due to catastrophic events (e.g., severe weather, epizootics).

Selecting the Founder Stock: Size and Source Populations

In an ideal reintroduction, the founder stock should harbor sufficient evolutionary potential to ensure the future of the population, and little inbreeding. To maximize both parameters, founders should be sourced from populations with similar environments to the reintroduction site, with the highest heterozygosity, allelic richness, and least inbreeding. Typically this translates to the remnant population(s), or to the population(s) with the least number of past bottlenecks. If more than one source population exists, sourcing from several populations may also help maintain higher levels of genetic diversity within reintroduced populations (Ransler et al. 2011, Biebach and Keller 2012,

Kennington et al. 2012). This even holds for populations that descended from a single source only a few generations ago (Box 8.2). When individuals from different source populations are admixed, the risk of outbreeding depression needs to be considered (Jamieson and Lacy 2012); however, the risk of outbreeding depression is low where potential source populations have the same karyotype, have been isolated for less than 500 years, and occupy similar environments (see Frankham et al. 2011). Under such conditions, admixing individuals in the founder group from different source populations is a suitable management option.

When the environment of the reintroduction site differs from that of the source populations, it is essential to maximize evolutionary potential at the time of establishment (Allendorf et al. 2013, 415). Sourcing animals from different source populations may be more costly, but is particularly important in these conditions (Jamieson and Lacy 2012). Sourcing individuals evenly from several source populations can have a greater positive effect on genetic variation than taking more individuals from a single source (Box 8.2). This applies especially to serially bottlenecked populations. In bottlenecked populations, rare alleles have been already lost and therefore fewer founders already carry most genetic variation present in a single population. Under such conditions, taking individuals from different source populations is the only way to increase genetic variation and, therefore, maximize the potential to adapt to the new environment. This approach has been successfully employed in the reintroduction of the peregrine falcon (*Falco peregrinus*) to the Midwestern United States (Tordoff and Redig 2001). In our view, maximizing evolutionary potential is more important than maintaining "purity" of subspecies or populations for the long-term survival of endangered species, a view substantiated by the experience with the peregrine falcon reintroduction.

To determine the ideal founder group size, it is useful to look at the loss of heterozygosity and allelic richness that will arise from groups of different sizes. In a founder group of 10 effective individuals, 95% heterozygosity of the source population is maintained (see "Genetic Drift" earlier in this chapter); however, the census size this corresponds to will depend on life-history traits and post-release mortality. All life-history characteristics that lower the ratio of the effective to the census population size (N_e/N_c, see Box 8.1) and that affect the loss of heterozygosity in the following generations must be considered. After reintroduction, subsequent generations lose heterozygosity at a higher rate if population growth is slow and if carrying capacity is small. Importantly, loss of allelic diversity during a reintroduction relates not only to the number of founders, but also to the number and frequency of alleles in the source population (Kimura 1955, Allendorf 1986, see "Genetic Drift" earlier in this chapter). In a source population with alleles at a frequency of 0.05, 30 founders have to be taken to have a 95% probability of capturing alleles at this frequency. With alleles at a frequency of 0.01, this number increases to 150 individuals (Allendorf et al. 2013).

Assuming that not all founders contribute genetically to the reintroduced population (see Box 8.2 for an example), we recommend reintroducing at least 60 individuals (Tracy et al. 2011, Groombridge et al. 2012). This should be higher in species with sex-biased breeding systems, low ratio of effective to census population size, low population growth rate, or low expected post-release survival (Allendorf et al. 2013). Notably, post-release mortality seems to be higher among individuals raised in captivity than in the wild (Parker et al. 2012).

The composition of the founder group with respect to age and sex will also influence the genetic constitution of a population and reintroduction success, and thus needs careful consideration (see Robert et al. 2004, for details).

Captive Breeding

The points raised above on managing a reintroduction also apply to establishing captive breeding programs. Thus, here we discuss only points that are specific to captive breeding programs.

Reintroductions cannot always avoid captive breeding. Hence, awareness of the genetic processes that operate in captivity and of the demographic and genetic consequences of captive breeding is paramount for an optimal design of the captive breeding phase (Robert 2009, Allendorf et al. 2013, 396–399).

Determining the size of a founder group for a captive population differs from that of a wild reintroduction in that a genetic goal is often formally stated. Typically a captive population has a specific goal for the amount of genetic diversity that will be captured from the source populations and maintained for a number of years. Common recommendations are to retain at least 90% of the heterozygosity for 100 years (Frankham et al. 2010). A defined genetic goal can be reached in several ways. First, fast population growth allows for a smaller target population size to retain variation. Likewise, a larger founder group size means a smaller target population size is needed. Smaller target population sizes are valuable in captivity in contrast to the wild, as they may save long-term costs, despite a higher upfront investment in sourcing more founders at the outset (Frankham et al. 2010).

Once a captive population has been founded, genetic management should be applied to reduce genetic problems that arise in captivity. The dependence of genetic adaptation to the captive environment on the effective population size (see equation 2) suggests that species with high N_e (e.g., fish) have the greatest risk of adaptation to captivity. In such cases, substantial adaptation to captivity can occur rapidly within the first captive generations (Heath et al. 2003, Box 8.3). As a general rule, the smaller N_e, the less risk of adaptation to captivity. However, small populations suffer from a higher rate of genetic drift. Thus for management, an intermediate population size with N_e small enough to prevent rapid adaptation to captivity but high enough to create a low rate of genetic drift should be considered (figure 8.2). Minimizing both processes is necessary to increase

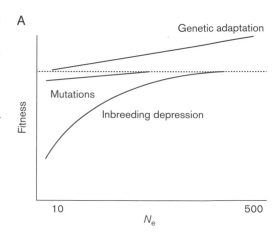

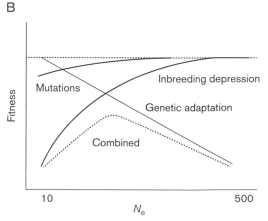

FIGURE 8.2. Expected relationships between fitness and population size (N_e) due to the consequences of inbreeding and inbreeding depression, genetic drift (accumulation of harmful mutations), and genetic adaptation to captivity in captive (A) and after translocation to wild environments (B). The effects of inbreeding depression and genetic drift do not change with the environment. In contrast, genetic adaptation to the captive environment increases fitness in captivity, but is detrimental in the wild environment. The dotted line in the wild environment shows the combined fitness curve for the effects of genetic drift and genetic adaptation, with an optimum at intermediate population sizes.

SOURCE: After Woodworth et al. (2002), with kind permission from Springer Science and Business Media.

reintroduction success, and has been experimentally tested in a study with fruit flies (Woodworth et al. 2002).

The risk of adaptation to captivity is also minimized when the selection differential is reduced. This can be achieved by mimicking natural conditions or by reducing differences in the fitness of individuals, for example, by regulating differences in survival and equalizing reproductive success (Rodríguez-Ramilo et al. 2006). Furthermore, immigration from wild populations into the captive population retards adaptation to captivity, particularly when animals are brought in from the wild in later generations (Frankham and Loebel 1992). However, this option is restricted to selected cases where individuals from the wild population can be safely and ethically removed.

Due to space constraints, we do not discuss more details of the genetic management of captive populations. They are discussed in detail in Frankham (2008), Robert (2009), Williams and Hoffman (2009), Frankham et al. (2010), and Ivy and Lacy (2012).

Post-release Management Decisions

Monitoring

Post-release monitoring is essential (see Chapter 12, this volume). Genetic methods can be helpful in monitoring populations, and sometimes they are the only way to access the necessary information. For instance, genetic paternity testing is the only way to assess reproductive success in species with nonmonogamous mating systems. The minimum level of genetic monitoring we recommend is an estimation of genetic drift. This allows timely responses to strong genetic drift, such as the release of additional founders. For the last decades, an N_e of 50 was considered sufficient to prevent inbreeding depression in the short term (Franklin 1980, Soulé 1980, Jamieson and Allendorf 2012); however, Frankham et al. (2014) recently revised this number upward to an N_e of 100. In addition, the N_e needed to maintain evolutionary potential in the long term is 10 times higher; therefore, managers should try to maximize N_e where possible (Franklin 1980, Frankham et al. 2014).

Supplemental Reintroductions

If existing populations (reintroduced or not) exhibit potential genetic problems, supplemental reintroductions should be considered. Here

we distinguish between supplemental reintroductions where the main goal is to increase census size (demographic supplementation) and those where the main goal is to increase evolutionary potential and reduce inbreeding (genetic supplementation).

Demographic supplementations are usually conducted in commercially exploited populations to prevent population collapse and extinction (Box 8.3). For supplementation, individuals should be sourced from similar populations to avoid maladaptation and where captive breeding is required, the breeding generation should be transferred into captivity anew each generation to avoid adaptation to captivity (Box 8.3). Demographic supplementations, especially supplementation from captive breeding, carry the risk of introducing maladapted genes (e.g., adapted to captivity) and of decreasing N_e, despite increasing the census size, N_c. This decrease of N_e is unlikely to be obvious, as it arises through a high number of introduced individuals representing few breeding individuals (Box 8.3). Thus, it is important to monitor the effects of demographic supplementations genetically to prevent damage to the remnant wild populations.

For genetic supplementations, new genetic material is introduced to decrease the harmful effects of genetic drift. Therefore, the source population should be one that is genetically different, but has a low risk of causing outbreeding depression when admixed with the recipient population (see "Selecting the Founder Stock: Size and Source Populations" earlier in this chapter). If individuals are taken from captive breeding, they should be selected to complement the genetic diversity in the wild. The sex of the supplemented individuals might influence the propagation of the new genes. For instance, in polygynous species females bring a relatively steady influx of new genes into the new population, whereas the contribution of males is highly variable and may be nil.

Hybridization

Managing hybridizing populations requires quantifying the extent of hybridization (see

"Detecting Hybridization" earlier in this chapter) and deciding whether hybridization is potentially threatening to a reintroduced population. Not all hybridization is deleterious, since hybridization can also provide new genetic variation that allows adaptation to new environments (Allendorf et al. 2013). Once it has been established that hybridization has sufficiently strong negative fitness consequences in a reintroduced population or is threatening the integrity of the gene pool, management actions to prevent further propagation of the introgressing alleles can be taken.

Such management action was implemented by the Red Wolf Adaptive Management Plan (RWAMP), which concluded that introgression from sympatrically occurring coyote (*Canis latrans*) posed the biggest threat to the restoration of reintroduced red wolf (*C. rufus*) populations in North Carolina. Thus, captured hybrids as well as pure coyote in the monitored management zones were either sterilized (and left in as "placeholders," preventing their territories freeing up for the colonization of immigrant non-sterile coyote and hybrids) or altogether removed from the population (Gese et al. 2015). Hybrids were defined as animals with >25% coyote ancestry. This threshold was modified to a more conservative 12.5% after 2002 (Miller et al. 2003).

If introgression has advanced to the point where a population has become a hybrid swarm, the focus of conservation efforts might shift to alternative populations that do not have a history of interspecific admixture. In the absence of such alternatives, managers have little choice but to attempt to preserve the residual genetic legacy of the displaced parental species retained in this hybridized population (Allendorf et al. 2001; Epifanio and Nielson 2001).

For instance, male American bison (*Bison bison*) were deliberately crossed with female cattle in the nineteenth century with the aim of boosting beef production. These "beefalo" subsequently mated with reintroduced bison herds, leading to widespread introgression of cattle alleles. Wild hybrids carrying the maternally inherited cattle mitochondrial DNA (mtDNA) have a 7.8% lower body mass on average than those carrying bison mtDNA (Derr et al. 2012). Thus, hybrids likely experience outbreeding depression. A screen of mtDNA and nuclear microsatellite markers showed that while introgression seemed pervasive across the species, its magnitude varied between herds, and four federal conservation herds were found to have no detectable cattle ancestry. There have been calls to selectively cull animals carrying mtDNA haplotypes (Hedrick 2009, Derr et al. 2012), or at least avoid translocating them into less hybridized populations (Halbert and Derr 2007). However, care must be taken to ensure that the benefits of hybrid removal outweigh the loss of unique bison genetic variation retained by these animals at autosomal loci.

There is no consensus on a hybridization threshold below which individuals or populations warrant conservation—this should be evaluated on a case-by-case basis. However, a key point is that genetic tools provide the more accurate means for quantifying hybridization and hence, wherever possible, should be favored over more ambiguous morphological assignments (Glover et al. 2013). The RWAMP provides one such cautionary tale: prior to the introduction of genetic testing, three red wolves were mistakenly classified as hybrids and sterilized, thereby inadvertently reducing the breeding population (Gese et al. 2015).

Of course, when extinction is imminent, managers might decide that compromising genetic integrity by reintroducing closely related individuals is of relatively minor consequence (Jachowski et al. 2015). This may be particularly true when low genetic diversity and inbreeding are contributing to extinction risk. These considerations were the motivation behind the translocation of Texas cougars (*Felis concolor*) to Florida to rescue the small population of Florida panthers (Land and Lacy 2000). In general, however, reintroduction biologists should determine whether closely related forms are present at recipient sites, evaluate whether hybridization is likely to occur, and determine whether this hybridization will put the reintroduction goals at risk.

FUTURE RESEARCH AND DEVELOPMENT

Studies in conservation genetics have increased strikingly in number over the last decades, improving our understanding of genetic problems faced in reintroductions. Fast developments of molecular techniques had a major role in this. For example, the advent of microsatellites made it possible to estimate population genetic parameters for many species. As molecular methods continue to develop, more longstanding questions in conservation genetics can be addressed and estimation of important population genetics parameters in endangered species will become more precise and affordable. With the latest developments in sequencing technology, accurate estimates of inbreeding coefficients without a pedigree are possible and these estimates may even outperform inbreeding estimates from pedigrees (Keller et al. 2011, Speed and Balding 2015). As long as genetic material can be sampled, precise inbreeding estimates of any population can be obtained. Similar arguments can be made for hybridization. However, to investigate the fitness consequences of inbreeding or hybridization, fitness data are necessary. Fitness data will likely be the limiting resource for research in conservation genetics in the future.

Ultimately, conservation biology is concerned with populations, not individuals. However, studies that assess the genetic consequences at the population level remain scarce (Saccheri and Hanski 2006, Keller et al. 2007). Future research should focus on the population-specific consequences of inbreeding, outbreeding, hybridization, and the other genetic issues outlined in this chapter. One important measure for the future persistence of populations is their adaptive evolutionary potential. However, while new molecular methods open a wide range of opportunities to study the genetics of threatened species in detail, it is still not clear how genetic variation needed for adaptation to new or changing environments is best quantified. Therefore, estimating the adaptive evolutionary potential remains a major challenge for future research in conservation biology.

SUMMARY

Reintroductions are an important conservation tool. Reintroduction biologists have to take many decisions, most of which will affect—directly or indirectly—the genetic composition of the released populations. This, in turn, can affect reintroduction success. To maximize reintroduction success in the long term, it is therefore important that reintroduction biologists are aware of the genetic consequences of the decisions they take.

In this chapter, we describe the most important genetic issues that arise in reintroduced populations: genetic drift, loss of evolutionary potential, inbreeding, and hybridization. Genetic drift, the random changes in allele frequencies from one generation to the next, leads to the loss of genetic variation within populations and to increased genetic divergence among populations. Since most reintroduced populations are small for a substantial length of time, genetic drift is often pronounced in reintroductions. The random changes in allele frequencies may lead to the proliferation of deleterious alleles, because selection against deleterious alleles is less efficient when genetic drift is strong. The increased frequency of harmful alleles can reduce the average fitness in these populations. The loss of genetic variation through genetic drift also reduces a population's evolutionary potential (i.e., its ability to adapt to changing environments). Through both processes, strong genetic drift may reduce reintroduced success.

Like genetic drift, inbreeding is a consequence of the small population sizes typical for certain phases of reintroductions. Inbreeding can affect reintroduction success because inbred individuals generally experience reduced fitness. The extent of such inbreeding depression varies greatly among populations and environments. However, on average, individual fitness reductions are pronounced, and evidence exists that inbreeding depression in individual fitness traits affects population dynamics and extinction risks of populations. Thus, reintroduction biologists should minimize inbreeding where possible.

Hybridization between reintroduced individuals and domestic or related wild species may also affect reintroduction success. Managing potential hybridization requires reintroduction biologists to determine whether closely related forms are present at recipient sites, evaluate whether hybridization is likely to occur, and determine whether this hybridization will put the reintroduction goals at risk. Not all hybridization is deleterious, since hybridization can also provide new genetic variation that allows adaptation to new environments. However, if hybridization has sufficiently strong negative fitness consequences or is threatening the genetic integrity of the reintroduced populations, management actions may become necessary. Genetic tools can provide accurate means for identifying hybrids and appropriate management actions.

On occasion, captive breeding is necessary prior to reintroduction. Captive-bred populations face the same genetic risks as wild populations, but they suffer from an additional threat: individuals may adapt to the captive environment and the traits that are favored in captivity may prove disadvantageous in the wild. Captive breeding programs thus should limit adaptation to captive conditions as much as possible, through such measures as limiting the time spent in ex situ breeding.

Reintroductions always affect the genetic composition of reintroduced populations. Reintroduction biologists need to manage these changes in a way that gives reintroduced populations the best long-term survival prospects. This includes giving populations the ability to adapt to future environmental change (Pierson et al. 2015). This is a challenging task, but recent developments in genomic techniques combined with the long-standing theoretical tradition in conservation genetics promise exciting new approaches that will help us to better understand and manage the genetic composition of reintroduced populations.

MANAGEMENT RECOMMENDATIONS

- Source populations should be selected to maximize genetic diversity of the reintroduced population or the captive breeding program.

- Genetic diversity is increased when the founders originate from more than one source population. Therefore, with limited resources it might be better to reintroduce fewer individuals from each of several different source populations than more individuals from a single source population.

- Where possible, founders should be sourced from populations with high heterozygosity, allelic richness, and low inbreeding.

- Where possible, founders should be sourced from populations with a similar environment as the future reintroduction site.

- At least 60 individuals should be released, preferably more.

- Age and sex of founders should be taken into account as they may influence reintroduction success.

- Use captive breeding only if it brings a great advantage in terms of number and genetic diversity of released individuals.

- If captive breeding is necessary, the breeding stock should be kept in captivity for as short a time as possible.

- After reintroduction, populations should be monitored both demographically and genetically.

- Supplemental reintroductions need to minimize the potential negative effects on the recipient population.

LITERATURE CITED

Allendorf, F. W. 1986. Genetic drift and the loss of alleles versus heterozygosity. Zoo Biology 5:181–190.

Allendorf, F. W., R. F. Leary, P. Spruell, and J. K. Wenburg. 2001. The problems with hybrids: setting conservation guidelines. Trends in Ecology and Evolution 16:613–622.

Allendorf, F. W., G. H. Luikart, and S. N. Aitken. 2013. Conservation and the Genetics of Populations. Wiley-Blackwell, Oxford.

Amato, G., R. DeSalle, O. A. Ryder, and H. C. Rosenbaum. 2009. Conservation genetics in the age of genomics. Columbia University Press, New York.

Anderson, E. C., and M. Slatkin. 2007. Estimation of the number of individuals founding colonized populations. Evolution 61:972–983.

Anderson, E. C., and E. A. Thompson. 2002. A model-based method for identifying species hybrids using multilocus genetic data. Genetics 160:1217–1229.

Angeloni, F., C. A. M. Wagemaker, M. S. M. Jetten, H. J. M. O. den Camp, E. M. Janssen-Megens, K. J. Francoijs, H. G. Stunnenberg, and N. J. Ouborg. 2011. De novo transcriptome characterization and development of genomic tools for *Scabiosa columbaria* L. using next-generation sequencing techniques. Molecular Ecology Resources 11:662–674.

Aparicio, J. M., J. Ortego, and P. J. Cordero. 2007. Can a simple algebraic analysis predict markers-genome heterozygosity correlations? Journal of Heredity 98:93–96.

Araki, H., W. R. Ardren, E. Olsen, B. Cooper, and M. S. Blouin. 2007b. Reproductive success of captive-bred steelhead trout in the wild: evaluation of three hatchery programs in the Hood River. Conservation Biology 21:181–190.

Araki, H., B. A. Berejikian, M. J. Ford, and M. S. Blouin. 2008. Fitness of hatchery-reared salmonids in the wild. Evolutionary Applications 1:342–355.

Araki, H., B. A. Berejikian, M. J. Ford, and M. S. Blouin. 2009. Carry-over effect of captive breeding reduces reproductive fitness of wild-born descendants in the wild. Biology Letters 5:621–624.

Araki, H., B. Cooper, and M. S. Blouin. 2007a. Genetic effects of captive breeding cause a rapid, cumulative fitness decline in the wild. Science 318:100–103.

Armbruster, P., and D. H. Reed. 2005. Inbreeding depression in benign and stressful environments. Heredity 95:235–242.

Baird, N. A., P. D. Etter, T. S. Atwood, M. C. Currey, A. L. Shiver, Z. A. Lewis, E. U. Selker, W. A. Cresko, and E. A. Johnson. 2008. Rapid SNP discovery and genetic mapping using sequenced RAD markers. PLoS ONE 3(10):e3376. doi:10.1371/journal.pone.0003376.

Ballou, J. D., M. Gilpin, and T. J. Foose. 1995. Population Management for Survival and Recovery: Analytical Methods and Strategies in Small Population Conservation. Columbia University Press, New York.

Biebach, I., and L. F. Keller. 2009. A strong genetic footprint of the re-introduction history of Alpine ibex (*Capra ibex ibex*). Molecular Ecology 18:5046–5058.

Biebach, I., and L. F. Keller. 2010. Inbreeding in reintroduced populations: the effects of early reintroduction history and contemporary processes. Conservation Genetics 11:527–538.

Biebach, I., and L. F. Keller. 2012. Genetic variation depends more on admixture than number of founders in reintroduced Alpine ibex populations. Biological Conservation 147:197–203.

Bozzuto, C., I. Biebach, S. Muff, A. R. Ives, and L. F. Keller. Inbreeding reduces growth rates of reintroduced Alpine ibex populations. Forthcoming.

Brambilla, A., I. Biebach, B. Bassano, G. Bogliani, and A. von Hardenberg. 2015. Direct and indirect causal effects of heterozygosity on fitness-related traits in Alpine ibex. Proceedings of the Royal Society B: Biological Sciences 282:20141873.

Brekke, P., P. M. Bennett, A. W. Santure, and J. G. Ewen. 2011. High genetic diversity in the remnant island population of hihi and the genetic consequences of re-introduction. Molecular Ecology 20:29–45.

Caballero, A., and A. Garcia-Dorado. 2013. Allelic diversity and its implications for the rate of adaptation. Genetics 195:1373–1384.

Carroll, S. P., and C. W. Fox. 2008. Conservation Biology: Evolution in Action. Oxford University Press, New York.

Chapman, J. R., S. Nakagawa, D. W. Coltman, J. Slate, and B. C. Sheldon. 2009. A quantitative review of heterozygosity-fitness correlations in animal populations. Molecular Ecology 18:2746–2765.

Charlesworth, D., and J. H. Willis. 2009. Fundamental concepts in genetics: the genetics of inbreeding depression. Nature Reviews Genetics 10:783–796.

Christie, M. R., R. A. French, M. L. Marine, and M. S. Blouin. 2014. How much does inbreeding contribute to the reduced fitness of hatchery-born steelhead (*Oncorhynchus mykiss*) in the wild? Journal of Heredity 105:111–119.

Christie, M. R., M. L. Marine, R. A. French, and M. S. Blouin. 2012a. Genetic adaptation to captivity can occur in a single generation. Proceedings of the National Academy of Sciences of the United States of America 109:238–242.

Christie, M. R., M. L. Marine, R. A. French, R. S. Waples, and M. S. Blouin. 2012b. Effective size of a wild salmonid population is greatly reduced by hatchery supplementation. Heredity 109:254–260.

Coutellec, M. A., and T. Caquet. 2011. Heterosis and inbreeding depression in bottlenecked populations: a test in the hermaphroditic freshwater snail Lymnaea stagnalis. Journal of Evolutionary Biology 24:2248–2257.

Crnokrak, P., and D. A. Roff. 1999. Inbreeding depression in the wild. Heredity 83:260–270.

Crow, J. W., and C. Denniston. 1988. Inbreeding and variance effective population numbers. Evolution 42:482–495.

Crow, J. F., and M. Kimura. 1970. An Introduction to Population Genetics Theory. Harper & Row, New York.

Denney, N. H., S. Jennings, and J. D. Reynolds. 2002. Life-history correlates of maximum population growth rates in marine fishes. Proceedings of the Royal Society B: Biological Sciences 269:2229–2237.

Denniston, C. 1978. Small population size and genetic diversity: implications for endangered species. Pp.281–289. In S. A. Temple, ed., Endangered Birds: Management Techniques for Preserving Threatened Species. University of Wisconsin Press, Madison.

Derr, J. N., P. W. Hedrick, N. D. Halbert, L. Plough, L. K. Dobson, J. King, C. Duncan, D. L. Hunter, N. D. Cohen, and D. Hedgecock. 2012. Phenotypic effects of cattle mitochondrial DNA in American Bison. Conservation Biology 26:1130–1136.

Epifanio, J., and J. Nielsen. 2001. The role of hybridization in the distribution, conservation and management of aquatic species. Reviews in Fish Biology and Fisheries 10:245–251.

Ewen, J. G., K. A. Parker, K. Richardson, D. Armstrong, and C. Smuts-Kennedy. 2011. Translocations of hihi *Notiomystis cincta* to Maungatautari, a mainland reserve protected by a predator-exclusion fence, Waikato, New Zealand. Conservation Evidence. 8:58–65.

Ewing, S. R., R. G. Nager, M. A. C. Nicoll, A. Aumjaud, C. G. Jones, and L. F. Keller. 2008. Inbreeding and loss of genetic variation in a reintroduced population of Mauritius Kestrel. Conservation Biology 22:395–404.

Falconer, D. S., and R. F. C. Mackay. 1996. Introduction to quantitative genetics. Longman, Essex, UK.

Fischer, J., and D. B. Lindenmayer. 2000. An assessment of the published results of animal relocations. Biological Conservation 96:1–11.

Forsman, A. 2014. Effects of genotypic and phenotypic variation on establishment are important for conservation, invasion, and infection biology. Proceedings of the National Academy of Sciences of the United States of America 111:302–307.

Forstmeier, W., H. Schielzeth, J. C. Mueller, H. Ellegren, and B. Kempenaers. 2012. Heterozygosity-fitness correlations in zebra finches: microsatellite markers can be better than their reputation. Molecular Ecology 21:3237–3249.

Fox, C. W., and D. H. Reed. 2011. Inbreeding depression increases with environmental stress: an experimental study and meta-analysis. Evolution 65:246–258.

Frankham, R. 1980. The founder effect and response to artificial selection in *Drosophila*. Pp.87–90. In A. Robsertson, ed. Selection Experiments in Laboratory and Domestic Animal. Commonwealth Agricultural Bureaux, Farnham Royal, UK.

Frankham, R. 1995. Effective population-size adult-population size ratios in wildlife: a review. Genetical Research 66:95–107.

Frankham, R. 2008. Genetic adaptation to captivity in species conservation programs. Molecular Ecology 17:325–333.

Frankham, R., J. D. Ballou, and D. A. Briscoe. 2010. Introduction to Conservation Genetics. Cambridge University Press, New York.

Frankham, R., J. D. Ballou, M. D. B. Eldridge, R. C. Lacy, K. Ralls, M. R. Dudash, and C. B. Fenster. 2011. Predicting the probability of outbreeding depression. Conservation Biology 25:465–475.

Frankham, R., C. J. A. Bradshaw, and B. W. Brook. 2014. Genetics in conservation management: Revised recommendations for the 50/500 rules, Red List criteria and population viability analyses. Biological Conservation 170:56–63.

Frankham, R., D. M. Gilligan, D. Morris, and D. A. Briscoe. 2001. Inbreeding and extinction: Effects of purging. Conservation Genetics 2:279–285.

Frankham, R., and D. A. Loebel. 1992. Modeling problems in conservation genetics using captive drosophila populations: rapid genetic adaptation to captivity. Zoo Biology 11:333–342.

Franklin, I. R. 1980. Evolutionary change in small populations. Pp.135–150. In M. E. Soulé and B. Wilcox, eds., Conservation Biology: An Evolutionary-Ecological Perspective. Sinauer Associates, Sunderland, MA.

Fredrickson, R. J., P. Siminski, M. Woolf, and P. W. Hedrick. 2007. Genetic rescue and inbreeding depression in Mexican wolves. Proceedings of the Royal Society B: Biological Sciences 274:2365–2371.

Gese, E. M., F. F. Knowlton, J. R. Adams, K. Beck, T. K. Fuller, D. L. Murray, T. D. Steury, M. K. Stoskopf, W. T. Waddell, and L. P. Waits. 2015. Managing hybridization of a recovering endangered species: the red wolf Canis rufus as a case study. Current Zoology 61:191–205.

Glover, K. A., C. Pertoldi, F. Besnier, V. Wennevik, M. Kent, and Ø. Skaala. 2013. Atlantic salmon populations invaded by farmed escapees: quantifying genetic introgression with a Bayesian approach and SNPs. BMC Genetics 14:74.

Griffith, B., J. M. Scott, J. W. Carpenter, and C. Reed. 1989. Translocation as a species conservation tool: status and strategy. Science 245:477–480.

Grodinsky, C., and M. Stüwe. 1987. The reintroduction of the Alpine ibex to the Swiss Alps. Smithsonian 18:68–77.

Groombridge, J. J., C. Raisin, R. Bristol, and D. S. Richardson. 2012. Genetic consequences of reintroductions and insights from population history. Pp.395–440. In J. G. Ewen, D. P. Armstrong, K. A. Parker, and P. J. Seddon, eds., Reintroduction Biology: Integrating Science and Management. Conservation Science and Practice No. 9. Wiley-Blackwell, Chichester, UK.

Halbert, N. D., and J. N. Derr. 2007. A comprehensive evaluation of cattle introgression into US federal bison herds. Journal of Heredity 98:1–12.

Hammerly, S. C., M. E. Morrow, and J. A. Johnson. 2013. A comparison of pedigree- and DNA-based measures for identifying inbreeding depression in the critically endangered Attwater's Prairie-chicken. Molecular Ecology 22:5313–5328.

Hartl, D. L., and A. G. Clark. 2007. Principles of population genetics. 4th ed., Sinauer, Sunderland, MA.

Heath, D. D., J. W. Heath, C. A. Bryden, R. M. Johnson, and C. W. Fox. 2003. Rapid evolution of egg size in captive salmon. Science 299:1738–1740.

Heber, S., A. Varsani, S. Kuhn, A. Girg, B. Kempenaers, and J. Briskie. 2012. The genetic rescue of two bottlenecked South Island robin populations using translocation of inbred donors.

Proceedings of the Royal Society B: Biological Sciences 280:2012–2228

Hedrick, P. W. 2009. Conservation genetics and North American bison (Bison bison). Journal of Heredity 100:411–420.

Hedrick, P. W., and R. J. Fredrickson. 2008. Captive breeding and the reintroduction of Mexican and red wolves. Molecular Ecology 17:344–350.

Hoeck, P. E. A., J. L. Bollmer, P. G. Parker, and L. F. Keller. 2010. Differentiation with drift: a spatio-temporal genetic analysis of Galapagos mockingbird populations (Mimus spp.). Philosophical Transactions of the Royal Society B: Biological Sciences 365:1127–1138.

Hogg, J. T., S. H. Forbes, B. M. Steele, and G. Luikart. 2006. Genetic rescue of an insular population of large mammals. Proceedings of the Royal Society B: Biological Sciences 273:1491–1499.

Höglund, J. 2009. Evolutionary Conservation Genetics. Oxford University Press, New York.

Ivy, J. A., and R. C. Lacy. 2012. A comparison of strategies for selecting breeding pairs to maximize genetic diversity retention in managed populations. Journal of Heredity 103:186–196.

Jachowski, D. S., D. C. Kesler, D. A. Steen, and J. R. Walters. 2015. Rethinking baselines in endangered species recovery. Journal of Wildlife Management 79:3–9.

James, J. W. 1970. Founder effect and response to artificial selection. Genetical Research 16:241–250.

Jamieson, I. G., and F. W. Allendorf. 2012. How does the 50/500 rule apply to MVPs? Trends in Ecology & Evolution 27:578–584.

Jamieson, I. G., and R. C. Lacy. 2012. Managing genetic issues in reintroduction biology. In J. G. Ewen, D. P. Armstrong, K. A. Parker, and P. J. Seddon, eds., Reintroduction Biology: Integrating Science and Management. Conservation Science and Practice No. 9. Wiley-Blackwell, Chichester, UK.

Jamieson, I. G., L. N. Tracy, D. Fletcher, and D. P. Armstrong. 2007. Moderate inbreeding depression in a reintroduced population of North Island robins. Animal Conservation 10:95–102.

Jule, K. R., L. A. Leaver, and S. E. G. Lea. 2008. The effects of captive experience on reintroduction survival in carnivores: a review and analysis. Biological Conservation 141:355–363.

Kardos, M., G. Luikart, and F. W. Allendorf. 2015. Measuring individual inbreeding in the age of genomics: marker-based measures are better than pedigrees. Heredity 115:63–72.

Keller, I., C. E. Wagner, L. Greuter, S. Mwaiko, O. M. Selz, A. Sivasundar, S. Wittwer, and O. Seehausen. 2013. Population genomic signatures of divergent adaptation, gene flow and hybrid speciation in the rapid radiation of Lake Victoria cichlid fishes. Molecular Ecology 22:2848–2863.

Keller, L. F., P. Arcese, J. N. M. Smith, W. M. Hochachka, and S. C. Stearns. 1994. Selection against inbred song sparrows during a natural-population bottleneck. Nature 372:356–357.

Keller, L. F., I. Biebach, S. R. Ewing, and P. E. A. Hoeck. 2012. The genetics of reintroductions: inbreeding and genetic drift. In J. G. Ewen, D. P. Armstrong, K. A. Parker, and P. J. Seddon, eds., Reintroduction Biology: Integrating Science and Management. Conservation Science and Practice No. 9. Wiley-Blackwell, Chichester, UK.

Keller, L. F., I. Biebach, and P. E. A. Hoeck. 2007. The need for a better understanding of inbreeding effects on population growth. Animal Conservation 10:286–287.

Keller, L. F., and D. M. Waller. 2002. Inbreeding effects in wild populations. Trends in Ecology & Evolution 17:230–241.

Keller, M. C., P. M. Visscher, and M. E. Goddard. 2011. Quantification of inbreeding due to distant ancestors and its detection using dense single nucleotide polymorphism data. Genetics 189:237–249.

Kennington, W. J., T. H. Hevroy, and M. S. Johnson. 2012. Long-term genetic monitoring reveals contrasting changes in the genetic composition of newly established populations of the intertidal snail Bembicium vittatum. Molecular Ecology 21:3489–3500.

Kimura, M. 1955. Random genetic drift in multi-allelic locus. Evolution 9:419–435.

Kirin, M., R. McQuillan, C. S. Franklin, H. Campbell, P. M. McKeigue, and J. F. Wilson. 2010. Genomic runs of homozygosity record population history and consanguinity. PLoS ONE 5:e13996.

Kostow, K. E. 2004. Differences in juvenile phenotypes and survival between hatchery stocks and a natural population provide evidence for modified selection due to captive breeding. Canadian Journal of Fisheries and Aquatic Sciences 61:577–589.

Koutsogiannouli, E. A., K. A. Moutou, T. Sarafidou, C. Stamatis, and Z. Mamuris. 2010. Detection of hybrids between wild boars (Sus scrofa scrofa) and domestic pigs (Sus scrofa f. domestica) in Greece, using the PCR-RFLP method on melanocortin-1 receptor (MC1R) mutations. Mammalian Biology 75:69–73.

Kristensen, T. N., and A. C. Sørensen. 2005. Inbreeding: lessons from animal breeding, evolutionary biology and conservation genetics. Animal Science 80:121–133.

Land, E. D., and R. C. Lacy. 2000. Introgression level achieved through Florida Panther genetic restoration. Endangered Species Update 17:100–105.

Luikart, G., N. Ryman, D. A. Tallmon, M. K. Schwartz, and F. W. Allendorf. 2010. Estimation of census and effective population sizes: the increasing usefulness of DNA-based approaches. Conservation Genetics 11:355–373.

Luquet, E., P. David, J. P. Lena, P. Joly, L. Konecny, C. Dufresnes, N. Perrin, and S. Plenet. 2011. Heterozygosity-fitness correlations among wild populations of European tree frog (Hyla arborea) detect fixation load. Molecular Ecology 20:1877–1887.

Luquet, E., J. P. Lena, P. David, J. Prunier, P. Joly, T. Lengagne, N. Perrin, and S. Plenet. 2013. Within- and among-population impact of genetic erosion on adult fitness-related traits in the European tree frog Hyla arborea. Heredity 110:347–354.

Lynch, M. 1996. A quantitative-genetic perspective on conservation issues. Pp. 471–501. In J. C. Avise and J. L. Hamrick, eds., Conservation Genetics: Case Histories from Nature. Chapman & Hall, New York.

Maran, T., and H. Henttonen. 1995. Why is the European mink (Mustela lutreola) disappearing? – A review of the process and hypotheses. Annales Zoologici Fennici 32:47–54.

Marsden, C. D., R. Woodroffe, M. G. L. Mills, J. W. McNutt, S. Creel, R. Groom, M. Emmanuel et al. 2012. Spatial and temporal patterns of neutral and adaptive genetic variation in the endangered African wild dog (Lycaon pictus). Molecular Ecology 21:1379–1393.

Maruyama, T., and P. A. Fuerst. 1985a. Population bottlenecks and nonequilibrium models in population genetics. II. Number of alleles in a small population that was formed by a recent bottleneck. Genetics 111:675–689.

Maruyama, T., and P. A. Fuerst. 1985b. Population bottlenecks and nonequilibrium models in population genetics. III. Genic homozygosity in populations which experience periodic bottlenecks. Genetics 111:691–703.

Maudet, C., C. Miller, B. Bassano, C. Breitenmoser-Wursten, D. Gauthier, G. Obexer-Ruff, J. Michallet, P. Taberlet, and G. Luikart. 2002. Microsatellite DNA and recent statistical methods in wildlife conservation management:

applications in Alpine ibex (*Capra ibex [ibex]*). Molecular Ecology 11:421–436.

McQuillan, R., A.-L. Leutenegger, R. Abdel-Rahman, C. S. Franklin, M. Pericic, L. Barac-Lauc, N. Smolej-Narancic et al. 2008. Runs of homozygosity in European populations. American Journal of Human Genetics 83:359–372.

Miller, C. R., J. R. Adams, and L. P. Waits. 2003. Pedigree-based assignment tests for reversing coyote (*Canis latrans*) introgression into the wild red wolf (*Canis rufus*) population. Molecular Ecology 12:3287–3301.

Miller, W., V. M. Hayes, A. Ratan, D. C. Petersen, N. E. Wittekindt, J. Miller, B. Walenz et al. 2011. Genetic diversity and population structure of the endangered marsupial *Sarcophilus harrisii* (Tasmanian devil). Proceedings of the National Academy of Sciences of the United States of America 108:12348–12353.

Morton, N. E., J. F. Crow, and H. J. Muller. 1956. An estimate of the mutational damage in man from data on consanguineous marriages. Proceedings of the National Academy of Sciences of the United States of America 42:855–863.

Nei, M., T. Maruyama, and R. Chakraborty. 1975. Bottleneck effect and genetic variability in populations. Evolution 29:1–10.

Nussberger, B., M. P. Greminger, C. Grossen, L. F. Keller, and P. Wandeler. 2013. Development of SNP markers identifying European wildcats, domestic cats, and their admixed progeny. Molecular Ecology Resources 13: 447–460.

O'Grady, J. J., B. W. Brook, D. H. Reed, J. D. Ballou, D. W. Tonkyn, and R. Frankham. 2006. Realistic levels of inbreeding depression strongly affect extinction risk in wild populations. Biological Conservation 133:42–51.

Parker, K. A., M. J. Dickens, R. H. Clarke, and T. G. Lovegrove. 2012. The theory and practice of catching, holding, moving and releasing animals. Pp.105–137. In J. G. Ewen, D. P. Armstrong, K. A. Parker, and P. J. Seddon, eds., Reintroduction Biology: Integrating Science and Management. Conservation Science and Practice No. 9. Wiley-Blackwell, Chichester, UK.

Pemberton, J. M. 2008. Wild pedigrees: the way forward. Proceedings of the Royal Society B: Biological Sciences 275:613–621.

Perry, G. H., E. E. Louis, Jr., A. Ratan, O. C. Bedoya-Reina, R. C. Burhans, R. Lei, S. E. Johnson, S. C. Schuster, and W. Miller. 2013. Aye-aye population genomic analyses highlight an important center of endemism in northern Madagascar. Proceedings of the National Academy of Sciences of the United States of America 110:5823–5828.

Pierson, J. C., S. R. Beissinger, J. G. Bragg, D. J. Coates, J. G. B. Oostermeijer, P. Sunnucks, N. H. Schumaker, M. V. Trotter, and A. G. Young. 2015. Incorporating evolutionary processes into population viability models. Conservation Biology 29:755–764.

Plard, F., J.-M. Gaillard, T. Coulson, A. J. M. Hewison, D. Delorme, C. Warnant, and C. Bonenfant. 2014. Mismatch between birth date and vegetation phenology slows the demography of roe deer. PLoS Biology 12(4):e1001828.

Põdra, M., A. Gómez, and S. Palazón. 2013. Do American mink kill European mink? Cautionary message for future recovery efforts. European Journal of Wildlife Research 59:431–440.

Purfield, D. C., D. P. Berry, S. McParland, and D. G. Bradley. 2012. Runs of homozygosity and population history in cattle. BMC Genetics 13:70.

Ralls, K., K. Brugger, and J. Ballou. 1979. Inbreeding and juvenile mortality in small populations of ungulates. Science 206:1101–1103.

Ransler, F. A., T. W. Quinn, and S. J. Oyler-McCance. 2011. Genetic consequences of trumpeter swan (*Cygnus buccinator*) reintroductions. Conservation Genetics 12:257–268.

Reed, D. H., and R. Frankham. 2001. How closely correlated are molecular and quantitative measures of genetic variation? A meta-analysis. Evolution 55:1095–1103.

Reid, J. M., L. F. Keller, A. B. Marr, P. Nietlisbach, R. J. Sardell, and P. Arcese. 2014. Pedigree error due to extra-pair reproduction substantially biases estimates of inbreeding depression. Evolution 68:802–815.

Robert, A. 2009. Captive breeding genetics and reintroduction success. Biological Conservation 142:2915–2922.

Robert, A., F. Sarrazin, D. Couvet, and S. Legendre. 2004. Releasing adults versus young in reintroductions: Interactions between demography and genetics. Conservation Biology 18:1078–1087.

Robertson, A. 1960. A theory of limits in artificial selection. Proceedings of the Royal Society B: Biological Sciences 153:234–249.

Rodríguez-Ramilo, S. T., P. Morán, and A. Caballero. 2006. Relaxation of selection with equalization of parental contributions in conservation programs: an experimental test with *Drosophila melanogaster*. Genetics 172:1043–1054.

Ross-Gillespie, A., M. J. O'Riain, and L. F. Keller. 2007. Viral epizootic reveals inbreeding depression in a habitually inbreeding mammal. Evolution 61:2268–2273.

Ruiz-Lopez, M. J., N. Ganan, J. Antonio Godoy, A. Del Olmo, J. Garde, G. Espeso, A. Vargas, F.

Martinez, E. R. S. Roldan, and M. Gomendio. 2012. Heterozygosity-fitness correlations and inbreeding depression in two critically endangered mammals. Conservation Biology 26:1121–1129.

Ruiz-Lopez, M. J., E. R. S. Roldan, G. Espeso, and M. Gomendio. 2009. Pedigrees and microsatellites among endangered ungulates: what do they tell us? Molecular Ecology 18:1352–1364.

Saccheri, I., and I. Hanski. 2006. Natural selection and population dynamics. Trends in Ecology & Evolution 21:341–347.

Saccheri, I., M. Kuussaari, M. Kankare, P. Vikman, W. Fortelius, and I. Hanski. 1998. Inbreeding and extinction in a butterfly metapopulation. Nature 392:491–494.

Scandura, M., L. Iacolina, B. Crestanello, E. Pecchioli, M. F. Di Benedetto, V. Russo, R. Davoli, M. Apollonio, and G. Bertorelle. 2008. Ancient vs. recent processes as factors shaping the genetic variation of the European wild boar: are the effects of the last glaciation still detectable? Molecular Ecology 17:1745–1762.

Shackleton, D. M., and IUCN/SSC Caprinae Specialist Group, eds. 1997. Wild Sheep and Goats and their Relatives. Status Survey and Conservation Action Plan for Caprinae. IUCN, Gland, Switzerland.

Shrimpton, J. M., and D. D. Heath. 2003. Census vs. effective population size in chinook salmon: large- and small-scale environmental perturbation effects. Molecular Ecology 12:2571–2583.

Smith, S., L. Bernatchez, and L. B. Beheregaray. 2013. RNA-seq analysis reveals extensive transcriptional plasticity to temperature stress in a freshwater fish species. BMC Genomics 14:375.

Soulé, M. E. 1980. Thresholds for survival: maintaining fitness and evolutionary potential. Pp.151–170. In M. E. Soulé and B. Wilcox, eds., Conservation Biology: An Evolutionary-Ecological Perspective. Sinauer Associates, Sunderland, MA.

Soulé, M. E. 1986. Conservation Biology: The Science of Scarcity and Diversity. Sinauer Associates, Sunderland, MA.

Speed, D., and D. J. Balding. 2015. Relatedness in the post-genomic era: is it still useful? Nature Reviews Genetics 16:33–44.

Stüwe, M., and B. Nievergelt. 1991. Recovery of alpine ibex from near extinction: the result of effective protection, captive breeding, and reintroductions. Applied Animal Behaviour Science 29:379–387.

Swindell, W. R., and J. L. Bouzat. 2005. Modeling the adaptive potential of isolated populations:

experimental simulations using Drosophila. Evolution 59:2159–2169.

Szulkin, M., N. Bierne, and P. David. 2010. Heterozygosity-fitness correlations: a time for reappraisal. Evolution 64:1202–1217.

Tavecchia, G., C. Viedma, A. Martinez-Abrain, M.-A. Bartolome, J. Antonio Gomez, and D. Oro. 2009. Maximizing re-introduction success: assessing the immediate cost of release in a threatened waterfowl. Biological Conservation 142:3005–3012.

Taylor, S. S., and I. G. Jamieson. 2008. No evidence for loss of genetic variation following sequential translocations in extant populations of a genetically depauperate species. Molecular Ecology 17:545–556.

Tordoff, H. B., and P. T. Redig. 2001. Role of genetic background in the success of reintroduced peregrine falcons. Conservation Biology 15:528–532.

Townsend, S. M., and I. G. Jamieson. 2013. Molecular and pedigree measures of relatedness provide similar estimates of inbreeding depression in a bottlenecked population. Journal of Evolutionary Biology 26:889–899.

Tracy, L. N., G. P. Wallis, M. G. Efford, and I. G. Jamieson. 2011. Preserving genetic diversity in threatened species reintroductions: how many individuals should be released? Animal Conservation 14:439–446.

Varvio, S. L., R. Chakraborty, and M. Nei. 1986. Genetic-variation in subdivided populations and conservation genetics. Heredity 57:189–198.

Vedder, O., S. Bouwhuis, and B. C. Sheldon. 2013. Quantitative assessment of the importance of phenotypic plasticity in adaptation to climate change in wild bird populations. PLoS Biology 11(7):e1001605.

Vitalis, R., K. Dawson, and P. Boursot. 2001. Interpretation of variation across marker loci as evidence of selection. Genetics 158:1811–1823.

Waples, R. S. 2004. Salmonid insights into effective population size. Heredity 109:254–260.

Waples, R. S., and C. Do. 2008. LDNE: a program for estimating effective population size from data on linkage disequilibrium. Molecular Ecology Resources 8:753–756.

Waples, R. S., G. Luikart, J. R. Faulkner, and D. A. Tallmon. 2013. Simple life-history traits explain key effective population size ratios across diverse taxa. Proceedings of the Royal Society Series B 280:20131339.

Weir, B. S., and C. C. Cockerham. 1984. Estimating F-Statistics for the analysis of population-structure. Evolution 38:1358–1370.

Westemeier, R. L., J. D. Brawn, S. A. Simpson, T. L. Esker, R. W. Jansen, J. W. Walk, E. L. Kershner, J. L. Bouzat, and K. N. Paige. 1998. Tracking the long-term decline and recovery of an isolated population. Science 282:1695–1698.

Whitlock, M. C., and R. Bürger. 2004. Fixation of new mutations in small populations. Pp.155–170. In R. Ferrière, U. Dieckmann, and D. Couvet, eds., Evolutionary Conservation Biology. Cambridge University Press, New York.

Willi, Y., P. Griffin, and J. Van Buskirk. 2013. Drift load in populations of small size and low density. Heredity 110:296–302.

Williams, S. E., and E. A. Hoffman. 2009. Minimizing genetic adaptation in captive breeding programs: a review. Biological Conservation 142:2388–2400.

Wolf, C. M., B. Griffith, C. Reed, and S. A. Temple. 1996. Avian and mammalian translocations: update and reanalysis of 1987 survey data. Conservation Biology 10:1142–1154.

Woodworth, L. M., M. E. Montgomery, D. A. Briscoe, and R. Frankham. 2002. Rapid genetic deterioration in captive populations: causes and conservation implications. Conservation Genetics 3:277–288.

Wright, S. 1922. Coefficients of inbreeding and relationship. American Naturalist 56:330–338.

Wright, S. 1931. Evolution in Mendelian populations. Genetics 16:0097–0159.

Wright, S. 1965. The interpretation of population structure by F-statistics with special regard to systems of mating. Evolution 19:395–420.

Wright, S. 1969. Evolution and the Genetics of Populations. University of Chicago Press, Chicago, IL.

NINE

Accounting for Potential Physiological, Behavioral, and Community-Level Responses to Reintroduction

*David S. Jachowski, Samantha Bremner-Harrison,
David A. Steen, and Kim Aarestrup*

ACCOUNTING FOR PHYSIOLOGICAL and behavioral responses by fish and wildlife to reintroduction is of great importance because they can have dramatic short- and long-term impacts on reintroduction success. Despite over 20 years of study, reintroduced populations still often fail to reach desired goals (Griffith et al. 1989, Fischer and Lindenmayer 2000). To enhance reintroduction success, a great deal of attention is typically placed on population dynamics (Chauvenet et al., Chapter 6, this volume) and on extrinsic characteristics of the release site itself (Dunham et al., Chapter 5, this volume). Yet to understand the mechanisms behind why a population fails to become established, scientific investigations into the behavioral and physiological responses of reintroduced animals often are needed (Parker et al. 2012). Correspondingly, to optimize reintroduction strategies and enhance probability of success prior to undertaking a reintroduction, knowledge of common physiological and behavioral issues is critical.

Investigations into animal physiology and behavior in the context of reintroduction biology began with captive-reared animals. One of the first and most extreme examples involved extensive behavioral preconditioning of captive-reared whooping cranes (*Grus americana*) (Archibald 1974). Care was taken to avoid imprinting on humans not only during captive rearing, but also by members of Operation Migration while guiding whooping cranes on a migration to overwintering grounds in the Southern United States (Ellis et al. 2003). It is also important that captive-reared individuals do not acclimatize to a captive setting. For example, the black-footed ferret (*Mustela nigripes*) was extirpated from the wild and required multiple generations of captive breeding before reintroduction. This species required behavioral conditioning to avoid predators and to limit post-release dispersal (Biggins et al. 1999). By contrast, in aquatic systems where dispersal is more often a desired behavior, hatchery managers tend to select for fish that exhibit more exploratory or risk-prone phenotypes (Sundström et al. 2004).

Practitioners working with wild-caught individuals translocated directly to reintroduction sites also might observe problematic behavioral and physiological patterns. For example, translocated individuals often are prone to disperse relatively long distances away from reintroduction areas, sometimes in an attempt to return to capture sites (Linnell et al. 1997). Physiological stress, related to capture and transport, has led to decreased reproductive output, individual injury, and death (Price 1989, Lloyd and Powlesland 1994). These have contributed to reintroduction failures and in many cases have led to the subsequent development of well-designed capture, transport, and release protocols within reintroduction programs (Parker et al. 2012; Moehrenschalger and Lloyd, Chapter 11, this volume).

Post-release, persistently high concentrations of stress hormones make reintroduced individuals less equipped to adapt to novel environmental conditions at release sites (Dickens et al. 2010). Through the sampling of stress hormones, practitioners can quantitatively evaluate not only the current physiological state of an organism, but also the longer-term acclimatization of that organism to its new environment (Box 9.1). Long-term elevations of stress hormone levels resulting from reintroduction have been correlated with several of the leading causes of reintroduction failure—including decreased immune function, increased susceptibility to predation, aberrant behaviors such as aggression toward and avoidance of humans, and death (Teixeira et al. 2007, Dickens et al. 2010).

Investigations of post-release behavior also have proven to be a key to understanding why reintroduction attempts often fail, particularly for highly social species. For example, reintroduction success for black-tailed prairie dogs (*Cynomys ludovicianus*) increased when family groups were translocated and released together at the same site (Truett et al. 2001). Interactions among species also are critical to understanding the limiting effects that competition and predation have on reintroduced populations (Biggins et al. 1999, Ward et al. 2008, Robinson and Ward 2011). This is particularly true in systems altered from historical conditions, where novel species and habitat conditions can limit population growth and establishment of self-sustaining populations (Osborne and Seddon 2012).

To address these challenges and to optimize reintroduction strategies, there is an increasing need for detailed post-release monitoring that integrates measures of physiology and behavior. For example, high overwinter mortality (44% more than normal) in captive-reared northern water snakes (*Nerodia sipedon sipedon*) was associated with abnormal movements and habitat use that resulted in failure to maintain appropriate body temperature and mass (Roe et al. 2010). Simultaneous monitoring of stress hormone concentrations and movement data through satellite tracking provided a mechanistic understanding of why reintroduced African elephants are reclusive and aggressive post-release, as well as how to mitigate potential human-elephant conflict (Box 9.1; Jachowski et al. 2012b, 2013b). Additionally, following reintroduction of gray wolves (*Canis lupus*) to Yellowstone National Park, integration of physiological monitoring with behavioral observations helped in understanding direct and indirect effects of wolves on ungulates and on broader ecosystem processes (Christianson and Creel 2010). Thus, an improved understanding of the physiological and behavioral responses that released animals are likely to undergo has relevance not only to individual animals or populations, but also to community-level interactions and ecological processes that are likely altered through reintroduction.

In this chapter, we summarize current knowledge regarding physiological and behavioral responses of fish and wildlife species to reintroduction. We review and provide examples of how physiology and behavior, both individually and interactively, have advanced our understanding of individual, population, and community-level responses to reintroduction. Finally, we provide guidance on how to establish reintroduction protocols that minimize physiological, behavioral, and community-level responses that are detrimental to achieving

9a Bull African elephant (*Loxodonta africana*) in Pilanesburg National Park, where prior to introduction of large bulls, young males had elevated musth cycles and exhibited aggressive behavior and goring of both black rhinoceros (*Diceros bicornis*) and white rhinoceros (*Ceratotherium simum*) (Slotow et al. 2000).

The reintroduction of African elephants has become a common enterprise in South Africa, with over 38 populations initiated over the past four decades (Garaï et al. 2004). However, restoration of these megaherbivores within fenced reserves has not been easy, with multiple behavioral and physiological problems encountered as a result of reintroduction attempts. Early experience revealed the importance of social networks and translocating entire family groups. Female elephants form social groups or herds that often remain intact for decades; thus, translocation of entire family groups was an early priority in elephant reintroduction programs (Garaï et al. 2004). Adult male elephants are typically solitary or form smaller bachelor groups. Due to the large size of old bulls, younger (but still sexually mature) males were selected for translocation. This led to younger bulls no longer being behaviorally and physiologically suppressed by older bulls following release into a new environment, and prolonged periods of musth when bulls exhibited elevated testosterone levels (figure 9a). These extended musth periods were found to be associated with disruptive, aberrant aggressive behavior by reintroduced males, including the goring and killing of rhinos (Slotow and van Dyk 2001). As a solution, once older bull

elephants were brought in, the prolonged musth cycles and aberrant behaviors by younger males were suppressed (Slotow et al. 2000).

Aberrant behavioral problems by reintroduced elephants were not restricted to males. Female elephants exhibited a number of problematic behaviors including fence breaking, aggression toward humans, and even mortality of some young calves that was linked to long-distance dispersal movements (Jachowski et al. 2013). Each of these behaviors was most evident in one or two reserves where elephants were relatively recently reintroduced and fecal glucocorticoid levels were chronically elevated (figure 9b). Furthermore, finer-scale evaluations of the driving role physiological state plays in elephant movement behavior post-release revealed close associations between stress hormone levels and incidences of refugia (Jachowski et al. 2012) and corridor use behavior (Jachowski et al. 2014); elephants were unlikely to leave refugia and streaked through corridors when in elevated physiological states (Jachowski et al. 2013). Resulting findings have informed managers on how to manage access by elephants to certain habitats and to mitigate human-elephant conflict through a variety of tools, including the use of virtual fences (Jachowski et al. 2014).

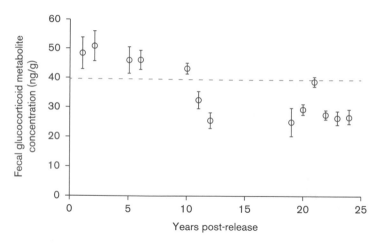

9b Fecal glucocorticoid metabolite (FGM) stress hormone concentration monitoring revealed that elephants reintroduced into five parks and reserves in South Africa required up to 10 years before physiologically acclimatizing and returning to FGM levels below average baseline levels of 40 ng/g, depicted here with the black dashed line (adapted from Jachowski et al. 2013a).

common recovery goals and conclude by providing recommendations for future research.

PHYSIOLOGICAL ISSUES

Understanding physiological responses by fish and wildlife to reintroduction is best approached under the broader field of animal physiology. While physiological measures are often integrated across multiple systems (including the simultaneous monitoring of physiological and behavioral responses [Boxes 9.1 and 9.2]), we break apart issues into four main fields of inquiry based on the critical roles they have played to date in reintroduction attempts. Fortunately, each of these areas of investigation has been well developed in the broader animal physiology literature, and a variety of invasive and noninvasive monitoring techniques are available. We specifically focus on physiological issues that take place during and following reintroduction, and thus do not touch on short-term responses associated with specific capture or release events such as capture myopathy or

reactions such as flight or fight in this review (see Chapter 11, this volume, for a review of these issues).

Energetics and Nutrition

At a minimum, a release site should be selected that provides opportunities to meet energetic needs of the reintroduced individual and larger population (Dunham et al., Chapter 5, this volume). As discussed in Chapter 11, it is essential to select appropriate times and locations to release animals so as to provide sufficient foraging opportunities for reintroduced populations (Davies-Mostert et al. 2009). While this is particularly true for endothermic animals that have higher energy requirements, releasing ectothermic animals in habitat that consistently fluctuates within a species' optimal thermal range is critical for digestion, and overall fitness and recruitment (Besson and Cree 2011). Furthermore, similar to endothermic animals, ectothermic animals can also be vulnerable to temperature extremes. For example, temperature

BOX 9.2 · Migration Timing Mismatch in Translocated Salmonids

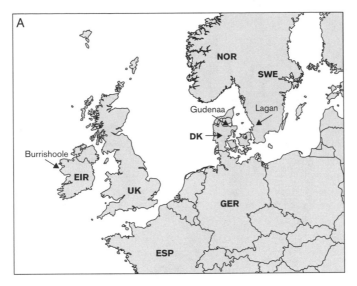

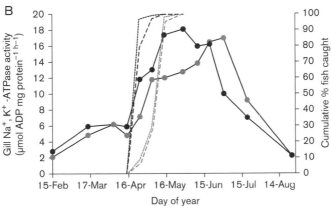

9c Atlantic salmon (*Salmo salar* L.) collected from Burrishoole, Scotland, and Lagan, Sweden, were released into the River Gudenaa in Denmark in 1996 (panel A). The graph in panel B illustrates gill Na+, K+ -ATPase activity in the two strains of Atlantic salmon (*Salmo salar* L.) held in fresh water in 1997. Black lines indicate the Lagan strain and gray lines indicate the Burrishoole strain. Each value is a mean based on sampling of 10 fish. On the right axis, cumulative percentage of Atlantic salmon caught (dashed lines) in River Gudenaa following release are shown at 3 km (dotted line) and 9 km (dashed line) upstream. When released in River Gudenaa, Burrishoole-strain fish exhibited a 15–19 day delay in migration compared to the Lagan strain, highlighting the importance of considering genetic and physiological drivers of timing and intensity of smolt migration (lower panel).

SOURCE: Adapted from Nielsen et al. (2001).

Successful migration is often a prerequisite of future reproduction success; thus, failing to account for migratory behavior might hamper or even impede successful reintroduction. This is particularly true for anadromous fish such as Atlantic salmon (*Salmo salar*). To understand migration ecology, a special focus on physiological adaptations is particularly important. Numer- ous populations of Atlantic salmon have been extirpated in the last several decades, and reintro- duction is attempted in many places with varying success. This species undertakes long migrations between freshwater and sea, requiring dramatic physiological changes. In spring, juvenile salmon in the freshwater phase preadapt for life at sea with physiological and morphological changes including increased salinity tolerance and changes in body coloration (Hoar 1988, Nielsen et al. 2006). The changes are necessary to survive in the new environment, and this migration often is associated with large mortalities (Thorstad et al. 2012). It becomes even more critical because hatchery-raised fish might have an even larger mortality rate than wild ones (Melnychuk et al 2014, Aarestrup et al. 2014).

The timing of this physiologically driven migration is at least partially under genetic control, and because specific chemical conditions differ between rivers, local adaptations to the prevailing conditions have evolved in many populations (Aarestrup et al 1999, Nielsen et al 2001). If a particular reintroduction involves individuals from a population with a different smolt migratory period, potential match/mismatch scenarios arise and are almost sure to end in a poor outcome of the reintroduction. For example, figure 9c shows the cumulative migration of five different introduced salmon populations in an experiment undertaken during a large reintroduction project in Denmark. The fish were raised under identical conditions in a hatchery and released at the same time and place. Subsequently, they were recaptured in a downstream trap. There was a clear difference between populations in their time of migration. Physiological developments were closely linked to future migratory behavior and physiological measurements could be used to predict whether an individual would migrate (Nielsen et al. 2004). For example, gill Na^+, K^+-ATPase activity rose during smolt migration and could reliably be used to indicate the potential migratory period for various populations (figure 9c, Nielsen et al. 2001). Matching the migratory pattern of the introduced population with the original population's migration during mid to late April was expected to produce better results in this particular river than might be achieved by releasing smolt from populations migrating predominantly in May. Hence, because the physiological information can be obtained before the actual introduction, the potential exists for using this information to find the best population to be used for introduction at a given place.

monitoring of cold-sensitive gopher tortoises (*Gopherus polyphemus*) at the northern extent of their distribution revealed that reintroduction success was enhanced by the availability of suitable burrows as refugia to avoid overwinter mortality (DeGregorio et al. 2012). Therefore, an understanding of both natural history and physiology is often necessary to address the energetic needs of a species proposed for reintroduction.

In addition to meeting basic energetic needs, reintroduction practitioners need to consider dietary preferences and nutritional needs of a given species. This is of concern in fish and wildlife reintroduction when there is a risk of mismatch in gut flora, morphology, and dietary preference with available prey items, or when captive individuals transition from human-modified feedstuffs to wild diets. For example, relative to wild brown teal (*Anas chlorotis*), cap-tive-reared teal exhibited small intestines that were 21% shorter and 33% lighter, which likely reduced their ability to digest a relatively diverse and fibrous wild diet, while simultaneously contributing to poor nutritional condition of teal found dead post-release (Moore and Battley 2006). Additionally, decreased foraging efficiency by captive-reared animals must be taken into account. For example, in comparison to resident brook trout (*Salvelinus fontinalis*) that maintained steady fitness levels, captive-reared brook trout were approximately 36% less successful when foraging and consumed about 44% fewer prey items—contributing to high mortality rates of released individuals (Ersbak and Haase 1983).

Meeting dietary and nutritional needs can be challenging because populations often are reintroduced into fragmented or fenced portions of their former range. For example, through simulation models, Armstrong et al. (2006) found successful growth of hihi (*Notiomystis cincta*) reintroduced to the island of Tiritiri Matangi, New Zealand, was most likely to be accomplished through provision of sugar water as supplemental food (in conjunction with predator control). Similarly, within a year of release, tule elk (*Cervus elaphus mammodes*) reintroduced to Point Reyes, California, exhibited copper deficiencies in the liver (−93%) and serum (−70%), requiring the provision of dietary supplements (Gogan et al. 1989). Additionally, dietary and nutritional needs can be major drivers of important movement behaviors such as dispersal and migration (see section on dispersal below). Thus, in some cases, dietary manipulation involving careful monitoring and experimentation might be necessary to enhance reintroduction success.

Stress Response

Given the fundamental role of the endocrine system in directing vertebrate response to environmental variation (Walker et al. 2005), it is increasingly apparent that an understanding of endocrinology is critical to understanding observed responses by fish and wildlife to reintroduction. In vertebrates, glucocorticoid stress hormones (cortisol and corticosterone) are released by the hypothalamus-pituitary-adrenal axis following exposure to a stressor. These hormones interact with internal receptors to facilitate the direction and prioritization of energy to differing processes within the organism (McEwen and Wingfield 2003, Romero and Butler 2007). An acute or short-term elevation in stress hormones has the beneficial effect of prioritizing energy away from basic physiological processes (digestion, immune system, etc.) to mount a response to a specific stressor (Sapolsky et al. 2000). However, persistent elevation in stress hormones (hereafter referred to as chronic stress) can have detrimental effects on individual fitness and can contribute to the failure of an individual or population to adapt to its environment post-release (Dickens et al. 2010).

A state of chronic stress following translocation has been observed across a wide variety of taxa, and has been associated with problematic behavioral and physiological responses post-release (Teixeira et al. 2007, Dickens et al. 2010). Individuals vary in how they perceive stressors, but a stressor is essentially any disturbance an individual perceives in its surrounding environment that is unpredictable, novel, or leads to a lack of control (Parker et al. 2012). In this sense, the process of reintroduction is clearly a stressor (or a series of stressors as discussed below) that in many cases is likely to elicit a chronic stress response (Dickens et al. 2010). Chronic stress is correlated with a number of issues that commonly limit reintroduction success, including increased susceptibility to disease (Glaser and Kiecolt-Glaser 2005), decreased reproductive output (Herzog et al. 2009), and starvation (García-Díaz et al. 2007, Herzog et al. 2009). Furthermore, a number of potentially problematic behaviors have been associated with chronic stress, including increased risk of predation (Cyr et al. 2009, Dickens and Romero 2009), dispersal away from the release site (Dickens et al. 2010), or refuge behavior within a portion of the release site (Jachowski et al. 2012), and even aggression toward humans (Jachowski et al. 2013c).

While a physiological stress response might be inevitable when reintroducing fish and wildlife populations, understanding what specific factors contribute to chronic stress and minimizing exposure to those stressors can be important parts of a successful reintroduction (Parker et al. 2013). Studying physiological stress responses requires knowledge of baseline physiological states or normal stress hormone levels of an individual, population, or species from which to compare subsequent hormone concentrations (figure 9.1). For wild-caught individuals, pre-capture measurements or measurements of a similar population will be invaluable. Prerelease sampling is likely to

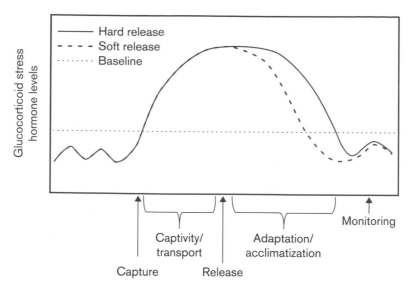

FIGURE 9.1. Conceptual diagram of stressors and stress responses expected to occur during reintroduction. Stress hormone concentrations go through natural daily or seasonal rhythms prior to disturbance, and are predicted to become elevated above baseline conditions following capture, and typically remain elevated during captivity, transport, and release, indicative of chronic stress. Ideally, soft-release techniques should be used when (1) it facilitates the more rapid decline of circulating stress hormone values and/or (2) captivity at the release site limits the risk of negative behavioral and physiological issues associated with chronic stress such as dispersal, predation, and disease. Managers also should consider the likelihood that duration and intensity of monitoring post-release can elicit a stress response.

be easier on captive-reared individuals, but interpreting those values as baseline physiological conditions is considerably more difficult due to the confounding effect of a captive setting. Conceptually, each phase of the reintroduction process should be assessed individually as a potential stressor, beginning with the capture of an individual, and on through to its holding, transport, and release (figure 9.1). Questions of whether to initiate a hard or a soft release (Chapter 11, this volume) can be thought of in a stress-related context, where managers should consider whether holding an animal in a captive setting at the release site is likely to protect an individual from factors associated with chronic stress (e.g., disease, predation, dispersal) and facilitate physiological acclimatization, or exacerbate stress hormone production (figure 9.1). Finally, there is value in maintaining a sampling protocol for stress hormones

following release to critically evaluate physiological acclimatization time and to provide information needed for altering management protocols to mitigate chronic stress (Box 9.1).

Contaminants and Immune System Function

Reintroduction efforts rely on the ability of individuals to mount immune responses to common and novel pathogens (Muths and McCallum, Chapter 10, this volume). For example, over 30 years ago, a combination of canine distemper (caused by *Morbillivirus*) and plague (*Yersinia pestis*) led to the extirpation of the last wild population of black-footed ferrets in the extreme western portion of their historical range (Thorne and Williams 1988). The extreme genetic bottleneck that ferrets experienced when surviving individuals were bred in captivity has led to fears that the species now is unable to mount effective responses to bacterial infections and other common pathogens (Marinari and Kreeger 2006). Furthermore, efforts to reintroduce black-footed ferrets over the past 21 years rarely have succeeded, in large part due to the spread and persistence of plague on the Great Plains over the past several decades (Jachowski 2014). Thus, in the case of the black-footed ferret and many other species, pathogens not only led to their decline in the wild, but also continue to limit reintroduction success for the foreseeable future regardless of other management practices (Muths and McCallum, Chapter 10, this volume).

In addition to pathogens, altered levels of trace elements and contaminants can negatively impact the physiology of reintroduced animals. This is particularly evident in long-term attempts to reintroduce the endangered California condor (*Gymnogyps californianus*). The California condor was largely extirpated from its range due to lead poisoning, which still poses a major threat to the species (Meretsky et al. 2000). Ongoing efforts to reduce lead exposure by restricting lead-based ammunition have so far been ineffective and it is estimated that 60% of the population maintains elevated levels of lead in their blood sufficient to decrease physiological function, and that at least 20% of the wild population needs to be regularly captured and treated to prevent mortality (Finkelstein et al. 2012). Thus, establishment of a self-sustaining population and full recovery of the species will only be possible with intensive, long-term management of contaminant loads in individual animals (Finkelstein et al. 2012).

Rather than simply measuring absolute levels of infection or contamination, the impact of these diseases has to be evaluated in relation to the overall physiological state of the individual. Because stress hormones facilitate a stress response that could include reprioritization of energetic demands away from immunological responses (Ricklefs and Wikelski 2002, Berger et al. 2005), chronic stress associated with reintroductions can limit an individual's ability to mount an immune response to these otherwise "common" disturbances. However, chronic stress is only indicative of an elevated physiological state and disease is the direct mechanism that leads to decreased fitness or mortality. Therefore, simultaneous measurements of stress hormones and diseases often are required to gain insight into the potential for decreased immunological response and the cause for reintroduction failure (Dickens et al. 2010).

Monitoring thyroid hormone levels in addition to stress hormones can provide insight into immunological function and general health of reintroduced animals. Thyroid hormone measurements have been used as an indicator of metabolic rate, thermoregulatory capacity, and growth and development (Wasser et al. 2010). While insight into each of these metrics can be individually important, the value of measuring thyroid hormones in animal reintroduction is likely greatest for identifying subtle or subclinical effects of disease (Mönig et al. 1999). Furthermore, the simultaneous monitoring of multiple hormones can be particularly instructive. For example, measuring elevated levels of stress hormones can indicate exposure to a stressor (such as a contaminant or pathogen) and the diversion of resources away from certain organs or functions, and thyroid hormones can be used to assess disease status and overall health (Tarszisz et al. 2014). The benefits of this type of integrated, multi-hormonal monitoring extend beyond contaminants and disease. In the case of glucocorticoids and thyroid hormones, the simultaneous monitoring of both hormones could lead to an increased understanding of the processes and potential mechanisms (e.g., exposure to a specific extrinsic persistent stressor and associated declining health) that contribute to reintroduction failure.

Reproduction

Ensuring high rates of reproductive success, along with survivorship, are the two primary requirements for demographic restoration of a species through reintroduction (Converse and Armstrong, Chapter 7, this volume). However, a number of reproductive issues have been encountered post-release, ranging from adjusting parturition (Whiting et al. 2011, 2012) to complete reproductive failure (Wolf et al 2000). While, in practice, these reproductive changes will become obvious to managers over time due to low or declining growth in numbers of wild-born individuals, both invasive and noninvasive techniques have been devised to directly assess reproductive status and preemptively identify problems that can lead to decreased reproductive output. First, when translocating sexually mature individuals, internal imagery and/or palpation for the presence of developing embryos are established techniques for most vertebrate taxa (Smith et al. 1989, Canon et al. 1997). Following release, mating and reproductive behavior can often be determined noninvasively through focal following (Sarrazin et al. 1996) or through remote devices and retrospective assessment of animal movement patterns (Barbknecht et al. 2009).

Molecular-level investigations also can be conducted to gain advanced insights into potential reproductive concerns in reintroduced animals. The field of reproductive biology provides a variety of tools that are relatively well developed for captive breeding purposes, such as pregnancy testing through blood sampling in females and collection and analysis of sperm to assess male fertility (Howard et al. 2006). Such intensive sampling has been valuable in determining potential causes of reproductive failure in reintroduced populations, particularly in species that have undergone severe population bottlenecks (Wolf et al. 2000). Genetic sampling, either invasively through capture or noninvasively through collection of scat, hair, or feathers, can be used not only to determine relatedness and dispersal patterns (Vernesi et al. 2002), but also to assess the extent of hybridization in reintroduced populations (Adams et al. 2003).

Monitoring variation in the production of sex hormones can provide an even more detailed and integrative assessment of reproductive function that can be used to enhance reintroduction success. For example, Slotow et al. (2000) found reintroduction of young bull African elephants (*Loxodonta africana*) without large adult male bulls present resulted in musth cycles that were prolonged for monthly periods, breeding attempts that occurred 7–10 years before normal, and aberrant aggressive behaviors driven by elevations in testosterone (Box 9.1). A much more common application of sex hormone monitoring involves female animals, whereby the monitoring of estrogen and progestagen metabolites in blood, urine, or feces can be used to assess reproductive function (Hodges et al. 2010). In addition, as mentioned above, chronic stress has been shown to influence animal reproduction, primarily through altered behavior and by delayed or abandoned reproduction. While the physiological mechanisms behind these phenomena are not well understood, elevated glucocorticoids likely inhibit production and/or responsiveness to reproductive hormones (Sapolsky et al. 2000). Thus, monitoring reproductive hormone levels (often in synchrony with other hormones and behavioral data) can be useful in understanding reproductive health following reintroduction.

BEHAVIORAL ISSUES

Behavior can influence reintroduction success at a variety of scales. A majority of attention has been placed on dispersal, yet other individual-based behavioral factors such as individual personality or readiness for release through prerelease training require careful management consideration. Further, interspecific interactions such as predation and competition can limit or enhance reintroduction success, and have broader cascading ecological effects.

Dispersal

Reintroduction practitioners often face contrasting goals of both limiting movement immediately following release and facilitating movement between isolated populations over the long term. In a general sense, dispersal is simply categorized as any long-distance movement outside of a predefined reintroduction zone, which could be a consequence of selecting an inappropriate spatial scale for reintroduction. We prefer here to define dispersal as a period of exploratory behavior post-release that is common in reintroduced taxa. Immediate post-release dispersal away from the release site is commonly observed in reintroduced fish and wildlife populations (Griffith et al. 1989, Saenz et al. 2002, Germano and Bishop 2009). As discussed below, the extent and duration of dispersal behavior varies based on a suite of factors, with the consistent theme that if post-release dispersal is not accompanied by homing behavior back to the release area, dispersal by reintroduced individuals does not contribute to the establishment of the target population and it thereby limits reintroduction success.

By leaving the release site where conditions are often the most optimal, post-release dispersal can lead to mortality, hybridization, human-wildlife conflict, and a variety of other conditions that contribute to reintroduction problems. The likelihood of losses from predation (Biggins et al. 1999, Hervas et al. 2010), disease (Singer et al. 2000), or incidental killing (Kramer-Schadt et al. 2004) is generally increased during dispersal events. Human-wildlife conflict is a particular concern in larger animals, where dispersal can lead individuals outside of often carefully selected and discretely defined reintroduction zones (Slotow 2012). Conversely, in some instances, dispersal outside of reintroduction areas can facilitate human interactions that lead to imprinting on humans and eventual human-wildlife conflict. For example, of four young Hawaiian monk seals that were exposed to humans (either through captivity or through areas of high human use in the wild), three became habituated and navigated away from translocation sites back toward humans 1–6 days post-release (Baker et al. 2011). Subsequently, these three

individuals were removed from an already small population. Finally, in many reintroduction attempts, maintaining genetic purity of the reintroduced population and limiting the risk of hybridization with other species or populations are key reintroduction goals (Rhymer and Simberloff 1996, Jakober et al. 1998). Limiting dispersal and potential genetic exchange is important for successful reintroduction and recovery of such species (Biebach et al., Chapter 8, this volume).

Because of the negative impact post-release dispersal can have on reintroduction success, managers should consider a suite of tools, often species or site specific, to minimize dispersal distance. These management tools essentially fall into three general categories: (1) physically restricting dispersal at the release site during a period of acclimatization, (2) manipulating environmental conditions and social interactions to encourage residency, and (3) use of conspecifics in highly social species. The use of conditioning pens or bomas at release sites is relatively widely used in reptile (Tuberville et al. 2005, Hyslop 2007), avian (Lockwood et al. 2005, Mitchell et al. 2011), and mammalian species (Letty et al. 2000), with the goal of allowing animals to acclimatize and imprint to a release site (often termed a "soft release") (figure 9.1). However, soft release strategies are rarely used in fish translocations (with the exception of delayed release in salmon; Novotny 1980), and numerous studies have shown no difference between immediate and delayed "soft release" in a variety of terrestrial avian and mammalian species (see Parker et al. 2012, Le Gouar et al. 2012). Thus, the effectiveness of this technique in mitigating dispersal, and decisions regarding the design, placement, and the amount of time animals are retained in an enclosure, often depends on species-specific attributes, animal welfare concerns, and logistical constraints (Chapter 11). As a general rule, we recommend an experimental approach to evaluate how differing conditions (e.g., the size and construction of an enclosure and the amount of time an animal is retained in it) facilitate physiological acclimatization (figure 9.1).

Post-release dispersal also can be mitigated through strategic placement of release sites and consideration of surrounding environmental conditions. Release sites selected should be those that optimally fit with knowledge of habitat use by that species and that are likely to serve as refugia post-release (Moehrenschlager and Lloyd, Chapter 11, this volume). For example, translocation of fully aquatic Ozark hellbender salamanders (*Cryptobranchus alleganiensis bishopi*) into areas with a relatively high density of large rocks increased site fidelity by 30% (Bodinof et al. 2012). Similarly, African elephants released into iSimangaliso Wetland Park navigated toward dense forests and intensively used them as refugia during periods of chronic stress following reintroduction (Jachowski et al. 2012). Timing of release also is critical to managing dispersal behavior, where seasonal weather and food availability could either increase or decrease an individual's tendency to disperse from a release site (Hardouin et al. 2014). Manipulation of release site conditions also can decrease exploratory behavior. For example, retaining tall grass or placing vinyl barriers near burrows where prairie dogs were released limited visibility and reduced dispersal tendencies of the animals following translocation (Truett et al. 2001). Finally, release of individuals near conspecific animals (Dullum et al. 2005, Støen et al. 2009) or artificial conspecific attraction signals such as acoustic playbacks (Kress 1997) can be used as tools to minimize movement post-release, particularly for animals that live in groups. However, close proximity of conspecifics for other, typically solitary animals can have the opposite effect (Dunham 2000, Weilenmann et al. 2011).

Over the long term, dispersal is an important behavior that maintains connectivity for restored fish and wildlife populations (McCormick et al. 1998, Trakhtenbrot et al. 2005). Further, dispersal of reintroduced species can allow individuals to fulfill lost ecological roles

such as long-distance seed dispersal (Polak et al. 2014). Thus, while immediate post-release dispersal can limit reintroduction success in the short term, dispersal is a critical component to population viability and ecosystem services provided by the reintroduced species over the long term. At the same time, management plans for reintroduced species increasingly assume that future dispersal and connectivity among populations will be limited, and sometimes call for hands-on management and active translocation of individuals over the long term to replicate lost immigration and emigration dispersal behaviors (Rohlf et al. 2014). In this sense, populations are not viewed as self-sustaining but rather "conservation reliant" (Scott et al. 2010). Regardless, monitoring dispersal over the short and long term either through genetic analyses, satellite or radio tracking, or mark-recapture (banding, biologgers, etc.) is critical to developing management strategies that facilitate connectivity among targeted populations (Hayward and Slotow, Chapter 13, this volume).

Restoring migratory behaviors by reintroduced fish and wildlife can be viewed as the ultimate form of facilitated post-release dispersal due to the critical role migrations play in the long-term survival and successful reproduction of some reintroduced species. All species that undergo migratory movements are of increased management concern due to their typically large spatial requirements (Singh and Milner-Guland 2011). To facilitate migratory behavior by reintroduced species, provided physical barriers to migration are addressed, the first key aspect to consider in reintroduction planning is the extent to which migratory behavior is innate or learned, and how the interaction of those factors is driven by environmental cues (Box 9.2). For many fish species where migratory behavior is innately timed based on environmental cues, managers must consider how reintroduced individuals are likely to perceive environmental cues at a release site where cues might differ from those elsewhere (Box 9.2). For animals in which migratory behavior relies on a large learning component, as is the case for many avian and mammalian species, restoring migratory behaviors can be even more complex (e.g., Ellis et al. 2003).

Individual Variation in Behavior

Selection of individual release candidates traditionally has considered the individual only in terms of its potential contribution to the overall released population, i.e., whether it contributes appropriately to the genetic variability and demographic profile required by a functioning founder population. However, because individual animals exhibit different behaviors (sometimes referred to as personalities) in their responses to environmental stimuli (Sih et al. 2004, Watters and Meehan 2007), it also is important to consider behavioral characteristics when choosing among individuals to reintroduce (Waples and Stagoll 1997, Dingemanse and Reale 2005).

One of the key aspects of behavioral variation of importance in reintroduction is the extent to which individual animals exhibit boldness in response to environmental stimuli. Individual variation in boldness/shyness is widespread across a range of species and taxa, including invertebrates (Pruitt et al. 2008), fish (Wilson et al. 1993), birds (Dingemanse et al. 2004), reptiles (Carter et al. 2012a), and mammals (Réale et al. 2000, Bremner-Harrison et al. 2004, 2011). Levels of boldness are subject to natural selection (Huntingford and Giles 1987), with evidence that relaxed selection can result in greater variability within a population and more animals being observed at the extremes of the shy-bold continuum (McPhee 2003). Individual variability in boldness is of particular importance in a reintroduction context because of clear linkages between boldness and four factors associated with reintroduction success: survival, growth, reproduction, and dispersal.

The correlation between boldness and mortality is known from a variety of species (Réale and Festa-Bianchet 2003, Biro and Stamps 2008, Bremner-Harrison et al. 2011), although it has infrequently been examined within reintroduced populations. A study of reintroduced swift fox (*Vulpes velox*) demonstrated that individuals that died following release were those with higher levels of boldness (Bremner-Harrison et al. 2004). This suggests that releasing individuals that are overly bold is to be avoided as they have a higher likelihood of being predated and thus are less likely to contribute to the development of a self-sustaining population. However, individuals that are too shy might not explore their new habitat, and thus might have difficulty in locating suitable dens, food, or mates (Sinn et al. 2014). Thus, incorporating high levels of variation in behavioral type is likely important when compiling a founding population for reintroduction that can adapt to selection pressures that change spatially and temporally over time (Watters and Meehan 2007, Bremner-Harrison et al. 2013).

Boldness can also influence growth of reintroduced individuals post-release. Assessments of growth post-release not only serve as approximate measures of fitness, but also indicate sexual maturation and fecundity in many vertebrate species (Blueweiss et al. 1978, Rowe and Thorpe 1990). Boldness has widely been associated with risk-prone personality traits that are positively associated with growth (Biro and Stamps 2008). However, because most studies are of animals in captivity, they do not account for changing environmental conditions and predation risk (Adriaenssens and Johnsson 2009, Biro and Stamps 2008). Recent evidence from studies involving wild fish (Adriaenssens and Johnsson 2011), birds (Both et al. 2005), and mammals (Boon et al. 2007) suggests that growth is negatively correlated with boldness. Thus, the relationship between behavioral types and growth rates can be significant, but it is likely to depend on fluctuations in selective pressures such as predation risk and food availability (Dingemanse et al. 2004, Boon et al. 2007).

Selecting of individuals for reintroduction based on boldness can have important impacts on reproduction and long-term population growth. Smith and Blumstein (2008) conducted a meta-analysis of behavioral studies on wild, captive, and domestic animals, and determined that although bolder animals showed shorter life spans than their shyer counterparts, they had higher reproductive output. The analysis did not find a significant effect of increased boldness on reproductive output when considering only wild animals; however, there were relatively few studies of wild animals in their analysis. More recent studies have supported the correlation between increased boldness and higher reproductive output for wild animals (Réale et al. 2009, Bremner-Harrison et al. 2013, Patrick and Weimerskirch 2014). Within a reintroduction program, individuals that are likely to produce more offspring at an earlier stage of their reproductive life span are desirable, as they will facilitate population growth (Converse and Armstrong, Chapter 7, this volume). However, given the propensity of bolder animals to show reduced survival, it might be optimal to release animals exhibiting a diversity of behavioral types to ensure that reproduction is occurring, whether at a fast rate through the breeding of bolder animals or at a slower but potentially more sustained rate by shyer individuals.

The final correlate with boldness that is of key importance to reintroduction is dispersal, or movement behavior. As discussed above, dispersal can either limit or enhance reintroduction success. Bremner-Harrison et al. (2004) showed that bolder reintroduced swift fox exhibited greater mean distances between telemetry fixes within the release site and moved farther from their point of release than shyer counterparts, leading to increased predation-linked mortality of bold individuals. Similarly, in an experimental release of freshwater killifish (*Rivulus hartii*), bolder individuals moved farther in the stream than lower ranked individuals (Fraser et al. 2001). Indeed managers of captively bred fish stocks often select for increased boldness, leading to hatchery-raised fish that consistently exhibit risk-prone and aggressive phenotypes (Sundström et al. 2004). However, in some cases, dispersal behavior and

linkages among populations might be detrimental to reintroduction success. For example, in many salmonids, local adaptations exist within individual populations to specific natal breeding streams (Box 9.2), so dispersal behavior and associated genetic linkages in this scenario could be detrimental to overall recovery.

While studies directly relating behavioral type to reintroduction are still relatively rare, boldness is an adaptive trait that can influence survival, reproduction, and movement across a variety of ecological contexts, as suggested in the early discussions of the shy-bold continuum and the context specificity of boldness (Wilson et al. 1993, Wilson and Stevens 2005). Various methods for assessing boldness in animals are available in both captive and field settings (Table 9.1). How the knowledge of variation in boldness can be utilized by reintroduction managers is not yet fully known, but reintroduction programs will benefit from having an awareness of the importance of behavioral diversity, its potential effects, and the ecological trade-offs that might occur, such as between boldness-survival and boldness-reproduction.

Intraspecific Interactions

Behavioral interactions between individuals of the same species can both enhance and limit reintroduction success. As discussed above, social cues and conspecific attraction can limit dispersal behavior and enhance fidelity to reintroduction sites. Considering social behavior of reintroduced species is particularly important for species with close social structure. For example, following numerous issues with African elephants exhibiting aberrant post-release behavioral problems, it is now common practice to attempt to translocate and reintroduce entire family groups of elephants when attempting a reintroduction (Garaï et al. 2004). Release strategies for highly social species must be adjusted to fit the species, population, and often the individual animal. For example, when releasing male chimpanzees (*Pan troglodytes*), it is critical to space males apart from other

TABLE 9.1
Behavioral Tests Used to Assess Boldness in Wild and Captive Animals

Assessment Approach	Methods	Tests	Examples	
			Captive	*Wild*
Rating	Questionnaire-based approach rating individuals on a scale in relation to a variety of behavioral descriptors	Likert scale (0–5 or 10) or visual analog scale (a mark placed by investigators on a 10-cm-length line)	Carlstead et al. (1999)	Lee and Moss (2012)
		Binary	None found	Bremner-Harrison et al. (2004)
Ethological coding	Recording of observed behavior of individuals in non-manipulated situations	N/A	Bergvall et al. (2011)	Bremner-Harrison and Cypher (2011) and Bremner-Harrison et al. (2013)
Experimental tests	Recording behavioral responses of individuals to particular stimuli or contexts	Novel object test	Korhonen and Niemela (1996)	Bremner-Harrison et al. (2004) and Bremner-Harrison and Cypher (2011)
		Open field test (novel environment)	Brydges et al. (2008)	Boon et al. (2007)
		Startle test	Rudin and Briffa (2012)	None found
		Mirror test	Blumstein et al. (2006)	None found
		Predator test	McPhee (2003)	Carter et al. (2012b)
		Handling test	Carere and van Oers (2004)	Réale et al. (2000)
		Trappability (often used in conjunction with handling test)	Not found	Carter et al. (2012a)

males because the released individuals are likely to be attacked and killed by resident males (Goossens et al. 2005).

Consideration of intraspecific interactions also is important for more solitary species that can be highly territorial and aggressive toward other individuals. For example, concerns over cannibalism contributed to the decision to release individuals of eastern indigo snakes apart from one another (Stiles 2013). Determining appropriate distribution or density of individuals to release within a site is compli-cated by the fact that territoriality often fluctuates based on season and habitat quality (Jachowski et al. 2010). Thus, it is evident that species- and site-specific release protocols should be designed to minimize territorial aggression (see Chapter 11). Conversely, practitioners must also consider how population density might lead to over-dispersion or long-distance dispersal to find mates (Le Gouar et al. 2008, 2012). Therefore, accounting for intraspecific interactions in reintroduction protocols requires a detailed knowledge of the behavioral ecology of the species being reintroduced. As discussed in Chapter 5, we encourage practitioners to undertake an adaptive, experimental approach to release site protocols that account for varying ways in which intraspecific interactions can limit population establishment and reintroduction success.

Interspecific Interactions

The presence and strength of various interspecific interactions are likely to influence not only the short-term survival of individual animals and long-term viability of reintroduced populations, but also broader ecosystem structure and function. Conceptually, these issues can be broken down by considering how the species targeted for reintroduction is likely to perform in its new environment (which has likely changed since extirpation) as a predator, and as prey, as well as the potential for competition, risk of hybridization, and cascading effects on other trophic levels following release.

Reintroduced Species as Predators

Suitable prey density often is necessary to achieve viability in reintroduced predator populations. For example, the principal force determining persistence of reintroduced Canada lynx (*Lynx canadensis*) populations was the density of snowshoe hares (*Lepus americanus*), their primary prey (Steury and Murray 2004). Similarly, lack of success in reintroduction of African wild dogs (*Lycaon pictus*) in portions of Southern Africa (Scheepers and Venzke 1995) and black-footed ferrets (*Mustela nigripes*) in western North America (Jachowski et al. 2011) was linked to releases occurring in areas of low prey abundance. Clearly, managers must consider, prior to reintroduction attempts, whether the prey base is sufficient to support a reintroduced population.

Ideally, reintroduced predatory animals have an innate ability to find, kill, and consume prey. However, this should not be assumed to be true, particularly for social animals or those that spend considerable time with their mothers before becoming independent (e.g., many carnivores). In these cases, it might be necessary to train individuals to support themselves. For example, captive-bred black-footed ferrets that were exposed to live prairie dogs while still in captivity were better able to kill prairie dogs after their release into the wild, although this prerelease exposure did not result in observable effects on ferret survival (Biggins et al. 1999).

Conversely, reintroduced predators often are naïve to the danger posed by recently arrived invasive prey species that can be poisonous and cause population declines. This phenomenon has been most studied in relation to the toxic and invasive cane toad (*Bufo marinus*) in Australia; predation attempts on cane toads caused mortality in multiple taxonomic groups (Letnic et al. 2008, Doody et al. 2009; reviewed in Shine 2010). Although some animals might learn to avoid these animals rapidly enough to prevent population declines (e.g., Llewelyn et al. 2010), the cane toad can cause additive mortality that hinders reintroduction attempts. Because it is unlikely that it will be possible to eliminate cane toads from their invaded habitats, researchers have attempted to condition predators to avoid cane toads prior to their release. For example, dead toads laced with thiabendazole, a chemical that causes nausea, were fed to northern quolls (*Dasyurus hallucatus*) in captivity prior to their release. Mortality rates for the released quolls that received this aversion training appeared to be lower than for quolls that did not receive it, particularly male quolls that more typically attack cane toads, for which training increased survival by 30% (O'Donnell et al. 2010). In a similar study, blue-tongued skinks (*Tiliqua scincoides*) learned to avoid toad bait when it was associated with lithium chloride, another chemical that induces nausea (Price-Rees et al. 2011). Thus, prior to reintroduction, predators can be trained to learn to avoid the poisonous amphibians. However, each successive generation of any reintroduced population would need to learn to avoid cane toads, and the efficacy of such a long-term commitment is untested.

Reintroduced Species as Prey

The presence of both native and introduced predators can have dramatic limiting effects on typically small initial populations of reintroduced fish and wildlife. For example, in the Arid Recovery Reserve in South Australia, three of five numbats (*Myrmecobius fasciatus*) were killed by native avian predators (Bester and Rusten 2009) and all released wombats (*Aspidites ramsayi*) were killed by native mulga snakes (*Pseudechis australis*) within four months (Moseby et al. 2011, Read et al. 2011). Similarly, a native fish and generalist predator, the slimy sculpin (*Cottus cognatus*) was thought to reduce recruitment of reintroduced Atlantic salmon (*Salmo salar*) (Ward et al. 2008). Collectively, there is evidence that predation is one of the primary causes of failure among reintroduction attempts (e.g., White et al. 2012), thus emphasizing the need for managers to account for and mitigate loss due to predation through the selection of individuals for release, prerelease training, or other management actions.

SELECTING SOURCE INDIVIDUALS FOR RELEASE
As a general rule, when animals are introduced into an area that contains predators, individuals from source populations that had previously been exposed to predators might be better suited to persist because they recognize the potential threat and behave accordingly (Maloney and McLean 1995, Parlato and Armstrong 2013). For example, experimentally released elk (*Cervus elaphus*) from populations that had experienced wolf predation or human hunting pressure had survival rates 1.0–2.2 times higher in their first year than elk that were naïve to these threats (Frair et al. 2007). However, this benefit might degrade after some time in a predator-free environment (Jamieson and Ludwig 2012), emphasizing the importance of continued vigilance and long-term monitoring of predator populations. In addition, although animals might exhibit behavior that reduces chances of predation (or shift their behavior in response to the presence of a novel predator), this behavior might not be sufficient to compensate demographically for high predation levels (Hudgens and Garcelon 2010).

PRERELEASE CONDITIONING If source populations that have been exposed to predator guilds similar to those at release sites are not available, animals destined for reintroduction can be trained or conditioned to avoid predators prior to release (Griffin et al. 2000, Berger-Tal et al. 2011). This type of training might be particularly important for social animals (Griffin 2004), but is relevant to multiple taxonomic groups (Brown and Laland 2001, Crane and Mathis 2011). For example, exposure to appropriate species-specific alarm calls conditioned black-tailed prairie dogs (*Cynomys ludovicianus*) to recognize predators and behave accordingly; this learned behavior was associated with greater survival in the year following release (Shier and Owings 2006). Correspondingly, active training of animals to demonstrate appropriate behavior when they detect a predator has been attempted for a wide variety of species destined for release into the wild, including numerous avian (McLean et al. 1999), mammalian (McLean et al. 1996, 2000, Short and Turner 2000, Moseby et al. 2012), and fish species (Suboski et al. 1989, Brown and Laland 2001).

When attempting prelease conditioning, training should ideally be tailored to the specific predator species or attack style (i.e., terrestrial or avian). For example, captive-born Vancouver Island marmots (*Marmota vancouverensis*) were approximately four times more vulnerable to predation from golden eagles (*Aquila chrysaetos*) and three times more vulnerable to predation by mammalian predators such as cougars (*Felis concolor*) than were wild-born individuals (Aaltonen et al. 2009). Thus, prerelease conditioning should be designed specifically for the type of predators (i.e., avian, fossorial, etc.) as well as the species of predators likely to be of most concern at a given reintroduction site.

Despite the implementation of a wide variety of methods for teaching animals to avoid predators (e.g., Brown and Laland 2001), there is limited evidence that this training actually increases survival after animals are released (Moseby et al. 2012). The lack of experimental evidence with which to evaluate prerelease strategies might relate to a reluctance of biologists to release control animals without any conditioning, given the value of each individual animal to reintroduction efforts. For example, although 21% of all captive-reared Puerto Rican parrots (*Amazona vittata*) were killed by raptors, all had received prerelease training that was thought to reduce their risk from these predators (White et al. 2005). In Spain, juvenile little owls (*Athene noctua*) that were conditioned to recognize potential predators appeared to be more likely to avoid predation than owls that received no training prior to release, but sample sizes were low and the two groups of owls were released in different years (Alonso et al. 2011). In general, we encourage managers to ensure at a minimum that appropriate antipredator refuge is sufficiently abundant in the wild to facilitate predator avoidance post-release. Further, given a lack of empirical evidence regarding the efficacy of various training protocols and resulting shifts in behavior, it might be important to ensure that a variety of anti-predatory behaviors exist among animals destined for reintroduction (e.g., Watters and Meehan 2007, Sinn et al. 2014).

PREDATOR CONTROL While often controversial, the control of native predators is a tool available to managers to attempt to enhance post-release survival. For example, high rates of mountain lion (*Puma concolor*) predation on desert bighorn sheep (*Ovis canadensis mexicana*) translocated into New Mexico led to culling of 34 lions over the span of 14 years, with the harvest targeting female lions that posed the greatest overall threat to sheep recovery (Rominger et al. 2004, Phillips 2013). This example highlights how attempts to control native predators can enhance restoration of prey populations, but such attempts must be approached cautiously with considerable public outreach to avoid tension and backlash within a restoration team as

well as the general public (Phillips 2013). Further, any attempt to control native predators should be viewed as a short-term experiment to enhance initial population growth, particularly because despite predation being a lead cause of mortality, predator control alone does not always lead to increased survival, recruitment, and overall reintroduction success (Breck et al. 2006).

Predation by exotic species often poses a significant threat to reintroduction success, particularly in island systems. A variety of examples come from islands where management attempts have focused on eradication of invasive mammals. Feral cats (*Felis catus*), rats (*Rattus* sp.), and red foxes (*Vulpes vulpes*) are particularly cosmopolitan invaders that often prey on reintroduced native species (Armstrong et al. 2006, Moseby et al. 2011). For example, rat eradication efforts via poison baits have been implemented with some success prior to the reintroduction of tuatara (*Sphenodon punctatus*) on Motuhora (Whale Island), New Zealand (Towns 2005), and Antiguan racers (*Alsophis antiguae*) on Antigua (Daltry 2006). If invasive mammal eradication is not achieved, long-term efforts are likely necessary to suppress their populations and ensure the best possible viability of a reintroduced population, as for the North Island robin (*Petroica longipes*) in New Zealand (Armstrong et al. 2006).

Where large-scale eradications are not successful or feasible, managers in terrestrial systems might attempt to limit the occurrence of predators within fenced areas where animals are to be released. This strategy has been effective at increasing survival of several reintroduced terrestrial reptile and mammalian species (e.g., Lettink et al. 2010, Moseby et al. 2011). However, because of persistent risk of reinvasion, fencing comes with considerable long-term maintenance costs (Richards and Short 2003). Therefore, we encourage careful consideration of the long-term costs and benefits of fencing prior to implementation (Scofield et al. 2011).

Invasive species can act in concert with native predators to impact reintroduction success. For example, predation by raccoons (*Procyon lotor*) together with feral housecats (*F. catus*) influenced the distribution of the Key Largo woodrat (*Neotoma floridana smalli*) in Florida, United States (Winchester et al. 2009), and the success of reintroduction programs (McCleery et al. 2013, 2014). Unfortunately, eradication or reduction of invasive predator populations might not be sufficient to benefit reintroduced animals because of compensatory predation by native predators—as was observed for gray partridge (*Perdix perdix*): hens present in areas where red fox (*V. vulpes*) were harvested relatively intensively were instead preyed upon by raptors (Parish and Sotherton 2007). Therefore, dynamic and often experimental treatments followed by intensive monitoring are needed to evaluate optimal short- and long-term predator management strategies.

Reintroduced Species and Competition

In addition to direct predation, competition between resident (either native or nonnative) and reintroduced species can limit reintroduction success. The management actions conducted to reduce competition with reintroduced animals typically focus on at least temporarily reducing the numbers of potential competitors, such as when invasive species have colonized a niche formerly occupied by the species subject to reintroduction and are likely to outcompete that species. For example, successful recruitment by reintroduced Eagle Lake rainbow trout (*Oncorhynchus mykiss aquilarum*) was only evident in streams where nonnative brook trout (*Salvelinus fontinalis*) had been actively removed (Carmona-Catot et al. 2010). Similarly, over 50% of reintroduction failures experienced during greenback cutthroat trout (*O. clarki*) reintroduction attempts were due to incomplete removal and reinvasion of nonnative salmonids that typically outcompete or are less vulnerable to harvest than the native species (Harig et al. 2000).

When multiple species are reintroduced to the same area at the same time, competition between them might limit reintroduction suc-

cess. Desert pupfish (*Cyprinodon macularius*) and Gila topminnows (*Poeciliopsis occidentalis*) are two rare species from the Southwestern United States that were reintroduced to the same habitats. Desert pupfish populations were more likely to become established when they were stocked before Gila topminnows (population increases during 50% of trials) or where the topminnows were never released at all (increases during 75% of trials)—patterns likely influenced by competition between the species for resources, as well as by predation of each species on the young of the other (Robinson and Ward 2011).

Cascading Ecological Effects of Reintroduction

In addition to direct comparisons between individual pairs of species, it is important to consider long-term and perhaps unexpected consequences of species reintroduction for broader ecological communities. Although we have largely focused here on single interspecific interactions in pairwise comparisons between species, actual relationships might be more complicated.

Management actions in preparation for reintroduction might, by themselves, be sufficient to cause cascading ecological effects. As discussed above, whether a resident species negatively influences reintroduced animals via predation, competition, disease transmission, or any other interaction, the preferred management action for the species typically is population control or, in the case of invasive species, attempted eradication. These types of efforts are likely to be essential prior to reintroduction attempts when a given species was responsible for the extirpation of the native species or is significantly limiting the success of reintroduction efforts (Jones and Murton 2012). However, the removal of an exotic species might benefit other exotic species with negative effects on native species on ecosystems (Myers et al. 2000). For example, a successful rat eradication program on Buck Island Reef National

Monument, Saint Croix, US Virgin Islands, conducted in advance of Saint Croix ground lizard (*Ameiva polops*) reintroduction, prompted an approximately 60% increase in the abundance of the invasive house mouse (*Mus musculus*), an organism also likely to influence native flora and fauna populations (Witmer et al. 2007).

Similarly, the presence of exotic species in a particular area might have changed the system such that their removal does not substantially influence the forces that threaten reintroduced populations. For example, the introduction of feral hogs (*Sus scrofa*) to the California Channel Islands provided a stable food source for golden eagles, a species that did not previously reside on the islands in substantial numbers. Once the golden eagles became associated with these islands, however, their predation on island gray fox (*Urocyon littoralis*) was not sustainable and caused severe declines (Bakker et al. 2009, Roemer et al. 2001). Therefore, following removal of the feral hogs (Morrison et al. 2007), management and removal of golden eagles was an essential component of successful island gray fox reintroduction attempts (Coonan and Schwemm 2009).

Reintroduction of top predators is widely known to have cascading ecological effects on ecosystem structure and function. This has been best demonstrated in (currently controversial) attempts to explain how wolf reintroduction in Yellowstone National Park has influenced elk populations, beaver dams, and aspen growth (Ripple et al. 2001, Mech 2012), involving not only the direct effects of predation, but also indirect effects through what has been termed "the ecology of fear" (Ripple and Beschta 2003). Most recently, in Australia there are increasing calls for reintroduction of dingos (*Canis lupus dingo*) to help restore native fauna through the suppression of invasive mesocarnivores (Dickman et al. 2009), while at the same time there is concern that dingos are a potential threat to native threatened species (Allen and Fleming 2012). Thus, the reintroduction of top predators, and associated direct (i.e., preda-

tion) and behavioral indirect effects (i.e., landscape of fear) can have important impacts on other trophic levels that require consideration prior to release, and careful monitoring post-release.

Herbivory similarly can have direct and indirect impacts on lower trophic levels. For example, herbivores often act as important dispersal agents for seeds, and the reintroduction of diverse herbivore assemblages often is needed to ensure dispersal and germination of many plant species (Polak et al. 2014). An analogous interaction is that of pollination. Further, in addition to enhancing germination and dispersal through direct herbivory, herbivores can alter broader habitat structure and plant community ecology (Lagendijk et al. 2011). These simultaneous effects were seen on Isla Española, Galápagos, where reintroduced giant tortoises (*Geochelone nigra hoodensis*) increased recruitment of an endangered cactus (*Opuntia megasperma*) by about 68% due to their patterns of foraging and seed dispersal. In turn, the cactus represented an important component of the ecosystem, facilitating the recovery and conservation of sympatric species and ecosystem processes (Gibbs et al. 2008).

Collectively, despite the small body of knowledge regarding complex ecological relationships of reintroduced species, there is compelling evidence that reintroductions can have cascading top-down or bottom-up ecological effects on ecosystem function. These multifaceted relationships among species require that managers have a basic knowledge of the broader ecosystem they wish to alter through reintroduction. To accomplish this, they must first conceptualize the ecological niche of species targeted for reintroduction and how the species can best fulfill its targeted role in a given system. Further, it is important to acknowledge that a reintroduced species might be dependent on other species or ecological processes no longer present within the ecosystem. Plants, for example, might require various dispersal agents and appropriate environmental conditions to ensure population viability (Primack

1996). Assuming a reintroduced species is able to occupy its previous niche, biologists and managers should consider the corresponding integrity of the ecosystem. In this sense, reintroductions should be viewed not only as a technique to recover a single species, but also as a potential mechanism to restore degraded communities and ecosystems.

INTEGRATION OF PHYSIOLOGICAL AND BEHAVIORAL MONITORING

While we have largely partitioned physiological and behavioral responses as separate phenomena thus far, they are actually necessarily intertwined. The considerable variation in behavioral responses that often is observed following reintroduction is not easily explained by environmental variables or physical adjustments in release strategy, but rather by differences in the physiological states of individuals or populations (Box 9.1). Further, physiological adaptations to particular environments that influence differing behaviors and responses to reintroduction often become evident only after detailed physiological and behavioral monitoring (Box 9.2). Therefore, we strongly advise the simultaneous monitoring and integration of data regarding both physiological and behavioral responses when undertaking animal reintroductions.

RECOMMENDATIONS FOR FUTURE RESEARCH AND DEVELOPMENT

As emphasized above, one of the brightest frontiers and most urgent needs is for practitioners to explicitly design monitoring programs that integrate physiological and behavioral data when conducting species reintroductions. In addition to collecting basic demographic data needed for population viability analyses, as often suggested to evaluate overall reintroduction success (Converse and Armstrong, Chapter 7, this volume), behavioral and physiological data should be gathered to address the mechanisms behind observed demographic responses. Specifically, we encourage practitioners to better

account for these patterns by using an experimental approach. This involves the identification of questions and hypotheses prior to release, and the subsequent monitoring and evaluating of these hypotheses post-release (Chauvenet et al., Chapter 6, this volume).

At the same time, there is a need for research to evaluate and refine new techniques for noninvasively monitoring physiological state and behavior post-release. For example, smaller, longer-lasting telemetry devices continue to be developed that provide invaluable data on post-release movement and physiology at increasingly finer spatial and temporal scales (Millspaugh et al. 2012). Perhaps most promising is the increasingly feasible application of noninvasive monitoring techniques. Provided appropriate collection techniques are followed, a suite of physiological measures are available through scat sampling and other noninvasive techniques that can be analyzed using well-validated methods. Further, the number of tools, diversity of information, and cost of each of these techniques will surely decline as technology advances, thus increasing the feasibility of incorporating these measures into reintroduction activities.

While we have attempted here to group fish and wildlife reintroductions, some key differences exist that are likely to influence how research, monitoring, and management are integrated to enhance recovery. First, the most obvious difference in the nature of aquatic *versus* terrestrial systems is that assessing habitat quality in aquatic systems is typically much more complex, requiring consideration of fine-scale changes in basic biogeochemical parameters such as dissolved oxygen or salinity. Second, in ocean, river, and stream systems the potential for unrestricted movement makes many aquatic reintroductions more vulnerable to a variety of post-release issues such as dispersal, hybridization, and reinvasion by nonnative species (Harig et al. 2000). Third, as a general rule, the fewer animals involved, the more important each individual becomes and the more that knowledge about those individuals is required. Mammal and bird reintroductions might use tens to hundreds of individuals, but typical fish stockings use several thousands, if not millions. In these situations, it might not be feasible to incorporate such fine-scale, integrated measures as are suggested for mammals and birds. For reintroduction of fishes, an emphasis on populations rather than on individual animals often can be more useful in designing research, monitoring, and management approaches (Box 9.2).

Finally, because reintroductions might involve intensive management such as fencing, supplemental feeding, and many other practices that will have to be continued over both the short and the long term, there is the possibility that such management might alter the behavior and physiology of the animals involved. The obvious sociological-political debate then must become, at what point are reintroduced populations managed so intensively as to alter behavioral characteristics of the species that made them unique. This should be a particular concern for attempts to use reintroduction in an effort to restore biological communities and functional relationships within ecosystems. It will become increasingly important for practitioners of restoration ecology and others to address the possibility that reintroduced species will no longer perform behavioral and physiological roles they maintained in historical states.

SUMMARY

There is clear evidence of the importance of accounting for behavioral and physiological responses by fish and wildlife following reintroduction. These measures are critical to understanding the mechanisms behind the demographic success or failure of reintroduction. To incorporate these measures requires prerelease planning that often is species and site specific, and based on historical knowledge of the species and an experimental approach to optimizing reintroduction strategies. Physiological monitoring employing a set of tools of proven utility is increasingly feasible, both logistically

and financially. Behavioral monitoring is similarly feasible, and can be integrated with physiological monitoring to provide deeper insights into animal responses. Reintroduced species are members of dynamic communities, and interactions within populations and with other species will influence the success of fish and wildlife reintroductions. Further, there are often cascading direct and indirect effects of species reintroduction on ecosystem functions that should be anticipated and monitored.

MANAGEMENT RECOMMENDATIONS

- To understand mechanisms behind observed demographic patterns that often determine reintroduction success or failure, it is necessary to have knowledge of the physiological and behavioral ecology of the target species.

- As a general rule, the fewer individual animals being reintroduced, the more important it is to have knowledge about fine-scale variation in behavior and physiological responses to reintroduction.

- Evaluation of personalities of individual release animals and release site parameters can help in predicting the likelihood of success and in determining the extent of management interventions required.

- Collection of baseline data on physiological and behavioral characteristics of a population (either the donor or an established representative population) is vital for subsequent comparisons post-release.

- A scientific approach to reintroduction requires both short- and long-term monitoring strategies that underpin an adaptive, experimental approach to optimizing reintroduction protocols.

LITERATURE CITED

Aaltonen, K., A. A. Bryant, J. A. Hostetler, and M. K. Oli. 2009. Reintroducing endangered Vancouver Island marmots: survival and cause-specific mortality rates of captive-born versus wild-born individuals. Biological Conservation 142:2181–2190.

Aarestrup, K., H. Baktoft, K. Koed, D. D. Villar-Guerra, and E. B. Thorstad. 2014. Comparison of the riverine and early marine migration behaviour and survival of wild and hatchery-reared sea trout (*Salmo trutta*) smolt. Marine Ecology Progress Series 496:197–206.

Aarestrup, K., N. Jepsen, and G. Rasmussen. 1999. Movements of two strains of radio tagged Atlantic salmon (*Salmo salar L.*) smolts through a reservoir. Fisheries Management and Ecology 6:97–107.

Adams, J. R., B. T. Kelly, and L. P. Waits. 2003. Using faecal DNA sampling and GIS to monitor hybridization between red wolves (*Canis rufus*) and coyotes (*Canis latrans*). Molecular Ecology 12:2175–2186.

Adriaenssens, B., and J. I. Johnsson. 2009. Personality and life-history productivity: consistent or variable association? Trends in Ecology & Evolution 24:179–180.

Adriaenssens, B., and J. I. Johnsson. 2011. Shy trout grow faster: exploring links between personality and fitness-related traits in the wild. Behavioral Ecology 22:135–143.

Allen, B. L., and P. J. S. Fleming. 2012. Reintroducing the dingo: the risk of dingo predation to threatened vertebrates of western New South Wales. Wildlife Research 39:35–50.

Alonso, R., P. Orejas, F. Lopes, and C. Sanz. 2011. Pre-release training of juvenile little owls *Athene noctua* to avoid predation. Animal Biodiversity and Conservation 34:389–393.

Archibald, G. W. 1974. Methods for breeding and rearing cranes in captivity. International Zoo Yearbook 14:147–155.

Armstrong, D. P., E. H. Raeburn, R. M. Lewis, and D. Ravine. 2006. Estimating the viability of a reintroduced New Zealand robin population as a function of predator control. Journal of Wildlife Management 70:1020–1027.

Baker, J. D., B. L. Becker, T. A. Wurth, T. C. Johanos, C. L. Littnan, and J. R. Henderson. 2011. Translocation as a tool for conservation of the Hawaiian monk seal. Biological Conservation 144:2692–2701.

Bakker, V. J., D. F. Doak, G. W. Roemer, D. K. Garcelon, T. J. Coonan, S. A. Morrison, C. Lynch, K. Ralls, and R. Shaw. 2009. Incorporating ecological drivers and uncertainty into a demographic population viability analysis for the island fox. Ecological Applications 79:77–108.

Barbknecht, A. E., W. Fairbanks, J. D. Rogerson, E. J. Maichak, and L. L. Meadows. 2009. Effectiveness of vaginal-implant transmitters for locating elk parturition sites. Journal of Wildlife Management 73:144–148.

Berger, S., L. B. Martin II, M. Wikelski, L. M. Romero, E. K. Kalko, M. N. Vitousek, and T. Rödl. 2005. Corticosterone suppresses immune activity in territorial Galapagos marine iguanas during reproduction. Hormones and Behavior 47:419–429.

Berger-Tal, O., T. Polak, A. Oron, Y. Lubin, B. P. Kotler, and D. Saltz. 2011. Integrating animal behavior and conservation biology: a conceptual framework. Behavioral Ecology 22:236–239.

Bergvall, U. A., A. Schäpers, P. Kjellander, and A. Weiss. 2011. Personality and foraging decisions in fallow deer, *Dama dama*. Animal Behaviour 81:101–112.

Besson, A. A., and A. Cree. 2010. Integrating physiology into conservation: an approach to help guide translocations of a rare reptile in a warming environment. Animal Conservation 14:28–37.

Bester, A. J., and K. Rusten. 2009. Trial translocation of the numbat (*Myrmecobius fasciatus*) into arid Australia. Australian Mammalogy 31:9–16.

Biggins, D. E., A. Vargas, J. L. Godbey, and S. H. Anderson. 1999. Influence of prerelease experience on reintroduced black-footed ferrets (*Mustela nigripes*). Biological Conservation 89:121–129.

Biro, P. A., and J. A. Stamps. 2008. Are animal personality traits linked to life-history productivity? Trends in Ecology & Evolution 23:361–368.

Blueweiss, L., H. Fox, V. Kudzma, D. Nakashima, R. Peters, and S. Sams. 1978. Relationships between body size and some life history parameters. Oecologia 37:257–272.

Blumstein, D. T., B. D. Holland, and J. C. Daniel. 2006. Predator discrimination and "personality" in captive Vancouver Island marmots (*Marmota vancouverensis*). Animal Conservation 9:274–282.

Bodinof, C. M., J. T. Briggler, R. E. Junge, J. Beringer, M. D. Wanner, C. D. Schuette, J. Ettling, R. A. Gitzen, and J. J. Millspaugh. 2012. Post-release movements of captive-reared Ozark hellbenders (*Cryptobranchus alleganiensis bishopi*). Herpetologica 68:160–173.

Boon, A. K., D. Réale, and S. Boutin. 2007. The interaction between personality, offspring fitness and food abundance in North American red squirrels. Ecology Letters 10:1094–1104.

Both, C., N. J. Dingemanse, P. J. Drent, and J. M. Tinbergen. 2005. Pairs of extreme avian personalities have highest reproductive success. Journal of Animal Ecology 74:667–674.

Breck, S. W., D. E. Biggins, T. M. Livieri, M. R. Matchett, and V. Kopcso. 2006. Does predator management enhance survival of reintroduced black-footed ferrets? Pp.203–209. In J. E. Rolle, B. J. Miller, J. L. Godbey, and D. E. Biggins, eds., Recovery of the Black-Footed Ferret: Progress and Continuing Challenges. Scientific Investigations Report 2005–5293. US Geological Survey, Reston, VA.

Bremner-Harrison, S., and B. L. Cypher. 2011. Reintroducing San Joaquin Kit Fox to Vacant and Restored Lands: Identifying Optimal Source Populations and Candidate Foxes. Final Report for US Bureau of Reclamation/Central Valley Program Conservation Project.

Bremner-Harrison, S., B. L. Cypher, and S. W. R. Harrison. 2013. An investigation into the effect of individual personality on reintroduction success, examples from three North American fox species: swift fox, California Channel Island fox and San Joaquin kit fox. Pp.152–158. In P. S. Soorae, ed., Global Reintroduction Perspectives: 2013. IUCN/SSC Reintroduction Specialist Group, Gland, Switzerland/Environment Agency, Abu Dhabi, UAE.

Bremner-Harrison, S., P. A. Prodohl, and R. W. Elwood. 2004. Behavioural trait assessment as a release criterion: boldness predicts early death in a reintroduction programme of captive-bred swift fox (*Vulpes velox*). Animal Conservation 7:313–320.

Brown, C., and K. Laland. 2001. Social learning and life skills training for hatchery reared fish. Journal of Fish Biology 59:471–493.

Brydges, N. M., N. Colegrave, R. J. P. Heathcote, and V. A. Braithwaite. 2008. Habitat stability and predation pressure affect temperament behaviours in populations of three-spined sticklebacks. Journal of Animal Ecology 77:229–235.

Canon, S. K., F. C. Bryant, K. N. Bretzlaff, and J. M. Hellman. 1997. Pronghorn pregnancy diagnosis using trans-rectal ultrasound. Wildlife Society Bulletin 25:832–834.

Carere, C., and K. van Oers. 2004. Shy and bold great tits (*Parsus* major): body temperature and breath rate in response to handling stress. Physiology and Behaviour 82:905–912.

Carlstead, K., J. Mellon, and D. G. Kleimam. 1999. Black rhinoceros (*Diceros bicornis*) in U. S. zoos: I. Individual behavior profiles and their relationship to breeding success. Zoo Biology 18:17–34.

Carmona-Catot, G., P. B. Moyle, E. Aparicio, P. K. Crain, L. C. Thompson, and E. García-Berthou. 2010. Brook trout removal as a conservation tool to restore Eagle Lake rainbow trout. North American Journal of Fisheries Management 30:1315–1323.

Carter, A. J., R. Heinsohn, A. W. Goldizen, and P. A. Biro. 2012a. Boldness, trappability and sampling bias in wild lizards. Animal Behaviour 83:1051–1058.

Carter, A. J., H. H. Marshall, R. Heinsohn, and G. Cowlishaw. 2012b. How not to measure boldness: novel object and antipredator responses are not the same. Animal Behaviour 84:603–609.

Christianson, D., and S. Creel. 2010. A nutritionally mediated risk effect of wolves on elk. Ecology 91:1184–1191.

Coonan, T. J., and C. A. Schwemm. 2009. Factors contributing to success of Island fox reintroductions on San Miguel and Santa Rosa Islands, California. Pp.363–376. In C. C. Damiana and D. K. Garcelon, eds., Proceedings of the 7th California Islands Symposium. Institute for Wildlife Studies, Arcata, CA.

Crane, A. L., and A. Mathis. 2011. Predator-recognition training: a conservation strategy to increase post-release survival of hellbenders in head-starting programs. Zoo Biology 30:611–622.

Cyr, N. E., M. J. Dickens, and L. M. Romero. 2009. Heart rate and heart rate variability responses to acute and chronic stress in a wild-caught passerine bird. Physiological and Biochemical Zoology 82:332–344.

Daltry, J. 2006. Control of the black rat *Rattus rattus* for the conservation of the Antiguan racer *Alsophis antiguae* on Great Bird Island, Antigua. Conservation Evidence 3:27–28.

Davies-Mostert, H. T., M. G. L. Mills, and D. W. Macdonald. 2009. A critical assessment of South Africa's managed metapopulation recovery strategy for African wild dogs. Pp.10–42. In M. W. Hayward and M. J. Somers, eds., Reintroduction of Top-Order Predators. Wiley-Blackwell, Oxford, UK.

DeGregorio, B. A., K. A. Buhlmann, and T. D. Tuberville. 2012. Overwintering of gopher tortoises (*Gopherus polyphemus*) translocated to the northern limit of their geographic range: temperatures, timing, and survival. Chelonian Conservation and Biology 11:84–90.

Dickens, M. J., and L. M. Romero. 2009. Wild European starlings (*Sturnus vulgaris*) adjust to captivity with sustained sympathetic nervous system drive and a reduced fight-or-flight response. Physiological and Biochemical Zoology 82:603–610.

Dickens, M. J., D. J. Delehanty, and L. M. Romero. 2010. Stress: an inevitable component of animal translocation. Biological Conservation 143:1329–1341.

Dickman, C. R., A. S. Glen, and M. Letnic. 2009. Reintroducing the dingo: can Australia's conservation wastelands be restored. Pp.238–269. In M. W. Hayward and M. J. Somers, eds., Reintroduction of Top-Order Predators. Wiley-Blackwell, Oxford, UK.

Dingemanse, N. J., C. Both, P. J. Drent, and J. M. Tinbergen. 2004. Fitness consequences of avian personalities in a fluctuating environment. Proceedings of the Royal Society of London, Series B: Biological Sciences 271:847–852.

Dingemanse, N. J., and D. Reale. 2005. Natural selection and animal personality. Behaviour 142:1159–1184.

Doody, J. S., B. Green, D. Rhind, C. M. Castellano, R. Sims, and T. Robinson. 2009. Population-level declines in Australian predators caused by an invasive species. Animal Conservation 12: 46–53.

Dullum, J. A. L., K. R. Foresman, and M. R. Matchett. 2005. Efficacy of translocations for restoring populations of black-tailed prairie dogs. Wildlife Society Bulletin 33:842–850.

Dunham, K. M. 2000. Dispersal pattern of mountain gazelles (Gazella gazella) released in central Arabia. Journal of Arid Environments 44:247–258.

Ellis, D. H., W. J. Sladen, W. A. Lishman, K. R. Clegg, J. W. Duff, G. F. Gee, and J. C. Lewis. 2003. Motorized migrations: the future or mere fantasy? BioScience 53:260–264.

Ersbak, K., and B. L. Haase. 1983. Nutritional deprivation after stocking as a possible mechanism leading to mortality in stream-stocked brook trout. North American Journal of Fisheries Management 3:142–151.

Finkelstein, M. E., D. F. Doak, D. George, J. Burnett, J. Brandt, M. Church, J. Grantham, and D. R. Smith. 2012. Lead poisoning and the deceptive recovery of the critically endangered California condor. Proceedings of the National Academy of Sciences 109:11449–11454.

Fischer, J., and D. B. Lindenmayer. 2000. An assessment of the published results of animal relocations. Biological Conservation 96:1–11.

Frair, J. L., E. H. Merrill, J. R. Allen, and M. S. Boyce. 2007. Know thy enemy: experience affects elk translocation success in risky landscapes. Journal of Wildlife Management 71:541–554.

Fraser, D. F., F. F. Gilliam, M. J. Daley, A. N. Le, and G. T. Skalski. 2001. Explaining leptokurtic movement distributions: intrapopulation variation in boldness and exploration. American Naturalist 158:124–135.

Garaï, M. E., R. Slotow, R. D. Carr, and B. Reilly. 2004. Elephant reintroductions to small fenced reserves in South Africa. Pachyderm 37:28–36.

García-Díaz, D. F., J. Campion, F. I. Milagro, A. Lomba, F. Marzo, and J. A. Martínez. 2007. Chronic mild stress induces variations in locomotive behavior and metabolic rates in high fat fed rats. Journal of Physiology and Biochemistry 63:337–346.

Germano, J. M., and P. J. Bishop. 2009. Suitability of amphibians and reptiles for translocation. Conservation Biology 23:7–15.

Gibbs, J. P., C. Marquez, and E. J. Sterling. 2008. The role of endangered species reintroduction in ecosystem restoration: tortoise-cactus interactions on Española Island, Galápagos. Restoration Ecology 16:88–93.

Glaser, R., and J. K. Kiecolt-Glaser. 2005. Stress-induced immune dysfunction: implications for health. Nature Reviews Immunology 5:243–251.

Gogan, P. J., D. A. Jessup, and M. Akeson. 1989. Copper deficiency in tule elk at Point Reyes, California. Journal of Range Management 42:233–238.

Goossens, B., J. M. Setchell, E. Tchidongo, E. Dilambaka, C. Vidal, M. Ancrenaz, and A. Jamart. 2005. Survival, interactions with conspecifics and reproduction in 37 chimpanzees released into the wild. Biological Conservation 123:461–475.

Griffin, A. S. 2004. Social learning about predators: a review and prospectus. Learning and Behavior 32:131–140.

Griffin, A. S., D. T. Blumstein, and C. S. Evans. 2000. Training captive-bred or translocated animals to avoid predators. Conservation Biology 14:1317–1326.

Griffith, B., J. M. Scott, J. W. Carpenter, and C. Reed. 1989. Translocation as a species conservation tool: status and strategy. Science 245:477–480.

Hardouin, L. A., A. Robert, M. Nevoux, O. Gimenez, F. Lacroix, and Y. Hingrat. 2014. Meteorological conditions influence short-term survival and dispersal in a reinforced bird population. Journal of Applied Ecology 51:1494–1503.

Harig, A. L., K. D. Fausch, and M. K. Young. 2000. Factors influencing success of greenback cutthroat trout translocations. North American Journal of Fisheries Management 20: 994–1004.

Hervas, S., K. Lorenzen, M. A. Shane, and M. A. Drawbridge. 2010. Quantitative assessment of a white seabass (Atractoscion nobilis) stock enhancement program in California: Post-release dispersal, growth and survival. Fisheries Research 105:237–243.

Herzog, C. J., B. Czéh, S. Corbach, W. Wuttke, O. Schulte-Herbrüggen, R. Hellweg, G. Flügge, and E. Fuchs. 2009. Chronic social instability stress in female rats: a potential animal model for female depression. Neuroscience 159:982–992.

Hoar, W. S. 1988. The physiology of smolting salmonids. Pp.275–343. In Hoar, W. S. and Randall, D. J., eds., Fish Physiology, Vol. XIB. Academic Press, New York.

Hodges, K., J. Brown, and M. Heistermann. 2010. Endocrine monitoring of reproduction and stress. Pp.447–468. In D. G. Kleiman, K. V. Thompson, and C. K. Baer, eds., Wild Mammals in Captivity: Principles and Techniques for Zoo Management. University of Chicago Press, Chicago, IL.

Howard, J., R. M. Santymire, P. E. Marinari, J. S. Kreeger, L. Williamson, and D. E. Wildt. 2006. Use of reproductive technology for black-footed ferret recovery. Pp.28–36. In J. E. Rolle, B. J. Miller, J. L. Godbey, and D. E. Biggins, eds., Recovery of the Black-Footed Ferret: Progress and Continuing Challenges. Scientific Investigations Report 2005–5293. US Geological Survey, Reston, VA.

Hudgens, B. R., and D. K. Garcelon. 2010. Induced changes in island fox (*Urocyon littoralis*) activity do not mitigate the extinction threat posed by a novel predator. Oecologia 165:699–705.

Huntingford, F. A., and N. Giles. 1987. Individual variation in anti-predator responses in the three-spined Stickleback (*Gasterosteus aculeatus* L.). Ethology 74:205–210.

Hyslop, N. 2007. Movements, habitat use, and survival of the threatened eastern indigo snake (*Drymarchon couperi*) in Georgia. PhD dissertation, University of Georgia, Athens.

Jachowski, D. S. 2014. Wild Again: The Struggle to Save the Black-footed Ferret. University of California Press, Berkeley.

Jachowski, D. S., N. L. Brown, M. Wehtje, D. W. Tripp, J. J. Millspaugh, and M. E. Gompper. 2012a. Mitigating plague risk in Utah prairie dogs: evaluation of a systemic flea-control product. Wildlife Society Bulletin 36:167–175.

Jachowski, D. S., R. A. Gitzen, M. B. Grenier, B. Holmes, and J. J. Millspaugh. 2011. The importance of thinking big: large-scale prey conservation drives black-footed ferret reintroduction success. Biological Conservation 144: 1560–1566.

Jachowski, D. S., J. J. Millspaugh, D. E. Biggins, T. M. Livieri, and M. R. Matchett. 2010. Home-range size and spatial organization of black-footed ferrets *Mustela nigripes* in South Dakota, USA. Wildlife Biology 16:66–76.

Jachowski, D. S., R. A. Montgomery, R. Slotow, and J. J. Millspaugh. 2013a. Unravelling complex associations between physiological state and movement of African elephants. Functional Ecology 27:1166–1175.

Jachowski, D. S., R. Slotow, and J. J. Millspaugh. 2012b. Physiological stress and refuge behavior by African elephants. PLoS ONE 7:e31818.

Jachowski, D. S., R. Slotow, and J. J. Millspaugh. 2013b. Delayed physiological acclimatization by African elephants following reintroduction. Animal Conservation 16:575–583.

Jachowski, D. S., R. Slotow, and J. J. Millspaugh. 2013c. Corridor use and streaking behavior by African elephants in relation to physiological state. Biological Conservation 167:276–282.

Jachowski, D. S., R. Slotow, and J. J. Millspaugh. 2014. Good virtual fences make good neighbors: opportunities for conservation. Animal Conservation 17:187–196.

Jakober, M. J., T. E. McMahon, R. F. Thurow, and C. G. Clancy. 1998. Role of stream ice on fall and winter movements and habitat use by bull trout and cutthroat trout in Montana headwater streams. Transactions of the American Fisheries Society 127:223–235.

Jamieson, I. G., and K. Ludwig. 2012. Rat-wise robins quickly lose fear of rats when introduced to a rat-free island. Animal Behaviour 84:225–229.

Jones, C. G., and D. V. Merton. 2012. A tale of two islands: the rescue and recovery of endemic birds in New Zealand and Mauritius. Pp.33–72. In J. G. Ewen, D. P. Armstrong, K. A. Parker, and P. J. Seddon, eds., Reintroduction Biology: Integrating Science and Management. Wiley-Blackwell, Oxford, UK.

Korhonen, H., and P. Niemela. 1996. Temperament and reproductive success in farmbred silver foxes housed with and without platforms. Journal of Animal Breeding and Genetics 113:209–218.

Kramer-Schadt, S., E. Revilla, T. Wiegand, and U. R. S. Breitenmoser. 2004. Fragmented landscapes, road mortality and patch connectivity: modelling influences on the dispersal of Eurasian lynx. Journal of Applied Ecology 41:711–723.

Kress, S. W. 1997. Using animal behavior for conservation: case studies in seabird restoration from the Maine coast, USA. Journal of the Yamashina Institute for Ornithology 29: 1–26.

Lagendijk, D. G., R. L. Mackey, B. R. Page, and R. Slotow. 2011. The effects of herbivory by a mega-and mesoherbivore on tree recruitment in Sand Forest, South Africa. PLoS ONE 6:e17983.

Le Gouar, P., A. Robert, J. P. Choisy, S. Henriquet, P. Lecuyer, C. Tessier, and F. Sarrazin. 2008. Roles of survival and dispersal in reintroduction success of griffon vulture (*Gyps fulvus*). Ecological Applications 18:859–872.

Le Gouar, P., Mihoub, J. B., and Sarrazin, F. (2012). Dispersal and habitat selection: behavioural and spatial constraints for animal translocations. Pp.138–164. In J. G. Ewen, D. P. Armstrong, K. A. Parker, and P. J. Seddon, eds., Reintroduction Biology: Integrating Science and Management. Wiley-Blackwell, Oxford, UK.

Lee, P. C., and C. J. Moss. 2012. Wild female African elephants (*Loxodonta africanus*) exhibit personality traits of leadership and social integration. Journal of Comparative Psychology 126:224–232.

Letnic, M., J. K. Webb, and R. Shine. 2008. Invasive cane toads (*Bufo marinus*) cause mass mortality of freshwater crocodiles (*Crocodylus johnstoni*) in tropical Australia. Biological Conservation 141:1773–1782.

Lettink, M., G. Norbury, A. Cree, P. J. Seddon, R. P. Duncan, and C. J. Schwarz. 2010. Removal of introduced predators, but not artificial refuge supplementation, increases skink survival in coastal duneland. Biological Conservation 143:72–77.

Letty, J., S. Marchandeau, J. Clobert, and J. Aubineau. 2000. Improving translocation success: an experimental study of anti-stress treatment and release method for wild rabbits. Animal Conservation 3:211–219.

Linnell, J. D., R. Aanes, J. E. Swenson, J. Odden, and M.E., Smith. 1997. Translocation of carnivores as a method for managing problem animals: a review. Biodiversity and Conservation 6:1245–1257.

Llewelyn, J., J. K. Webb, L. Schwarzkopf, R. Alford, and R. Shine. 2010. Behavioural responses of carnivorous marsupials (*Planigale maculata*) to toxic invasive cane toads (*Bufo marinus*). Austral Ecology 35:560–567.

Lloyd, B. D., and R. G. Powlesland. 1994. The decline of kakapo (*Strigops habroptilus*) and attempts at conservation by translocation. Biological Conservation 69:75–85.

Lockwood, M. A., C. P. Griffin, M. E. Morrow, C. J. Randel, and N. J. Silvy. 2005. Survival, movements, and reproduction of released captive-reared Attwater's prairie-chicken. Journal of Wildlife Management 69:1251–1258.

Maloney, R. F., and I. G. McLean. 1995. Historical and experimental learned predator recognition in free-living New Zealand robins. Animal Behaviour 50:1193–1201.

Marinari, P. E., and J. S. Kreeger. 2006. An adaptive management approach for black-footed ferrets in captivity. Pp.23–27. In J. E. Rolle, B. J. Miller, J. L. Godbey, and D. E. Biggins, eds., Recovery of the Black-footed Ferret: Progress and Continuing Challenges. Scientific Investigations Report 2005–5293. US Geological Survey, Reston, VA.

McCleery, R., J. A. Hostetler, and M. K. Oli. 2014. Better off in the wild? Evaluating a captive breeding and release program for the recovery of an endangered rodent. Biological Conservation 169:198–205.

McCleery, R., M. K. Oli, J. A. Hostetler, B. Karmacharya, D. Greene, C. Winchester, J. Gore, S. Sneckenberger, S. B. Castleberry, and M. T. Mengak. 2013. Are declines of an endangered mammal predation-driven, and can a captive-breeding and release program aid their recovery? Journal of Zoology 291:59–68.

McCormick, S. D., L. P. Hansen, T. P. Quinn, and R. L. Saunders. 1998. Movement, migration, and smolting of Atlantic salmon (*Salmo salar*). Canadian Journal of Fisheries and Aquatic Sciences 55:77–92.

McEwen, B. S., and J. C. Wingfield. 2003. The concept of allostasis in biology and biomedicine. Hormones and Behavior 43:2–15.

McLean, I. G., C. Holzer, and B. J. S. Studholne. 1999. Teaching predator-recognition to a naïve bird. Implications for management. Biological Conservation 87:123–130.

McLean, I. G., G. Lundie-Jenkins, and P. J. Jarman. 1996. Teaching an endangered mammal to recognize predators. Biological Conservation 75:51–62.

McLean, I. G., N. T. Schmitt, P. J. Jarman, C. Duncan, and C. D. L. Wynne. 2000. Learning for life: training marsupials to recognize introduced predators. Animal Behaviour 137:1361–1376.

McPhee, M. E. 2003. Generations in captivity increases behavioural variance: considerations for captive breeding and reintroduction programs. Biological Conservation 115:71–77.

Mech, D. L. 2012. Is science in danger of sanctifying the wolf? Biological Conservation 150:143–149.

Melnychuk, M. C., J. Korman, S. Hausch, D. W. Welch, D. J. F. McCubbing, and C. J. Walters. 2014. Marine survival difference between wild and hatchery-reared steelhead trout determined during early downstream migration. Canadian Journal of Fisheries and Aquatic Sciences 71:831–846.

Meretsky, V. J., N. F. Snyder, S. R. Beissinger, D. A. Clendenen, and J. W. Wiley. 2000. Demography of the California Condor: implications for reestablishment. Conservation Biology 14:957–967.

Millspaugh, J. J., D. C. Kesler, R. W. Kays, R. A. Gitzen, J. H. Schulz, J. L. Belant, C. T. Rota, B. J. Keller, and C. M. Bodinof. 2012. Wildlife radio-tracking and remote monitoring. Pp.258–283. In N. Silvy, ed., Wildlife Techniques Manual. John Hopkins Press, Baltimore, MD.

Mitchell, A. M., T. I. Wellicome, D. Brodie, and K. M. Cheng. 2011. Captive-reared burrowing owls show higher site-affinity, survival, and reproductive performance when reintroduced using a soft-release. Biological Conservation 144:1382–1391.

Mönig, H., T. Arendt, M. Meyer, S. Kloehn, and B. Bewig. 1999. Activation of the hypothalamo-pituitary-adrenal axis in response to septic or non-septic diseases: implications for the euthyroid sick syndrome. Intensive Care Medicine 25:1402–1406.

Moore, S. J., and P. F. Battley. 2006. Differences in the digestive organ morphology of captive and wild Brown Teal Anas chlorotis and implications for releases. Bird Conservation International 16:253–264.

Morrison, S. A., N. Macdonald, K. Walker, L. Lozier, and M. R. Shaw. 2007. Facing the dilemma at eradication's end: uncertainty of absence and the Lazarus effect. Frontiers in Ecology and the Environment 5:271–276.

Moseby, K. E., A. Cameron, and H. A. Crisp. 2012. Can predator avoidance training improve reintroduction outcomes for the greater bilby in arid Australia? Animal Behaviour 83:1011–1021.

Moseby, K. E., J. L. Read, D. C. Paton, P. Copley, B. M. Hill, and H. A. Crisp. 2011. Predation determines the outcome of 10 reintroduction attempts in arid South Australia. Biological Conservation 144:2863–2872.

Myers, J. H., D. Simberloff, A. M. Kuris, and J. R. Carey. 2000. Eradication revisited: dealing with exotic species. Trends in Ecology and Evolution 15:316–320.

Nielsen, C., K. Aarestrup, and S. S. Madsen. 2006. Comparison of physiological smolt status in descending and nondescending wild brown trout (Salmo trutta) in a Danish stream. Ecology of Freshwater Fish 15:229–236.

Nielsen, C., K. Aarestrup, U. Nørum, and S. S. Madsen. 2004. Future migratory behaviour predicted from premigratory levels of gill Na+/K+-ATPase activity in individual wild brown

trout (Salmo trutta). Journal of Experimental Biology 207:527–533.

Nielsen, C., G. Holdensaard, H. C. Petersen, B. T. Björnsson, and S. S. Madsen. 2001. Genetic differences in physiology, growth hormone levels and migratory behaviour of Atlantic salmon smolts. Journal of Fish Biology 59:28–44.

Novotny, A. J. 1980. Delayed release of salmon. Pp.325–369. In J. E. Thorpe, ed., Salmon Ranching. Academic Press, London.

O'Donnell, S., J. K. Webb, and R. Shine. 2010. Conditioned taste aversion enhances the survival of an endangered predator imperiled by a toxic invader. Journal of Applied Ecology 47:558–656.

Osborne, P. E., and P. J. Seddon. 2012. Selecting suitable habitats for reintroductions: variation, change and the role of species distribution modelling. Pp.73–104. In J. G. Ewen, D. P. Armstrong, K. A. Parker, and P. J. Seddon, eds., Reintroduction Biology: Integrating Science and Management. Wiley-Blackwell, Oxford, UK.

Parish, D. M. B., and N. W. Sotherton. 2007. The fate of released captive-reared grey partridges Perdix perdix: implications for reintroduction programmes. Wildlife Biology 13:140–149.

Parker, K. A., M. J. Dickens, R. H. Clarke, and T. G. Lovegrove. 2012. The theory and practice of catching, holding, moving and releasing animals. Pp.105–137. In J. G. Ewen, D. P. Armstrong, K. A. Parker, and P. J. Seddon, eds., Reintroduction Biology: Integrating Science and Management. Wiley-Blackwell, Oxford, UK.

Parlato, E. H., and D. P. Armstrong. 2013. Predicting post-release establishment using data from multiple reintroductions. Biological Conservation 160:97–104.

Patrick, S., and H. Wiemerskirch. 2014. Personality, foraging and fitness consequences in a long lived seabird. PLoS ONE 9:e87269.

Phillips, M. 2013. Establishment of a desert bighorn sheep population to the Fra Cristobal Mountains, New Mexico, USA. Pp.198–203. In P. S. Soorae, ed., Global Reintroduction Perspectives: 2013. IUCN/SSC Reintroduction Specialist Group, Gland, Switzerland/Environment Agency, Abu Dhabi, UAE.

Polak, T., Y. Gutterman, I. Hoffman and D. Saltz. 2014. Redundancy in seed dispersal by three sympatric ungulates: a reintroduction perspective. Animal Conservation 17:565–572.

Price, M. R. S. 1989. Animal Re-introductions: The Arabian oryx in Oman. Cambridge University Press, Cambridge, UK.

Price-Rees, S. J., J. K. Webb, and R. Shine. 2011. School for skinks: can conditioned taste aversion enable blue-tongue lizards (*Tiliqua scincoides*) to avoid toxic cane toads (*Rhinella marina*) as prey? Ethology 117:749–757.

Primack, R.B. 1996. Lessons from ecological theory: dispersal, establishment, and population structure. Pp.209–213. In D.A. Falk, C.I. Millar, and M. Olwell, eds., Restoring Diversity. Island Press, Washington, DC.

Pruitt, J.N., S.E. Riechert, and T.C. Jones. 2008. Behavioural syndromes and their fitness consequences in a socially polymorphic spider, *Anelosimus studiosus*. Animal Behaviour 76:871–879.

Read, J.L., G.R. Johnston, and T.P. Morley. 2011. Predation by snakes thwarts trial reintroduction of the endangered woma python *Aspidites ramsayi*. Oryx 45:505–512.

Réale, D., and M. Festa-Bianchet. 2003. Predator-induced natural selection on temperament in bighorn ewes. Animal Behaviour 65:463–470.

Réale, D., B.Y. Gallant, M. LeBlanc, and M. Festa-Bianchet. 2000. Consistency in temperament in bighorn ewes and correlates with behaviour and life history. Animal Behaviour 60:589–597.

Réale, D., J. Martin, D.W. Coltman, J. Poissant, and M. Festa-Bianchet. 2009. Male personality, life-history strategies and reproductive success in a promiscuous mammal. Journal of Evolutionary Biology 22:1599–1607

Rhymer, J.M., and D. Simberloff. 1996. Extinction by hybridization and introgression. Annual Review of Ecology and Systematics 27:83–109.

Richards, J.D., and J. Short. 2003. Reintroduction and establishment of the western barred bandicoot *Perameles bougainville* (Marsupialia: Peramelidae) at Shark Bay, Western Australia. Biological Conservation 109:181–195.

Ricklefs, R.E., and M. Wikelski. 2002. The physiology/life-history nexus. Trends in Ecology and Evolution 17:462–468.

Ripple, W.J., and R.L. Beschta. 2003. Wolf reintroduction, predation risk, and cottonwood recovery in Yellowstone National Park. Forest Ecology and Management 184:299–313.

Ripple, W.J., E.J. Larsen, R.A. Renkin, and D.W. Smith. 2001. Trophic cascades among wolves, elk and aspen on Yellowstone National Park's northern range. Biological Conservation 102:227–234.

Robinson, A.T., and D.L. Ward. 2011. Interactions between desert pupfish and Gila topminnow can affect reintroduction success. North American Journal of Fisheries Management 31:1093–1099.

Roe, J.H., M.R. Frank, S.E. Gibson, O. Attum, and B.A. Kingsbury. 2010. No place like home: an experimental comparison of reintroduction strategies using snakes. Journal of Applied Ecology 47:1253–1261.

Roemer, G.W., T.J. Coonan, D.K. Garcelon, J. Bascompte, and L. Laughrin. 2001. Feral pigs facilitate hyperpredation by golden eagles and indirectly cause the decline of the island fox. Animal Conservation 4:307–318.

Rohlf, D.J., C. Carroll, and B. Hartl. 2014. Conservation-reliant species: toward a biology-based definition. BioScience 64:601–611.

Romero, M.L., and L.K. Butler. 2007. Endocrinology of stress. International Journal of Comparative Psychology 20:89–95.

Rominger, E.M., H.A. Whitlaw, D.L. Weybright, W.C. Dunn, and W.B. Ballard. 2004. The influence of mountain lion predation on bighorn sheep translocations. Journal of Wildlife Management 68:993–999.

Rowe, D.K., and J.E. Thorpe. 1990. Differences in growth between maturing and non-maturing male Atlantic salmon, *Salmo salar L.*, parr. Journal of Fish Biology 36:643–658.

Rudin, F.S., and M. Briffa. 2012. Is boldness a resource-holding potential trait? Fighting prowess and changes in startle response in the sea anemone, *Actinia equina*. Proceedings of the Royal Society Series B 279:1904–1910.

Saenz, D., K.A. Baum, R.N. Conner, D.C. Rudolph, and R. Costa. 2002. Large-scale translocation strategies for reintroducing red-cockaded woodpeckers. Journal of Wildlife Management 66:212–221.

Sapolsky, R.M., L.M. Romero, and A.U. Munck. 2000. How do glucocorticoids influence stress responses? Integrating permissive, suppressive, stimulatory, and preparative actions. Endocrine Reviews 21:55–89.

Sarrazin, F., C. Bagnolinp, J.L. Pinna, and E. Danchin. 1996. Breeding biology during establishment of a reintroduced griffon vulture *Gyps fulvus* population. Ibis 138:315–325.

Scheepers, J.L., and K.A.E. Venzke. 1995. Attempts to reintroduce African wild dogs *Lycaon pictus* into Etosha National Park, Namibia. South African Journal of Wildlife Research 25:138–140.

Scofield, R.P., R. Cullen, and M. Wang. 2011. Are predator-proof fences the answer to New Zealand's terrestrial faunal biodiversity crisis? New Zealand Journal of Ecology 35:312–317.

Scott, J.M., D.D. Goble, A.M. Haines, J.A. Wiens, and M.C. Neel. 2010. Conservation-reliant species and the future of conservation. Conservation Letters 3:91–97.

Shier, D.M., and D.H. Owings. 2006. Effects of predator training on behavior and post-release survival of captive prairie dogs (*Cynomys ludovicianus*). Biological Conservation 132:126–135.

Shine, R. 2010. The ecological impact of invasive cane toads (*Bufo marinus*) in Australia. The Quarterly Review of Biology 85:253–291.

Short, J., and B. Turner. 2000. Reintroduction of the burrowing bettong *Bettongia lesueur* (Marsupialia: Potoroidae) to mainland Australia. Biological Conservation 96:185–196.

Sih, A., A. Bell, and J.C. Johnson. 2004. Behavioral syndromes: an ecological and evolutionary overview. Trends in Ecology and Evolution 19:372–378.

Singer, F.J., C.M. Papouchis, and K.K. Symonds. 2000. Translocations as a tool for restoring populations of bighorn sheep. Restoration Ecology 8:6–13.

Singh, N.J., and E.J. Milner-Gulland. 2011. Conserving a moving target: planning protection for a migratory species as its distribution changes. Journal of Applied Ecology 48:35–46.

Sinn, D.L., L. Cawthen, S.M. Jones, C. Pukk, and M.E. Jones. 2014. Boldness towards novelty and translocation success in captive-raised, orphaned Tasmanian devils. Zoo Biology 33:36–48.

Slotow, R. 2012. Fencing for purpose: a case study of elephants in South Africa. Pp.91–104. In M.J. Somers and M.W. Hayward, eds., Fencing for Conservation. Springer, New York.

Slotow, R., and G. van Dyk. 2001. Role of delinquent young "orphan" male elephants in high mortality of white rhinoceros in Pilanesberg National Park, South Africa. Koedoe 44:85–94.

Slotow, R., G. van Dyk, J. Poole, B. Page, and A. Klocke. 2000. Older bull elephants control young males. Nature 408:425–426.

Smith, C.R., R.E. Cartee, J.T. Hathcock, and D.W. Speake. 1989. Radiographic and ultrasonographic scanning of gravid eastern indigo snakes. Journal of Herpetology 23:426–429.

Steury, T.D., and D.L. Murray. 2004. Modeling the reintroduction of lynx to the southern portion of its range. Biological Conservation 117:127–141.

Stiles, J.A. 2013. Evaluating the use of enclosures to reintroduce eastern indigo snakes. PhD dissertation, Auburn University, Auburn, AL.

Støen, O.G., M.L. Pitlagano, and S.R. Moe. 2009. Same-site multiple releases of translocated white rhinoceroses *Ceratotherium simum* may increase the risk of unwanted dispersal. Oryx 43:580–585.

Suboski, M.D., and J.J. Templeton. 1989. Life skills training for hatchery fish: social learning and survival. Fisheries Research 7:343–352.

Sundström, L.F., E. Petersson, J. Höjesjö, J.I. Johnsson, and T. Järvi. 2004. Hatchery selection promotes boldness in newly hatched brown trout (*Salmo trutta*): implications for dominance. Behavioral Ecology 15:192–198.

Tarszisz, E., C.R. Dickman, and A.J. Munn. 2014. Physiology in conservation translocations. Conservation Physiology 2:cou054.

Teixeira, C.P., C.S. De Azevedo, M. Mendl, C.F. Cipreste, and R.J. Young. 2007. Revisiting translocation and reintroduction programmes: the importance of considering stress. Animal Behaviour 73:1–13.

Thorne, E., and E.S. Williams. 1988. Disease and endangered species: the black-footed ferret as a recent example. Conservation Biology 2:66–74.

Thorstad, E.B., F. Whoriskey, I. Uglem, A. Moore, A.H. Rikardsen, and B. Finstad. 2012. A critical life stage of the Atlantic salmon *Salmo salar*: behaviour and survival during the smolt and initial post-smolt migration. Journal of Fish Biology 81:500–542.

Towns, D. 2005. Eradication of introduced mammals and reintroduction of the tuatara *Sphenodon punctatus* to Motuhora/Whale Island, New Zealand. Conservation Evidence 2:92–93.

Trakhtenbrot, A., R. Nathan, G. Perry, and D.M. Richardson. 2005. The importance of long-distance dispersal in biodiversity conservation. Diversity and Distributions 11:173–181.

Truett, J.C., J.A.L. Dullum, M.R. Matchett, E. Owens, and D. Seery. 2001. Translocating prairie dogs. Wildlife Society Bulletin 29:863–872.

Tuberville, T.D., E.E. Clark, K.A. Buhlmann, and J.W. Gibbons. 2005. Translocation as a conservation tool: site fidelity and movement of repatriated gopher tortoises (*Gopherus polyphemus*). Animal Conservation 8:349–358.

Vernesi, C., E. Pecchioli, D. Caramelli, R. Tiedemann, E. Randi, and G. Bertorelle. 2002. The genetic structure of natural and reintroduced roe deer (*Capreolus capreolus*) populations in the Alps and central Italy, with reference to the mitochondrial DNA phylogeography of Europe. Molecular Ecology 11:1285–1297.

Walker, B.G., P.D. Boersma, and J.C. Wingfield. 2005. Field endocrinology and conservation biology. Integrative and Comparative Biology 45:12–18.

Waples, K.A., and C.S. Stagoll. 1997. Ethical issues in the release of animals from captivity. BioScience 47:115–121.

Ward, D. M., K. H. Nislow, and C. L. Folt. 2008. Do native species limit survival of reintroduced Atlantic salmon in historic rearing streams? Biological Conservation 141:146–152.

Wasser, S. K., J. C. Azkarate, R. K. Booth, L. Hayward, K. Hunt, K. Ayres, C. Vynne, K. Gobush, D. Canales-Espinosa, and L. E. Rodriguez-luna. 2010. Non-invasive measurement of thyroid hormone in feces of a diverse array of avian and mammalian species. General and Comparative Endocrinology 168:1–7.

Watters, J. L., and C. L. Meehan. 2007. Different strokes: can managing behavioral types increase post-release success? Applied Animal Behaviour Science 102:364–379.

Weilenmann, M., M. Gusset, D. R. Mills, T. Gabanapelo, and M. Schiess-Meier. 2011. Is translocation of stock-raiding leopards into a protected area with resident conspecifics an effective management tool? Wildlife Research 37:702–707.

White, T. H., N. J. Collar, R. J. Moorhouse, V. Sanz, E. D. Stolen, and D. J. Brightsmith. 2012. Psittacine reintroductions: common denominators of success. Biological Conservation 148:106–115.

White, T. H., J. A. Collazo, and F. J. Vilella. 2005. Survival of captive-reared Puerto Rican parrots released in the Caribbean National Forest. The Condor 107:424–432.

Whiting, J. C., R. T. Bowyer, J. T. Flinders, and D. L. Eggett. 2011. Reintroduced bighorn sheep: fitness consequences of adjusting parturition to local environments. Journal of Mammalogy 92:213–220.

Whiting, J. C., D. D. Olson, J. M. Shannon, R. T. Bowyer, R. W. Klaver, and J. T. Flinders. 2012. Timing and synchrony of births in bighorn sheep: implications for reintroduction and conservation. Wildlife Research 39:565–572.

Wilson, A. D. M., and E. D. Stevens. 2005. Consistency in context-specific measures of shyness and boldness in rainbow trout, *Oncorhynchus mykiss*. Ethology 111:849–862.

Wilson, D. S., K. Coleman, A. B. Clark, and L. Biederman. 1993. Shy-bold continuum in pumpkinseed sunfish (*Lepomis gibbosus*): an ecological study of a psychological trait. Journal of Comparative Psychology 107:250–260.

Winchester, C., S. B. Castleberry, and M. T. Mengak. 2009. Evaluation of factors restricting distribution of the endangered key largo woodrat. Journal of Wildlife Management 73:374–379.

Witmer, G. W., F. Boyd, and Z. Hillis-Starr. 2007. The successful eradication of introduced roof rats (*Rattus rattus*) from Buck Island using diphacinone, followed by an irruption of house mice (*Mus musculus*). Wildlife Research 34:108–115.

Wolf, K. N., D. E. Wildt, A. Vargas, P. E. Marinari, M. A. Ottinger, and J. G. Howard. 2000. Reproductive inefficiency in male black-footed ferrets (*Mustela nigripes*). Zoo Biology 19:517–528.

TEN

Why You Cannot Ignore Disease When You Reintroduce Animals

Erin Muths and Hamish McCallum

ALL WILD ANIMAL POPULATIONS harbor a range of parasitic organisms, ranging from viruses and bacteria to multicellular parasites such as helminths and arthropods. While some of these are mutualists and some are commensal, others cause infectious disease in at least some members of the population and some may have substantial population-level impacts. The reintroduction[1] of animals requires an immense amount of effort, and considering known and potential diseases is crucial to all phases of a well-formulated reintroduction plan (site and donor selection, implementation, and monitoring). While disease has been recognized as a potential factor in reintroductions for decades (e.g., Griffith et al. 1993), heightened conservation concerns and increasing numbers of emerging pathogens bring a consideration of disease to the forefront of many projects.

Infectious disease may have been the primary factor leading to the disappearance or decline of a species that has necessitated its reintroduction. Alternatively, the decline may have occurred for other reasons and the question may be how best to reintroduce the species, taking into account issues raised by parasites and pathogens. This latter question has been extensively reviewed previously (Ballou 1993, Viggers et al. 1993, Mathews et al. 2006, Sainsbury and Vaughan-Higgins 2012), but we will summarize the main issues. How to approach reintroduction when disease has been the cause of extinction or near extinction has been discussed less fully; we therefore concentrate on this aspect of the problem.

One under-considered aspect of reintroduction within the context of disease is how the various levels of uncertainty can affect decision-making (e.g., Ewen et al. 2014, Jakob-Hoff et al. 2014, Canessa 2015). We briefly address this topic of adapting strategies as new information arises, but other chapters (Chapters 5

---

1 There are a variety of terms referring to the movement of wildlife by humans including translocation, repatriation, and reintroduction among others. We use reintroduction throughout to indicate the movement of wildlife by humans into an area where they are currently absent (possibly for numerous reasons).

and 6, this volume) consider uncertainty and methods to deal with it more explicitly.

Another aspect that is under-considered is the specificity of the disease organism to the target species. The role of disease and possible management actions in reintroductions depend strongly on whether the parasite or pathogen in question is specific to the species being reintroduced. Most infectious diseases that are important in conservation biology affect more than one host species (de Castro and Bolker 2005). There is usually one or more reservoir species on which the pathogen or parasite has a limited effect, in contrast to the species of conservation concern that is far more susceptible. For example, although the amphibian chytrid fungus (*Batrachochytrium dendrobatidis*) has caused the extinction of numerous frog species worldwide, particularly in the Americas and Australia (Skerratt et al. 2007, Heard et al. 2011), some amphibian species are tolerant of infection and can maintain significant levels of infection without greatly increased mortality (e.g., bullfrogs, Schloegel et al. 2010). The existence of a reservoir species allows a high force of infection to be maintained even as susceptible species decline toward extinction. It is less common for a parasite or pathogen to threaten the extinction of species for which it is the sole host, because once the host has become sufficiently rare, transmission will decrease and the disease is likely to become extinct before the host. However, this may not be the case if transmission depends weakly on host density.

REINTRODUCTION FOLLOWING DISEASE-INDUCED DECLINE OR EXTINCTION

One of the basic principles of conservation biology, and reintroductions in particular, is that if you have not identified, and mitigated or removed, the agent or situation that is responsible for the decline in the first place, simply adding animals to a landscape is likely to fail (Caughley 1994). We present a tree (figure 10.1) to guide decision-making following dis-

ease-induced local extinction or decline[2] (Boxes 10.1 and 10.2).

Decision Point One: Is Reintroduction Necessary Because of Disease-Induced Extinction or Decline?

The first decision point is to confirm whether reintroduction is necessary because of disease-induced extinction, or whether extinction occurred primarily for another reason. Extinction of an entire species due to disease has rarely been unequivocally demonstrated—an analysis based on the 2004 IUCN Red List found that infectious disease had been identified as contributing to extinction in only 31/833 listed extinctions of animals and plants (Smith et al. 2006). However, it is likely that this may be a substantial underestimate (McCallum 2012). The most likely situation in which reintroduction may be feasible will be one in which disease has caused local or regional extirpation rather than global extinction. If extinction has occurred primarily for a reason other than infectious disease, it is also important to consider disease issues when planning reintroductions (see chapters in Part 3 of this volume). The remainder of the decision tree, however, is concerned with cases where there is confidence that infectious disease was the primary factor in the disappearance of the population where reintroduction is proposed as a strategy.

Decision Point Two: Is Resistant/Tolerant Stock Available?

The second decision point, given that reintroduction is necessary because of disease, is governed by whether a stock of animals resistant or tolerant to the disease in question is available. These may be from another wild population that has been exposed to this pathogen for some time and has evolved resistance, a population or

---

2 In the case of a decline, reintroduction of animals would be to augment the existing population to promote persistence.

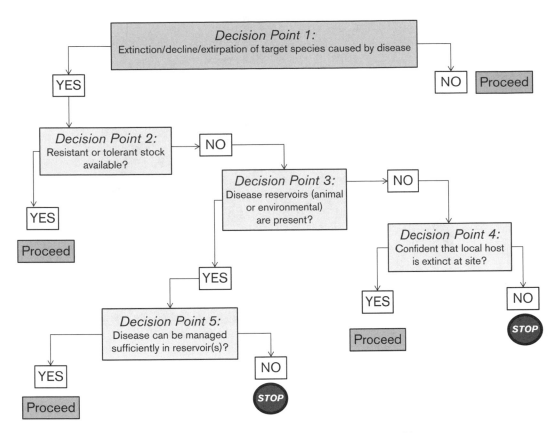

FIGURE 10.1 Simple decision tree to guide reintroduction decisions within the context of disease.

individual that has been artificially selected for resistance or tolerance, or a population or individual that may have been treated with a probiotic that induces resistance (see, e.g., Woodhams et al. 2011). Questions of resistance and tolerance to pathogens have been addressed minimally in the reintroduction literature and are a crucial piece of missing information for conservation efforts of this kind. If information is not acquired before species are in need of reintroduction, it is often difficult to acquire because there may be too few individuals or too few populations to yield applicable information. However, a number of possibilities exist within the realm of resistance and tolerance to pathogens that may be useful in reintroductions. A laboratory study using guppies (*Poecilia reticulata*) as a model organism showed that loads of the parasite *Gyrodactylus turnbulli* were lower in guppies "reintroduced" to tanks with resident infected fish, if the reintroduced guppies had been preexposed to the parasite (Faria et al. 2010). This experiment suggests that preexposure to those parasites that reintroduced animals will encounter in the wild may be advisable. Innate tolerances may differ among subspecies (e.g., white-tailed deer, *Odocoileus virginianus*; Gaydos et al. 2002) with negative implications if subspecies with lower innate resistance are substituted in translocation projects as well as positive implications if the converse is tried. In amphibians, some populations of the boreal toad in the Rocky Mountains, United States, persist with disease (Pilliod et al. 2010), suggesting an innate resistance (Box 10.1). Other research on this species shows that some genetic lines have higher survival when challenged by disease (Murphy et al. 2009). There are several examples of the evolution of resistance or tolerance of pathogens following

10a Boreal toads *(Anaxyrus [Bufo] boreas)* in amplexus. Photo by
D. Herasimtschuk.

There is clearly no single cause for the global phenomenon of amphibian decline, but disease (including, but not limited to, the amphibian chytrid fungus *Batrachochytrium dendrobatidis* [Bd]) has been labeled as one of the "enigmatic" causes (Stuart et al. 2004) and in a number of cases is culpable for observed declines. This case study (Muths et al. 2014) describes a reintroduction effort that was begun with a limited understanding of the disease. It serves as a cautionary tale and provides lessons for future amphibian reintroductions and data to inform management decisions. This effort was initiated through collaboration among researchers at the US Geological Survey, and managers at Rocky Mountain National Park (RMNP), Colorado Parks and Wildlife (CPW), and Colorado State University.

Context: The Target Species

Boreal toads (*Anaxyrus [Bufo] boreas*; figure 10a) inhabit high-elevation wetlands, ponds, and wet meadows (Hammerson 1999) that are located primarily in protected areas considered to be "pristine" (e.g., national parks and other federal lands). Despite this, major declines in boreal toads in the Southern Rocky Mountains (SRM, southern Wyoming, Colorado, and northern New Mexico) were

observed in the 1970s (Carey 1993) and 1980s (Corn et al. 1989). In RMNP, the number of populations of toads declined in the 1980s and 1990s (Corn et al. 1997). Significant population declines, coincident with the identification of Bd, occurred in the mid-1990s (Muths et al. 2003). The identification of Bd coincident with population die-offs provided evidence for the likely, or at least proximate, cause of the observed declines of toads in the SRM and precipitated a reintroduction effort in RMNP that began in 2004. The boreal toad in the SRM is part of the eastern clade of boreal toads in the Western United States. Evidence suggests that toads in the SRM may be a distinct population segment although reports are somewhat equivocal (Goebel et al. 2009, Switzer et al. 2009). The boreal toad is considered to be a species of concern in Wyoming, an endangered species in Colorado, and is extirpated in New Mexico, except for one population of reintroduced individuals.

The initial plan for reintroduction was formulated within the context of the Boreal Toad Recovery Plan (Loeffler 2001) that provided basic requirements for reintroductions including acknowledgment of the importance of disease. However, our understanding of Bd—the focal disease when the recovery plan was written—was limited.

Context: The Disease

Chytridiomycosis is caused by Bd, a fungal pathogen that thickens the skin (Berger et al. 1998)—an important component of an amphibian's respiratory system—and thus interferes with water intake and electrolyte balance (Voyles et al. 2009). Chytridiomycosis can cause death in individuals (Pessier et al. 1999) and appears to reduce survival in populations (Muths et al. 2011). It has been associated with population-level declines and extirpations around the world in a variety of species (e.g., Central America [*Atelopus* species], Lips et al. 2008; Australia [*Taudactylus acutirostris*], Berger et al. 1998; United States, the mountain yellow-legged frog [*Rana muscosa*], Vredenburg et al. 2010). The origin of the Bd in the Western United States is debated (Oeullet et al. 2005, Rachowicz et al. 2005, Fisher et al. 2009).

The virulence of Bd varies among strains (Berger et al. 2005, Carey et al. 2006) and susceptibility to the disease varies among amphibians (Daszak et al. 2005, Ouellet et al. 2005, Adams et al. 2007, Longcore et al. 2007, Murphy et al. 2009). For example, in contrast to observed pathogenicity in some species, other species like the American bullfrog (*Lithobates catesbeiana*, Schloegel et al. 2010) and the Pacific tree frog (*Pseudacris regilla*, Reeder et al. 2012) appear to be carriers of this fungal disease and apparently suffer no ill effects. The broad range of Bd (identified on all continents except Antarctica) makes it an important factor in amphibian decline and in some cases, a primary factor to consider in reintroductions.

Although Bd has been the focus of extensive research since it was identified in the late 1990s (Berger et al. 1998, Longcore et al. 1999), and despite the need to consider the potential effect of Bd relative to management actions, there is still little definitive information on predicting the effects of infection on populations or individual species (but see e.g., Briggs et al. 2005, 2010). In contrast to many declining or extirpated populations, some populations appear to be surviving with Bd. Recent work from California suggests that some populations of yellow-legged frogs are persisting with Bd at an endemic state, that is, infected adults have low fungal loads, survive between years, and frequently lose and regain infection (Briggs et al. 2010). This suggests that the host and pathogen dynamic is variable.

Research in Colorado and Wyoming suggests that under favorable environmental conditions, populations of western toads may be able to compensate for lower survival with increases in recruitment (Muths et al. 2011), thus offsetting or at least slowing population declines associated with Bd. These bits of information, and indeed the equivocality of the magnitude of the effects of disease on long-term persistence, only serve to highlight the importance of considering disease when designing reintroduction activities.

Among those considerations is mitigation, aimed at ameliorating the disease or its impacts on species or habitats after it has been identified. Mitigation efforts directed at Bd have been implemented. In Europe, removal and treatment of amphibians coupled with a dry down and disinfection of the habitat has been successful (Woodhams et al. 2011). Such methods may not be practical on a large scale, but may be effective for isolated habitats that contain species of considerable risk and conservation value. Bioaugmentation (the addition of beneficial microbiota to individuals or habitats) is a feasible form of mitigation (Bletz et al. 2013). Bioaugmentation has been tested in the Sierra Nevada where yellow-legged frogs were inoculated with a specific bacterium (*Janthinobacterium lividum*) that produces a metabolite toxic to Bd (the bacterium occurs naturally on some amphibians), but with no long-term success (Woodhams et al. 2011, Vredenburg et al. 2011).

Despite constant vigilance and attention to both preventative and mitigative measures, emerging infectious diseases are part of the conservation landscape. Illustrating this is the recent identification of *Batrachochytrium salamandrivorans* sp. nov., in fire salamanders (*Salamandra salamandra*) (Martel et al. 2013, 2014). The recognition of another pathogenic chytrid fungus within the nearly ubiquitous (Pessier et al. 1999) Chytridiomycota has bearing on mitigation efforts directed at amphibian decline including efforts at reintroduction.

Reintroduction Effort

This reintroduction is described in Muths et al. (2014) and is summarized here to lay the basis for the following critique. The effort began with in-depth site consideration and selection based on both logistical and biological concerns, and historical and contemporary data. Prerelease surveys revealed no disease (i.e., Bd), no fish and, initially, no other amphibians. While not a boreal toad breeding site according to previous surveys (Corn et al. 1997), the selected site was within the area of toad occurrence within a national park. As specified in the conservation plan (Loeffler 2001), the site was >5 km from other known breeding sites. Site selection took 3 years, which enabled adequate surveys (using visual encounter surveys, Dodd 2012) to determine what amphibians were present. Later surveys during this 3-year period revealed the boreal chorus frog (*Pseudacris maculata*), but in low numbers. Water was assessed for Bd (Kirshtein et al. 2007) in only 1 year and Bd was not assessed in animals until the chorus frogs were detected and sentinel boreal toads (wild caught, but in captivity for multiple years) were released. Donor animals (tadpoles) were chosen from nearest neighbor populations and were either from wild-collected eggs that were reared in captivity or eggs laid and reared in captivity by captive stock from those nearest neighbor populations. Animals were released each year from 2007 to 2015, but the number of tadpoles reintroduced varied by year (minimum 175, pilot year; maximum >13,000) and also by size within and between years. Quantification of the reintroduction was undertaken with data collected from 2010 to 2013 and found to be feasible, resulting in defensible measures of success at that point in the reintroduction. This quantification resulted in survival estimates for tadpoles from release to metamorphosis (i.e., over summer survival) and information on hatching rate and survival differences between wild- and hatchery-laid eggs (Muths et al. 2014). Juvenile toads were observed in 2009 through 2013 but were not quantified. No evidence of breeding has been observed; however, in 2014, one male adult boreal toad was observed at the site but subsequently found dead. The animal had a significant Bd infection but exhibited normal body condition (e.g., adequate fat bodies; Muths et al., unpublished data). This animal had survived 2 years post-metamorphosis, suggesting that it had either not encountered the pathogen (e.g., low prevalence at the site) or some level of resistance or tolerance had been achieved in animals from that cohort. Boreal toads tend to live in

excess of 10 years and only those from the initial years would have reached reproductive age. The return of only one animal is somewhat bleak, but the effort will continue through 2016 with reintroductions, assessments, and monitoring.

A Critique

Initial planning for this reintroduction was accomplished prior to the availability of most of the information about this disease in the Rocky Mountains. For example, the prescription of a minimum of 5 km between reintroduced and extant populations was made when Bd was considered to be a pathogen that could be potentially spread to disease-free systems and was based on the maximum dispersal distance of boreal toads known at the time (Muths et al. 2003) in an effort to isolate the population from disease. Our realization that Bd seems to be everywhere we look, albeit undetectable or at very low prevalence, suggests that selecting a site that had a higher potential for immigration, and thus gene flow, may have been a better choice. A more concerted effort to determine disease status in potential reservoirs (chorus frogs) could have been undertaken. An option that could have been considered in targeting donor populations is using a genetically similar and more resistant toad for reintroductions. This option was not considered for two reasons: first, management preferred to maintain the genetic integrity of resident toads, and, second, while there is some evidence that different populations are less susceptible than toads in the SRM (Muths et al. 2011), the data to determine this quantitatively are lacking.

Because we have taken an adaptive management approach to toad conservation, some of the decisions made early in the process can be reconsidered. Such reconsideration is, in fact, what is currently underway as the SRM population of boreal toad is considered for federal endangered status with a listing decision due in 2017 (Greenwald et al. 2011). The boreal toad recovery team is in the process of a structured decision-making exercise. This exercise (sensu Runge 2011, Converse et al. 2016) provides a venue where the experience of this reintroduction effort as well as newly acquired data about Bd and population dynamics of toads with and without the disease can be assimilated into management decisions. Applying figure 10.1 to this case study, the answer to the first decision point (is reintroduction necessary because of disease-induced extinction?) is a clear yes. The next decision points were unanswerable at the time of planning and initial implementation but have been incorporated into the current structured decision-making exercise where risk tolerance by signatory agencies will factor into identifying overarching objectives and the means to achieve them. Only decision point 4 was answered in the affirmative and was supported by field data.

This case study presents a situation that is likely fairly typical in reintroductions. There is much information lacking, but managers are under pressure to implement actions. We were fortunate to have the cooperation of supportive state and federal agencies who shared in all aspects of the cost and implementation of the reintroduction. In contrast to many reintroductions, this effort has been well documented from inception to implementation. Data were collected with specific questions in mind and were thus able to be used in rigorous analyses (Muths et al. 2014). Despite the commitment to this project (10 years, reintroductions of tadpoles 2007–2016), it is unlikely that we will see a self-sustaining population of boreal toads at this site. The amphibian chytrid fungus is present and whether it was latent in the environment, existed at very low prevalence, or was carried only by the chorus frog is unknown. If some adult toads do return to the "natal" pond and breeding occurs, a second iteration of reintroductions might be in order to augment those returning individuals. The discovery of resistant or tolerant stock to use for this effort would be key and would force a decision: choosing to champion genetic "purity" in SRM populations with the likely outcome of extirpation versus releasing resistant or tolerant animals into existing populations or as part of reintroductions with the expected outcome of maintaining toad populations in the SRM. This decision about whether to engage in assisted evolution of resistance or tolerance, in species and populations challenged by disease, is likely to become more common.

their introduction into wildlife (Vander Wal et al. 2014), and strategies can be developed to accelerate these evolutionary responses (Kilpatrick 2006). Improved resistance to viral and bacterial infections has been achieved through artificial selection in livestock (e.g., Stear et al. 2001, Borriello et al. 2006) and in fish (Henryon et al. 2005). In addition, resistance or tolerance may also be location specific, as with white-nose syndrome in bats, for which there appear to be intercontinental differences in bat mortality that are related more to the susceptibility of the host or the habitat rather than differences in the pathogen (Cryan et al. 2013). Strategies based on these examples may not provide a complete solution to the problem, but may be worthwhile in the sense of "buying time" while natural selection operates (sensu Grant and Grant 1993), perhaps in the direction of more resistant populations. If a stock of resistant or tolerant animals exists, their reintroduction is worth considering, whether or not reservoir hosts are present. Such populations may exist in the wild or animals could be vaccinated prior to reintroduction, although the success of this strategy will likely depend on the size and degree of isolation of the target population. As the population grows, resilience should begin to increase naturally through epidemiological processes. Vaccination against anthrax has been used with some measure of success for wildlife in Kruger National Park (Kock et al. 2010).

Decision Point Three: Are Disease Reservoirs Present?

If resistant or tolerant stock are not available, the third critical decision point relies on the presence or absence of reservoir hosts or environmental reservoirs in the area where the reintroduction is proposed. If the pathogen is host specific (no reservoir species) and has caused the local extinction of the species being reintroduced, it must be a pathogen able to persist at very low host densities (McCallum and Dobson 1995). A possible example of this is Tasmanian devil facial tumor disease (McCallum et al. 2009).

Some wildlife pathogens can persist for extended periods in the environment in the absence of a host species. For example, *Pseudogymnoascus destructans*, the causative agent of white-nose syndrome in bats, can persist for months in hibernacula without bats being present (Lorch et al. 2013, Hoyt et al. 2014), the amphibian chytrid fungus is able to persist for months in moist sand without the presence of host organisms (Johnson and Speare 2005), and deer can acquire chronic wasting disease from paddocks for more than 2 years after infected animals have been removed (Miller et al. 2004). If pathogens can persist in the environment, some form of decontamination will be necessary before reintroduction. If there is any chance that the pathogen might persist or recolonize the site, adding more susceptible hosts is likely to simply feed a potential epidemic and cause the reintroduction to fail. This situation may be equally plausible in reintroductions where the decline occurred because of factors unrelated to disease. If one or more reservoirs for the pathogen in question do exist, then reintroduction of a highly susceptible species will almost certainly fail unless the disease can be managed in the reservoirs (see "Decision Point Five").

Decision Point Four: Is the Local Host Extinct?

Is the host species (reintroduction target species) really extinct in the area where the reintroduction is proposed? It is extremely difficult to determine when the final extinction of any declining species may have occurred (Lee et al. 2014). However, if there is a high level of confidence that the species being reintroduced is locally extinct and that no potentially infected wild members of the species are able to migrate into the reintroduction area, then successful reintroduction may be feasible. If not, then reintroduction should be delayed until it is determined that local extinction has occurred unless disease-resistant or disease-tolerant reintroduction stock are available.

Decision Point Five: Can the Disease Be Effectively Managed in the Reservoir(s)?

Decision point five is the need to determine whether disease can be managed sufficiently in the reservoir population(s) to enable persistence of a more susceptible species targeted for reintroduction. What management is "sufficient" is not a straightforward question. One critical issue is whether or not transmission chains of the pathogen can be maintained in "target" species. A classification of zoonotic pathogens developed by Wolfe et al. (2007) and Lloyd-Smith et al. (2009) is important here. Some pathogens (e.g., rabies in humans) spill over from their reservoir host to a susceptible target, without further transmission chains within the susceptible host population, meaning that R_0, the basic reproductive number, i.e., "the expected number of secondary cases produced by a typical infectious individual in a wholly susceptible population" (Lloyd-Smith et al. 2009), in the spillover host population is 0. These are described as "stage 2 spillover pathogens." In this case, disease-induced mortality in the target, resulting from spillover, might be sufficiently high to prevent population growth. Reducing disease prevalence in the reservoir, or reducing contact between reservoir and target species, may reduce mortality resulting from spillover sufficiently, and it may not be necessary to entirely eliminate disease in the reservoir population. Other pathogens, described as "stage 3 spillover pathogens" have stuttering chains of transmission in the target host. This means that target-to-target transmission does occur, but is insufficient to maintain infection within the target population without continuing spillover from the reservoir. In this case, R_0, is less than 1 in the target population of hosts. As with stage 2 spillover pathogens, reduction of infection within the reservoir and management of contact rates between reservoir and target may be effective in allowing a reintroduction to succeed.

The most difficult situation to manage is when there is a stage 4 spillover pathogen, in which continuing transmission within the target population occurs. In this case, R_0 exceeds 1 in the target population. Once transmission from the reservoir to the reintroduced population occurs, it is then highly likely that the pathogen will become established within the reintroduced host, and given that it has previously caused extinction (or extirpation), this is likely to reoccur. A successful reintroduction will therefore require either elimination of the pathogen from the reservoir population or reintroduction of disease-tolerant or disease-resistant hosts. Strategies to reduce host-to-host transmission within the reintroduced population are likely to contribute to the success of a reintroduction regardless of the degree of spillover.

In certain cases, it may be worthwhile to continue replenishing a population knowing that it is not self-sustaining. It may be considered desirable to maintain a wild population of a species extirpated by a pathogen even though the population is not self-maintaining in the absence of continuing reintroduction. For example, Canessa et al. (2014) describe optimal release strategies for the southern corroboree frog (*Pseudophryne corroboree*) in Australian alpine areas. This species has been driven close to extinction in the wild by the amphibian chytrid fungus, but it is possible to maintain wild populations by continual introduction of either eggs or subadults, which can be mass-reared in captivity. Fishes and some amphibians are relatively inexpensive to rear from eggs to larvae (Canessa et al. 2014, Muths et al. 2014). Reasons behind adopting this strategy often include human values attributed to the species or to their contribution to the environment (i.e., intrinsic value; sensu Leopold 1949) or may be driven by directives that require managers to retain species extant to a particular park or forest (e.g., "The National Park Service will strive to understand, maintain, restore, and protect the inherent integrity of the natural resources, processes, systems, and values of the parks," National Park Service 2006). Additionally, federal mandates (e.g., the US Endangered Species Act, http://www.fws.gov/endangered/esa-library/pdf/ESA_basics.pdf) may dictate recovery of a species that would not be possible without reintroductions (e.g., the black-footed ferret; Biggins et al. 2012).

REINTRODUCTION FOLLOWING DECLINE THAT WAS CAUSED PRIMARILY BY FACTORS OTHER THAN DISEASE

Infectious disease can be an important consideration in reintroductions even if disease was not the primary cause of the initial decline. We discuss four possible scenarios where disease is an important player in the system, but not the initial driver of decline in the target organism. These include examples of situations to be considered when attempting a reintroduction.

Disease Is Present at the Reintroduction Site at the Time of Reintroduction

In many cases, a disease can occur at low prevalence in an environment but erupt in the presence of a population of naïve and susceptible hosts. There may be genetic predisposition or unique physiologies that facilitate infection and adding large numbers of hosts—made more susceptible by stresses associated with reintroduction (transportation, crowding) that can facilitate disease and disease transfer (Teixeira et al. 2007, Dickens et al. 2010). At low densities, a disease may be elusive. Pre-reintroduction screening methods for disease in the environment may lack sensitivity, be flawed (e.g., a high rate of false positives or low detection probability), or may not exist. For example, the amphibian chytrid fungus has been difficult to detect in the environment (i.e., water) although methods are now being developed (Kirshtein et al. 2007, Walker et al. 2007, Chestnut et al. 2014). This issue could be addressed by releasing sentinel organisms (e.g., excess genetic stock from captive-rearing programs) to see if the disease manifests when the species intended for reintroduction (or an epidemiologically linked species) is present (e.g., Martin et al. 2011). Another promising method for pre-reintroduction detection of disease organisms is using environmental DNA (eDNA) screening to detect disease organisms in environmental samples (e.g., water). This has been used successfully on the detection of invasive species (Sepulveda et al. 2012) and disease (Chestnut et al. 2014), although it is not a panacea (Roussel et al. 2015.

Disease Is Introduced with the Reintroduced Animals

In this scenario, a disease may be either latent or not identified in animals before reintroduction. This can be the case for animals reared in captivity or for animals from donor sites where parasites or pathogens are different from those at the targeted release site. Captive-rearing facilities are potential sources for novel cross-species infections, with animals at relatively high densities often being maintained in close proximity to species they would not encounter in the wild. Amateur wildlife caretakers and rehabilitators play an important role in rearing orphaned animals, but frequently keep wild animals in close association with other species, particularly domestic pets. These situations pose a serious threat of introduction of exotic diseases. High densities and cohousing of same-species individuals in captivity or during transport can amplify parasite burdens, and increase prevalence and intensity of infection (e.g., the amphibian chytrid fungus; Fisher and Garner 2007). Kock et al. (2010) provide examples of infectious diseases being introduced into release areas by translocated wildlife, with pathogens ranging from viruses such as rabies, to helminths and warble flies, and in hosts ranging from mammals to birds, reptiles, and fish. A specific example of a disease being co-reintroduced to a site occurred when midwife toads (*Alytes mulentensis*) were reintroduced into Mallorca, Spain, and captive-source individuals carried the amphibian chytrid fungus, compromising the reintroduction effort (Walker et al. 2008). Quarantine procedures, disease surveillance, and ongoing efforts to ensure proper diets and husbandry protocols for animals brought in from the wild as well as those bred in captivity (sensu Pessier 2008, Green et al. 2009) will reduce the probability of disease being brought to reintroduction sites by released animals. Similar situations can occur when individuals are reintroduced to augment already reintroduced populations (also termed reinforcement when applied to extant but declining populations). This practice has the potential of confounding the disease dynamics that exist or that are developing in the target population (Aiello et al. 2014).

Disease Is Brought to Reintroduced Populations by a Third Party after Reintroduction

Pathogens might derive from researchers monitoring the population, supplemental feeding in "soft" releases (e.g., proximity of individuals), or from other species that reside in or move through the area. Immigrants of the target species might also carry disease. Addressing the human aspect is straightforward: care in disinfection of equipment, including clothing and footwear, can control inadvertent contamination by researchers or other persons. This approach has been implemented in areas that harbor invasive species (including disease) and in areas where invasion is a potential threat. For example, the Australian threat abatement plan for the amphibian chytrid fungus requires disinfection of all field equipment used in sampling frogs (Australian Department of the Environment and Heritage 2006); and in the United States, anglers are encouraged to disinfect equipment, including waders, in areas where whirling disease is present (Rocky Mountain National Park, http://www.nps.gov/romo/planyourvisit/upload/Fishing-2012.pdf). Whirling disease has been a concern in reintroductions of greenback cutthroat trout in Colorado (US Fish and Wildlife Service 2009). The level of compliance from the general public varies and while public outreach programs have raised awareness, results are mixed (e.g., Strayer 2009).

Disease Associated with the Reintroduction Effort Can Affect Resident Species

Disease introduced inadvertently with reintroduced animals, or by researchers whose presence is related to the reintroduction effort, can pose a nuisance to, or have catastrophic effects on, resident species (i.e., naïve hosts with no prior evolutionary exposure to a novel pathogen). The inadvertent introduction of the rinderpest virus to Africa in the 1890s was associated with cattle that were part of a military campaign, rather than a conservation reintroduction, but the resulting cascading effects on the numbers of wildebeest, lion, and hyena (Dobson and Hudson 1995) illustrate the potential ecosystem-wide impacts of a disease introduced along with its host species. Similarly, harlequin frogs (*Atelopus* spp) are declining across their range, most species are endangered, and reintroduction is suggested as a necessary management action (Lötters et al. 2005, Lampo et al. 2012). However, *A. zeteki* is considered to be a "supershedder" of the amphibian chytrid fungus (DiRenzo et al. 2014) such that reintroductions of this species would likely affect other anuran species present at reintroduction sites. Reintroduced animals would likely be "clean," but if the fungus were present in the environment, there is the potential for infection and elevated transmission. The decline of the Mallorcan midwife toad (*Alytes muletensis*) was attributed to introduced species (e.g., the viperine snake, *Natrix maura*) (Moore et al. 2004), but reintroductions of captive-reared midwife toads inadvertently included the amphibian chytrid fungus (Walker et al. 2008). Any number of scenarios can be imagined where novel disease, or even slightly different strains of disease from those associated with the original residents, might be introduced during reintroductions and have deleterious consequences for resident species. Stringent testing of animals slated for release and a working knowledge of shared susceptibilities among native species and the species to be reintroduced are necessary (figure 10.2).

STRATEGIES TO MANAGE DISEASE

Managing Infectious Agents in the Individuals Being Released

Appropriate screening of reintroduced stock for parasites and pathogens that could be threats to the reintroduced population or to other species in the reintroduction area is essential (e.g., Pessier 2008, OIE/IUCN 2014). The value of such screening is illustrated in two case studies on water voles (*Arvicola terrestris*) and dibblers (a species of marsupial carnivore: *Parantechinus apicalis*) where health surveys were implemented prior to and after reintroductions (Mathews et al. 2006). Although treatment or prophylaxis for major pathogens in reintroduced stock has become routine in most reintroductions (Woodford 2000), removal of all parasites from stock for reintroduction is not necessarily desirable. Parasites are normal components of the biology of all wild populations and immunologically naïve reintroduced hosts may be particularly susceptible to "normal" parasites or pathogens present in the reintroduction environment (Jule et al. 2008, Almberg et al. 2012). Dilemmas can also arise when the parasites or pathogens are endangered species themselves (e.g., lice; Rózsa and Vas 2015, Gompper and Williams 1998).

Managing Infectious Agents in the Ecological Community Following Reintroduction

There are essentially five strategies available to manage infectious diseases in wild populations (Wobeser 2002, McCallum and Jones 2006): (1) isolation of uninfected populations; (2) treatment or vaccination; (3) culling of either all hosts or infected hosts only; (4) assisted evolution of tolerance or resistance; and (5) management of the environment to reduce transmission. Each of these could potentially be applied to manage infectious disease in one or more reservoir species in an area where reintroduction is attempted and, depending on the scenario, they may be applicable to managing

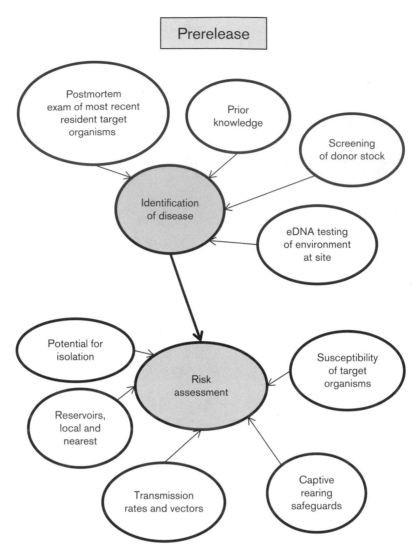

Postmortem exam of most recent resident target organisms

Prior knowledge

Screening of donor stock

Identification of disease

eDNA testing of environment at site

Potential for isolation

Risk assessment

Susceptibility of target organisms

Reservoirs, local and nearest

Transmission rates and vectors

Captive rearing safeguards

FIGURE 10.2 Prerelease considerations (gray circles) and potential information sources to address these concerns.

disease in a host population following reintroduction (figure 10.3).

Isolation of Uninfected Populations

Perhaps the most attractive attribute of this strategy is that there is no requirement for any knowledge of the etiological agent of disease (common with point 2). This strategy has been implemented by moving disease-free individuals to islands; it has been used in the management of Tasmanian devil facial tumor disease (Box 10.2) and toxoplasmosis (as well as other causes of death) in the eastern barred bandicoot (Kemp et al. 2015). Reintroducing uninfected animals to islands is probably the most feasible means of isolating wild populations, although fencing has also been used to isolate uninfected populations (Bode and Wintle 2009). Fences have been used in Africa to limit the spread of pathogens, particularly foot and mouth disease (Bruckner et al. 2002, Hayward and Kerley 2009), but usually require an associated buffer zone in which other disease control strategies such as culling or

10b Tasmanian devil *(Sarcophilus harrisii)*. Photo by H. McCallum.

The Tasmanian devil *(Sarcophilus harrisii;* figure 10b) is threatened with extinction by an infectious cancer, Tasmanian devil facial tumor disease (McCallum et al. 2009). The tumor cells themselves are the infectious agent (Pearse and Swift 2006) and appear not to be rejected by recipient devils at least partially because of the very low genetic diversity in Tasmanian devil populations (Siddle et al. 2007, 2010), although recent work shows that the tumors have evolved the ability to "hide" from an infected devil's immune system by down-regulating major histocompatibility complex (MHC) expression on the tumor cell surface (Kreiss et al. 2011, Siddle and Kaufman 2013). It is remarkable enough that cancer cells can develop within other members of the same species, and the disease is certainly host specific. It has already caused an overall population decline of at least 80% of the entire population, with declines in excess of 90% in the areas where disease has been present the longest (DSEWPaC 2013). High prevalence of infection is maintained even as populations have declined dramatically, suggesting that transmission depends weakly, if at all, on host density (McCallum et al. 2009). In the face of this potential threat of extinction of the largest surviving marsupial carnivore, insurance populations have been established on the Australian mainland with the objective of reintroducing devils to Tasmania should they become extinct. In addition, devils have recently been translocated to Maria Island, approximately 100 km², off the east coast of Tasmania (http://www.tassiedevil.com.au/tasdevil.nsf/Insurance-population). This is technically a translocation rather than a reintroduction, as Tasmanian devils do not occur on the island and probably have not since the last ice age (Rounsevell 1989). The habitat and recipient ecological community are very similar to adjacent areas on the Tasmanian mainland less than 5 km distant, which supported thriving Tasmanian devil populations until the arrival of the disease around 2000. There is an obvious potential risk of introducing the disease along with any reintroduced devils, particularly as the latent or incubation period is poorly known (Beeton and McCallum 2011). However, there has been no evidence of vertical transmission of the tumor (McCallum et al. 2007) and therefore there is a very low risk of inadvertent introduction of disease provided captive-born animals maintained in complete isolation from wild devils are used (as is the case for captive colonies on the Australian mainland).

Following the decision tree (figure 10.1), a reintroduction strategy is being developed because of the risk of disease-induced extinction (McCallum et al. 2009), although no population has yet become locally extinct. This brings us to the next point in the decision tree, the availability of resistant or tolerant stock. Despite suggestions that Tasmanian devils from the northwest of Tasmania might have MHC genotypes resistant to disease (Siddle et al. 2010) and some field evidence consistent with the suggestion (Hamede et al. 2012), no evidence of resistant or tolerant host genotypes has yet been found. At the next decision point, Tasmanian devil facial tumor disease is clearly host specific. The critical question regarding the feasibility of reintroduction is thus whether we can be confident of local host extinction in areas where reintroduction is proposed.

Given that Tasmanian devil facial tumor can be transmitted at low density and remains at high prevalence even in small populations (McCallum et al. 2009), reintroducing disease-free Tasmanian devils to an area in which disease has caused decline would simply feed the epidemic if any diseased animals remained present. There have also been some suggestions that the few remaining devils could be extirpated from selected areas that could be isolated from immigration and then the areas could be restocked with disease-free devils (http://www.tassiedevil.com.au/tasdevil .nsf/Insurance-population). This proposal is problematic for at least three reasons. First, extirpating all remaining devils would be extremely difficult. Second, constructing an effective barrier fence to prevent immigration of diseased devils would likely be impractical. There have been examples of predator-proof fences in Australia and elsewhere being used effectively for conservation purposes (Hayward and Kerley 2009, Moseby et al. 2009), but fencing to prevent spread of disease is more difficult because onward transmission could occur before an animal breaching the barrier is removed. Third, despite the lack of current evidence for resistance or tolerance, the few devils that may remain in an area in which disease has caused decline might be either resistant or tolerant to infection and therefore could serve as the nucleus of a recovering population. Removing them and replacing them with naïve devils could thus be counterproductive.

A final, if very ambitious, reintroduction possibility for Tasmanian devils would be to reintroduce them to the Australian mainland. Tasmanian devils were widespread on the mainland of Australia until their extinction approximately 3,000 years ago, following the introduction of the dingo (Canis lupus dingo), although it is not entirely clear that the dingo itself was responsible. An alternative hypothesis is that an increase in aboriginal hunting pressure may be responsible (Prowse et al. 2014). Both these factors that may have led to the original extinction would no longer be major limiting factors in parts of southern Australia. Reintroduction of disease-free Tasmanian devils might therefore succeed and be the best prospect for the long-term survival of the species. In addition, feral cats and introduced European red foxes are major threats for much of Australia's medium-sized mammal fauna (Johnson 2006). There is empirical evidence that Tasmanian devils have played a role in suppressing feral cat populations in Tasmania, with cat numbers increasing as devils decline due to disease (Hollings et al. 2014), and devils may have been responsible for preventing foxes from becoming established in Tasmania until recently (Sarre et al. 2013). These instances suggest that there may be substantial biodiversity benefits, in addition to direct benefits to the devil, resulting from their reintroduction to the mainland of Australia (Hunter et al. 2015).

vaccination are implemented. Fences have been used to protect livestock rather than for conservation purposes under a number of scenarios with varying results. However, fences are generally inapplicable to managing infectious disease among free-living animals (Wobeser 2002), and we know of no case where a fence has been used to successfully protect a wildlife species from infectious disease.

Treatment or Vaccination

Mass vaccination of domestic dogs (reservoir population) has been used in attempts to protect wild carnivore populations (e.g., lions, Ethiopian wolves, and African wild dogs) from the transmission of rabies and canine distemper virus. Although empirical evidence and modeling suggest that vaccinating domestic dogs should reduce and possibly eliminate rabies and canine distemper in wildlife in the Serengeti, outbreaks of both continue despite a domestic dog vaccination program (Lembo et al. 2008, Viana et al. 2015). Vaccinating wildlife hosts has also been suggested as an option for mitigating these infectious diseases in Africa (Cleveland et al. 2006). In the United States, vaccination through ingestion of baits can reduce plague in black-tailed prairie dogs, which will aid in the reduction of population declines of this critical prey item for endangered black-footed ferrets (Biggins et al. 2012, Tripp et al. 2014). While we do not know of a case where these strategies have prevented the incursion of a disease, they can be effective in combating an outbreak (e.g., the case of the prairie dogs and ferrets). Tangentially, Woodhams et al. (2011) discuss bioaugmentation and biostimulation in amphibians as strategies to reduce susceptibility to infection and disease by adding probiotic microbiota to the epidermis or adding prebiotics to encourage beneficial microbiota.

Culling

Culling may target the removal of all hosts, irrespective of disease status, or target infected animals only following a "test and cull" protocol (e.g., Wolfe et al. 2004). Despite the poten-

tial appeal of culling and some evidence for its effectiveness with incursions of exotic livestock diseases (Ferguson et al. 2001), its use is somewhat equivocal, particularly in wildlife, and there are limited instances where it is a successful strategy for conservation. Examples include the substantial reduction in bovine tuberculosis in cattle and deer in New Zealand by culling brush-tailed possums (Ramsey and Efford 2010) and simulations from Australia indicating that targeted culling assists in controlling swine fever in wild pigs (Cowled et al. 2012). In contrast, both empirical studies and modeling efforts suggest that culling is rarely successful in managing infectious disease in a conservation context (e.g., Hallam and McCracken 2011); adequate modeling of the efficacy of culling can be complex (McCallum 2015). For example, Lachish et al. (2010) found that culling in Tasmanian devils failed to reduce the population-level impacts of facial tumor disease and did not affect the rate of disease progression. These results were attributed to a number of parameters including the frequency-dependent nature of the disease, long latent period, and the potential for immigration of diseased animals (Lachish et al. 2010, Beeton and McCallum 2011). Wild badgers carry tuberculosis and can transmit the disease to cattle. Bielby et al. (2014) reported that culling badgers in Britain changed the remaining animals' behavior such that the spread of disease increased rather than decreased. Culling appeared to increase badger movement (Pope et al. 2007), thus compromising control of bovine tuberculosis (Donnelly et al. 2006, Pope et al. 2007, Vial and Donnelly 2012, Bielby et al. 2014). Chronic wasting disease, a serious concern in North American ungulates, was not effectively managed by culling (Wolfe et al. 2004, Wasserberg et al. 2009).

Assisted Evolution of Resistance or Tolerance

When animals must be reintroduced into areas where the disease status is either acknowledged as a concern and unavoidable, or unknown, this prophylactic approach may be called for. There are numerous permutations to this strategy. Animals to be reintroduced could be challenged with specific diseases before release (Faria et al. 2010) or manipulated genetically for increased resistance and thus increased likelihood of withstanding disease(s) predicted or known to be encountered in the wild upon release (e.g., Jacobs et al. 2012, Venesky et al. 2012). An extreme case might be to incorporate genetic stock from different subspecies or sibling species that exhibit desirable characteristics such as lower susceptibility to a specific disease, or greater fecundity. For example, the *Anaxyrus (Bufo) boreas* species group is widespread (Goebel 2005, Goebel et al. 2009), but populations in the Southern Rocky Mountains (SRM) have been petitioned for listing under the US Endangered Species Act as a distinct population segment because of precipitous declines and extirpations (Muths et al. 2003, Carey et al. 2005) and putative genetic distinctness from other boreal toads in the region (Goebel 2005, Goebel et al. 2009). In Wyoming, outside the SRM but within 500 mi of extirpated toad populations in Colorado (part of the SRM), there are populations of *A. boreas* in which survival is affected by chytridiomycosis, but declines are occurring more slowly (about 6% per year; Muths et al. 2011). Incorporating individuals from the possibly less susceptible population might improve survival in the SRM. Although there are putative genetic differences between boreal toads in Wyoming and those in the SRM (Goebel 2005), maintaining a "boreal toad" in the SRM could be of more value than maintaining genetic distinctness. We might argue that in the world where diversity is declining at an alarming rate, maintaining presence of the boreal toad in Colorado preserves the fabric of the ecosystem that it resides in and provides value to humans in that it is present, an observable part of nature, and thus, a better option.

Management of the Environment to Reduce Transmission

This strategy depends on understanding the environmental requirements of the pathogen as well as the host species. For example, prevalence of brucellosis in elk (*Cervus elaphus*) in the Greater Yellowstone Ecosystem can be reduced by shortening the length of the period in early spring during which wildlife managers provide supplemental feeding to elk (Joseph et al. 2013). The high rates of contact produced by aggregation around supplementary feeding sites increase transmission of the pathogen. Where excessive congregation, because of either climatic conditions (e.g., reindeer, *Rangifer tarandus*; Parkinson et al. 2014) or behavior (kangaroo rats, *Dipodomys stephensi*; Shier and Swaisgood 2012), may facilitate disease transmission, it could be manipulated with different release strategies (e.g., attention to the arrangement of supplemental feeding, or the spatial placement of animal releases on the landscape). Wobeser (2002) described a number of cases in which habitat modification was effective in reducing transmission of wildlife diseases: prescribed burning in the reduction of tick populations, and draining of standing water around nesting sites in reducing the transmission of avian cholera among eider ducks (*Somateria mollissima*). Habitat management to limit transmission will inevitably be specific to the particular combination of pathogen, host, and environment, and will need to be considered carefully in the context of potential impacts on nontarget species. Managing the environment can go beyond addressing the pathogen itself and extend to addressing the general health (and possibly susceptibility) of the reintroduced individuals to disease, or by scrupulous selection of sites to limit disease or the effects of disease (e.g., Hansen and Budy 2011).

While not a strategy per se, the political landscape of disease also is a component of the management of reintroductions (see Chapters 3 and 5, this volume). There is utility in acknowledging the political impacts of disease, which can have a greater financial cost to the project than the direct (i.e., biological) impacts. Perception of disease can be more significant than the reality. For example, the notion that avian influenza is introduced into new landscapes by wild

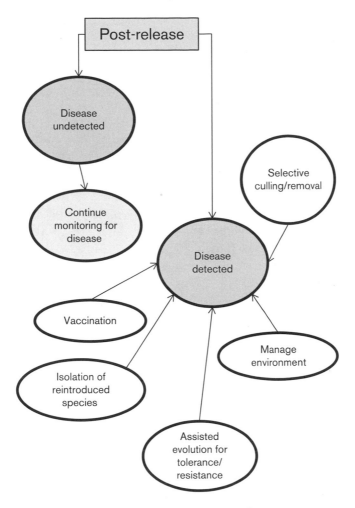

FIGURE 10.3 Post-release considerations to manage disease.

birds has been repeated in scientific literature for over a decade with little confirmatory evidence that wild birds either carry this disease or introduce it into domestic poultry (e.g., Feare 2007, Simms and Jeggo 2014). Misconceptions may abound especially when the public is interested in or affected by reintroductions and represents one more angle to be considered when disease is a consideration in wildlife reintroductions.

FUTURE RESEARCH AND DEVELOPMENT

An understanding of the etiology of diseases, disease ecology, and among-species transmission probabilities is vital to effective reintroduc-

tion efforts, yet these topics require substantial research. As habitat is lost and more species are imperiled, it is likely that reintroduction efforts will increase (Seddon et al. 2007) despite equivocal outcomes in reintroductions (including other movements of animals, such as supplementing existing populations or introducing species into habitat not formerly occupied) (e.g., Dodd 2005, Ricciardi and Simberloff 2009, Muths et al. 2014). Similarly, as emergent diseases continue to make headlines (e.g., Lips and Mendelson 2014), disease will increasingly become a focal concern in reintroductions, as will the resilience of reintroduced populations to perturbations such as disease outbreaks. A related focal concern of reintroductions should be the collection of information at each stage in the process (e.g., Muths et al. 2014, Box 10.1). Preplanning and thoughtful execution (see Moehrenschlager and Lloyd, Chapter 11, this volume) will lead to increased understanding of the organism and its interactions with its "new" habitat, and also of the reintroduction process itself, such that no effort is wasted even if the final outcome is deemed unsuccessful.

In addition to knowledge gaps related directly to disease, demographic parameters (i.e., survival, recruitment, population size) are often lacking for populations before reintroduction is considered (i.e., baseline information) or after reintroduced individuals are in place (but see, e.g., Sarrazin and Legendre 2000, Schaub et al. 2009, Swanack et al. 2009). Understanding vital rates and how disease is affecting them (e.g., Briggs et al. 2005, Lambert et al., in review) is important in assessing the success of a reintroduction and in developing mitigation strategies. For example, the density of a host population can influence the behavior of infectious disease (Rachowicz and Briggs 2007). To acquire disease and demographic data applicable to understanding the outcomes of the project, long-term and well-planned research efforts are required (Muths and Dreitz 2008, Kock et al. 2010).

Ewen et al. (2014) emphasize the criticality of clearly stated and prioritized objectives and constraints—a priori—if reintroductions are to succeed. The presence of disease adds an additional layer of concern and highlights the uncertainty that is inherent in manipulating natural systems. Increased attention has been accorded to addressing uncertainty, and advances in dealing with uncertainty have been made (e.g., Chapters 5 and 6, this volume, Runge 2011, Ewen et al. 2014, Canessa 2015). From the perspective of disease, population and epidemiological modeling can provide data on which to build predictions (Restif et al. 2012, McCallum 2015). Structured decision-making (Runge 2011) can facilitate an integration of the science (including the ramifications of disease and models that describe it) and human dimensions (e.g., costs and politics) that figure into reintroduction projects. There is also a body of literature on risk assessment (e.g., Covello and Merkhofer 1993, Kock et al. 2010, Sainsbury et al. 2012, Jakob-Hoff et al. 2014) to turn to when information is lacking at the decision points that we have illustrated, but when action is required due to other conservation concerns or environmental catastrophes.

Regardless of the successes (e.g., Stanley-Price 1991, Denton et al. 1997, Walters et al. 2010), failures (e.g., Seigel and Dodd 2002; Griffith et al.1989, Wimburger et al. 2009, Soorae 2011, Germano et al. 2015), or uncertainties (e.g., disease) in reintroduction efforts, the continued use of reintroduction as a management tool is certain and the methods applied to assess efforts deserve a critical look (e.g., Burton and Rivera-Milan 2014). Reintroduction projects are not necessarily selected based on conservation need (Seddon et al. 2005) and are not always cost effective (Germano et al. 2015). In the face of ongoing biodiversity loss and the need for triage decisions as conservation resources diminish, the validity of arguably Sisyphean reintroduction efforts is questionable. Emerging infectious disease adds complexity and an additional sense of urgency to the dilemma. When considering reintroductions, especially when disease is a component, the considerations noted in this chapter are worth examining. At the same time, it is worth remembering that most management actions are likely to be host, pathogen, and site specific, making overarching prescriptions difficult.

SUMMARY

Disease is one of the obvious risks involved in moving animals around a landscape, but shifting animals from place to place without their associated biota (including potential or realized pathogens) is nearly impossible. Addressing risks related to disease (pathogens and parasites) in planning reintroductions is complex but critical. It requires knowledge of the etiologies of potential diseases, potential effects on target animals and nontarget animals, the development of responses to a variety to potential scenarios as well as attention to potential "political fallout" from concerns (both legitimate and misconceived) about disease. Disease is a legitimate concern in reintroductions, as determined by recent outbreaks as well as historical information that suggests that extinctions due to disease are underrepresented in the literature (Abbott 2006, McCallum 2012). If disease was a primary contributing factor to the initial extirpation that necessitates reintroduction and is not addressed, then the reintroduction will almost certainly fail. If disease was not the original cause of extinction, it is still necessary to take it into account, to ensure the success of a reintroduction and to ensure that there are no detrimental effects on the recipient community through inadvertent introduction of a parasite or pathogen. The decision tree we have presented can guide managers in determining whether or not to reintroduce in the face of disease. The two case studies concerning boreal toads and Tasmanian devils provide on-the-ground examples of reintroduction programs grappling with different systems and different diseases. It is unlikely that the number of attempted reintroductions will decrease as the need to address declining fish and wildlife populations with direct conservation actions escalates. Reintroduction science has begun to incorporate several new frontiers in its application. Decision science, social science, and advances in biology (e.g., genomics, transmission pathways, genetic engineering, and vaccines) are all avenues of research that have direct bearing on the multiple aspects of disease that need to be addressed for successful reintroductions.

MANAGEMENT RECOMMENDATIONS

- Inattention to disease aspects of a reintroduction program can be disastrous from several perspectives: (a) the target animal to be reintroduced; (b) other native organisms that are susceptible to disease that is inadvertently introduced with the reintroduced animals; and (c) funding agencies, as reintroduction is costly and failure to account for disease can compromise an entire program. The actions below should be incorporated into the development and planning stages for any reintroduction.

- Determine if the cause of extinction is disease related.

- If that is so, determine if disease-resistant or disease-tolerant stock of the species to be reintroduced (target species) is available or can be bred from existing stock.

- Determine the degree of confidence in the knowledge that (a) the local host (target species) is extinct at the site and (b) the probability of recolonization by that host from other areas.

- Determine if disease reservoirs (other than the target species) are present at the targeted site or if disease could be present but unrecognized (e.g., latent) in reintroduced animals.

- If reservoirs other than target species are identified, can they be sufficiently managed in that reservoir?

ACKNOWLEDGMENTS

We thank D. Jachowski for the invitation to participate in this book. We thank D. Jachowski and two anonymous reviewers for helpful comments on earlier drafts. EM thanks L. Bailey and M. K. Watry for their generosity in allowing the use of our paper (Muths et al. 2014) as the basis for the case study and critique on toads, and the Weasel for support. EM's research is supported by the US Geological Survey (USGS) and the USGS Amphibian Research and Monitoring Initiative (ARMI). HM's research on Tasmanian devil facial tumor disease is supported by the Australian Research Council (ARC DP110102656) and the National Science Foundation (Award 1316549). His research on parasites and reintroductions is supported by the

Australian Research Council (ARC ARC LP130101073). This is contribution no. 520 of ARMI. Any use of trade, firm, or product names is for descriptive purposes only and does not imply endorsement by the US government.

LITERATURE CITED

Abbott, I. 2006. Mammalian faunal collapse in Western Australia, 1875–1925: the hypothesized role of epizootic disease and a conceptual model of its origin, introduction, transmission, and spread. Australian Zoologist 33:530–561.

Adams, M. J., S. Galvan, D. Reinitz, R. C. Cole, S. Pyare, M. Hahr, and P. Govindarajulu. 2007. Incidence of the fungus *Batrachochytrium dendrobatidis* in amphibian populations along the northwest coast of North America. Herpetological Review 38:430–431.

Aiello, C. M., K. E. Nussear, A. D. Walde, T. C. Esque, P. G. Emblidge, P. Sah, S. Bansal, and P. J. Hudson 2014. Disease dynamics during wildlife translocations: disruptions to the host population and potential consequences for transmission in desert tortoise contact networks. Animal Conservation 17:27–39.

Almberg, E. S., P. C. Cross, A. P. Dobson, D. W. Smith, and P. J. Hudson. 2012. Parasite invasion following host reintroduction: a case study of Yellowstone's wolves. Philosophical Transactions of the Royal Society B: Biological Sciences 367:2840–2851.

Australian Department of the Environment and Heritage. 2006. Threat abatement plan: Infection of amphibians with chytrid fungus results in Chytridomycosis. Department of the Environment and Heritage, Commonwealth of Australia. Available online at: http://www .amphibians.org/wp-content/uploads/2013/09 /Australian-chytrid-threat-abatement-plan.pdf.

Ballou, J. D. 1993. Assessing the risks of infectious disease in captive breeding and reintroduction programs. Journal of Zoo and Wildlife Medicine 24:327–335.

Beeton, N., and H. McCallum. 2011. Models predict that culling is not a feasible strategy to prevent extinction of Tasmanian devils from facial tumour disease. Journal of Applied Ecology 48:1315–1323.

Berger, L., R. Speare, P. Daszak, D. E. Green, A. A. Cunningham, L. Goggin, R. Slocombe et al. 1998. Chytridiomycosis causes amphibian mortality associated with population declines in the rain forests of Australia and Central America. Proceedings of the National Academy of Science of the United States of America 95:9031–9036.

Berger, L., G. Marantelli, L.F. Skerratt, and R. Speare. 2005. Virulence of the amphibian chytrid fungus Batrachochytrium dendrobatidis varies with the strain. Diseases of Aquatic Organisms 68:47–50.

Bielby, J., C. A. Donnelly, L. C. Pope, T. Burke, and R. Woodroffe. 2014. Badger responses to small-scale culling may compromise targeted control of bovine tuberculosis. Proceedings of the National Academy of Sciences of the United States of America 111:9193–9198.

Biggins, D. E., T. M. Livieri, and S. W. Breck. 2012. Interface between black-footed ferret research and operational conservation. Journal of Mammalogy 92:699–704.

Bletz, M. C., A. H. Loudon, M. H. Becker, S. C. Bell, D. C. Woodhams, K. P. C. Minbiole, and R. N. Harris. 2013. Mitigating amphibian chytridiomycosis with bioaugmentation: characteristics of effective probiotics and strategies for their selection and use. Ecology Letters 16:807–820.

Bode, M., and B. Wintle. 2010. How to build an efficient conservation fence. Conservation Biology 24:182–188.

Borriello, G., R. Capperelli, M. Bianco, D. Fenizia, F. Alfano, F. Capuano, D. Ercolini, A. Parisi, S. Roperto, and D. Iannelli. 2006. Genetic resistance to *Brucella abortus* in the water buffalo (*Bubalus bubalis*). Infection and Immunity 74:2115–2120.

Briggs, C. J., R. A. Knapp, and V. T. Vredenburg. 2010. Enzootic and epizootic dynamics of the chytrid fungal pathogen of amphibians. Proceedings of the National Academy of Sciences of the United States of America 107.21:9695–9700.

Briggs, C. J., V. T. Vredenburg, R. A. Knapp, and L. J. Rachowicz. 2005. Investigating the population-level effects of chytridiomycosis: an emerging infectious disease of amphibians. Ecology 86:3149–3159.

Bruckner, G. K., W. Vosloo, B. J. A. Du Plessis, P. Kloeck, L. Connoway, M. D. Ekron, D. B. Weaver et al. 2002. Foot and mouth disease: the experience of South Africa. Revue Scientifique et Technique De L Office International Des Epizooties 21:751–764.

Burton, F. J., and F. F. Rivera-Milan. 2014. Monitoring a population of translocated Grand Cayman blue iguanas: assessing the accuracy and precision of distance sampling and repeated counts. Animal Conservation 17:40–47.

Canessa, S. 2015. Structured decision making for

designing complex release strategies. Pp.17–28. In D. P. Armstrong, M. W. Haywarad, D. Moro, and P. J. Seddon, eds., Advances in Reintroduction Biology of Australian and New Zealand Fauna. CSIRO, Clayton, VIC.

Canessa, S., D. Hunter, M. McFadden, G. Marantelli, and M. A. McCarthy. 2014. Optimal release strategies for cost-effective reintroductions. Journal of Applied Ecology 51:1107–1115.

Carey, C. 1993. Hypothesis concerning the causes of the disappearance of boreal toads from the mountains of Colorado. Conservation Biology 7:355–362.

Carey, C., J. E. Bruzgul, L. J. Livo, M. L. Walling, K. A. Kuehl, B. F. Dixon, A. P. Pessier, R. A. Alford, and K. B. Rogers. 2006. Experimental exposures of boreal toads (Bufo boreas) to a pathogenic chytrid fungus (*Batrachochytrium dendrobatidis*). EcoHealth 3:5–21.

Carey, C., P. S. Corn, M. S. Jones, L. J. Livo, E. Muths, and C. W. Loeffler. 2005. Environmental and life history factors that limit recovery in southern Rocky Mountain populations of boreal toads (*Bufo boreas*). Pp.222–236. In M. Lannoo, ed., Amphibian Declines: The Conservation Status of United States Species. University of California Press, Berkley.

Caughley, G. 1994. Directions in Conservation Biology. Journal of Animal Ecology 3:215–244.

Chestnut, T., C. W. Anderson, R. Popa, A. R. Blaustein, M. Voytek, D. H. Olson, and J. Kirshtein. 2014. Heterogeneous occupancy and density estimates of the pathogenic fungus *Batrachochytrium dendrobatidis* in waters of North America. PLoS ONE 9:e106790.

Cleveland, S., M. Kaare, D. Knobel, and M. K. Laurenson. 2006. Canine vaccination: providing broader benefits for disease control. Veterinary Microbiology 117:43–50.

Converse, S. J., L. L. Bailey, B. Mosher, E. Muths, and Funk, W. C. 2016. Building a management model to inform reintroduction and other actions as a response to chytridiomycosis-associated decline. Ecohealth. doi:10.1007/s10393-016-1117-9.

Corn, P. S., Jennings, M. L., and Muths, E., 1997. Survey and assessment of amphibian populations in Rocky Mountain National Park. Northwestern Naturalist 78:34–55.

Corn, P. S., W. Stolzenburg, and R. B. Bury. 1989. Acid Precipitation Studies in Colorado and Wyoming: Interim Report of Surveys of Montane Amphibians and Water Chemistry. Report 80. US Fish and Wildlife Service, Washington, DC.

Covello, V. T., and M. W. Merkhofer. 1993. Risk Assessment Methods. Springer, New York.

Cowled, B. D., M. G. Garner, K. Negus, and M. P. Ward. 2012. Controlling disease outbreaks in wildlife using limited culling: modelling classical swine fever incursions in wild pigs in Australia. Veterinary Research 43:3.

Cryan, P. M., C. U. Meteyer, J. G. Boyle, and D. S. Blehert. 2013. White-nose syndrome in bats: Illuminating the darkness. BMC Biology. 11:1–4.

Daszak, P., D. E. Scott, A. M. Kilpatrick, C. Faggioni, J. W. Gibbons, and D. Porter. 2005. Amphibian population declines at Savannah River site are linked to climate, not chytridiomycosis. Ecology 86:3232–3237.

de Castro, F., and B. Bolker. 2005. Mechanisms of disease-induced extinction. Ecology Letters 8:117–126.

Denton, J. S., S. P. Hitchings, T. J. C. Beebee, and A. Gent. 1997. A recovery program for the natterjack toad (*Bufo calamita*) in Britain. Conservation Biology, 11:1329–1338.

Dickens, M. J., D. J. Delehanty, and L. M. Romero. 2010. Stress: an inevitable component of animal translocation. Biological Conservation 143:1329–1341.

DiRenzo, G. V., P. F. Langhammer, K. R. Zamudio, and K. R. Lips. 2014. Fungal infection intensity and zoospore output of *Atelopus zeteki*, a potential acute chytrid supershedder. PloS ONE 9:e93356. doi:10.1371/journal.pone.0093356.

Dobson, A. P., and P. J. Hudson. 1986. Parasites, disease and the structure of ecological communities. Trends in Ecology & Evolution 1:11–15.

Dodd, C. K., Jr. 2005. Population manipulations. Pp.265–270. In M. Lannoo, ed., Amphibian Declines: The Conservation Status of United States Species. University of California Press, Berkley.

Dodd, C. K., Jr. 2012. Amphibian Ecology and Conservation. Oxford University Press. Oxford, UK.

Donnelly, C. A., R. Woodroffe, D. R. Cox, F. J. Bourne, C. L. Cheeseman, R. S. Clifton-Hadley, G. Wei et al. 2006. Positive and negative effects of widespread badger culling on tuberculosis in cattle. Nature 439:843–846.

DSEWPaC (Department of Sustainability Environment Water Population and Communities). 2013. *Sarcophilus harrisii*. Species Profile and Threats Database. Department of Sustainability, Environment, Water, Population and Communities, Canberra. Available online at: http://www.environment.gov.au/sprat (accessed June 20, 2016.

Ewen, J.G., P. Soorae, and S. Canessa. 2014. Reintroduction objectives, decisions and outcomes: global perspectives from the herpetofauna. Animal Conservation 17(Suppl. 1):74–81.

Faria, P.J., C. VanOosterhout, and J. Cable. 2010. Optimal release strategies for captive-bred animals in reintroduction programs: experimental infections using the guppy as a model organism. Biological Conservation 143:35–41.

Feare, C.J. 2007. The role of wild birds in the spread of HPAI H5N1. Avian Diseases 51(s1):440–447.

Ferguson, N.M., C.A. Donnelly, and R.M. Anderson. 2001. The foot-and-mouth epidemic in Great Britain: pattern of spread and impact of interventions. Science 292:1155–1160.

Fisher, M.C., and T.W.J. Garner. 2007. The relationship between the emergence of *Batrachochytrium dendrobatidis*, the international trade in amphibians and introduced amphibian species. Fungal Biology Reviews 21:2–9.

Fisher, M.C., T.J.W. Garner, and S. Walker. 2009. Global emergence of *Batrachochytrium dendrobatidis* and amphibian chytridiomycosis in space, time and host. Annual Review of Microbiology 63:291–310.

Gaydos, J.K., W.R. Davidson, F. Elvinger, D.G. Mead, E.W. Howerth, and D.E. Stallknecht. 2002. Innate resistance to epizootic hemorrhagic disease in white-tailed deer. Journal of Wildlife Diseases 38:713–719.

Germano, J.M., K.J. Field, R.A. Griffiths, S. Clulow, J. Foster, G. Harding, and R.R. Swaisgood. 2015. Mitigation-driven translocations: are we moving wildlife in the right direction? Frontiers in Ecology and the Environment 13.2:100–105.

Goebel, A.M. 2005. Conservation systematics: the *Bufo boreas* species group. Pp.210–221. In M. Lannoo, ed., Amphibian Declines: The Conservation Status of United States Species. University of California Press, Berkley.

Goebel, A.M., T. Ranker, and P.S. Corn. 2009. Mitochondrial DNA evolution in the *Anaxyrus boreas* species group. Molecular Phylogenetics and Evolution 50:209–225.

Gompper, M.E., and E.S. Williams. 1998. Parasite conservation and the black-footed ferret recovery program. Conservation Biology 12:730–732.

Grant, B.R., and P.R. Grant. 1993. Evolution of Darwin's finches caused by a rare climatic event. Proceedings of the Royal Society, London, Biology 251:111–117.

Green, D.E., M.J. Gray, and D.L. Miller. 2009. Disease monitoring and biosecurity. Forestry, Wildlife, and Fisheries Publications and Other Works. Available online at: http://trace.tennessee.edu/utk_forepubs/2 (accessed August 2015).

Greenwald, D.N., C.L. Adkins Giese, M. Mueller, and E. Molvar. 2011. Petition to List a Distinct Population Segment of the Boreal Toad (*Anaxyrus boreas boreas*) as Endangered or Threatened Under the Endangered Species Act. Center for Biological Diversity, Center for Native Ecotystems, and Biodiversity Conservation Alliance, 72pp.

Griffith, B., J.M. Scott, J.W. Carpenter, and C. Reed. 1989. Translocation as a species conservation tool: status and strategy. Science 245:477–480.

Griffith, B., J.M. Scott, and J.W. Carpenter. 1993. Animal translocations and potential disease transmission. Journal of Zoo and Wildlife Medicine 24:231–236.

Hallam, T.G., and G.F. McCracken. 2011. Management of the panzootic white-nose syndrome through culling of bats. Conservation Biology 25:189–194.

Hamede, R., S. Lachish, K. Belov, G. Woods, A. Kreiss, A.-M. Pearse, B. Lazenby, M. Jones, and H. McCallum. 2012. Reduced effect of Tasmanian devil facial tumor disease at the disease front. Conservation Biology 26:124–134.

Hammerson, G.A. 1999. Amphibians and Reptiles in Colorado. 2nd ed. University Press of Colorado/Colorado Division of Wildlife, Denver.

Hansen, E.S., and P. Budy. 2011. The potential of passive stream restoration to improve stream habitat and minimize the impact of fish disease: a short-term assessment. Journal of the North American Benthological Society 30:573–588.

Hayward, M.W., and G.I.H. Kerley. 2009. Fencing for conservation: Restriction of evolutionary potential or a riposte to threatening processes? Biological Conservation 142:1–13.

Heard, M., K.F. Smith, and K. Ripp. 2011. Examining the evidence for chytridiomycosis in threatened amphibian species. PLoS ONE 6:e23150.

Henryon, M., P. Berg, N.J. Olesen, T.E. Kjaer, W.J. Slierendrecht, A. Jokumsen, and I. Lund. 2005. Selective breeding provides an approach to increase resistance of rainbow trout (*Onchorhynchus mykiss*) to the diseases, enteric redmouth disease, rainbow trout fry syndrome, and viral haemorrhagic septicaemia. Aquaculture 250:621–636.

Hollings, T., M. Jones, N. Mooney, and H. McCallum. 2014. Trophic cascades following the disease-induced decline of an apex predator, the Tasmanian devil. Conservation Biology 28:63–75.

Hoyt, J. R., K. E. Langwig, J. Okoniewski, W. F. Frick, W. B. Stone, and A. M. Kilpatrick. 2014. Long-term persistence of *Pseudogymnoascus destructans*, the causative agent of white-nose syndrome, in the absence of bats. EcoHealth 2014:1–4.

Hunter, D. O., T. Britz, M. Jones, and M. Letnic. 2015. Reintroduction of Tasmanian devils to mainland Australia can restore top-down control in ecosystems where dingoes have been extirpated. Biological Conservation 191:428–435.

Jacobs, D. F., H. J. Dalgleish, and C. D. Nelson. 2012. A conceptual framework for restoration of threatened plants; the effective model of American chestnut (*Castanea dentata*) reintroduction. New Phytologist 197:378–393.

Jakob-Hoff, R. M., S. C. MacDiarmid, C. Lees, P. S. Miller, D. Travis, and R. Kock. 2014. Manual of Procedures for Wildlife Disease Risk Analysis. World Organisation for Animal Health, Paris.

Johnson, C., 2006. Australia's mammal extinctions: a 50,000-year history. Cambridge University Press, Cambridge, UK.

Johnson, M. L., and R. Speare. 2005. Possible modes of dissemination of the amphibian chytrid *Batrachochytrium dendrobatidis* in the environment. Diseases of Aquatic Organisms 65:181–186.

Joseph, M. B., J. R. Mihaljevic, A. L. Arellano, J. G. Kueneman, D. L. Preston, P. C. Cross, and P. T. J. Johnson. 2013. Taming wildlife disease: bridging the gap between science and management. Journal of Applied Ecology 50:702–712.

Jule, K. R., L. A. Leaver, and S. E. Lea. 2008. The effects of captive experience on reintroduction survival in carnivores: a review and analysis. Biological Conservation 141:355–363.

Kemp, L., G. Norbury, R. Groenewegen, and S. Comer. 2015. The roles of trials and experiments in fauna reintroduction programs. Pp.74–89. In D. Armstrong, M. W. Hayward, D. Moro, and P. J. Seddon, eds., Advances in Reintroduction Biology of Autralian and New Zealand Fauna. CSIRO, Clayton, VIC.

Kilpatrick, A. M. 2006. Facilitating the evolution of resistance to avian malaria in Hawaiian birds. Biological Conservation 128:475–485.

Kirshtein, J. D., C. W. Anderson, J. S. Wood, J. E. Longcore, and M. A. Voytek. 2007. Quantitative PCR detection of *Batrachochytrium dendrobatidis* DNA from sediments and water. Diseases of Aquatic Organisms 77:11–15.

Kock, R. A., M. H. Woodford, and P. B. Rossiter. 2010. Disease risks associated with the translocation of wildlife. Revue Scientifique et Technique-Office International des Epizooties 29:329–350.

Kreiss, A., Y. Y. Cheng, F. Kimble, B. Wells, S. Donovan, K. Belov, and G. M. Woods. 2011. Allorecognition in the Tasmanian devil (*Sarcophilus harrisii*), an endangered marsupial species with limited genetic diversity. PLoS ONE 6:e22402–e22402.

Lachish, S., H. McCallum, D. Mann, C. E. Pukk, and M. E. Jones. 2010. Evaluation of selective culling of infected individuals to control Tasmanian devil facial tumor disease. Conservation Biology 24:841–851.

Lambert, B. A., R. A. Schorr, S. C. Schneider, and E. Muths. In Review. Influence of demography and environment on persistence in toad populations. Journal of Wildlife Management.

Lampo, M., S. J. Celsa, A. Rodríguez-Contreras, F. Rojas-Runjaic, and C. Z. García. 2012. High turnover rates in remnant populations of the harlequin frog *Atelopus cruciger* (Bufonidae): low risk of extinction? Biotropica 44:420–426.

Lee, T. E., M. A. McCarthy, B. A. Wintle, M. Bode, D. L. Roberts, M. A. Burgman, and J. Matthiopoulos. 2014. Inferring extinctions from sighting records of variable reliability. Journal of Applied Ecology 51:251–258.

Lembo, T., K. Hampson, D. T. Haydon, M. Craft, A. Dobson, J. Dushoff, E. Ernest, R. Hoare, M. Kaare, T. Mlengeya, and C. Mentzel. 2008. Exploring reservoir dynamics: a case study of rabies in the Serengeti ecosystem. Journal of Applied Ecology 45:1246–1257.

Leopold, A. 1949. A Sand County Almanac and Sketches Here and There. Oxford University Press, New York.

Lips, K. R., J. Diffendorfer, J. R. Mendelson, III, and M. W. Sears. 2008. Riding the wave: reconciling the roles of disease and climate change in amphibian declines. PLoS ONE 6:e72.

Lips, K. R., and J. R. Mendelson, III. 2014. Stopping the next amphibian apocalypse. New York Times, Op-Ed. November 15, 2014, A21.

Lloyd-Smith, J. O., D. George, K. M. Pepin, V. E. Pitzer, J. R. C. Pulliam, A. P. Dobson, P. J. Hudson, and B. T. Grenfell. 2009. Epidemic dynamics at the human-animal interface. Science 326:1362–1367.

Loeffler, C., 2001. Conservation Plan and Agreement for the Management and Recovery of the Southern Rocky Mountain Population of the Boreal Toad (*Bufo boreas boreas*). Colorado Division of Wildlife, Denver.

Longcore, J. E., A. P. Pessier, and D. K. Nichols. 1999. *Batrachochytrium dendrobatidis* gen. et sp. nov., a chytrid pathogenic to amphibians. Mycologia 91:219–227.

Longcore, J. R., J. E. Longcore, A. P. Pessier, and W. A. Halteman. 2007. Chytridiomycosis widespread in anurans of northeastern United States. Journal of Wildlife Management 71:435–444.

Lorch, J. M., J. K. Muller, R. E. Russell, M. O'Connor, D. I. Lindner, and D. S. Blehert. 2013. Distribution and environmental persistence of the causative agent of white-nose syndrome, *Geomyces destructans*, in bat hibernacula of the Eastern United States. Applied and Environmental Microbiology 79:1293–1301.

Lötters, S., R. Schulte, J. H. Córdova, and M. Veith. 2005. Conservation priorities for harlequin frogs (*Atelopus* spp.) of Peru. Oryx 39:343–346.

Martel, A., M. Blooi, C. Adriaensen, P. Van Rooij, W. Beukema, M. C. Fisher, R. A. Farrer et al. 2014. Recent introduction of a chytrid fungus endangers Western Palearctic salamanders. Science 346:630–631.

Martel, A., A. Spitzen-van der Sluijs, M. Blooia, W. Bert, R. Ducatelle, M. C. Fisher, A. Woeltjes et al. 2013. *Batrachochytrium salamandrivorans* sp. nov. causes lethal chytridiomycosis in amphibians. Proceedings of the National Academy of Sciences of the United States of America 110:15325–15329.

Martin, C., P. Pastoret, B. Brochier, M. Humblet, and C. Saegerman. 2011. A survey of the transmission of infectious diseases/infections between wild and domestic ungulates in Europe. Veterinary Research 42:70.

Mathews, F., D. Moro, R. Strachan, M. Gelling, and N. Buller. 2006. Health surveillance in wildlife reintroductions. Biological Conservation 131:338–347.

McCallum, H. I. 2012. Disease and the dynamics of extinction. Philosophical Transactions of the Royal Society of London B: Biological Sciences 367:2828–2839.

McCallum, H. I. 2015. Models for managing wildlife disease. Parasitology 18:1–16. doi:http://dx.doi.org/10.1017/S0031182015000980.

McCallum, H. I., and A. P. Dobson. 1995. Detecting disease and parasite threats to endangered species and ecosystems. Trends in Ecology and Evolution 10:190–194.

McCallum, H., and M. Jones. 2006. To lose both would look like carelessness: Tasmanian devil facial tumour sisease. PLoS ONE 4:1671–1674.

McCallum, H., M. Jones, C. Hawkins, R. Hamede, S. Lachish, D. L. Sinn, N. Beeton, and B. Lazenby. 2009. Transmission dynamics of Tasmanian devil facial tumor disease may lead to disease-induced extinction. Ecology 90:3379–3392.

McCallum, H., D. M. Tompkins, M. Jones, S. Lachish, B. Lazenby, G. Hocking, J. Wiersma, and C. Hawkins. 2007. Distribution and impacts of Tasmanian devil facial tumor disease. EcoHealth 4:318–325.

Miller, M. W., E. S. Williams, N. T. Hobbs, and L. L. Wolfe. 2004. Environmental sources of prion transmission in mule deer. Emerging Infectious Diseases 10:1003–1006.

Moore, R. D., R. A. Griffiths, and A. Roman. 2004. Distribution of the Mallorcan midwife toad (*Alytes muletensis*) in relation to landscape topography and introduced predators. Biological Conservation 116:327–332.

Moseby, K. E., B. M. Hill, and J. L. Read. 2009. Arid recovery: a comparison of reptile and small mammal populations inside and outside a large rabbit, cat and fox-proof exclosure in arid South Australia. Austral Ecology 34:156–169.

Murphy, P. J., S. St-Hilaire, S. Bruer, P. S. Corn, and C. R. Peterson. 2009. Distribution and pathogenicity of *Batrachochytrium dendrobatidis* in boreal toads from the Grand Teton area of western Wyoming. EcoHealth 6:109–120.

Muths, E., L. L. Bailey, and M. K. Watry. 2014. Animal reintroductions: an innovative assessment of survival. Biological Conservation 172:200–208.

Muths, E., P. S. Corn, A. P. Pessier, and D. E. Green. 2003. Evidence for disease-related amphibian decline in Colorado. Biological Conservation 110:357–365.

Muths, E., and V. Dreitz. 2008. Monitoring programs to assess reintroduction efforts: a critical component in recovery. Animal Biodiversity and Conservation 31:47–56.

Muths, E., R. D. Scherer, and D. S. Pilliod. 2011. Compensatory effects of recruitment and survival when amphibian populations are perturbed by disease. Journal of Applied Ecology 48:873–879.

National Park Service. 2006. Management Policies: The Guide to Managing the National Park System. NPS Office of Policy, Washington, DC.

OIE/IUCN (World Organisation for Animal Health and International Union for Conservation of Nature). 2014. Guidelines for Wildlife Disease Risk Analysis. OIE, Paris.

Ouellet, M., I. Mikaelian, B. D. Pauli, J. Rodriquez, and D. M. Green. 2005. Historical evidence of widespread chytrid infection in North American amphibian populations. Conservation Biology 19:1431–1440.

Parkinson, A. J., B. Evengard, J. C. Semenza, N. Ogden, M. L. Børresen, J. Berner, M. Brubaker et al. 2014. Climate change and infectious diseases in the Arctic: establishment of a circumpolar working group. International Journal of Circumpolar Health 73:25163.

Pearse, A.-M., and K. Swift. 2006. Allograft theory: transmission of devil facial-tumour disease. Nature 439:54e9.

Pessier, A. P. 2008. Management of disease as a threat to amphibian conservation. International Zoo Yearbook 42:30–39.

Pessier, A. P., D. K. Nichols, J. E. Longcore, and M. S. Fuller. 1999. Cutaneous chytridiomycosis in poison dart frogs (*Dendrobates* spp.) and White's tree frogs (*Litoria caerulea*). Journal of Veterinary Diagnostic Investigations 11:194–199.

Pilliod, D. S., E. Muths, R. D. Scherer, P. E. Bartelt, P. S. Corn, B. R. Hossack, B. A. Lambert, R. Mccaffery, and C. Gaughan. 2010. Effects of amphibian chytrid fungus on individual survival probability in wild boreal toads. Conservation Biology 24:1259–1267.

Pope, L. C., R. K. Butlin, G. J. Wilson, R. Woodroffe, K. Erven, C. M. Conyers, T. Franklin et al. 2007. Genetic evidence that culling increases badger movement: implications for the spread of bovine tuberculosis. Molecular Ecology 16:4919–4929.

Prowse, T. A., C. N. Johnson, C. J. Bradshaw, and B. W. Brook. 2014. An ecological regime shift resulting from disrupted predator-prey interactions in Holocene Australia. Ecology 95:693–702.

Rachowicz, L. J., and C. J. Briggs. 2007. Quantifying the disease transmission function: effects of density on *Batrachochytrium dendrobatidis* transmission in the mountain yellow-legged frog *Rana muscosa*. Journal of Animal Ecology 76:711–721.

Rachowicz, L. J., J-M. Hero, R. A. Alford, J. W. Taylor, J. A. T. Morgan, V. T. Vredenburg, J. P. Collins, and C. J. Briggs. 2005. The novel and endemic pathogen hypothesis: competing explanations for the origin of emerging infectious diseases of wildlife. Conservation Biology 19:1441–1448.

Ramsey, D. S. L., and M. G. Efford. 2010. Management of bovine tuberculosis in brushtail possums in New Zealand: predictions from a spatially explicit, individual-based model. Journal of Applied Ecology 47:911–919.

Reeder, N. M., A. P. Pessier, and V. T. Vredenburg. 2012. A reservoir species for the emerging amphibian pathogen *Batrachochytrium dendrobatidis* thrives in a landscape decimated by disease PLoS ONE 7:e33567.

Restif, O., D. T. S. Hayman, J. R. C. Pulliam, R. K. Plowright, D. B. George, A. D. Luis, A. A. Cunningham et al. 2012. Model-guided field-work: practical guidelines for multidisciplinary research on wildlife ecological and epidemiological dynamics. Ecology Letters 15:1083–1094.

Ricciardi, A., and D. Simberloff. 2009. Assisted colonization: good intentions and dubious risk assessment. Trends in Ecology and Evolution 24:476–477.

Rounsevell, D. E. 1989. Managing offshore island reserves for nature conservation in Tasmania. Pp.157–161. In A Burbidge, ed., Australian and New Zealand Islands: Nature Conservation Values and Management. Department of Conservation and Land Management, Perth, WA.

Roussel, J. M., J. M. Paillisson, A. Treguier, and E. Petit. 2015. The downside of eDNA as a survey tool in water bodies. Journal of Applied Ecology 52:823–826.

Rózsa, L., and Z. Vas. 2015. Co-extinct and critically co-endangered species of parasitic lice, and conservation-induced extinction: should lice be reintroduced to their hosts? Oryx 49:107–110.

Runge, M. C. 2011. An introduction to adaptive management for threatened and endangered species. Journal of Fish and Wildlife Management 2:220–233.

Sainsbury, A. W., D. P. Armstrong, and J. G. Ewen. 2012. Methods of disease risk analysis for reintroduction programmes. Pp.344–359. In J. G. Ewen, D. P. Armstrong, K. A. Parker, and P. J. Seddon, eds., Reintroduction Biology: Integrating Science and Management. Wiley-Blackwell, London, UK.

Sainsbury, A. W., and R. J. Vaughan-Higgins. 2012. Analyzing disease risks associated with translocations. Conservation Biology 26:442–452.

Sarrazin, F., and S. Legendre. 2000. Demographic approach to releasing adults versus young in reintroductions. Conservation Biology 14:488–500.

Sarre, S. D., A. J. MacDonald, C. Barclay, G. R. Saunders, and D. S. L. Ramsey. 2013. Foxes are now widespread in Tasmania: DNA detection defines the distribution of this rare but invasive carnivore. Journal of Applied Ecology 50:459–468.

Schaub, M., R. Zink, H. Beissmann, F. Sarrazin, and R. Arlettaz. 2009. When to end releases in reintroduction programmes: demographic rates and population viability analyses of bearded vultures in the Alps. Journal of Applied Ecology. 46:92–100.

Schloegel, L. M., C. M. Ferreira, T. Y. James, M. Hipolito, J. E. Longcore, A. D. Hyatt, M. Yabsley et al. 2010. The North American bullfrog as a reservoir for the spread of *Batrachochytrium dendrobatidis* in Brazil. Animal Conservation 13:53–61.

Seddon, P. J., D. P. Armstrong, and R. F. Maloney. 2007. Developing the science of reintroduction biology. Conservation Biology 21:303–312.

Seddon, P. J., P. Soorae, and F. Launay. 2005. Taxonomic bias in reintroduction projects. Animal Conservation 8:51–58.

Seigel, R. A., and C. K. Dodd, Jr., 2002. Translocations of amphibian: proven management method or experimental technique? Conservation Biology 16:552–554.

Sepulveda, A., A. Ray, R. Al-Chokhachy, C. Muhlfeld, R. Gresswell, J. Gross, and J. Kershner. 2012. Aquatic invasive species: lessons from cancer research. American Scientist 100:234–242.

Shier, D. M., and R. R. Swaisgood. 2012. Fitness costs of neighborhood disruption in translocations of a solitary mammal. Conservation Biology 26:116–123.

Siddle, H. V., and J. Kaufman. 2013. A tale of two tumours: comparison of the immune escape strategies of contagious cancers. Molecular Immunology 55:190–193.

Siddle, H. V., A. Kreiss, M. D. B. Eldridge, E. Noonan, C. J. Clarke, S. Pyecroft, G. M. Woods, and K. Belov. 2007. Transmission of a fatal clonal tumor by biting occurs due to depleted MHC diversity in a threatened carnivorous marsupial. Proceedings of the National Academy of Sciences of the United States of America 104:16221–16226.

Siddle, H. V., J. Marzec, Y. Cheng, M. Jones, and K. Belov. 2010. MHC gene copy number variation in Tasmanian devils: implications for the spread of a contagious cancer. Proceedings of the Royal Society B: Biological Sciences 277:2001–2006.

Simms, L., and M. Jeggo. 2014. Avian influenza from an ecohealth perspective. EcoHealth 11.1:4–14.

Skerratt, L. F., L. Berger, R. Speare, S. Cashins, K. R. McDonald, A. D. Phillott, H. B. Hines, and N. Kenyon. 2007. Spread of chytridiomycosis has caused the rapid global decline and extinction of frogs. EcoHealth 4:125–134.

Smith, K. F., F. S. Dov, and K. D. Lafferty. 2006. Evidence for the role of infectious disease in species extinction and endangerment. Conservation Biology 20:1349–1357.

Soorae, P. S., ed. 2011. Global Re-introduction Perspectives: 2011. More Case Studies from around the Globe. IUCN/SSC Re-introduction Specialist Group, Gland, Switzerland/Environment Agency-Abu Dhabi, Abu Dhabi, UAE.

Stanley-Price, M., 1991. A review of mammal re-introductions, and the role of the Reintroduction Specialist Group of IUCN/SSC. Pp.9–23. In J. H. W. Gipps, ed., Beyond Captive Breeding: Re-introducing Endangered Mammals to the Wild. Symposium of the Zoological Society of London. Oxford University Press, New York.

Stear, M. J., S. C. Bishop, B. A. Mallard, and H. Raadsma. 2001. The sustainability, feasibility and desirability of breeding livestock for disease resistance. Research in Veterinary Science 71:1–7.

Strayer, D. L. 2009. Twenty years of zebra mussels: lessons from the mollusk that made headlines. Frontiers in Ecology and the Environment 7:135–141.

Stuart, S. N., J. S. Chanson, N. A. Cox, B. E. Young, A. S. L. Rodrigues, D. L. Fischman, and R. W. Waller. 2004. Status and trends of amphibian declines and extinctions worldwide. Science 306:1783–1786.

Swanack, T. M., W. E. Grant, and M. R. J. Forstner. 2009. Projecting population trends of endangered amphibian species in the face of uncertainty: a pattern-oriented approach. Ecological Modelling 220:148.

Switzer, J. F., R. Johnson, B. A. Lubinski, and T. L. King. 2009. Genetic Structure in the *Anaxyrus boreas* Species Group (Anura, Bufonidae): An Evaluation of the Southern Rocky Mountain Population. Other Government Series. A Final Report Submitted to the US Fish and Wildlife Service, Mountain-Prairie Region.

Teixeira, C. P., C. S. De Azevedo, M. Mendl, C. F. Cipreste, and R. J. Young. 2007. Revisiting translocation and reintroduction programmes: the importance of considering stress. Animal Behaviour 73:1–13.

Tripp, D. W., T. E. Rocke, S. P. Streich, N. L. Brown, J. Rodriquez-Ramos Fernandez, and M. W. Miller. 2014. Season and application rates affect vaccine bait consumption by prairie dogs in Colorado and Utah, USA. Journal of Wildlife Diseases 50:224–234.

US Fish and Wildlife Service. 2009. Greenback Cutthroat Trout (*Oncorhynchus clarki stomias*) 5-Year Review: Summary and Evaluation. US Fish and Wildlife Service Colorado Field Office, Lakewood.

Vander Wal, E., D. Garant, S. Calmé, C. A. Chapman, M. Festa-Bianchet, V. Millien, S. Rioux-Paquette, and F. Pelletier. 2014. Applying evolutionary concepts to wildlife disease ecology and management. Evolutionary Applications 7:856–868.

Venesky, M. D., J. R. Mendelson, III, B. F. Sears, and P. Stiling. 2012. Selecting for tolerance against pathogens and herbivores to enhance success of reintroduction and translocation. Conservation Biology 26:586–592.

Vial, F., and C. A. Donnelly. 2012. Localized reactive badger culling increases risk of bovine tuberculosis in nearby cattle herds. Biology Letters 8:50–53.

Viana, M., S. Cleaveland, J. Matthiopoulos, J. Halliday, C. Packer, M. E. Craft, K. Hampson et al. 2015. Dynamics of a morbillivirus at the domestic–wildlife interface: Canine distemper virus in domestic dogs and lions. Proceedings of the National Academy of Sciences 112:1464–1469.

Viggers, K. L., D. B. Lindenmayer, and D. M. Spratt. 1993. The importance of disease in reintroduction programs. Wildlife Research 20:687–698.

Voyles, J., S. Young, L. Berger, C. Campbell, W. F. Voyles, A. Dinudom, D. Cook et al. 2009. Pathogenesis of chytridiomycosis, a cause of catastrophic amphibian declines. Science 326:582–585.

Vredenburg, V. T., C. J. Briggs, and R. N. Harris. 2011. Host-pathogen dynamics of amphibian chytridiomycosis: the role of the skin microbiome in health and disease. Pp.342–354. In L. Olsen, E. R. Choffnes, D. A. Relman, and L. Pray, eds., Fungal Diseases: An Emerging Threat to Human, Animal and Plant Health. National Academies Press, Washington, DC.

Vredenburg, V. T., R. A. Knapp, T. S. Tunstall, and C. J. Briggs. 2010. Dynamics of an emerging disease drive large-scale amphibian population extinctions. Proceedings of the National Academy of Sciences of the United States of America 107:9689–9694.

Walker, S. F., S. M. Baldi, D. Jenkins, T. W. Garner, A. A. Cunningham, A. D. Hyatt, J. Bosch, and M. C. Fisher. 2007. Environmental detection of *Batrachochytrium dendrobatidis* in a temperate climate. Diseases of Aquatic Organisms 77:105–112.

Walker, S. F., J. Bosch, T. Y. James, A. P. Litvintseva, J. Antonio, O. Valls, S. Pina, G. Garcia et al. 2008. Invasive pathogens threaten species recovery programs. Current Biology 18:R853–R854.

Walters, J. R., S. R. Derrickson, D. M. Fry, S. M. Haig, J. M. Marzluff, and J. M. Wunderle, Jr. 2010. Status of the California condor (*Gymnogyps californianus*) and efforts to achieve its recovery. Auk 127:969–1001.

Wasserberg, G., E. E. Osnas, R. E. Rolley, and M. D. Samuel. 2009. Host culling as an adaptive management tool for chronic wasting disease in white-tailed deer: a modelling study. Journal of Applied Ecology 46:457–466.

Wobeser, G. 2002. Disease management strategies for wildlife. Revue Scientifique Et Technique Office International Des Epizooties 21:159–178.

Wolfe, L. L., M. W. Miller, and E. S. Williams. 2004. Feasibility of "test-and-cull" for managing chronic wasting disease in urban mule deer. Wildlife Society Bulletin 32:500–505.

Wolfe, N. D., C. P. Dunavan, and J. Diamond. 2007. Origins of major human infectious diseases. Nature 447:279–283.

Woodford, M. H. 2000. Quarantine and Health Screening Protocols for Wildlife Prior to Translocation and Release into the Wild. IUCN Species Survival Commisson's Veterinary Specialist Group, Gland, Switzerland.

Woodhams, D. C., J. Bosch, C. J. Briggs, S. Cashins, L. R. Davis, A. Lauer, E. Muths et al. 2011. Mitigating amphibian disease: strategies to maintain wild populations and control chytridiomycosis. Frontiers in Zoology 8:8.

ELEVEN

Release Considerations and Techniques to Improve Conservation Translocation Success

Axel Moehrenschlager and Natasha A. Lloyd

DECISIONS REGARDING THE RELEASE of animals are numerous, and can make or break a reintroduction program. They can make the difference between individual survival or death, successful or failed population establishment, returning or lost ecological processes, and supportive or opposing public opinion. Releases stand out among thematic and chronological considerations of the broader reintroduction spectrum through their pragmatic reality: at some point, someone will be physically handling, moving, and setting free individual organisms into an environment that may or may not be suitable for survival or reproduction. Such actions will not only be repeated for tens, hundreds, or thousands of individuals, but also produce ripple effects in the receiving ecosystem that could be desirable or damagingly irreversible.

Reintroductions are fraught with uncertainty. Identifying and reducing original factors of population decline or newly emerging threats is an important and often difficult challenge. Using all available information to restore or create ideal ecological conditions in landscapes where habitats have been lost, fragmented, or altered may be crucial, but actual suitability may be unclear until releases are actually attempted and evaluated. Decisions about what individuals are released, where, when, and how are of immediate consequence. Perhaps at no other point does one have as much control to react to emerging monitoring data and to ultimately affect demographic change as through decisions in the release process.

Building upon the previous set of chapters on obstacles to successful reintroductions (Chapters 8–10, this volume), in this chapter we seek to provide a synthetic problem- or challenge-based framework to guide reintroduction decision-making through iterative planning, preparation, releases, monitoring, and evaluation. We then describe in detail the four primary release management actions that can be undertaken to address these prerelease and post-release management challenges for release candidates and destinations: (1) selection, (2) preparation, (3) support, and (4) addressing human dimensions. Finally, we conclude with a review of recommendations for future research and development.

Keeping to the theme of this book, we concentrate exclusively on the conservation translocation of animals that aims to restore populations within their indigenous range (Dalrymple and Moehrenschlager 2013). As such, we focus on reintroduction, which is "the intentional movement and release of an organism inside its indigenous range from which it has disappeared," and reinforcement, which is "the intentional movement and release of an organism into an existing population of conspecifics" (IUCN 2013; Jachowski et al., Chapter 1, this volume). Further, many release considerations we overview here could also be more broadly relevant to conservation translocations that focus on conservation introductions, namely ecological replacement, which is "the intentional movement and release of an organism outside its indigenous range to perform a specific ecological function," and assisted colonization, which is "the intentional movement and release of an organism outside its indigenous range to avoid extinction of populations of the focal species" (IUCN 2013).

PLANNING, MODELING, AND THE RELEASE MANAGEMENT MATRIX

Release decisions lead to management actions that can be optimized through sound alignment within the broader reintroduction framework. The IUCN Guidelines for Reintroductions and Other Conservation Translocations (IUCN 2013) recommend that for adequate planning "Any conservation translocation should follow a logical process from initial design, feasibility, risk assessment, decision-making, implementation, monitoring, adjustment and evaluation"; and "Actions are precise statements of what should be done to meet the objectives; they should be capable of measurement, have time schedules attached, indicate the resources needed and who is responsible and accountable for their implementation." While release decisions may change over time, they should always be aligned to achieve goals defined for individual reintroduction programs.

Ecological models can help optimize decision-making for releases. Models can range from rudimentary diagrammatic concepts to highly quantitative, statistical assessments (Chapters 6 and 7, this volume) that attempt to predict translocation outcomes under various scenarios at a population, metapopulation, or ecosystem level. Respective examples include determining how establishment probability is affected by size and composition of the release group, what the optimal allocation of translocated individuals is among release sites, or in what sequence reintroductions of different species should occur (Armstrong and Seddon 2008). The utility and expansion of the use of models in reintroduction decision-making is reflected in a recent review by Armstrong and Reynolds (2012), which found population models to be used with increasing frequency, taxonomic breadth, and geographic spread for either proposed or initiated reintroductions. Examples included modeling various release scenarios for the Wyoming toad (*Bufo baxteri*) in the United States (Muths and Dreitz 2008), white-tailed eagle (*Haliaeetus albicilla*) in Scotland (Evans et al. 2009), European lynx (*Lynx lynx*) in Germany (Kramer-Schadt et al. 2005), Przewalski's horse (*Equus ferus przewalskii*) in Mongolia (Slotta-Bachmayr et al. 2004), four skink species in New Zealand (Towns and Ferreira 2001), and the bridled nailtail wallaby (*Onychogalea fraenata*) as well as trout cod (*Maccullochella macquariensis*) in Australia (McCallum et al. 1995, Todd et al. 2004).

While there has been great academic progress in developing models to enhance decision-making, in practice, when managers of reintroduction programs are asked how they will deal with uncertainty, a common response we encounter is that it will "just be dealt with using adaptive management over time." Unfortunately, this almost always refers to random trial and error, without sound mechanisms to respond to emerging knowledge. In reality, proper adaptive management is an iterative process of structured decision-making (Chapters 6 and 7, this volume) that can align progress assessment, learning, and state-dependent decisions (Nichols and Armstrong 2012). This approach has rarely been published for reintroductions (Armstrong and Reynolds 2012), but bears great promise for the ongoing refinement of such programs (Chapters 6 and 7, this volume).

To enhance adoption of adaptive, model-based approaches to release strategies, it is important to define what monitoring outcomes could or should be linked to subsequent management actions. In particular, while almost all reintroduction programs utilize repeated releases, many reviews describe prerelease considerations as if they were a single set of considerations preceding a single step of subsequent release. Here we propose a correlative "release management matrix" that aims to iteratively integrate various forms of monitoring data (Gitzen et al., Chapter 12, this volume) with potential management responses that can improve reintroduction programs over time. In general, we propose that reintroductions should involve monitoring for six forms of potential prerelease or post-release management challenges: (1) social, cultural, and economic (hereafter referred to as "socioeconomic"); (2) health and mortality (hereafter referred to as "health"); (3) ecological; (4) demographic; (5) genetic; and (6) behavioral (figure 11.1). Further, in response to these challenges, potential release management actions can be categorized into four broad categories, namely to (1) select release candidates/destinations, (2) prepare release candidates/destinations, (3) support release candidates/destinations, and (4) address human dimensions (figures 11.1 and 11.2).

While not all release management actions can be used to address all monitoring-based challenges, every challenge is met by numerous options within at least three of the broad release management action categories (figure 11.2). Below we provide examples of such potential connections, and thereby also set the stage for detailed sections that follow, which strive to investigate how a diversity of management actions can assist the restoration of various animal taxa.

Demographic Monitoring

Demographic monitoring may identify low survival, reproduction, population sizes, or evidence of poaching. These can then be addressed through release candidate management actions (figures 11.1 and 11.2), such as choices of release numbers, sex ratios, or prerelease training techniques; changes in release area management actions such as the selection of release areas, soft-release techniques, or predator management; or changes in human-condition actions such as proactive education/incentive programs to prevent possible increases in poaching (for further discussion, see Chapter 13, this volume).

Behavioral Monitoring

Behavioral monitoring may reveal an inability to forage, inability to evade predators, excessive dispersal, Allee effects, or inadequate migratory ability. This, in turn, can stimulate release management actions regarding the selection or preparation of individuals for release, or the selection of particular release sites where suboptimal behaviors may have less serious effects (for further discussion, see Chapter 9, this volume).

Genetic Monitoring

Genetic monitoring may detect high levels of inbreeding or outbreeding among post-release survivors, as well as potential risks of genetic isolation, which can trigger subsequent choices as to which release candidates should be chosen from wild or captive populations, where they are released, or in the most extreme instance, whether genetically deleterious individuals should be removed from release populations (for further discussion, see Chapter 8, this volume).

Health Monitoring

Health monitoring may reveal that released animals have contracted, or transmitted, deleterious

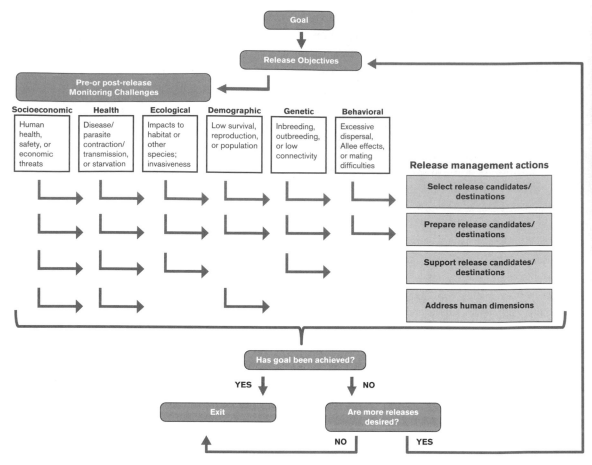

FIGURE 11.1. Flowchart for achieving release objectives in a reintroduction program. Following six key categories of "prerelease and post-release monitoring challenges" described in the IUCN Guidelines for Reintroductions and Other Conservation Translocations (2013) (i.e., socio economic, health, ecological, demographic, genetic, and behavioral), the flowchart illustrates feedback through "Release management actions" to correct the outstanding challenge(s) identified by prerelease risk assessments and/or monitoring.

disease or parasites. Depending on circumstance and effect, this may result in vaccination or parasite-treatment of captive or wild populations, quarantining of captive populations, capture and quarantining of wild animals, or in more extreme cases, the culling of released animals to alleviate welfare concerns or impacts upon conspecifics and other species at release sites (for further discussion, see Chapter 10, this volume).

Ecological or Socioeconomic Monitoring

Ecological or socioeconomic monitoring may unveil threats imposed by released animals on

the recipient ecosystem (for further discussion, see Chapter 9, this volume) or on human priorities such as personal safety and economic opportunity (for further discussion, see Chapter 4, this volume). This, in turn, can cause release actions that may pertain to direct management of individual animals, release sites, or effects imposed by humans (figure 11.2).

RELEASE MANAGEMENT ACTIONS

As overviewed and highlighted in the matrix presented above (and in figure 11.1), release management actions should be directly linked to management challenges that are observed through prerelease and post-release monitoring.

PREPARE RELEASE CANDIDATES/DESTINATIONS
- Behavioral training: antipredator, foraging, locomotion, mating
- Vaccinations, parasite treatment, reduce stressful stimuli
- Restore, create habitat, control predators/competitors

SELECT RELEASE CANDIDATES/DESTINATIONS
- Founders, sex ratio, genetics, life stage, social grouping, number, health, and behavior
- Suitable habitat, symbiotic species, sites with low risk to ecosystem
- Select/avoid conspecifics, competitors, predators

MONITORING
- Demographic, genetic, behavior, health, ecological, socioeconomic aspects
- Ongoing, iterative, long term
- Adaptive management context

SUPPORT RELEASE CANDIDATES/DESTINATIONS
- Soft release, foster, head start
- Food supplementation, refuge provision, olfactory cues,
- Physical protection, health checks, remove and reallocate, euthanize

ADDRESS HUMAN DIMENSIONS
- Education, stewardship
- Compensation and incentive programs
- Enforcement, policy

FIGURE 11.2. Release management actions (Address Human Dimensions, Select Release Candidates/Destinations, Prepare Release Candidates/Destinations, and Support Release Candidates/Destinations) for prerelease risk assessments and monitoring-dependent refinements that enable the adaptive management of release objectives.

Below we present the four major categories of release management actions available to reintroduction practitioners. It should be noted that the order in which we present these actions is not meant to imply their importance (figure 11.2). Rather, actions should be directly linked to observed challenges and prioritized accordingly (figure 11.1). Further, the relationship between monitoring outcomes, and release management actions, should continually be revisited to optimize subsequent decision-making. Finally, while this process aims to increase the efficiency and success of a reintroduction, not every monitoring result necessitates further releases. Releases can be terminated because restoration targets have been achieved. On the other hand, one of the most difficult, but also one of the most responsible, decisions is to abandon releases in circumstances where goal achievement becomes unlikely or where associated risks begin to outweigh potential benefits (figure 11.1).

Selecting Release Candidates and Destinations

During establishment and growth phases of reintroductions, multiple criteria underlie critical decisions to select source populations, individuals for release, release destination areas/sites, and appropriate release timing. Optimal decisions for selecting source population(s) will depend on the species and the goals of the given translocation, and will involve a trade-off between needs for the source and recipient populations (IUCN 2013). General considerations for determining source populations and suitable founder individuals include potential adverse effects on source populations, welfare implications for remaining animals at source sites or for moved individuals, diseases and parasites, and genetic considerations (IUCN 2013). From the source population(s), whether wild or captive, decisions as to the most appropriate number of individuals, life stage, age, sex and/or social group to be translocated must be determined prior to release (figure 11.1). For a conservation translocation of the Laysan duck (*Anas laysanensis*), selection of individuals involved multiple criteria, including age, sex, body condition, and reproductive status. Selected individuals were fledged but not of breeding age, representative of an even sex ratio, and in good body condition. Due to the specific selection criteria for founder individuals coupled with the fact that the source population for the Laysan duck was near carrying capacity, no negative consequences were incurred by the source population for the translocation (Reynolds and Klavitter 2006). Therefore, the viability of both the source and release populations need to be incorporated into translocation scenarios, and the optimal situation should be determined through a detailed risk assessment utilizing decision theory as well as population modeling (IUCN 2013).

Select Appropriate Genetics from Wild or Captive Populations

Selection of founders must consider the genetic diversity of the focal species or subspecies (Biebach et al., Chapter 8, this volume), but it can vary slightly based on the type of release being attempted. For reintroduction or reinforcements, proximate populations should be considered to incorporate locally adaptive traits (IUCN 2013). For conservation introductions, it may be beneficial to use multiple founder groups to maximize genetic diversity. For ecological replacements, the closest subspecies or species in terms of phylogeny, ecology, and behavior to the previously present form should be selected (IUCN 2013). Regardless of the conservation translocation approach being attempted, the number and demography of released individuals influence the probability of retaining alleles, the ability to minimize genetic loss in the subsequent release population (Tracy et al. 2011), and the potential for Allee effects (Deredec and Courchamp 2007, Armstrong and Wittmer 2011). Further, founder genetic diversity can

also promote release site establishment. For example, translocated populations of color polymorphic pygmy grasshoppers (*Tetrix subulata*) with greater genetic variation had more offspring the following year relative to those with low variation in founder individuals (Forsman et al. 2012).

The availability of wild source populations, or lack thereof, often determines whether captive breeding is necessary prior to releasing individuals. Captive-breeding programs are typically initiated when wild populations cannot provide a source of animals for translocation without compromising sustainability. Sometimes, wild populations have decreased to the point that the capture of all wild individuals needs to be considered, as was the case for the Guam rail (*Hypotaenidia owstoni*; Dillon 1988).

Captive breeding may enable managers to breed selectively for optimal release group composition, but managers must balance the risk of inbreeding depression (i.e., by increasing genetic variation) and the risk of outbreeding depression (i.e., by preventing mixing or dilution of genetic variation) to maintain the genetic purity of populations (Edmands 2007; Biebach et al., Chapter 8, this volume). Studbooks in captive-breeding programs assist managers in selecting certain individuals as breeding pairs to aid in genetic management. For example, in a captive-breeding program for Atlantic salmon (*Salmo salar*), a comparison of conservation risks between inbreeding and outbreeding determined that fitness consequences due to outbreeding were more variable. Inbreeding consistently resulted in negative fitness consequences, whereas outbreeding had both positive and negative fitness consequences for juvenile salmon (Rollinson et al. 2014). Avoiding inbreeding can be particularly difficult in captive-breeding programs when the diversity of wild founders is limited and multiple successive generations are bred. Additional techniques, such as cryopreservation of gametes, can aid in genetic management and ex situ conservation breeding (Figiel 2013). As a general rule, we suggest that genetic risk assessments and experimental procedures should be incorporated in captive-breeding programs for species involved in reintroductions.

Minimize Health Concerns in the Selection of Individuals

Disease and parasite considerations are important not only for the translocated individuals, but also to minimize potential transmission to the release site ecosystem (Muths and McCallum, Chapter 10, this volume). Therefore, risk assessments encompassing potential disease/parasite issues prior to translocation are critical (OIE and IUCN 2014). Surveying the source and recipient population(s) and ecosystem(s) for potential diseases and parasites should be completed prior to moving animals, while recognizing that some parasites may be beneficial for the released species. Insufficient risk assessments for disease considerations can lead to selection of sick individuals or carriers of disease that may introduce novel or damaging pathogens into populations and greatly increase financial costs of the translocation program (Woodford and Rossiter 1994). For example, translocations in 1925 of plains bison (*Bison bison bison*) to Wood Buffalo National Park, Canada, led to the transmission of bovine tuberculosis to resident wood bison (*B. b. athabascae*) populations (Carbyn and Watson 2001). Similarly, chytridiomycosis (Muths and McCallum, Chapter 10, this volume) has led to worldwide declines of amphibians (Fisher et al. 2009, Baláž et al. 2014). In response, detailed protocols have been developed for translocations of amphibians in the face of this disease (Pessier and Mendelson 2010) and should be strictly followed to reduce transmission. In addition to disease issues, parasites can also impact the success of a translocation (Cunningham 1996, Woodroffe 1999), and parasite introduction or transmission from translocations is a concern for resident species conservation (Viggers et al. 1993).

Select Appropriate Behaviors, Demographics, and Release Numbers

Social relationships and dominance hierarchies should be considered when selecting individuals to minimize stress in source or release populations (Jachowski et al., Chapter 9, this volume). Founder-group familiarity can increase post-release survival (Shier 2006); however, familiarity may have a greater post-release effect for social species compared to more solitary ones (Armstrong 1995). Group-living and social interactions (i.e., alarm calling, social foraging, communal nesting, allogrooming, etc.) confer fitness benefits for multiple taxa (Sherman 1980, Griffin 2004, Caro 2005) and can influence translocation success (Kleiman 1989, Shier 2006). Juveniles of social species, such as primates, often learn critical behaviors through group-living processes (Rabin 2003). Therefore, releasing young individuals along with experienced conspecific group members may increase post-release survival (Griffin et al. 2000, Shier and Owings 2007). Shier (2006) found that when family groups of black-tailed prairie dogs (*Cynomys ludovicianus*) were translocated together, post-release dispersal was lower and post-release survival was up to five times higher than in randomly related translocation groups. Fish that live in schools or harems may also benefit from releases in large groups that reflect demographic compositions suitable for species-specific mating strategies (Brown and Day 2002). Even for comparatively solitary animals, behavioral considerations may be valuable to consider when selecting individuals for release. Behavioral studies of swift foxes determined that "bold" individuals in captivity tended to disperse more than "shy" individuals, thereby suggesting that selection of "shy" individuals in captivity may increase post-release survival (Bremner-Harrison et al. 2004, Box 11.1).

Demographic considerations for release group candidates should optimize the probability of establishment, survival, and population growth post-release (Converse and Armstrong, Chapter 7, this volume). Demographic modeling was used to predict the most appropriate age-class to increase establishment of reintroductions and avoid extinction for griffon vulture (*Gyps fulvus*; Sarrazin and Legendre 2000) and helmeted honeyeater (*Lichenostomus melanops cassidix*; McCarthy 1994). Sometimes, large release numbers can contribute to translocation success (Fischer and Lindenmayer 2000, IUCN 2013). However, population growth can become asymptotic with increased numbers of release animals (Griffith et al. 1989, Wolf et al. 1996) and will depend on the life history of the species. For amphibians, translocation success was significantly greater when over 1,000 individuals were released (Germano and Bishop 2009). For Artiodactyla, population growth increased with up to 15–20 release individuals (Komers and Curman 2000, Matson et al. 2004), after which growth rate did not increase with additional group size (Komers and Curman 2000).

Selection of appropriate individuals or life stages requires species-specific considerations. Muths et al. (2014) found late-stage boreal toad (*Anaxyrus boreas*) tadpoles released to the wild had a greater chance of surviving compared to younger tadpoles. For European wild rabbits (*Oryctolagus cuniculus*), survival was greater for females compared to males, likely due to "sex-specific social behavior" (Letty et al. 2000). Although certain life stages can have particular advantages in terms of post-release success, combinations of several life stages for releases may maximize overall success and aid in cost-effectiveness for programs (Canessa et al. 2014). Canessa et al. (2014) found that trade-offs between the source captive population of the southern corroboree frog (*Pseudophryne corroboree*) and the release cohorts allowed for a cost-effective optimal release strategy which included both subadult (high success in wild but high cost in captivity) and egg (lower success in wild but lower cost in captivity) life stages. Experimental age- and sex-specific releases should be investigated whenever possible to determine the effect of these variables on site fidelity, release site establishment, and survival for different species and taxa (Hardman and Moro 2006).

Potential post-release dispersal generally, or of particular individuals, is rarely incorporated into release candidate selection criteria (van Heezik et al. 2009; Jachowski et al., Chapter 9, this volume) despite potential influences on post-release fitness. Swift fox (*Vulpes velox*) post-release movements were highest during the acclimation and establishment phases, when both the survival and likelihood of mating decreased with increasing dispersal distances (Moehrenschlager and MacDonald 2003, Box 11.1). Similarly, translocated young kaki (*Himantopus novaezelandiae*) had low site fidelity with over 30% of released individuals dispersing from their intended release sites (van Heezik et al. 2009). Interestingly, juvenile kaki tended to disperse to a greater extent compared to subadults (van Heezik et al. 2009), suggesting that age-specific selection for releases may decrease post-release movements for this species. Adult tiger beetles (*Cicindela dorsalis dorsalis*) provide perhaps the most extreme example, where dispersal shortly after release caused an experimental translocation to fail (Knisley et al. 2005). However, subsequent translocations using larvae resulted in adult emergence, survival, and wild recruitment of tiger beetles, showing initial signs of success for this species (Knisley et al. 2005).

Select Appropriate Release Areas and Release Sites

Released animals require suitable habitats for survival, reproduction, and recruitment to found or support a population capable of long-term growth, stability, or desired ecological function. Choices regarding release destinations should be considered on two spatial scales, namely in terms of release areas and release sites. A release area is a habitat where animals are hoped to remain in order to found a population. A release site is a smaller area within the release area where actual releases occur (Le Gouar et al. 2012).

Release area selection consists of finding sites where the conditions for life-stage requirements are suitable, survival or reproduction may be optimal, ecological or human threats are minimal, and the risk of released animals causing damage to ecological conditions or

BOX 11.1 · Return of the Swift Fox to Canada and Montana

The swift fox is a 2.5 kg small canid unique to North America that races across grasslands at speeds exceeding 60 km/h in pursuit of prey, to avoid the talons of golden eagles, or to reach a small escape burrow that pursuing coyotes cannot enter. Once numbering in the hundreds of thousands in Canada, the species disappeared from the country by 1938 due to habitat loss, trapping, and poisoning. Reintroductions involving releases of 942 animals in Canada's provinces of Alberta and Saskatchewan spanned from 1983 to 1997, and this represents the greatest numerical release effort for canid reintroductions to date (Boitani et al. 2004). Early successes gave rise to other reintroduction attempts across the Great Plains of North America, including sites on aboriginal land in Montana, United States.

Captive-bred animals received basic support at release sites, as den boxes from captive-breeding facilities were temporarily installed on the prairie to provide a familiar site for acclimation and return (Ausband and Foresman 2007). Moreover, behavioral monitoring before release to Montana determined that individuals exhibiting bold behavior in captivity were most likely to die, which prompted recommendations to select shy individuals for release (Bremner-Harrison et al. 2004). One of the most controversial aspects of the Canadian release program dealt with release management decisions regarding the selection of individuals for release. Originally, captive-bred animals were used, but subsequent comparisons revealed that direct translocations of wild-caught individuals from extant populations in the United States yielded higher survival rates (Carbyn et al. 1994, Moehrenschlager and Macdonald 2003). Ultimately, translocated animals were preferred, which was objected to by the primary captive-breeding organization.

Demographic monitoring revealed that coyotes and golden eagles were prevalent predators (Moehrenschlager et al. 2007). Surveys conducted for these predators days before release were combined with concurrent tracking of established swift foxes, to select areas where the likelihood of finding mates was high but predation was low (Moehrenschlager and Macdonald 2003).

Mark-recapture surveys of the contiguous Canadian/Montana population in 1996/1997, 2000/2001, and 2005/2006 documented a growing and expanding reintroduced population where 100% of 196 swift foxes captured in 2005/2006 in Canada and contiguous areas of Montana were wild-born. Such demographic monitoring had policy implications as habitat analyses tied to the Canadian national recovery yielded recommendations for critical habitat protection in Canada (Pruss et al. 2008). Originally listed as "extirpated" in Canada, swift foxes were down-listed from a subsequent "endangered" status to "threatened" by 2012. Canadian surveys were also successful in capturing a fox that had travelled nearly 200 km from another Montana release site, and this potential for connectivity yielded support for prairie reclamation strategies to connect additional populations across the international border (Ausband and Moehrenschlager 2009).

In Canada, over 90% of the restored swift fox range lies outside protected areas. Over the duration of the release program and subsequent years of monitoring, local landowners have been involved in fox releases, foxes have been named after local ranchers and their children, presentations have been made at local schools, landowners have been hired when appropriate, and every effort has been made to invest in the local community by purchasing supplies locally and by paying for accommodation on local ranches whenever possible. As a result, local landowners have generally been neutral or supportive of the reintroduction effort.

Recent genetic monitoring of the contiguous population has revealed that genetic diversity in the reintroduced population is satisfactory and that approximations of effective population size corroborate population growth. Nevertheless apparent genetic structuring of the population, where gene exchange between eastern and western areas is limited by cropland, has raised the possibility of future translocations between these zones to maximize genetic connectivity (Cullingham and Moehrenschlager 2013).

human circumstances are minimal. This can be achieved through the adequate selection or improvement of potential release habitats. Relevant choices are scale-dependent, ranging from the scale of continents to centimeters, as one examines, chooses, avoids, or alters environmental conditions on a taxon-specific basis. In general, however, IUCN guidelines (IUCN 2013) suggest that a release area should (1) meet all the species' biotic and abiotic requirements, (2) be appropriate habitat for the life stage released and all life stages of the species, (3) be adequate for all seasonal habitat needs, (4) be large enough to meet the required conservation benefit, (5) have adequate connectivity to suitable habitat if that habitat is fragmented, and (6) be adequately isolated from suboptimal or non-habitat areas that might be sink areas for the population.

Choosing appropriate release areas can be difficult, because it requires a thorough understanding of ecological and anthropogenic factors in potential release habitats. Given these challenges, we frequently see managers reach for useful "rules of thumb," such as that the center of a previously used species range or the last place the species survived in the wild must be suitable for potential release. Unfortunately, it is not that simple because individuals from across a species' range may not be equally suited to the chosen release site, and because habitat suitability varies across a species' range and may need to be engineered (Osborne and Seddon 2012). To address these challenges, Osborne and Seddon (2012) provide the following useful checks to prevent erroneous conclusions regarding release area suitability: (1) historical locations of a species' presence may not indicate a present-day suitable habitat, (2) present-day locations of a species' presence may not indicate a currently suitable habitat, (3) present-day locations where a species is absent may not indicate an unsuitable habitat, and (4) present-day locations of a species' presence may not indicate a future suitable habitat.

The suitability of release areas may depend upon potentially beneficial or harmful relationships with conspecifics or other species (Jachowski et al., Chapter 9, this volume). The value of conspecifics was apparent for Columbian sharp-tailed grouse (*Tympanuchus phasianellus columbianus*) released in Nevada, United States, as individuals released near nest sites of previously released birds exhibited the fewest risky post-release movements (Coates et al. 2006). Moreover, abundant prey may be important, particularly for highly specialized predators. For example, food addition experiments aimed at bolstering the density or colony expansion of black-tailed prairie dogs (*Cynomys ludovicianus*) in Canada (Lloyd et al. 2013) were motivated not only to sustain this imperiled species, but also to provide a necessary food source for recently reintroduced black-footed ferrets (*Mustela nigripes*). For species that rely on symbiotic relationships like the Karner blue butterfly (*Lycaeides melissa samuelis*), identifying suitable reintroduction areas in Ontario, Canada, included searches for symbiotic ant species because the presence of ants elicits butterfly larvae to secrete a nectar-like solution that the ants feed upon, and in turn larvae survivorship increases, presumably because of ant protection against predators and parasitoids (Chan and Packer 2006).

Release area selection needs to be particularly sensitive to the prevalence of competitors and predators, especially when these are hyperabundant invasive species. On O'ahu, Hawaii, the only potential release site for a damselfly (*Megalagrion xanthomelas*) was one where a 6-m waterfall had precluded access to invasive predatory poeciliid fish, prawns, and crayfish (Preston et al. 2007). Even though species may have coexisted historically, the return of an extirpated species is not necessarily of negligible consequence to other imperiled species. For example, Channel Island foxes (*Urocyon littoralis*) had been extirpated from over half of their indigenous range and reintroduction programs were ongoing on several islands. However, population reinforcements were deemed impossible on San Clemente Island because of potential impacts on the endangered San Clemente loggerhead shrike (*Lanius ludovicianus mearnsi*; Roemer and Wayne 2003).

Considerations regarding future habitat suitability should be taken into account. For example, current IUCN reintroduction guidelines suggest that "the climate requirements of the focal species should be understood and matched to the current and/or future climate" and "the occurrence and severity of episodic or unpredictable events that are extreme and adverse for the species should be assessed" (IUCN 2013). Release area considerations encompass not only fauna and flora, but also human-dependent parameters, such that "the release area and essential habitat for the translocated organisms should be secured from incompatible land-use change before the conservation goal is reached, and ideally in perpetuity" (IUCN 2013). For example, modeling habitat distributions of the Italian spadefoot toad (*Pelobates fuscus insubricus*) aided in selecting suitable reintroduction release areas that differed from recent or currently occupied sites (Giovannini et al. 2014). Models assessing habitat suitability for future releases of Eurasian black vultures (*Aegypius monachus*) not only determined potential areas of breeding habitat, but also highlighted that the extent, location, and timing of forest logging should be considered to increase reintroduction effectiveness (Mihoub et al. 2014).

Once a release area is identified, additional choices are pertinent to release sites within release areas. As outlined in the IUCN guidelines (IUCN 2013), a release site should also (1) meet all practical needs for effective release with least stress for the released organisms, (2) enable released organisms to exploit the surrounding release area quickly, and (3) be suitable for media and public awareness needs, and any community involvement.

Select Appropriate Release Timing

Timing of releases should be selected to optimize seasonal resource availability and take into account the life history of the translocated species (Steury and Murray 2004). For example, various life stages can differ in their susceptibility to seasonal weather conditions at the time of release. Juveniles and adult females had lower survival than male Utah prairie dogs (*Cynomys parvidens*) when translocations were done in cold, wet spring weather (Truett et al. 2001). Activity patterns should also be considered; if the species is nocturnal, then releases could occur at dusk or during the night. In contrast, releases could be scheduled during times of low activity, which may potentially aid in reducing stress-induced dispersal in species highly prone to post-release movements (Jachowski et al., Chapter 9, this volume). Timing releases either before or after major life-history events, such as reproduction or dispersal, can aid with establishment and overall translocation success (Lloyd and Powlesland 1994, Bloxam and Tonge 1995, Sarrazin et al. 1996). Conducting multiple releases over time at the same location can also aid with translocation success (IUCN 2013). Indeed, Harrington et al. (2013) found over 80% of projects involved multiple releases of species over time and in multiple locations.

Preparing Release Candidates and Destinations

After choosing release individuals, sites and areas, conservation practitioners can increase the likelihood of translocation success, particularly during the establishment and growth phase, by further preparing release candidates and destinations through actions such as behavior training or habitat restoration.

Optimize Behavioral Considerations for Release Candidates

Ensuring that released individuals are adequately fit for life in the wild is of utmost importance, especially when they are obtained from captive-bred populations. Releasing individuals without adequate survival skills may be considered unethical and unacceptable to individual welfare, the status of the post-release population, and the overall conservation of the species (McLean 1997 in Teixeira and Young 2014, Rabin 2003). Although the primary focus of animal welfare differs from species conservation (Harrington et al. 2013), the call to align these disciplines more closely is growing (Swaisgood 2010, Linklater and Gedir 2011, Harrington et al. 2013).

"Conservation behavior" is an emerging field that utilizes animal behavior to aid conservation (Blumstein and Fernandez-Juricic 2004). Integrating behavior-based research (Jachowski et al., Chapter 9, this volume) into conservation strategies may help success rates (Linklater 2004, Buchholz 2007, Moore et al. 2008, Berger-Tal et al. 2011). Reintroductions can benefit from behavior-based tools as released individuals must be able to avoid predators, locate suitable food, find refuge/shelter, effectively locomote, and interact with conspecifics for mate choice, parental care, and territorial defense (Kleiman 1989). Retaining or cultivating natural behaviors is critical, but learned or socially transmitted behaviors are particularly prone to being lost or altered over time in captivity (Rabin 2003, Mathews et al. 2005, McPhee and McPhee 2012, de Mestral and Herbinger 2013). Thus, prerelease training should be considered for captive-bred or head-started animals (Griffin et al. 2000, Alberts 2007) for which the first step should be to establish a species-specific natural behavior baseline (Mathews et al. 2005).

Ideally, environmental conditions in captivity should closely mimic those in the wild (Stamps and Swaisgood 2007). For example, hibernating species may need to experience fluctuating or extreme temperatures in captivity prior to release. For example, northern water snakes (*Nerodia sipedon sipedon*) reared in captivity under constant temperature and photoperiods had lower body temperatures and growth rates than wild translocated individuals after release and experienced high mortalities (50%) through the overwinter period compared to resident snakes (0%; Roe et al. 2010).

Environmental enrichment can be used as a tool to help induce natural behaviors for species while in captivity (Swaisgood 2007). When steelhead trout (*Oncorhynchus mykiss*) were provided with environmental enrichment in captivity, such as cover and underwater feeders, they had increased growth rate and social rank compared to fish in non-enriched enclosures (Berejikian et al. 2000). Additional enrichment for captive fish can include in-water structures, overhead cover, seminatural streambeds, differential flow rates, and areas with dark backgrounds (Maynard et al. 1995, Brown and Day 2002). Conversely, to avoid imprinting on humans, human activity, including voices or sounds associated with equipment, should generally be kept to a minimum around prerelease captive individuals. For example, in the captive-breeding program for whooping cranes (*Grus americana*), caretakers dress in a crane-like costume to reduce the likelihood that hand-raised chicks imprint on humans. Crane puppet heads and stuffed crane models with devices to emit brooding calls further aid in developing natural behaviors in captive-bred cranes (Horwich 1996).

One of the major challenges to reintroductions is high predation rates after release (Beck et al. 1994, Griffin et al. 2000, Aaltonen et al. 2009), particularly for captive-bred animals (Harrington et al. 2013). Antipredator training has been utilized for a diverse number of species and taxa, including: arthropods (Davis et al. 2004, Young et al. 2008), fish (Manassa and McCormick 2012), amphibians (Crane and Mathis 2011, Ferrari and Chivers 2011, Teixeira and Young 2014), birds (McLean et al. 1999, Quinn and Cresswell 2005, Alonso et al. 2011, de Azevedo et al. 2012, Saunders et al. 2013), and mammals (Hendrie et al. 1998, Hollén and Manser 2007, Shier and Owings 2007, Moseby et al. 2012). Prerelease antipredator training commonly involves associative learning processes. A predator cue may be presented with an aversive stimulus (Griffin et al. 2000, de Azevedo and Young 2006, Mesquita and Young 2007) to elicit an unpleasant experience and fearful behavioral response (Moseby et al. 2012). Aversive stimuli, such as a loud noise, water spray, elastic bands, or temporary capture, paired with a visual model of a predator, have increased antipredator behaviors in some species (Miller et al. 1990, McLean et al. 1996). Although antipredator training may compromise short-term welfare (Mench and Mason 1997 in Rabin 2003), it will likely increase long-term welfare and natural behavioral diversity (Rabin 2003).

Conservation managers should refine training techniques for individuals according to the type of predators they will face in the wild (e.g., terrestrial or aerial) and the sensory system(s) likely used for predator detection (e.g., visual or olfactory). Thus, antipredator training should be targeted to the species' natural predator-detection physiology systems, i.e., olfactory (Jędrzejewski et al. 1993, Hollén and Manser 2007, Crane and Mathis 2011, Ferrari and Chivers 2011, Ferrero et al. 2011, Manassa and McCormick 2012), visual (McLean et al. 1999, van Heezik et al. 1999, Quinn and Cresswell 2005, Moseby et al 2012, Teixeira and Young. 2014), and/or auditory (Hendrie et al. 1998, Saunders et al. 2013). Experiments testing post-release survival between prerelease trained and untrained control groups have shown higher survival rates for trained groups of some species in the wild (van Heezik et al. 1999, Shier and Owings 2007). However, others conclude that even if captive individuals can be trained in antipredator behaviors, this training does not necessarily increase survival in the wild (Moseby et al. 2012). Long-term studies and post-release monitoring are crucial to determine the conditions under which prerelease training for a particular species will be beneficial.

In addition to avoiding predators, released animals need to learn natural forage and foraging techniques while in captivity. For example, a reintroduction of captive-bred cheetahs (*Acinonyx jubatus*) failed to expose individuals to the appropriate prey before release, resulting in the released cheetahs attempting to hunt prey species that were too large for successful capture, thus increasing their risk of injury

and starvation in the wild (Pettifer 1981). Similarly, reintroduced bank voles (*Clethrionomys glareolus*) were able to find similar rations of food compared to wild voles, but captive-bred individuals were unable to open hazelnuts, a natural food item found in the wild but not part of the captive diet (Mathews et al. 2005). Therefore, provision of all types of natural food items while in captivity may help increase natural foraging behaviors post-release. For example, Brown and Day (2002) recommend sporadic provision of live prey to captive fish to aid in learning natural foraging techniques prior to release while also minimizing feed costs.

Effective locomotion is generally connected to efficient foraging techniques. Dietz et al. (1987) found that prerelease training that involved searching for food in natural areas aided in post-release survival for golden lion tamarins (*Leontopithecus rosalia*). In addition, post-released tamarins that were trained in family groups exhibited increased searching, locomotion, and foraging behavior relative to untrained groups (Dietz et al. 1987). Long-distance locomotion, such as migration and migration routes, may also need to be taught to semi-captive individuals prior to their full release. As discussed in Chapter 9, the reintroduction of whooping cranes to the eastern United States has involved extensive prerelease training techniques. Operation Migration used ultralight aircraft and flight conditioning to teach migration routes to captive-bred whooping cranes. While still in the egg, crane chicks were preconditioned with brood calls and aircraft engine sounds. Juvenile cranes were trained to follow the aircraft in flight for short distances to feeding grounds and eventually on a long-distance migration (Ellis et al. 2003, Converse and Armstrong, Chapter 7, this volume).

In general, prerelease training is most effective if conducted during critical times of development and throughout the individuals' time in captivity, and not left until shortly before the release date (Rabin 2003). Most species elicit natural behaviors after only a few training sessions (Chivers and Smith 1994, Maloney and McLean 1995, Griffin et al. 2000, Mesquita and Young 2007); however, there may be a threshold beyond which individuals become habituated to training (Magurran 1990 in Teixeira and Young 2014). Individuals in the captive program that do not show the necessary behaviors can undergo additional training or might be deemed inappropriate for release. Therefore, behavior studies and training in captivity need to be specific to the species, and even to the individual, prior to the release stage (Mathews et al. 2005).

Optimize Health Considerations of Release Candidates

In addition to preserving natural behaviors, conservation managers can improve the health of release animals initially, and then make adjustments for future releases based on monitoring data (figure 11.1). Most parasites and pathogens have negative effects on species fitness and post-release survival (Pedersen et al. 2007, Smith et al. 2009). In particular, reduced exposure to natural pathogens while in captivity can result in lower post-release survival of captive-bred individuals (Viggers et al. 1993, Cunningham 1996). For example, exposure of captive-bred guppies (*Poecilia reticulata*) to a parasite (*Gyrodactylus turnbulli*) prior to release, led to decreased parasite loads post-release (Faria et al. 2010). In addition, prerelease health checks, quarantine procedures, and disease monitoring can inform management interventions to mitigate disease risk. For example, radiography to assess skeletal development in mountain chicken frogs (*Leptodactylus fallax*) was used to evaluate the presence of metabolic bone disease and to guide the development of mitigation strategies such as dietary supplementation with vitamins and mineral supplements, and UV-B radiation while the animals were in captivity (Tapley et al. 2015). Immediately prior to release, vaccinations can help to reduce the likelihood of disease or parasite transmission following release. For example, black-footed ferrets were typically vaccinated for canine distemper in captivity before release (Williams et al. 1996), and more recently, a vaccine that protects ferrets from sylvatic plague was also developed (Rocke et al. 2008) and administered to reintroduced ferrets (Matchett et al. 2010).

Optimize Transportation

Moving or transporting animals is a fundamental component of reintroductions, and thus proper care and risk assessments need to be determined prior to any transfer. During transportation, individuals may be subjected to variations in temperature, humidity, light, vibrations, and noise to which they are unaccustomed. Thus, poor transport methods, additional time in transport, and an increase in novel stimuli can all lead to increased stress, stress-induced disease, or mortality during transfer (Jachowski et al., Chapter 9, this volume). Because transportation can be stressful for animals, precautions should be taken to minimize stress as much as possible during this time (Harrington et al. 2013). Stress levels during confinement and transport may be different for wild-caught and captive-bred animals (Swaisgood 2010; Jenni et al. 2015). In addition, stress can be elevated for solitary or territorial animals if confined and transported with multiple individuals. In contrast, social species may have increased stress levels if not transported with family or kin members (Shier 2006). Tranquilizers to reduce stress should be considered during transport, especially for long transit times and stress-prone species (Teixeira et al. 2007, Dickens et al. 2010). Capture myopathy is a noninfectious disease associated with stress from capture, handling, and transfer that can result in death of individuals (Nuvoli et al. 2014). For deer species in particular, rates of stress and mortality associated with capture can vary with different handling techniques, and using chemical immobilization may help reduce stress during transport compared to physical restraint (Haulton et al. 2001).

To mitigate transport-related stress, species- and life-stage specific modifications of transport containers should be developed and tested prior to use. Proper ventilation and protection from weather should be incorporated into the design of transport containers as well as transport vehicles to prevent the risk of overheating or hyperthermia of individuals (Truett et al. 2001). Also, as a general rule, dark transport containers likely reduce stress for species (CITES Secretariat 1981). Transport containers should be large enough for the animal to move around in but not so large that the individual could injure itself. The size of the container also depends on the length of time transportation will likely take. Asher et al. (2009) found that a small transport bag may be adequate for starlings (*Sturnus vulgaris*) in transit for less than an hour, whereas a larger transport box should be used if transportation time is greater than an hour. Depending on the species and the length of time in transport, food, water, and/or bedding should be included in the transfer container. Food can sometimes be wedged into wire mesh cages to avoid spillage (Truett et al. 2001). For example, translocations of kakapo (*Strigops habroptilus*) used helicopters or planes to transfer individuals to release sites (Lloyd and Powlesland 1994). Wire mesh cages with vegetation covering the bottom were used in transit. The cages were covered externally with rubber foam to reduce noise and vibrations during transfer. Transit time was kept to less than 12 hours. Once at the release area, each kakapo was transferred to the release site via open-weave hessian sack (Lloyd and Powlesland 1994). When transporting eggs, movement due to transport may impact embryo development; therefore, materials and packaging surrounding eggs are critical to reduce vibrations (Mortimer 1999, Vazquez-Sauceda et al. 2008). Transport of aquatic species can be especially difficult since the water environment needs to be transported with the individuals. Transportation experiments for the mangrove crab (*Ucides cordatus*) involved different stocking densities of megalopae that were placed in plastic bags filled with marine water with specific salinity and inflated with oxygen

prior to being vibrated for varying lengths of time (Ventura et al. 2010). Compared to control groups, all experimental groups had relatively high survival; however, after 24 hours the low-density groups showed greater survival, suggesting that stocking rates for transport should be low for this species during transport (Ventura et al. 2010). Overall, species-specific planning for the transportation phase can help prepare individual animals for release by reducing stress, to the extent possible, prior to the release stage.

Optimize Release Areas and Release Sites

Release areas and the release sites therein may appear suitable, but may be lacking in certain characteristics. Depending upon the context and organism, a number of management options are available to improve conditions that will maximize the likelihood of survival, reproduction, and population establishment in desired regions.

RESTORE, CONNECT, CREATE, OR PREPARE RELEASE HABITAT Habitat restoration may be a crucial precursor for successful reintroductions. For example, habitat assessments for the Karner blue butterfly (*Lycaeides melissa samuelis*) in Ontario at each of four potential release sites determined that several sites needed nectar source plants to be restored, while others required changes to spatial heterogeneity that would allow for variable light conditions (Chan and Packer 2006). Similarly, in Switzerland, reclaiming cropland into wildflower strips and increasing hedgerow densities yielded preferred habitats for released gray partridges (*Perdix perdix*) that were crucial for establishing a reintroduced population (Buner et al. 2005).

For some taxa, improving habitat connectivity is critical to reintroduction success. In Nepal, one-horned rhinoceros (*Rhinoceros unicornis*) were reintroduced to two national parks between 1986 and 2003. This subsequently resulted in successful reproduction and population growth. However, subsequent increases in poaching during a period of civil strife threatened to drive these populations to extinction. Consequently, an initiative was launched using community forestry projects to restore connectivity among 11 protected areas in southern Nepal and northwestern India, where protection by community-led antipoaching teams allowed for the dispersal of reintroduced populations (Thapa et al. 2013). Corridors are particularly critical for migratory species, such as the European sturgeon (*Acipenser sturio*). Construction of a weir in 1961 at the mouth of a river blocked the migration route for this species. A fish-passage facility was built in 2010 to allow access, and habitat improvements were made along the river to provide clean gravel beds for spawning sites as well as to reduce sediment load and flood risk (Kirschbaum et al. 2011).

In some cases, the suitability of available habitat may be poor, and the likelihood of habitat restoration equally limited, but it may still be possible to create the necessary habitat. For example, modified concrete blocks were placed and prepared in shallow waters on Pulau Weh, Indonesia, to host two subsequently translocated coral species. These artificial reefs not only succeeded in establishing the target species, but subsequently also supported 23 other coral taxa from 8 families and 29 fish species from 11 families (Fadli et al. 2012). Release sites for the reintroduction of the Mallorcan midwife toad (*Alytes muletensis*) were limited by multiple threats at all naturally occurring permanent bodies of water (Bloxam and Tonge 1995). Creation of new pools using cement, to ensure they did not drain prior to toad metamorphosis, provided additional adequate release sites (Bloxam and Tonge 1995).

INCREASE OR REDUCE THE IMPACT OF OTHER SPECIES Sometimes reintroductions benefit from, or require, restoration of other species. For example, the burrowing owl (*Athene cunicularia*), which became extirpated in British Columbia, Canada, has been the subject of reintroduction efforts for over 20 years. One limiting factor is the availability of burrows

(Mitchell et al. 2011), which necessitated the annual creation of artificial burrows (Smith et al. 2011). Although the numbers of reestablished owls are modestly growing, the rate of increase is primarily a function of release numbers, while the proportion of successful migration returns is minimal. This has triggered questions regarding the efficacy of the entire program, and a call for more comprehensive ecosystem restoration including the potential reintroduction of American badgers (*Taxidea taxus*), which would naturally create burrows for burrowing owl releases and subsequent breeding (Smith et al., unpublished data). Limitations of burrows also were accompanied by related concerns over prey availability for potential burrowing owl reintroductions in California. There, recent restoration efforts for the California ground squirrel (*Otospermophilus beecheyi*) show signs of success as manipulated habitat areas have increased squirrel and burrow densities (Deutschman et al. 2014).

In contrast to the restoration of additional species to aid reintroductions, species reintroduction success can also benefit from the reduction of predatory or competitive species. Species interactions are important to consider, particularly in reintroductions of multiple species (Robinson and Ward. 2011). As discussed by Jachowski et al. (Chapter 9, this volume), the desert pupfish (*Cyprinodon macularius*) and the Gila topminnow (*Poeciliopsis occidentalis*) were generally released together in previous reintroductions. However, experiments found that pupfish releases were more successful either when released without topminnows or when allowed to establish prior to releases of topminnows (Robinson and Ward 2011). Active control of predators prior to and following release is also a viable management tool for many prey species. For example, fencing and/or the control of mammalian predators such as rats benefitted reintroduction attempts for the North Island robin (*Petroica longipes*) in New Zealand (Parlato and Armstrong 2013). Similarly, swift foxes, which have been subject to a number of reintroduction programs in

North America, had significantly higher survival rates following the killing of 227 coyotes in Texas (Kamler et al. 2003, Box 11.1). However, predator control is not always a successful predation mitigation strategy. Reduction of predation pressure through the killing of red foxes (*Vulpes vulpes*) and coyotes (*Canis latrans*) may have marginally improved the survival of translocated greater sage grouse (*Centrocercus urophasianus*) in Utah, United States, between 1998 and 2011 (Baxter et al. 2013).

Support for Release Candidates and Destinations

Post-release supportive actions can increase the probability of survival and reproduction of released individuals, thus benefiting population establishment, growth, or regulation phases of reintroductions.

Select or Develop Optimal Release Methods

Release techniques are often classified as "soft" or "hard" (Beck et al. 1994, Campbell and Croft 2001, Thompson et al. 2001, Clarke et al. 2002, Hardman and Moro 2006). A "hard" or "immediate" release involves setting individuals free directly to the wild without previous experience in the area, preadaptation, or further support (Beck et al. 1994, Hardman and Moro 2006, Attum et al. 2013). For a "soft" or "delayed" release, individuals are confined for a set amount of time at the release site, usually in a predator-proof enclosure, and provided with food and water (Scott and Carpenter 1987, Beck et al. 1994, Hardman and Moro 2006, Attum et al. 2013). This conditioning period, in theory, enables the animal to become acclimatized to the new release site environment (i.e., temperature, light, landmarks, etc.) without the stress of needing to forage or avoid predators (Bright and Morris 1994, Campbell and Croft 2001, Hardman and Moro 2006).

Low release site fidelity, high mortality, and increased movements are common issues plaguing translocation post-release success (Truett et al. 2001, Moehrenschlager and Mac

donald 2003, Attum et al. 2013). Soft-release methods have been developed to increase success during the establishment period by increasing site fidelity and post-release survival (Switzer 1993, Bright and Morris 1994, Attum et al. 2013). For example, burrowing owls (*Athene cunicularia hypugaea*) released into artificial burrows using soft-release acclimatization enclosures had greater survival, site fidelity, and higher reproduction post-release in the wild than hard-released owls during the same time period (Mitchell et al. 2011). In addition to allowing animals to acclimatize to novel environments, a secondary benefit of soft-release protocols is the ability to closely monitor released individuals. Body condition, health, behavior, and stress can all be studied during the soft-release stage more easily than with a hard release (Wanless et al. 2002). However, soft-release techniques, in comparison to hard release techniques, are more expensive and can be logistically complicated (Thompson et al. 2001).

Decisions as to whether a translocation should use soft or hard releases should consider risk-benefit trade-offs, species-specific biology, and source of the animals. Captive-bred animals may rely more heavily on supplemental food or soft releases than wild-caught individuals (Jefferies et al. 1986, Kleiman et al. 1986, Zwank and Wilson 1987, Bright and Morris 1994). Captive-bred soft-released dormice had higher survival rates, greater site fidelity, and dispersed less relative to hard-released dormice. However, soft-released captive-bred dormice showed greater dependency on supplemental food post-release than hard-released individuals (Bright and Morris 1994). Many traditional aquatic reintroductions tend to hard-release numerous individuals directly into bodies of water (Brown and Day 2002). In such cases, the sheer volume of individuals released is thought to compensate for the likely high post-release mortality. However, soft-release acclimation periods can be beneficial for fish translocations as well (Jonssonn et al. 1999).

While many species translocations have benefited from a soft-release protocol (Gatti 1981, Davis 1983, Price 1989, Bright and Morris 1994, Carbyn et al. 1994, Ellis et al. 2000, Wanless et al. 2002, Lockwood et al. 2005, Tuberville et al. 2005, Mitchell et al. 2011), this is not always the case. Other studies have shown limited benefits (Short et al. 1992, Lovegrove 1996, Campbell and Croft 2001, Moro 2001, Thompson et al. 2001, Clarke et al. 2002, Hardman and Moro 2006), and some translocations even report negative effects of soft-release (Castro et al. 1994, Christensen and Burrows 1994). However, controlled studies with experimental comparisons between soft and hard releases are limited (Davis 1983, Bright and Morris 1994, Carbyn et al. 1994, Mitchell et al. 2011).

The benefit/cost ratio of a soft release likely depends on a combination of species-specific characteristics along with the length of the acclimation period. Too long of an acclimation period may affect subsequent territory establishment, while an acclimation period that is too short may result in confusion and increased dispersal (Kaya and Jeanes 1995). Depending on the species and situation, the duration of a soft release can vary from a few days, to weeks (Letty et al. 2000, Mitchell et al. 2011), months (Brightsmith et al. 2005), or years (Tuberville et al. 2005). For example, a simulated translocation of pygmy bluetongue lizards (*Tiliqua adelaidensis*) found that a 1-day soft release was preferential to a 5-day soft release because lizards decreased basking time and increased dispersal with increased containment time (Ebrahimi and Bull 2013). In addition, other support measures such as supplemental food and artificial burrows aided in reducing post-release dispersal for bluetongue lizards (Ebrahimi and Bull 2012, Ebrahimi and Bull 2014). Further, the behavioral ecology of the target species must be taken into account because territorial or aggressive behavior can lead to increased stress of individuals within acclimatization enclosures (Letty et al. 2000). Thus, the optimal release strategy will depend on the species

and situation, and we encourage an experimental approach to developing release protocols in species translocations, particularly in uncertain or novel situations.

Supporting or Replacing Parental Care

Fostering or adoption of young individuals into wild family groups may also increase post-release survival. Wallace and Buchholz (2001) used reciprocal fostering to increase genetic diversity of red-cockaded woodpecker (*Picoides borealis*), whereby unrelated nestlings were paired by age and switched into foster nests. They found that fostered nestlings were accepted into the foster nests and fledged at rates similar to unfostered young (Wallace and Buchholz 2001). Gray partridge chicks (*Perdix perdix*) fostered into wild nests had higher post-release survival compared to adult birds translocated from either the wild or captivity (Buner and Schaub 2008). As a general rule, when choosing individual nestlings for fostering, the species, age and weight should be similar to the brood in the foster nest to reduce nest abandonment. Fostering among conspecifics allows released individuals to learn species-specific behaviors without prerelease training by program managers. When intraspecific fostering is not possible (such as is the case of some endangered species), interspecific or cross fostering can be an important tool for conservation of imperiled bird taxa (Fyfe et al. 1978, Bunin and Jamieson 1996).

Fostering is generally easier for bird taxa compared to mammals. However, young mammals can be fostered into wild litters (Holmes and Sherman 1982). For example, juvenile Columbian ground squirrels (*Spermophilus columbianus*) that were of relatively similar size, age, and weight were cross-fostered and found to have similar survival rates compared to non-fostered young (Murie et al. 1998). For terrestrial mammals in particular, fostering can be a successful strategy for increasing gene flow among isolated populations, as demonstrated with African wild dogs (*Lycaon pictus*; McNutt et al. 2008).

In taxa or species without much parental care, head starting is a technique used to increase survival of young individuals. Head starting involves temporarily holding animals in a captive or semi-captive setting during their most vulnerable stage(s), in order to increase their survival and recruitment into the population (Wilson et al. 2004). For example, a head-starting program for green sea turtles (*Chelonia mydas*) in the Cayman Islands had turtles survive and return to breeding sites, showing positive results toward success (Bell et al. 2005). Similarly, King and Stanford (2006) found that head-started plains garter snakes (*Thamnophis radix*) reproduced successfully and had growth rates similar to those of wild conspecifics.

Assist Released Individuals

Additional supportive measures to aid translocated individuals can include techniques such as supplemental feeding, shepherding (protection from predators via human or domestic dog presence), sensory manipulation, and/or health checks/additional vaccines. Out of 199 translocation projects reviewed, 79% used post-release supportive measures and those measures often increased the success of translocations (Harrington et al. 2013). However, trade-offs between increased program costs and logistical constraints must be weighed when planning additional supportive measures for species translocations.

Supplemental feeding is often used as a supportive measure for post-release populations (figure 11.1) and has increased conservation success for many critically imperiled species (Komdeur 1996, Elliott et al. 2001, Gonzalez et al. 2006). Providing critical sustenance and nutrition for released individuals can increase their post-release survival, reproductive productivity, and social interactions, and reduce post-release dispersal (Finlayson and Moseby 2004, Brightsmith et al. 2005, Jones and Merton 2012). Feeding stations can aid monitoring of released populations and facilitate recapture of individuals for health assessments

(Brightsmith et al. 2005, Jones and Merton 2012). In addition, food supplementation can be beneficial to released populations at critical times of year. For example, translocated populations of eastern wild turkeys (*Meleagris gallopavo silvestris*) provided with supplemental food had higher survival over winter compared to populations without supplemental food (Kane et al. 2007). Decisions regarding supplemental feeding should include the optimal location for feeding stations (i.e., accounting for species home range size, territoriality), how to exclude nontarget species, and the most optimal method of feeding (e.g., scatter feeding, delivery into burrow/den, or local feeding stations).

As with other management techniques, the benefits of supplemental feeding should be weighed against potential costs. Predators can take advantage of feeding stations (Dunn and Tessaglia-Hymes 1999 in Schoech et al. 2008), and released individuals may become reliant on the supplemented food (Taylor and Castro 2000). Food stations can also increase disease transmission (Townsend et al. 1999, Turner et al. 2008), and their nutritional content can influence reproductive sex ratios (Clout et al. 2002). In addition, if the nutritional value of the supplemented food and the energy requirements of the focal species are not taken into account, the body condition of released individuals can decrease (Unangst and Wunder 2004).

A critical issue with supplemental feeding is deciding when to stop. The ultimate goal of many reintroductions is to achieve a self-sustaining population (Converse and Armstrong, Chapter 7, this volume). However, some reintroductions are still managed years after releases through continued supplemental feeding (Brightsmith et al. 2005, Chauvenet et al. 2012). For example, supportive measures were employed for over 30 years following the reintroduction of the white stork (*Ciconia ciconia*) to Switzerland (Schaub et al. 2004). Long-term monitoring, adaptive management, and population monitoring can all help identify how long supportive measures, such as food supplementation, should continue (Chauvenet et al. 2012).

Physical protection (shepherding or bodyguarding) of released individuals can also improve post-release survival if high predation or poaching is a concern. For example, in the first few years of reintroductions of the Vancouver Island marmot (*Marmota vancouverensis*), field crews were dispatched to colonies of released marmots to provide physical protection and deter predation events (Vancouver Island Marmot Recovery Team 2008). Due to the high predation risk at release sites and the high financial investment in each individual marmot, shepherding was determined to be necessary at least during initial establishment. However, shepherding can lead to habituation to humans and increased stress on released individuals. These potential consequences, as well as the financial costs of constant monitoring, must be weighed against the potential increased survival of release individuals.

Manipulating sensory cues, such as olfaction, at the release site area can further aid in site establishment, habitat selection, and overall reintroduction success (Reed 2004, Campbell-Palmer and Rosell 2010, 2011). Various signals within the habitat can encourage individuals to select suitable areas. Cues in the species' natal habitat can be used to determine quality of new habitats (Stamps and Swaisgood 2007). Presenting certain olfactory cues to animals as part of a captive-breeding program and subsequently scent marking the release site with the same odor may help individuals remain close to the release site (Campbell-Palmer and Rosell 2011). Furthermore, the addition of "self"-scent into the release area may increase post-release establishment through "territorial confidence" (Mykytowycz et al. 1976, Swaisgood et al. 1999). To reduce conspecific conflict post-release, "virtual territories" can be established prior to release via scent marking to allow resident individuals time to become "familiarized" with the scent of released individuals (Swaisgood et al. 1999, Swaisgood 2007). Olfactory cues may also aid in moving individuals through habitat corridors, thus connecting fragmented released populations (Swaisgood et al. 1999). The use of olfactory and chemical cues can also help to decrease human-wildlife conflicts. The use of scented oils caused an aversion to certain foods for wolverines (*Gulo gulo*) in captivity and may be useful to reduce targeted predation on domestic animals (Landa and Tømmerås 1997).

Post-release health checks may be helpful but should be weighed against impacts of recapture. Many vaccines require multiple boosters and thus may require re-trapping (e.g., Matchett et al. 2010). In other cases, serious but treatable health issues may arise where post-release care can help. Indeed, for reintroduced populations of California condors (*Gymnogyps californianus*), lead toxicosis was the primary cause of death for adult birds (Johnson et al. 2013). Mitigation measures such as provision of lead-free supplemental food and campaigns/regulations for lead-free ammunition have been ongoing. Further, individual condors with high amounts of lead poisoning are hospitalized and treated with chelation (Johnson et al. 2013). In fact, nearly 50% of the released population has been treated, showing that sometimes intensive post-release health management is required for reintroduced species (Finkelstein et al. 2012). Unfortunately, despite these mitigating measures, lead poisoning remains the greatest challenge to attaining self-sustaining populations of California condors to date (Kelly et al. 2014).

Remove Released Animals

While apparently counterintuitive as a release management technique, options for the removal of released animals or their offspring should be considered to assist individuals or to mitigate potential damage to recipient ecosystems or ecosystem services (also see Hayward and Slotow, Chapter 13, this volume). Removal of animals from the wild should be considered as a humane action, if releases are clearly ineffective or released animals are suffering. Such management actions were heavily debated for lynx *(Lynx canadensis)* reintroductions in Colorado, when some individuals apparently starved after release (Bekoff 1999). Conversely, in some cases, culling or euthanasia could also be considered as a humane management alternative for highly successful, overabundant restored populations (Hayward and Slotow, Chapter 13, this volume). For example, a reintroduced population of 219 African elephants *(Loxodonta africana)* was established in South Africa's Madikwe Game Reserve, although the carrying capacity was likely only 100–150. Negative impacts on tree species such as the marula *(Sclerocarya birrea)*, prompted subsequent elephant translocations from the reserve to other areas. Translocations of elephants to Swaziland from South Africa were also thought to have been excessive, with subsequent negative competitive effects on black rhinoceros *(Diceros bicornis)*, and increased damage to nesting trees that were impacting local vulture populations. As a result, efforts were initiated to remove elephants from the wild to zoos in the United States (Dublin and Niskanen 2003). Finally, stakeholder opposition to reintroductions of controversial species, such as in the case of gray wolf *(Canis lupus)* reintroductions to Yellowstone National Park, where wolves that endanger humans or livestock are lethally controlled outside of the park (Wilson 1997), illustrate that concessions are sometimes required for reintroduction programs to proceed.

Addressing Human Dimensions

Human dimensions in species reintroductions can be both a blessing and a curse. In practice, human dimensions transcend all reintroduction phases and need to be incorporated prior to releases, during the release event, and continue on post-release through the establishment, growth, and regulation phases. Dedicated and passionate people can positively influence conservation biology and species recovery through reintroduction (George and Sandhaus, Chapter 14, this volume). However, human opposition to

reintroducing or releasing species in certain habitats or areas can greatly impact the overall success of a translocation and in some cases halt releases all together (Riley and Sandström, Chapter 4, this volume). Even releases that appear to have profound benefits for imperiled species, restore ecological functions, or garner public attention can be undesirable. For example, releases of lions *(Panthera leo)* or elephants in areas where they would benefit from reintroductions or reinforcements would nevertheless be undesirable if they pose a significant hazard to the livelihood or safety of local communities. In our opinion, many reintroduction reviews omit or underrepresent the importance of human dimensions compared to biological considerations. Here we present it as a primary consideration that should not only be addressed before reintroductions are initiated, but also be revisited throughout the program (figures 11.1 and 11.2).

Listen To, Educate, and Involve Local Communities

The reintroduction of the Northern Rocky Mountain wolf (*Canis lupus irremotus*) into Yellowstone National Park has been one of the most controversial species translocations in North America. Opposition to the release of wolves resulted in significant project delays. The intense controversy around wolf reintroductions highlighted the importance of balancing biological and sociopolitical dimensions in reintroduction planning (Bath 1989, Thompson 1993). Thompson (1993) suggested that social assessment, public participation, and conflict management are key concepts to integrating human dimensions into wildlife translocations. In Yellowstone National Park, a multi-stakeholder committee was arguably established too late in the reintroduction planning process, when conflict levels were already high and compromise between members was unattainable (Mader 1991 in Thompson 1993). Frustrations also arose as public consultations were seen by some as allowing one-way communication tools that precluded public input and participation. (Thompson 1993).

Prerelease and post-release education regarding the threats that species face and the need for reintroductions can help to increase understanding and support for reintroduction programs (George and Sandhaus, Chapter 14, this volume). Following years of modeling, prey base population dynamic assessments, and public consultation, public participation was incorporated into considerations and actions for black-footed ferret reintroductions that were initiated in Canada in 2009 (Lloyd et al. 2013). At the initial reintroduction, a prerelease event involved all stakeholders in speeches and a traditional peace-pipe ceremony by the Lakota First Nation which blessed all groups, the land, and the ferrets themselves. Release of the first ferret involved representatives of the local community, government, and nongovernment organizations. Subsequent ferret releases were conducted solely by conservation managers and biologists involved in the project. Similarly, rehabilitation of an urban pond and translocation of blue mussels (*Mytilus edulis*) were accomplished through community volunteers, which resulted in local awareness of the lagoon ecosystem and improvements in the aquatic habitat (McDermott et al. 2008). Stewardship was increased by involving volunteers from local school groups, community groups, and the local government (McDermott et al. 2008).

Social media and marketing can play an important role in reintroduction success post-release. RARE, a nonprofit organization, partners with local organizations to launch extensive social marketing campaigns (called Pride Campaigns) to educate and inspire local communities to protect their natural resources. For example, RARE's "Go Native" campaign, on Cocos Island, Guam, used the Guam rail as a mascot to educate the public about reintroductions of this native species (Fortin 2012).

Consider Incentives and Compensation for Local Communities

Financial investments in behavioral training of animals to be released can help reduce conflict when potential predation on domestic animals, including livestock, could be a challenge. For example, European lynx (*Lynx lynx*) reintroduced to Poland began hunting domestic animals instead of native prey and had to be recaptured. After additional investments to expose the recaptured animals to natural prey in captivity, hunting of appropriate prey ensued after release and conflicts with local people diminished (Boer et al. 2000).

Financial incentives provided directly to the public can also motivate support for conservation programs such as reintroductions (Bromley 1994, Mishra et al. 2003). For example, an incentive program in India to ease human-wildlife conflict for the snow leopard (*Panthera uncia*) involved reimbursement for the designation of livestock-free areas. This was balanced with a communal herding insurance program, developed and managed by local people, whereby herders with low predation rates were paid to recognize "safe herding" (Mishra et al. 2003). Direct compensation to those impacted by the actions of reintroduced animals strives to increase the tolerance among local people and should also be considered in the planning process (Ciucci and Boitani 1998). For wolves reintroduced to Yellowstone National Park, compensation was sought for monetary loss (sheep and cattle predation and loss of time by local ranchers) that in one year exceeded $300,000 (Thompson 1993). While tolerance to species like wolves varies based on a variety of factors (level of education, direct loss from wolves, and gender), all groups surveyed regarding wolves in Wisconsin agreed that compensation was a good strategy for conservation managers (Naughton-Treves et al. 2003). Even when direct financial incentives are not possible, we suggest that practitioners should attempt to expend project costs within local communities and to employ local residents in the study to protect or support or release animals whenever possible (Box 11.1). These efforts could have multiple tangible benefits to reintroduction, such as the development of supportive public policy or legislation, and in some cases this may result in enforcement activities or antipoaching patrols.

RECOMMENDATIONS FOR FUTURE RESEARCH AND DEVELOPMENT

We encourage further research pertaining to each of the preceding release management action sections. Here we elaborate on section-specific questions that we believe deserve additional attention. These are subsequently followed by four broader thematic issues that would benefit from increased emphasis and resolution.

Research Questions

Selecting Release Candidates and Destinations

Which genetic differences are adaptive and how should such genetic differences influence release decisions?

We have presented genetic considerations of primary relevance. However, we increasingly encounter situations where genetic differences arc found between populations at a finer scale than subspecies or even ecotypes. Without a clear understanding of whether genetic structure is necessarily adaptive, managers are often uncertain whether movements among wild populations or between captive and wild populations are appropriate.

How can the transfer of gametes among wild populations or between captive and wild populations be further developed to preclude the need of releasing individuals for genetic outcomes?

In our review, we mentioned that released populations may need genetic supplementation, and traditionally this is done by releasing suitable individuals that are transferred from captivity or other populations. However, to contribute to the genetics of destination sites, release candidates must survive and reproduce, and many may perish before successful reproduction occurs. Therefore, we encourage additional research into techniques that seek to transfer gametes instead of individuals, such as artificial insemination at release sites using sperm garnered from other wild or captive populations.

How can the selection of release areas be better informed by climate change projections?

We discussed that release area selection needs to incorporate not only current, but also projected future habitat suitability. We emphasize this again here because the conservation ramifications are large, but the science to make adequate predictions regarding release decisions based on climate change projections is still in its infancy.

Preparing Release Candidates and Destinations

How can zoo populations be better aligned with the needs of wild populations to contribute more species and more individuals per species for release?

Captive breeding is a primary tool for reintroductions, but many captive-bred zoo populations are deemed as unsuitable for release because of genetic, behavioral, or demographic issues. Closer examination of reintroduction needs in the wild may allow zoos to focus on species with which they can align breeding resources for the maximal benefit of release populations.

What levels of stress among captive-bred animals link to post-release survival, reproduction, and recruitment?

We described the need to consider animal welfare, not only for ethical purposes, but also to benefit conservation outcomes. While examples exist of stress impacting post-release survival, it is still unclear to what extent captive breeders should allocate space and resources toward the improvement of procedures and enclosures, and for which species this is most critical.

Can behavioral training techniques be adequately developed beyond birds and mammals?

We outline innovative approaches in terms of behavioral training, but notice that the relevant literature is lacking for many species. Is this because there has been little interest or because many species are seen as unsuitable for such approaches?

Support for Release Candidates and Destinations

How does the efficacy of different release methods differ among populations for the same species?

We reviewed advantages and disadvantages of different release methods for a variety of taxa. For most species where information is available, conclusions are drawn from singular studies without any understanding of broader relevance behind conditions uniquely associated with individual release programs. We encourage research to determine whether supportive mechanisms for some species may be applicable in some contexts but not in others.

Can released animals that are removed from certain areas be successfully released in other areas?

While released animals that are suffering or problematic in some release areas may need to be removed, questions arise regarding their potential suitability for other release locations.

What ecosystem ramifications result from efforts to increase or decrease other species for the benefit of released species?

We reviewed the advantage or necessity of reestablishing or controlling other species for the benefit of release candidates. A central, and often unresolved, question in reintroduction biology is whether reintroduced species have a pronounced ecological effect in the ecosystem to which they are reestablished. Reaching one step further, we encourage additional investigations of ecosystem effects beyond the released species, to determine if additional restorations or limits of predators and competitors have more wide-reaching consequences at targeted release sites.

Addressing Human Dimensions

How can traditional knowledge be better integrated to align cultural and conservation goals in release phases of reintroductions?

We presented strategies to garner public support for reintroductions through consultation, education, incentives, and compensation. However, we generally see a paucity of information in the literature in terms of authentic alignment between cultural considerations and reintroduction approaches. A better alignment of traditional knowledge with conservation approaches could help to further support from communities, and potentially yield valuable advice in terms of the selection of release sites or post-release support mechanisms.

How can incentives and compensation be better aligned in developing countries?

Most examples we present on incentives and compensation are aligned with alleviating opposition or political pressure in richer countries. Similar approaches in relatively poor global regions, where even modest financial contributions could have big impacts for local people, likely have tremendous potential to garner public support.

How can reintroductions better benefit local communities?

While most human dimension approaches attempt to limit opposition by local communities, seemingly few actually attempt to use reintroductions as focused revenue opportunities for local communities. Few conservation tools capture the imagination like releases of imperiled species, and while ecotourism is often advocated for conservation benefits generally, we believe opportunities are currently missed in terms of the marketing potential of releases for the benefit of local communities. Of course, such initiatives would need to cautiously prevent any adverse effects on released animals and populations.

Thematic Issues

Beyond research questions, we believe that four thematic issues require additional attention.

Developing Taxon-Specific Guidelines

The IUCN Guidelines for Reintroductions and Other Conservation Translocations (IUCN 2013) were written to provide general recommendations that would be applicable for any species involved in any conservation translocation globally. However, taxon-specific knowledge can further improve reintroduction practice and pertinent release techniques. For example, detailed taxon-specific IUCN guidelines have been developed to add to the global guidelines for primates generally, gibbons specifically, African elephants, and Galliformes. We encourage practitioners to share pertinent information for similar taxa and contexts to further develop the toolbox of taxon-specific reintroduction guidelines. Moreover, we encourage practitioners to look for commonalities in release strategies for ecologically or functionally similar species across taxa, for example, among fossorial species.

Resolving Animal Welfare Dilemmas

Conservation biologists are primarily driven by population, species, or ecosystem outcomes such as preventing extinction, restoring ecological function, or retaining ecosystem services. However, for motivations of compassion, religion, or culture, many people care primarily about the well-being of individual animals. Good animal welfare can be beneficial for conservation outcomes, but ethical and practical implications for welfare and conservation are not always aligned (Harrington et al. 2013). Some of these pertain to releases, and need further reflection for optimal decision-making. Examples are the following: (1) removal of animals from wild source populations could impact the social structure of those individuals left behind; (2) prerelease training for carnivores may use live prey, which may suffer before dying; and (3) large release numbers could be used to overcome high mortality rates, which would successfully establish populations but involve potential suffering or ethical implications for many released individuals. Practitioners should grapple with evaluation efforts that determine whether the ends justify the means in all cases, or whether some animal welfare implications are too great to justify reintroduction release outcomes.

Determine Approaches toward Populations That Are No Longer "Wild"

Ideally, reintroductions should aim to yield self-sustaining populations (IUCN 2013). However, in an increasing number of circumstances, such outcomes are not realized because impeding threats can be impossible to control, such as invasive species that prey on released animals or economic forces that drive the poaching of valuable animals. Decisions sometimes need to be made on whether we accept the marooning of released animals on islands where they are safe from harm, whether their perpetual confinement within fences on semi-wild landscapes is desirable, or whether extinction is seen as an acceptable alternative to releases in such circumstances. This would call for an increased acceptance that sustainable populations in indigenous range, with minimal management support, should be seen as acceptable conservation endpoints.

Resolve Risk Tolerance in a Riskier World

Decision theory and risk analyses can help to elucidate the benefits, risks, and trade-offs of decisions regarding the release of animals. Population models can help to determine the likelihood of persistence, extinction, or the projected effects on viability associated with respective release actions. But these tools do not necessarily assess whether people are willing to take risks that could affect them individually, their organization, the release organism, or species and other affected components of the ecosystem. Understanding not just what we could decide, but indeed how we decide given evolving value-contexts dependent upon cultural, sociological, and economic drivers globally would help us to more predictably achieve beneficial outcomes for released animals specifically and reintroductions in particular.

SUMMARY

Decisions regarding release techniques may make or break the whole reintroduction program.

Selection of potential candidates and potential release areas/sites, as well as the preparation of individuals and release sites, allow conservation managers to make sound decisions for optimal release strategies. Human dimensions play a critical role in species reintroductions and need to be integrated throughout all release stages. Release strategies should also include risk assessments involving multiple criteria (i.e., genetics, disease, etc.) and exit strategies, including considerations for the potential removal of released animals. Reintroductions are fraught with uncertainty and dilemmas; however, proper planning, preparation, and monitoring involving an iterative process allow conservation managers to utilize decision-theory-based adaptive management processes to reduce uncertainty. Perhaps at no other point in the reintroduction process, does one have as much control to react to emerging monitoring data and ultimately to affect demographic change as through release decisions. Throughout this chapter we draw upon many examples to demonstrate that integration and refinement of release decisions within planning, preparation, monitoring, reflection, and informed management phases can yield benefits for reintroductions, and ultimately for conservation.

MANAGEMENT RECOMMENDATIONS

- Selecting individuals with the appropriate demographics, genetics, behavior, and good health will greatly enhance the likelihood of survival post-release and overall success of the release program. Thorough risk assessments on the viability of both source and recipient populations, genetics, behavior, and disease will enable the most appropriate selection of potential release candidates.

- Careful selection of the release area and site will increase the likelihood of success and potentially reduce costly financial investments. Over time, the suitability of habitat areas for specific species may change.

Therefore, release areas should not be chosen before the past and present distributions of a species, as well as potential future habitat suitability, are considered.

- When selecting a release area and site, life history and all life-stage requirements of the species need to be adequately taken into account and incorporated to the extent possible.

- Release timing may be critical to the success and establishment of released individuals and should be optimized to the season that will give the release individuals the greatest likelihood of survival.

- Successful population establishment and growth will depend on how each released individual responds to conditions in the wild post-release. By anticipating the potential health and behavioral challenges likely to be experienced by reintroduced populations, conservation practitioners can influence the likelihood that released individuals will survive to reproduce, thus increasing population establishment and growth.

- The act of transporting animals to release sites requires considerable planning and specific knowledge of species life history and welfare needs. We suggest that this topic within reintroduction science should receive much more attention (both in specific translocation guidelines and reintroduction publications), in order to advance our knowledge of this critical period in a reintroduction.

- It is crucial to recognize that even after optimal release sites are selected, some habitat requirements may not be met. Thus, preparation of the release site prior to releases (e.g., via habitat restoration or predator removal) may help to improve the likelihood of survival and reintroduction success.

- Reintroductions do not end at the release stage, and post-release supportive actions, such as supplementary feeding, provision of shelter, predator control, and health checks can assist released individuals to increase the likelihood of survival and population growth. In addition to providing assistance to individuals in the wild, post-release removal of certain individuals, for either humane or mitigative reasons, should also be considered as aspects of post-release support.

- Human dimensions are a fundamental aspect of reintroduction programs. All stakeholders and local communities can be engaged in a multitude of ways, from festivities on the actual release day to involvement in meetings and ongoing monitoring through citizen science programs to designing of incentive or compensation programs if the release species is controversial or could put human livelihoods at risk.

LITERATURE CITED

Aaltonen, K., A. A. Bryant, J. A. Hostetler, and M. K. Oli. 2009. Reintroducing endangered Vancouver Island marmots: survival and cause-specific mortality rates of captive-born versus wild-born individuals. Biological Conservation 142:2181–2190.

Alberts, A. C. 2007. Behavioral considerations of headstarting as a conservation strategy for endangered Caribbean rock iguanas. Applied Animal Behavior Science 102:380–391.

Alonso, R., P. Orejas, F. Lopes, and C. Sanz. 2011. Pre-release training of juvenile little owls *Athene noctua* to avoid predation. Animal Biodiversity and Conservation 34:389–393.

Armstrong, D. P. 1995. Effects of familiarity on the outcome of translocations, II. A test using New Zealand robins. Biological Conservation 71:281–288.

Armstrong, D. P., and M. H. Reynolds. 2012. Modelling reintroduced populations: the state of the art and future directions. Pp.165–222. In J. G. Ewen, D. P. Armstrong, K. A. Parker, and P. J. Seddon, eds., Reintroduction Biology: Integrating Science and Management. John Wiley & Sons, West Sussex, UK.

Armstrong, D. P., and P. J. Seddon. 2008. Directions in reintroduction biology. Trends in Ecology & Evolution 23:20–25.

Armstrong, D. P., and H. U. Wittmer. 2011. Incorporating Allee effects into reintroduction strategies. Ecological Research 26:687–695.

Asher, L., G. T. Davies, C. E. Bertenshaw, M. A. Cox, and M. Bateson. 2009. The effects of cage volume and cage shape on the condition and behavior of captive European starlings (*Sturnus vulgaris*). Applied Animal Behavior Science 116:286–294.

Attum, O., C. D. Cutshall, K. Eberly, H. Day, and B. Tietjen. 2013. Is there really no place like home? Movement, site fidelity, and survival probability of translocated and resident turtles. Biodiversity and Conservation 22:3185–3195.

Ausband, D. E., and K. R. Foresman. 2007. Swift fox reintroductions on the Blackfeet Indian Reservation, Montana, USA. Biological Conservation 136:423–430.

Ausband, D. E., and A. Moehrenschlager. 2009. Long-range juvenile dispersal and its implication for conservation of reintroduced swift fox *Vulpes velox* populations in the USA and Canada. Oryx 43:73–77.

Baláž, V., M. Kubečková, P. Civiš, R. Rozínek, and J. Vojar. 2014. Fatal chytridiomycosis and infection loss observed in captive toads infected in the wild. Acta Veterinaria Brno 82:351–355.

Bath, A. J. 1989. The public and wolf reintroduction in Yellowstone National Park. Society & Natural Resources 2:297–306.

Baxter, R. J., R. T. Larsen, and J. T. Flinders. 2013. Survival of resident and translocated greater sage-grouse in Strawberry Valley, Utah: A 13-year study. The Journal of Wildlife Management 7:802–811.

Beck, B., L. Rapaport, M. S. Price, and A. C. Wilson. 1994. Reintroduction of captive-born animals. Pp.265–286. In P. J. S. Olney, G. M. Mace, and A. T. C. Feistner, eds., Creative Conservation: Interactive Management of Wild and Captive Animals. Chapman & Hall, London.

Bekoff, M. 1999. Jinxed lynx? Journal of Applied Animal Welfare Science 2:239–242.

Bell, C. D., J. Parsons, T. J. Austin, A. C. Broderick, G. Ebanks-Petrie, and B. J. Godley. 2005. Some of them came home: the Cayman Turtle Farm headstarting project for the green turtle *Chelonia mydas*. Oryx 39:137–148.

Berejikian, B. A., E. P. Tezak, T. A. Flagg, A. L. LaRae, E. Kummerow, and C. V. Mahnken. 2000. Social dominance, growth, and habitat use of age-0 steelhead (*Oncorhynchus mykiss*) grown in enriched and conventional hatchery rearing environments. Canadian Journal of Fisheries and Aquatic Sciences 57:628–636.

Berger-Tal, O., T. Polak, A. Oron, Y. Lubin, B. P. Kotler, and D. Saltz. 2011. Integrating animal behavior and conservation biology: a conceptual framework. Behavioral Ecology 22:236–239.

Bloxam, Q. M., and S. J. Tonge. 1995. Amphibians: suitable candidates for breeding-release programmes. Biodiversity & Conservation 4:636–644.

Blumstein, D. T., and E. Fernandez-Juricic. 2004. The emergence of conservation behavior. Conservation Biology 18:1175–1177.

Boer, M., J. Reklewski, P. Tyrala, and J. Smielowski. 2000. Reintroduction of the European lynx (*Lynx lynx*) to the Kampinoski National Park/Poland: a field experiment with zooborn individuals. Part III. Demographic development of the population from December 1993 until January 2000. Zoologische Garten 70:304–312.

Boitani, L., Asa, C., and A. Moehrenschlager. 2004. Tools for canid conservation. Pp.143–159. In D. W. Macdonald and C. Sillero-Zubiri, eds., Biology and Conservation of Wild Canids. Oxford University Press, Oxford.

Bremner-Harrison, S., P. A. Prodohl, and R. W. Elwood. (2004). Behavioral trait assessment as a release criterion: boldness predicts early death in a reintroduction programme of captive-bred swift fox (*Vulpes velox*). Animal Conservation 7:313–320.

Bright, P., and P. Morris. 1994. Animal translocation for conservation: performance of dormice in relation to release methods, origin and season. Journal of Applied Ecology 31:699–708.

Brightsmith, D., J. Hilburn, A. Del Campo, J. Boyd, M. Frisius, R. Frisius, D. Janik, and F. Guillen. 2005. The use of hand-raised psittacines for reintroduction: a case study of scarlet macaws (*Ara macao*) in Peru and Costa Rica. Biological Conservation 121:465–472.

Bromley, D. W. 1994. Economic dimensions of community-based conservation. Pp.428–447. In D. Western, R. M. Wright, and S. C. Strum, eds., Natural Connections: Perspectives in Community Based Conservation. Island Press, Washington, DC.

Brown, C., and R. L. Day. 2002. The future of stock enhancements: lessons for hatchery practice from conservation biology. Fish and Fisheries 3:79–94.

Buchholz, R. 2007. Behavioral biology: an effective and relevant conservation tool. Trends in Ecology & Evolution 22:401–407.

Buner, F., and M. Schaub. 2008. How do different releasing techniques affect the survival of reintroduced grey partridges *Perdix perdix*? Wildlife Biology 14:26–35.

Buner, F., M. Jenny, N. Zbinden, and B. Naef-Daenzer. 2005. Ecologically enhanced areas–a key habitat structure for re-introduced grey partridges *Perdix perdix*. Biological Conservation 124:373–381.

Bunin, J. S., and I. G. Jamieson. 1996. Responses to a model predator of New Zealand's endangered takahe and its closest relative, the pukeko. Conservation Biology 10:1463–1466.

Campbell, L., and D. Croft. 2001. Comparison of hard and soft release of hand reared eastern grey kangaroos. Veterinary Conservation Biology, Wildlife Health and Management in Australasia. Proceedings of International Joint Conference, Taronga Zoo, Sydney, 173–180.

Campbell-Palmer, R., and F. Rosell. 2010. Conservation of the Eurasian beaver Castor fiber: an olfactory perspective. Mammal Review 40:293–312.

Campbell-Palmer, R., and F. Rosell. 2011. The importance of chemical communication studies to mammalian conservation biology: a review. Biological Conservation 144: 1919–1930.

Canessa, S., D. Hunter, M. McFadden, G. Marantelli, and M. A. McCarthy. 2014. Optimal release strategies for cost-effective reintroductions. Journal of Applied Ecology 51:1107–1115.

Carbyn, L. G. N., H. J. Armbruster, and C. Mamo. 1994. The swift fox reintroduction program in Canada from 1983 to 1992. Pp.247–271. In M. L. Bowles and C. J. Whelan, eds., Restoration of Endangered Species: Conceptual Issues, Planning and Implementation. Cambridge University Press, Cambridge, UK.

Carbyn, L. N., and D. Watson. 2001. Translocation of plains bison to Wood Buffalo National Park: economic and conservation implications. Pp.189–204. In D. S. Maehr, R. F. Noss, and J. L. Larkin, eds., Large Mammal Restoration: Ecological and Sociological Challenges in the 21st Century. Island Press, Washington, DC.

Caro, T. 2005. Antipredator Defenses in Birds and Mammals. University of Chicago Press, Chicago, IL.

Castro, I., J. Alley, R. Empson, and E. Minot. 1994. Translocation of hihi or stitchbird *Notiomystis cincta* to Kapiti Island, New Zealand: transfer techniques and comparison of release strategies. Pp.113–120. In M. Serena, ed., Reintroduction Biology of Australian and New Zealand Fauna, Surrey Beatty and Sons, Chipping Norton, NSW.

Chan, P. K., and L. Packer. 2006. Assessment of potential Karner Blue butterfly (*Lycaeides melissa samuelis*) (family: Lycanidae) reintroduction sites in Ontario, Canada. Restoration Ecology 14:645–652.

Chauvenet, A., J. Ewen, D. Armstrong, T. Coulson, T. Blackburn, L. Adams, L. Walker, and N. Pettorelli. 2012. Does supplemental feeding affect the viability of translocated populations? The example of the hihi. Animal Conservation 15:337–350.

Chivers, D., and R. Smith. 1994. The role of experience and chemical alarm signalling in predator recognition by fathead minnows, *Pimephales promelas*. Journal of Fish Biology 44:273–285.

Christensen, P., and N. Burrows. 1994. Project desert dreaming: experimental reintroduction of mammals to the Gibson Desert, Western Australia. Pp.199–207. In M. Serena, eds., Reintroduction Biology of Australian and New Zealand Fauna. Surrey Beatty and Sons, Chipping Norton, NSW.

CITES Secretariat. 1981. Guidelines for Transport and Preparation for Shipment of Live Wild Animals and Plants. CITES, Geneva.

Ciucci, P., and L. Boitani. 1998. Wolf and dog depredation on livestock in central Italy. Wildlife Society Bulletin 26:504–514.

Clarke, R. H., R. L. Boulton, and M. F. Clarke. 2002. Translocation of the socially complex black-eared miner *Manorina melanotis*: a trial using hard and soft release techniques. Pacific Conservation Biology 8:223–234.

Clout, M. N., G. P. Elliott, and B. C. Robertson. 2002. Effects of supplementary feeding on the offspring sex ratio of kakapo: a dilemma for the conservation of a polygynous parrot. Biological Conservation 107:13–18.

Coates, P. S., S. J. Stiver, and D. J. Delehanty. 2006. Using sharp-tailed grouse movement patterns to guide release-site selection. Wildlife Society Bulletin 34:1376–1382.

Crane, A. L., and A. Mathis. 2011. Predator-recognition training: a conservation strategy to increase postrelease survival of hellbenders in head-starting programs. Zoo Biology 30:611–622.

Cullingham, C. I., and A. Moehrenschlager. 2013. Temporal analysis of genetic structure to assess population dynamics of reintroduced swift foxes. Conservation Biology 27:1389–1398.

Cunningham, A. A. 1996. Disease risks of wildlife translocations. Conservation Biology 10:349–353.

Dalrymple, S. E., and A. Moehrenschlager. 2013. "Words matter." A response to Jørgensen's treatment of historic range and definitions of reintroduction. Restoration Ecology 21:156–158.

Davis, J. L. D., A. C. Young-Williams, R. Aguilar, B. L. Carswell, M. R. Goodison, A. H. Hines, M. A. Kramer, Y. Zohar, and O. Zmora. 2004. Differences between hatchery-raised and wild blue crabs: implications for stock enhancement potential. Transactions of the American Fisheries Society 133:1–14.

Davis, M. H. 1983. Post-release movements of introduced marten. The Journal of Wildlife Management 47:59–66.

de Azevedo, C. S., and R. J. Young. 2006. Do captive-born greater rheas *Rhea americana* Linnaeus (Rheiformes, Rheidae) remember antipredator training? Revista Brasileira de Zoologia 23:194–201.

de Azevedo, C. S., R. J. Young, and M. Rodrigues. 2012. Failure of captive-born greater rheas (*Rhea americana*, Rheidae, Aves) to discriminate between predator and nonpredator models. Acta Ethologica 15:179–185.

de Mestral, L., and C. Herbinger. 2013. Reduction in antipredator response detected between first and second generations of endangered juvenile Atlantic salmon *Salmo salar* in a captive breeding and rearing programme. Journal of Fish Biology 83:1268–1286.

Deredec, A., and F. Courchamp. 2007. Importance of the Allee effect for reintroductions. Ecoscience 14:440–451.

Deutschman, D. H., R. R. Swaisgood, D. M. Shier, L. A. Nordstrom, S. McCullough, C. M. Lenihan, J. P. Montagne, and C. L. Wisinski. 2014. Habitat structure and ecosystem engineers: Critical components of sustainable habitat restoration for burrowing owls. Paper presented at the 99th Annual Meeting of the Ecological Society of America, Sacramento, CA.

Dickens, M. J., D. J. Delehanty, and L. M. Romero. 2010. Stress: an inevitable component of animal translocation. Biological Conservation 143:1329–1341.

Dietz, J. M., M. I. Castro, B. B. Beck, and D. G. Kleiman. 1987. The effects of training on the behavior of captive born golden lion tamarins reintroduced into natural habitat. International Journal of Primatology 8:425–425.

Dillon, R. S. 1988. The reptile and the rail: a war in the Pacific. Smithsonian 22:112.

Dublin, H. T., and L. S. Niskanen. 2003. IUCN/SSC AfESG Guidelines for the in situ Translocation of the African Elephant for Conservation Purposes. IUCN, Gland, Switzerland/Cambridge, UK.

Dunn, E. H., and D. L. Tessaglia-Hymes. 1999. Birds at Your Feeder: A Guide to Feeding Habits, Behavior, Distribution, and Abundance. WW Norton & Company, New York.

Edmands, S. 2007. Between a rock and a hard place: evaluating the relative risks of inbreeding and outbreeding for conservation and management. Molecular Ecology 16:463–475.

Elliott, G. P., D. V. Merton, and P. W. Jansen. 2001. Intensive management of a critically endangered species: the kakapo. Biological Conservation 99:121–133.

Ellis, D. H., G. F. Gee, S. G. Hereford, G. H. Olsen, T. D. Chisolm, J. M. Nicolich, K. A. Sullivan, N. J. Thomas, M. Nagendran, and J. S. Hatfield. 2000. Post-release survival of hand-reared and parent-reared Mississippi Sandhill Cranes. The Condor 102:104–112.

Ellis, D. H., W. J. Sladen, W. A. Lishman, K. R. Clegg, J. W. Duff, G. F. Gee, and J. C. Lewis. 2003. Motorized migrations: the future or mere fantasy? BioScience 53:260–264.

Ebrahimi, M., and C. M. Bull. 2012. Food supplementation reduces post-release dispersal during simulated translocation of the endangered pygmy bluetongue lizard Tiliqua adelaidensis. Endangered Species Research 18:169–178.

Ebrahimi, M., and C. M. Bull. 2013. Determining the success of varying short-term confinement time during simulated translocations of the endangered pygmy bluetongue lizard (Tiliqua adelaidensis). Amphibia-Reptilia 34:31–39.

Ebrahimi, M., and C. M. Bull. 2014. Resources and their distribution can influence social behaviour at translocation sites: lessons from a lizard. Applied Animal Behaviour Science 156: 94–104.

Evans, R. J., J. D. Wilson, A. Amar, A. Douse, A. MacLennan, N. Ratcliffe, and D. P. Whitfield. 2009. Growth and demography of a re-introduced population of white-tailed eagles Haliaeetus albicilla. Ibis 151:244–254.

Fadli, N., S. J. Campbell, K. Ferguson, J. Keyse, E. Rudi, A. Riedel, and A. H. Baird. 2012. The role of habitat creation in coral reef conservation: a case study from Aceh, Indonesia. Oryx 46:501–507.

Faria, P. J., C. van Oosterhout, and J. Cable. 2010. Optimal release strategies for captive-bred animals in reintroduction programs: experimental infections using the guppy as a model organism. Biological Conservation 143:35–41.

Ferrari, M. C., and D. P. Chivers. 2011. Learning about non-predators and safe places: the forgotten elements of risk assessment. Animal Cognition 14:309–316.

Ferrero, D. M., J. K. Lemon, D. Fluegge, S. L. Pashkovski, W. J. Korzan, S. R. Datta, M. Spehr, M. Fendt, and S. D. Liberles. 2011. Detection and avoidance of a carnivore odor by prey. Proceedings of the National Academy of Sciences of the United States of America 108:11235–11240.

Figiel, C. R. 2013. Cryopreservation of sperm from the axolotl Ambystoma mexicanum: implications for conservation. Herpetological Conservation and Biology 8:748–755.

Finkelstein, M. E., D. F. Doak, D. George, J. Burnett, J. Brandt, M. Church, J. Grantham, and D. R. Smith. 2012. Lead poisoning and the deceptive recovery of the critically endangered California condor. Proceedings of the National Academy of Sciences of the United States of America 109:11449–11454.

Finlayson, G. R., and K. E. Moseby. 2004. Managing confined populations: the influence of density on the home range and habitat use of reintroduced burrowing bettongs (Bettongia lesueur). Wildlife Research 31:457–463.

Fischer, J., and D. Lindenmayer. 2000. An assessment of the published results of animal relocations. Biological Conservation 96:1–11.

Fisher, M. C., T. W. Garner, and S. F. Walker. 2009. Global emergence of Batrachochytrium dendrobatidis and amphibian chytridiomycosis in space, time, and host. Annual Review of Microbiology 63:291–310.

Forsman, A., L. Wennersten, M. Karlsson, and S. Caesar. 2012. Variation in founder groups promotes establishment success in the wild. Proceedings of the Royal Society B: Biological Sciences 279:2800–2806.

Fortin, A. 2012. Restoring a missing piece of Guam's culture. US Fish & Wildlife Endangered Species Program. Available online at: http://www.fws.gov/endangered/map/ESA_success_stories/HI/HI_story1/all_stories.html (accessed June 15, 2015).

Fyfe, R. W., H. Armbruster, U. Banasch, and L. J. Beaver. 1978. Fostering and cross-fostering of birds of prey. Pp.183–193. In S. A. Temple, ed., Endangered Birds: Management Techniques for Preserving Threatened Species. University of Wisconsin Press, Madison.

Gatti, R. C. 1981. A comparison of two hand-reared mallard release methods. Wildlife Society Bulletin 9:37–43.

Germano, J. M., and P. J. Bishop. 2009. Suitability of amphibians and reptiles for translocation. Conservation Biology 23:7–15.

Giovannini, A., D. Seglie, and C. Giacoma. 2014. Identifying priority areas for conservation of spadefoot toad, Pelobates fuscus insubricus using a maximum entropy approach. Biodiversity and Conservation 23:1427–1439.

Gonzalez, L. M., A. Margalida, R. Sánchez, and J. Oria. 2006. Supplementary feeding as an effective tool for improving breeding success in the Spanish imperial eagle (*Aquila adalberti*). Biological Conservation 129:477–486.

Griffin, A. S. 2004. Social learning about predators: a review and prospectus. Animal Learning & Behavior 32:131–140.

Griffin, A. S., D. T. Blumstein, and C. S. Evans. 2000. Training captive-bred or translocated animals to avoid predators. Conservation Biology 14:1317–1326.

Griffith, B., J. M. Scott, J. W. Carpenter, and C. Reed. 1989. Translocation as a species conservation tool: status and strategy. Science (Washington) 245:477–480.

Hardman, B., and D. Moro. 2006. Optimising reintroduction success by delayed dispersal: is the release protocol important for hare-wallabies? Biological Conservation 128:403–411.

Harrington, L. A., A. Moehrenschlager, M. Gelling, R. P. Atkinson, J. Hughes, and D. W. Macdonald. 2013. Conflicting and complementary ethics of animal welfare considerations in reintroductions. Conservation Biology 27:486–500.

Haulton, S. M., W. F. Porter, and B. A. Rudolph. 2001. Evaluating 4 methods to capture white-tailed deer. Wildlife Society Bulletin 29:255–264.

Hendrie, C. A., S. M. Weiss, and D. Eilam. 1998. Behavioral response of wild rodents to the calls of an owl: a comparative study. Journal of Zoology 245:439–446.

Hollén, L. I., and M. B. Manser. 2007. Persistence of alarm-call behavior in the absence of predators: a comparison between wild and captive-born meerkats (*Suricata suricatta*). Ethology 113:1038–1047.

Holmes, W. G., and P. W. Sherman. 1982. The ontogeny of kin recognition in two species of ground squirrels. American Zoologist 22:491–517.

Horwich, R. H. 1996. Imprinting, attachment and behavioral development in cranes. Pp.117–122. In D. H. Ellis, G. F. Gee, C. M. Mirande, eds., Cranes: Their Biology Husbandry and Conservation. National Biological Service, Washington, DC/International Crane Foundation, Baraboo, WI.

IUCN. 2013. Guidelines for Reintroductions and Other Conservation Translocations. Version 1.0. International Union for Conservation of Nature Species Survival Commission, Gland, Switzerland.

Jędrzejewski, W., L. Rychlik, and B. Jędrzejewska. 1993. Responses of bank voles to odours of seven species of predators: experimental data and their relevance to natural predator-vole relationships. Oikos 68:251–257.

Jefferies, D., P. Wayre, R. Jessop, and A. Mitchell-Jones. 1986. Reinforcing the native otter *Lutra lutra* population in East Anglia: an analysis of the behavior and range development of the first release group. Mammal Review 16:65–79.

Jenni, L., N. Keller, B. Almasi, J. Duplain, B. Homberger, M. Lanz, F. Korner-Nievergelt, M. Schaub, and S. Jenni-Eiermann. 2015. Transport and release procedures in reintroduction programs: stress and survival in grey partridges. Animal Conservation 18:62–72.

Johnson, C. K., T. R. Kelly, and B. A. Rideout. 2013. Lead in ammunition: a persistent threat to health and conservation. EcoHealth 10:455–464.

Jones, C. G., and D. V. Merton. 2012. A tale of two islands: the rescue and recovery of endemic birds in New Zealand and Mauritius. Pp.12–33. In J. G. Ewen, D. P. Armstrong, K. A. Parker, and P. J. Seddon, eds., Reintroduction Biology: Integrating Science and Management. John Wiley & Sons, West Sussex, UK.

Jonssonn, S., E. Brønnøs, and H. Lundqvist. 1999. Stocking of brown trout, *Salmo trutta* L.: effects of acclimatization. Fisheries Management and Ecology 6:459–473.

Kamler, J. F., W. B. Ballard, R. L. Gilliland, P. R. Lemons, and K. Mote. 2003. Impacts of coyotes on swift foxes in northwestern Texas. The Journal of Wildlife Management 67:317–323.

Kane, D. F., R. O. Kimmel, and W. E. Faber. 2007. Winter survival of wild turkey females in central Minnesota. The Journal of Wildlife Management 71:1800–1807.

Kaya, C. M., and E. D. Jeanes.1995. Notes: retention of adaptive rheotactic behavior by F1 fluvial Arctic grayling. Transactions of the American Fisheries Society 124:453–457.

Kelly, T. R., J. Grantham, D. George, A. Welch, J. Brandt, L. J. Burnett, K. J. Sorenson et al. 2014. Spatiotemporal patterns and risk factors for lead exposure in endangered California condors during 15 years of reintroduction. Conservation Biology 28:1721–1730.

King, R. B., and K. M. Stanford. 2006. Headstarting as a management tool: a case study of the plains gartersnake. Herpetologica 62:282–292.

Kirschbaum, F., P. Williot, F. Fredrich, R. Tiedemann, and J. Gessner. 2011. Restoration of the European sturgeon *Acipenser sturio* in Germany. Pp.309–333. In P. Williot, E. Rochard, N. Desse-Berset, F. Kirschbaum, and J. Gessner, eds., Biology and Conservation of the European

Sturgeon *Acipenser sturio L. 1758*. Springer-Verlag, Berlin/Heidelberg.

Kleiman, D.G. 1989. Reintroduction of captive mammals for conservation. BioScience 39:152–161.

Kleiman, D.G., B.B. Beck, J.M. Dietz, L.A. Dietz, J.D. Ballou, and A.F. Coimbra-Filho. 1986. Conservation program for the golden lion tamarin: captive research and management, ecological studies, educational strategies, and reintroduction. Pp.959–979. In K. Benirschke, ed., Primates: The Road to Self-sustaining Populations. Springer, New York.

Knisley, C.B., J.M. Hill, and A.M. Scherer. 2005. Translocation of threatened tiger beetle *Cicindela dorsalis dorsalis* (Coleoptera: Cicindelidae) to Sandy hook, New Jersey. Annals of the Entomological Society of America 98:552–557.

Komdeur, J. 1996. Breeding of the Seychelles magpie robin *Copsychus sechellarum* and implications for its conservation. Ibis 138:485–498.

Komers, P.E., and G.P. Curman. 2000. The effect of demographic characteristics on the success of ungulate re-introductions. Biological Conservation 93:187–193.

Kramer-Schadt, S., E. Revilla, and T. Wiegand. 2005. Lynx reintroductions in fragmented landscapes of Germany: projects with a future or misunderstood wildlife conservation? Biological Conservation 125:169–182.

Landa, A., and B.Å. Tømmerås. 1997. A test of aversive agents on wolverines. The Journal of Wildlife Management 61:510–516.

Le Gouar, P., J.-B. Mihoub, and F. Sarrazin. 2012. Dispersal and habitat selection: behavioral and spatial constraints for animal translocations. Pp.138–164. In J.G. Ewen, D.P. Armstrong, K.A. Parker, and P.J. Seddon, eds., Reintroduction Biology: Integrating Science and Management. John Wiley & Sons, West Sussex, UK.

Letty, J., S. Marchandeau, J. Clobert, and J. Aubineau. 2000. Improving translocation success: an experimental study of anti-stress treatment and release method for wild rabbits. Animal Conservation 3:211–219.

Linklater, W.L. 2004. Wanted for conservation research: behavioral ecologists with a broader perspective. BioScience 54:352–360.

Linklater, W.L., and J.V. Gedir. 2011. Distress unites animal conservation and welfare towards synthesis and collaboration. Animal Conservation 14: 25–27.

Lloyd, B.D., and R.G. Powlesland. 1994. The decline of kakapo *Strigops habroptilus* and attempts at conservation by translocation. Biological Conservation 69:75–85.

Lloyd, N., A. Moehrenschlager, D. Smith, and D. Bender. 2013. Food limitation at species range limits: Impacts of food availability on the density and colony expansion of prairie dog populations at their northern periphery. Biological Conservation 161:110–117.

Lockwood, M.A., C.P. Griffin, M.E. Morrow, C.J. Randel, and N.J. Silvy. 2005. Survival, movements, and reproduction of released captive-reared Attwater's prairie-chicken. Journal of Wildlife Management 69:1251–1258.

Lovegrove, T.G. 1996. Island releases of saddlebacks *Philesturnus carunculatus* in New Zealand. Biological Conservation 77:151–157.

Mader, T.R. 1991. Wrong direction forces wolf into Yellowstone. Casper Star Tribune, 25 April, 1991, p. A9.

Magurran, A.E. 1990. The inheritance and development of minnow anti-predator behavior. Animal Behavior 39:834–842.

Maloney, R.F., and I.G. McLean. 1995. Historical and experimental learned predator recognition in free-living New-Zealand robins. Animal Behavior 50:1193–1201.

Manassa, R., and M. McCormick. 2012. Social learning and acquired recognition of a predator by a marine fish. Animal Cognition 15:559–565.

Matchett, M.R., D.E. Biggins, V. Carlson, B. Powell, and T. Rocke. 2010. Enzootic plague reduces black-footed ferret (*Mustela nigripes*) survival in Montana. Vector-Borne and Zoonotic Diseases 10:27–35.

Mathews, F., M. Orros, G. McLaren, M. Gelling, and R. Foster. 2005. Keeping fit on the ark: assessing the suitability of captive-bred animals for release. Biological Conservation 121: 569–577.

Matson, T.K., A.W. Goldizen, and P.J. Jarman. 2004. Factors affecting the success of translocations of the black-faced impala in Namibia. Biological Conservation 116:359–365.

Maynard, D.J., T.A. Flagg, and C.V.W. Mahnken. 1995. A review of innovative culture strategies for enhancing the postrelease survival of anadromous salmonids. American Fisheries Society Symposium 15:307–314.

McCallum, H., P. Timmers, and S. Hoyle. 1995. Modelling the impact of predation on reintroductions of bridled nailtail wallabies. Wildlife Research 22:163–171.

McCarthy, M.A. 1994. Population viability analysis of the Helmeted Honeyeater: risk assessment of captive management and reintroduction.

Pp.21–25. In M. Serena, ed., Reintroduction Biology of Australian and New Zealand Fauna. Surrey Beatty and Sons, Chipping Norton, NSW.

McDermott, S., D. Burdick, R. Grizzle, and J. Greene. 2008. Restoring ecological functions and increasing community awareness of an urban tidal pond using blue mussels. Ecological Restoration 26:254–262.

McLean, I.G. 1997. Conservation and the ontogeny of behavior. Pp.132–156. In J.R. Clemmons and R. Buchholz, eds., Behavioral Approaches to Conservation in the Wild. Cambridge University Press, Cambridge, UK.

McLean, I.G., C. Hölzer, and B.J. Studholme. 1999. Teaching predator-recognition to a naive bird: implications for management. Biological Conservation 87:123–130.

McLean, I.G., G. Lundie-Jenkins, and P.J. Jarman. 1996. Teaching an endangered mammal to recognise predators. Biological Conservation 75:51–62.

McNutt, J.W., M.N. Parker, M.J. Swarner, and M. Gusset. 2008. Adoption as a conservation tool for endangered African wild dogs (Lycaon pictus). South African Journal of Wildlife Research 38:109–112.

McPhee, M., and N. McPhee. 2012. Relaxed selection and environmental change decrease reintroduction success in simulated populations. Animal Conservation 15:274–282.

Mench, J., and G. Mason. 1997. Behavior. Pp.127–141. In M.C. Appleby and B.O. Hughes, eds., Animal Welfare. Center for Agriculture and Bioscience International, New York.

Mesquita, F.d.O., and R.J. Young. 2007. The behavioral responses of Nile tilapia (Oreochromis niloticus) to anti-predator training. Applied Animal Behavior Science 106:144–154.

Mihoub, J.-B., F. Jiguet, P. Lécuyer, B. Eliotout, and F. Sarrazin. 2014. Modelling nesting site suitability in a population of reintroduced Eurasian black vultures Aegypius monachus in the Grands Causses, France. Oryx 48:116–124.

Miller, B., D. Biggins, C. Wemmer, R. Powell, L. Calvo, L. Hanebury, and T. Wharton. 1990. Development of survival skills in captive-raised Siberian polecats (Mustela eversmanni) II: predator avoidance. Journal of Ethology 8:95–104.

Mishra, C., P. Allen, T. McCarthy, M. Madhusudan, A. Bayarjargal, and H.H. Prins. 2003. The role of incentive programs in conserving the snow leopard. Conservation Biology 17:1512–1520.

Mitchell, A.M., T.I. Wellicome, D. Brodie, and K.M. Cheng. 2011. Captive-reared burrowing owls show higher site-affinity, survival, and reproduc-

tive performance when reintroduced using a soft-release. Biological Conservation 144:1382–1391.

Moehrenschlager, A., R. List, and D.W. Macdonald. 2007. Escaping intraguild predation: Mexican kit foxes survive while coyotes and golden eagles kill Canadian swift foxes. Journal of Mammalogy 88:1029–1039.

Moehrenschlager, A., and D.W. Macdonald. 2003. Movement and survival parameters of translocated and resident swift foxes Vulpes velox. Animal Conservation 6:199–206.

Moore, J.A., B.D. Bell, and W.L. Linklater. 2008. The debate on behavior in conservation: New Zealand integrates theory with practice. BioScience 58:454–459.

Moro, D. 2001. Update of translocation of Thevenard Island mice to Serrurier Island, Western Australia: success and excess. Reintroduction News 20:17–18.

Mortimer, J.A. 1999. Reducing threats to eggs and hatchlings: hatcheries. Pp.175–178. In K.L. Eckert, K.A. Bjorndal, F.A. Abreu-Grobois, and M. Donnelly, eds., Research and Management Techniques for the Conservation of Sea Turtles. IUCN/SSC Marine Turtle Specialist Group Publication No. 4.

Moseby, K.E., A. Cameron, and H.A. Crisp. 2012. Can predator avoidance training improve reintroduction outcomes for the greater bilby in arid Australia? Animal Behavior 83:1011–1021.

Murie, J., S. Stevens, and B. Leoppky. 1998. Survival of captive-born cross-fostered juvenile Columbian ground squirrels in the field. Journal of Mammalogy 79:1152–1160.

Muths, E., L.L. Bailey, and M.K. Watry. 2014. Animal reintroductions: an innovative assessment of survival. Biological Conservation 172:200–208.

Muths, E., and V.J. Dreitz. 2008. Monitoring programs to assess reintroduction efforts: a critical component in recovery. Animal Biodiversity and Conservation 31:47–56.

Mykytowycz, R., E. Hesterman, S. Gambale, and M. Dudziński. 1976. A comparison of the effectiveness of the odors of rabbits, Oryctolagus cuniculus, in enhancing territorial confidence. Journal of Chemical Ecology 2:13–24.

Naughton-Treves, L., R. Grossberg, and A. Treves. 2003. Paying for tolerance: rural citizens' attitudes toward wolf depredation and compensation. Conservation Biology 17:1500–1511.

Nichols, J.D., and D.P. Armstrong. 2012. Monitoring for reintroductions. Pp.223–255. In J.G. Ewen, D.P. Armstrong, K.A. Parker, and P.J. Seddon, eds., Reintroduction Biology: Integrating

Science and Management. John Wiley & Sons, West Sussex, UK.

Nuvoli, S., G. P. Burrai, F. Secci, N. Columbano, G. M. Careddu, L. Mandas, M. A. Sanna, S. Pinno, and E. Antuofermo. 2014. Capture myopathy in a corsican red deer *Cervus elaphus corsicanus* (Ungulata: Cervidae). Italian Journal of Zoology 81:457–462.

OIE and IUCN. 2014. Guidelines for Wildlife Disease Risk Analysis. World Organisation for Animal Health and International Union for Conservation of Nature, Paris.

Osborne, P. E., and P. J. Seddon. 2012. Selecting suitable habitats for reintroductions: variation, change and the role of species distribution modelling. Pp.73–104. In J. G. Ewen, D. P. Armstrong, K. A. Parker, and P. J. Seddon, eds., Reintroduction Biology: Integrating Science and Management. John Wiley & Sons, West Sussex, UK.

Parlato, E. H., and D. P. Armstrong. 2013. Predicting post-release establishment using data from multiple reintroductions. Biological Conservation 160:97–104.

Pedersen, A. B., K. E. Jones, C. L. Nunn, and S. Altizer. 2007. Infectious diseases and extinction risk in wild mammals. Conservation Biology 21:1269–1279.

Pessier, A., and J. Mendelson. 2010. A manual for control of infectious diseases in amphibian survival assurance colonies and reintroduction programs. IUCN/SSC Conservation Breeding Specialist Group, Apple Valley, MN.

Pettifer, H. 1981. The experimental release of captive-bred cheetah (*Acinonyx jubatus*) into the natural environment. Worldwide Furbearer Conference Proceedings 2:1001–1024.

Preston, D. J., R. A. Englund, and M. K. K. McShane. 2007. Translocation and monitoring efforts to establish a second population of the rare *Megalagrion xanthomelas* (Sélys-Longchamps) on O'ahu, Hawai'i (Zygoptera: Coenagrionidae). Bishop Museum Bulletin in Cultural and Environmental Studies 3:261–276.

Price, M. R. S. 1989. Animal Re-introductions: The Arabian Oryx in Oman. Cambridge University Press, Cambridge, UK.

Pruss, S., P. Fargey, and A. Moehrenschlager. 2008. National Swift Fox Recovery Strategy. Prepared in consultation with the Canadian Swift Fox Recovery Team. Species at Risk Act Recovery Strategy Series. Parks Canada Agency, 23pp.

Quinn, J. L., and W. Cresswell. 2005. Personality, anti-predation behavior and behavioral plasticity in the chaffinch *Fringilla coelebs*. Behavior 142:1377–1402.

Rabin, L. 2003. Maintaining behavioral diversity in captivity for conservation: natural behavior management. Animal Welfare 12:85–94.

Reed, J. M. 2004. Recognition behavior based problems in species conservation. Annales Zoologici Fennici 41:859–877.

Reynolds, M., and J. Klavitter. 2006. Translocation of wild Laysan duck *Anas laysanensis* to establish a population at Midway Atoll National Wildlife Refuge, United States and US Pacific Possession. Conservation Evidence 3:6–8.

Robinson, A. T., and D. L. Ward. 2011. Interactions between desert pupfish and Gila topminnow can affect reintroduction success. North American Journal of Fisheries Management 31:1093–1099.

Rocke, T. E., S. Smith, P. Marinari, J. Kreeger, J. T. Enama, and B. S. Powell. 2008. Vaccination with F1-V fusion protein protects black-footed ferrets (*Mustela nigripes*) against plague upon oral challenge with *Yersinia pestis*. Journal of Wildlife Diseases 44:1–7.

Roe, J. H., M. R. Frank, S. E. Gibson, O. Attum, and B. A. Kingsbury. 2010. No place like home: an experimental comparison of reintroduction strategies using snakes. Journal of Applied Ecology 47:1253–1261.

Roemer, G. W., and R. K. Wayne. 2003. Conservation in conflict: the tale of two endangered species. Conservation Biology 17:1251–1260.

Rollinson, N., D. M. Keith, A. L. S. Houde, P. V. Debes, M. C. Mcbride, and J. A. Hutchings. 2014. Risk assessment of inbreeding and outbreeding depression in a captive-breeding program. Conservation Biology 28:529–540.

Sarrazin, F., C. Bagnolinp, J. L. Pinna, and E. Danchin. 1996. Breeding biology during establishment of a reintroduced Griffon Vulture *Gyps fulvus* population. Ibis 138:315–325.

Sarrazin, F., and S. Legendre. 2000. Demographic approach to releasing adults versus young in reintroductions. Conservation Biology 14:488–500.

Saunders, S. P., T. W. Y. Ong, and F. J. Cuthbert. 2013. Auditory and visual threat recognition in captive-reared Great Lakes piping plovers (*Charadrius melodus*). Applied Animal Behavior Science 144:153–162.

Schaub, M., R. Pradel, and J.-D. Lebreton. 2004. Is the reintroduced white stork (*Ciconia ciconia*) population in Switzerland self-sustainable? Biological Conservation 119:105–114.

Schoech, S. J., E. S. Bridge, R. K. Boughton, S. J. Reynolds, J. W. Atwell, and R. Bowman. 2008.

Food supplementation: a tool to increase reproductive output? A case study in the threatened Florida Scrub-Jay. Biological Conservation 141:162–173.

Scott, J. M., and J. W. Carpenter. 1987. Release of captive-reared or translocated endangered birds: what do we need to know? The Auk 104: 544–545.

Sherman, P. W. 1980. The meaning of nepotism. American Naturalist 116:604–606.

Shier, D. M. 2006. Effect of family support on the success of translocated black-tailed prairie dogs. Conservation Biology 20:1780–1790.

Shier, D. M., and D. Owings. 2007. Effects of social learning on predator training and postrelease survival in juvenile black-tailed prairie dogs, *Cynomys ludovicianus*. Animal Behavior 73:567–577.

Short, J., S. Bradshaw, J. Giles, R. Prince, and G. R. Wilson. 1992. Reintroduction of macropods (Marsupialia: Macropodoidea) in Australia: a review. Biological Conservation 62:189–204.

Slotta-Bachmayr, L., R. Boegel, P. Kaczensky, C. Stauffer, and C. Walzer. 2004. Use of population viability analysis to identify management priorities and success in reintroducing Przewalski's horses to Southwestern Mongolia. Journal of Wildlife Management 68:790–798.

Smith, D., T. Everest, A. Moehrenschlager, and D. Brodie. 2011. Reintroducing a migratory raptor to the edge of its former range: are we feeding a sink? International Conference for Conservation Biology, December 5–9, Auckland, New Zealand.

Smith, K., K. Acevedo-Whitehouse, and A. Pedersen. 2009. The role of infectious diseases in biological conservation. Animal Conservation 12:1–12.

Stamps, J. A., and R. R. Swaisgood. 2007. Someplace like home: experience, habitat selection and conservation biology. Applied Animal Behavior Science 102:392–409.

Steury, T. D., and D. L. Murray. 2004. Modeling the reintroduction of lynx to the southern portion of its range. Biological Conservation 117:127–141.

Swaisgood, R. R. 2007. Current status and future directions of applied behavioral research for animal welfare and conservation. Applied Animal Behavior Science 102:139–162.

Swaisgood, R. R. 2010. The conservation-welfare nexus in reintroduction programmes: a role for sensory ecology. Animal Welfare 19:125–137.

Swaisgood, R. R., D. G. Lindburg, and X. Zhou. 1999. Giant pandas discriminate individual differences in conspecific scent. Animal Behavior 57:1045–1053.

Switzer, P. V. 1993. Site fidelity in predictable and unpredictable habitats. Evolutionary Ecology 7:533–555.

Tapley, B., M. Rendle, F. M. Baines, M. Goetz, K. S. Bradfield, D. Rood, J. Lopez, G. Garcia, and A. Routh. 2015. Meeting ultraviolet B radiation requirements of amphibians in captivity: a case study with mountain chicken frogs (*Leptodactylus fallax*) and general recommendations for pre-release health screening. Zoo Biology 34:46–52.

Taylor, S., and I. Castro. 2000. Translocation history of hihi (stitchbird), an endemic New Zealand honeyeater. IUCN/SSC Reintroduction Specialist Group Reintroduction News 19:28–30.

Teixeira, B., and R. J. Young. 2014. Can captive-bred American bullfrogs learn to avoid a model avian predator? Acta Ethologica 17:15–22.

Teixeira, C. P., C. S. De Azevedo, M. Mendl, C. F. Cipreste, and Young, R. J. 2007. Revisiting translocation and reintroduction programmes: the importance of considering stress. Animal Behavior 73:1–13.

Thapa, K., S. Nepal, G. Thapa, S. R. Bhatta, and E. Wikramanayake. 2013. Past, present and future conservation of the greater one-horned rhinoceros *Rhinoceros unicornis* in Nepal. Oryx 47:345–351.

Thompson, J. G. 1993. Addressing the human dimensions of wolf reintroduction: an example using estimates of livestock depredation and costs of compensation. Society & Natural Resources 6:165–179.

Thompson, J. R., V. C. Bleich, S. G. Torres, and G. P. Mulcahy. 2001. Translocation techniques for mountain sheep: does the method matter? The Southwestern Naturalist: 87–93.

Todd, C. R., S. J. Nicol, and J. D. Koehn. 2004. Density-dependence uncertainty in population models for the conservation management of trout cod, *Maccullochella macquariensis*. Ecological Modelling 171:359–380.

Towns, D. R., and S. M. Ferreira. 2001. Conservation of New Zealand lizards (Lacertilia: Scincidae) by translocation of small populations. Biological Conservation 98:211–222.

Townsend, D. E., R. L. Lochmiller, S. J. DeMaso, D. M. Leslie, Jr., A. D. Peoples, S. A. Cox, and E. S. Parry. 1999. Using supplemental food and its influence on survival of northern bobwhite (*Colinus virginianus*). Wildlife Society Bulletin 27:1074–1081.

Tracy, L., G. Wallis, M. Efford, and I. Jamieson. 2011. Preserving genetic diversity in threatened species reintroductions: how many individuals should be released? Animal Conservation 14:439–446.

Truett, J. C., J. A. L. Dullum, M. R. Matchett, E. Owens, and D. Seery. 2001. Translocating prairie dogs: a review. Wildlife Society Bulletin 29:863–872.

Tuberville, T. D., E. E. Clark, K. A. Buhlmann, and J. W. Gibbons. 2005. Translocation as a conservation tool: site fidelity and movement of repatriated gopher tortoises (*Gopherus polyphemus*). Animal Conservation 8:349–358.

Turner, A. S., L. Conner, and R. J. Cooper. 2008. Supplemental feeding of Northern bobwhite affects Red-tailed hawk spatial distribution. The Journal of Wildlife Management 72:428–432.

Unangst, Jr., E. T., and B. A. Wunder. 2004. Effect of supplemental high-fat forage on body composition in wild meadow voles (*Microtus pennsylvanicus*). The American Midland Naturalist 151:146–153.

Vancouver Island Marmot Recovery Team. 2008. Recovery Strategy for the Vancouver Island Marmot (*Marmota vancouverensis*) in British Columbia. Prepared for the B. C. Ministry of Environment, Victoria, BC.

van Heezik, Y., R. F. Maloney, and P. J. Seddon. 2009. Movements of translocated captive-bred and released Critically Endangered kaki (black stilts) *Himantopus novaezelandiae* and the value of long-term post-release monitoring. Oryx 43:639–647.

van Heezik, Y., P. J. Seddon, and R. F. Maloney. 1999. Helping reintroduced houbara bustards avoid predation: effective anti-predator training and the predictive value of pre-release behavior. Animal Conservation 2:155–163.

Vazquez-Sauceda, M. L., G. Aguirre-Guzman, R. Perez-Castaneda, J. G. Sanchez-Martinez, M. D. Campo, J. Loredo-Osti, and J. L. Rábago-Castro. 2008. Evaluation of the influence of two transport boxes on the incubation, hatching and emergence of Kemp's ridley turtle (*Lepidochelys kempii*) eggs. Ciencias Marinas 34:101–105.

Ventura, R., U. A. da Silva, A. Ostrensky, K. Cottens, and G. Perbiche-Neves. 2010. Survival of *Ucides cordatus* (Decapoda: Ocypodidae) megalopae during transport under different conditions of density and duration. Zoologia (Curitiba) 27:845–847.

Viggers, K., D. Lindenmayer, and D. Spratt. 1993. The importance of disease in reintroduction programmes. Wildlife Research 20:687–698.

Wallace, M. T., and R. Buchholz. 2001. Translocation of red-cockaded woodpeckers by reciprocal fostering of nestlings. The Journal of Wildlife Management 65:327–333.

Wanless, R. M., J. Cunningham, P. A. Hockey, J. Wanless, R. W. White, and R. Wiseman. 2002. The success of a soft-release reintroduction of the flightless Aldabra rail (*Dryolimnas [cuvieri] aldabranus*) on Aldabra Atoll, Seychelles. Biological Conservation 107:203–210.

Williams, E., S. Anderson, J. Cavender, C. Lynn, K. List, C. Hearn, and M. Appel. 1996. Vaccination of black-footed ferret (*Mustela nigripes*) × Siberian polecat (*M. eversmanni*) hybrids and domestic ferrets (*M. putorius furo*) against canine distemper. Journal of Wildlife Diseases 32:417–423.

Wilson, B. S., A. C. Alberts, K. S. Graham, R. D. Hudson, R. K. Bjorkland, D. S. Lewis, N. P. Lung, R. Nelson, N. Thompson, and J. L. Kunna. 2004. Survival and reproduction of repatriated Jamaican iguanas. Pp. 220–231. In A. Alberts, ed., Iguanas: Biology and Conservation. University of California Press, Los Angeles.

Wilson, M. A. 1997. The wolf in Yellowstone: science, symbol, or politics? Deconstructing the conflict between environmentalism and wise use. Society and Natural Resources 10:453–468.

Wolf, C. M., B. Griffith, C. Reed, and S. A. Temple. 1996. Avian and mammalian translocations: update and reanalysis of 1987 survey data. Conservation Biology 10:1142–1154.

Woodford, M. H., and P. B. Rossiter. 1994. Disease risks associated with wildlife translocation projects. Pp. 178–200. In P. J. S. Olney, G. M. Mace, and A. T. C. Feistner, eds., Creative Conservation: Interactive Management of Wild and Captive Animals. Chapman & Hall, London.

Woodroffe, R. 1999. Managing disease threats to wild mammals. Animal Conservation 2: 185–193.

Young, A. C., E. G. Johnson, J. L. Davis, A. H. Hines, O. Zmora, and Y. Zohar. 2008. Do hatchery-reared blue crabs differ from wild crabs, and does it matter? Reviews in Fisheries Science 16:254–261.

Zwank, P. J., and C. D. Wilson. 1987. Survival of captive, parent-reared Mississippi sandhill cranes released on a refuge. Conservation Biology 1:165–168.

PART FOUR

Managing Reintroduced Populations

TWELVE

Effective and Purposeful Monitoring of Species Reintroductions

Robert A. Gitzen, Barbara J. Keller, Melissa A. Miller, Scott M. Goetz, David A. Steen, David S. Jachowski, James C. Godwin, and Joshua J. Millspaugh

ECOLOGICAL MONITORING IS A COM-plex topic, often involving strong differences of opinion regarding the purpose (and even the definition) of monitoring and difficult challenges in deciding what to monitor, how to monitor, and how much to invest in the effort. Although there can be a temptation to view the statistical design of ecological monitoring surveys as a straightforward task, in reality it must deal with a difficult combination of design issues, e.g., yearly temporal variation, lack of experimental control, and the likely need for the monitoring program to adapt its design over time as information needs change. Most of these challenges arise when developing monitoring programs focused on species reintroductions. As in other monitoring contexts, reintroduction monitoring involves a myriad of difficult decisions, including decisions about how much to invest in monitoring versus other components of a reintroduction program. These decisions require careful thought, given the importance of monitoring data for guiding reintroduction decisions and assessing whether the reintroduction effort is achieving success (IUCN 1998, Fischer and Lindenmayer 2000, Muths and Dreitz 2008, Nichols and Armstrong 2012).

Our purposes in this chapter are to provide an overview of some key considerations and decisions faced in reintroduction monitoring, and to illustrate our discussion with selected real-world examples. We define "monitoring" simply as systematic, repeated data collection intended to assess, and reduce uncertainty about, parameters of interest such as population size and the effects of alternative reintroduction strategies. Our discussion is geared toward reintroduction practitioners such as agency biologists and program managers. Although monitoring design and the role of monitoring in structured decision-making contexts can involve complex quantitative issues and tools (Conroy and Peterson 2013), we do not focus heavily on quantitative details. Readers are referred to other literature for more in-depth technical treatments (e.g., Chapters 6 and 7, this volume; Gitzen et al. 2012). Our focus is on animal reintroductions, but most of our discussion is equally applicable to invertebrates and plants.

In developing reintroduction monitoring (Nichols and Armstrong 2012), as with broader ecological monitoring (Johnson et al. 2012, MacKenzie 2012, McDonald 2012), critical general questions that need to be addressed are "Why monitor?," "What should be monitored?," and "How (including where and when) to monitor?" Our overview of reintroduction monitoring is partly organized around these fundamental questions. Our discussion of "how" to monitor summarizes key aspects of developing any monitoring survey design. We provide additional discussion focusing on demographic monitoring because population establishment and persistence are core objectives of any reintroduction program. Throughout our overview, we emphasize that careful identification of reintroduction management objectives, and uncertainty affecting ongoing management decisions, guide the type and quality of information needed from monitoring (Box 12.1).

To illustrate the potential value of monitoring and to demonstrate some of our key recommendations, we refer frequently to published examples (e.g., Table 12.1) and we offer more extensive discussion of three examples with which the authors of this chapter have been involved, collectively. Throughout the chapter, we refer repeatedly to the reintroduction of black-footed ferrets (*Mustela nigripes*) in North America for illustrating potential benefits of multifaceted monitoring efforts across a collection of individual reintroductions (Box 12.2). As an extended case study of an intensive research and monitoring program involving close collaboration of scientists and managers, the latter part of this chapter focuses on the reintroduction of elk (*Cervus elaphus*) in the US state of Missouri. A third example focuses on reintroduction of eastern indigo snakes (*Drymarchon couperi*) in Alabama (Box 12.3).

MONITORING SERVES DECISION-MAKING IN REINTRODUCTION PROGRAMS

The "Why?" of Monitoring in the Reintroduction Context

The history of ecological monitoring, including reintroduction monitoring, involves countless efforts in which data collection was initiated in the absence of any carefully defined purpose and objectives for monitoring (Lindenmayer and Likens 2010). Without clear agreement on why monitoring is to be conducted and how the information will be used, there is no framework for developing the monitoring effort, leading to ad hoc choices about what and how to monitor (MacKenzie 2012, Reynolds 2012).

Stated most generally, the purpose of monitoring should be to produce useful information, not simply to spend money allocated for monitoring or to check off a box indicating that monitoring has occurred. In the broader field of ecological monitoring, there has been significant debate about the relative merits of surveillance monitoring intended to assess broadscale status and trends of selected ecological attributes versus targeted monitoring focused on clear management-related questions and scientific hypotheses (e.g., Nichols and Williams 2006, Johnson 2012). The reintroduction context eliminates the relevance of much of this debate. Reintroductions are decisive management actions that typically require a strong commitment and high investment in time, money, and often political capital (Jachowski et al., Chapter 1, this volume). It seems uncontroversial to assert that reintroduction monitoring should be a targeted effort aimed at assessing reintroduction success, helping increase the likelihood of success by the current and other reintroduction efforts, and assessing effects of the reintroduction, such as whether undesirable impacts are occurring. Monitoring in this context focuses on reducing core sources of uncertainty relevant to reintroduction decision-making.

BOX 12.1 · Steps in Quantitative Study Design for Monitoring Reintroductions

Book-length treatments are needed simply to provide an overview of issues and approaches for designing effective monitoring and other ecological studies (e.g., Thompson et al. 1998, de Gruijter et al. 2006, Gitzen et al. 2012). Here, we highlight a few key steps.

1. Determine the purpose of monitoring. The foundation for designing a monitoring program consists of quantitative reintroduction objectives, decision-focused models, and objectives for broader scientific learning, as this foundation identifies and prioritizes specific parameters for monitoring (e.g., population growth rate; odds ratio for survival of wild-caught vs. captive-reared animals).

2. Clearly define the statistical target population of interest about which monitoring is intended to provide information during the expected life of the monitoring program. In the reintroduction context, the target population may directly correspond to the reintroduced population of animals, or it may focus on a spatial target population, e.g., on the collection of patches of habitat that ultimately may be occupied by the reestablished biological population.

3. For the focal parameters, determine what quality of information (precision, statistical power, ability to correctly determine relative support for competing models) is needed to be useful (Reynolds 2012). In decision-focused monitoring, assess how precise the information needs to be to affect the predicted utility of alternative decision choices or rate of learning about the system (e.g., Kendall and Moore 2012).

4. Determine the specific analyses that provide direct estimates of the focal parameters, and then determine survey designs that will provide the data required by these analyses

(Reynolds 2012). This analysis-focused framework helps ensure that decisions faced during design of monitoring stay focused on the desired end result.

5. For these candidate survey designs, explicitly define, quantify, and carefully consider the potential impacts of sources of variability that affect expected precision and potential sources of bias (e.g., incomplete and variable detectability, or any aspect of data collection that produces a systematic mismatch between the subpopulation being monitored and the broader population of interest).

6. Quantify the expected precision (or power, etc.) of candidate survey designs in light of quantitative monitoring objectives. This step is essential both for understanding whether the monitoring program has any chance of producing useful information and for examining how effort should be allocated to optimize the effort (e.g., trade-offs between number of survey sites monitored vs. amount of effort spent when visiting each site).

7. If necessary, the scope of the monitoring program may need to be reduced such that effort can be focused where it is most needed. If monitoring is unlikely to support useful quantitative inference, then managers need to decide whether exploratory monitoring still is worth the cost, e.g., to decide whether there is value in monitoring just to assess whether some reintroduced animals are still alive, or as a program-development step in building a more effective longer-term monitoring program.

8. As early as possible, consider how to build in flexibility to expand, contract, or otherwise adjust the monitoring design as objectives change, as the target population expands, and as the magnitude of variation and sources of bias are better quantified.

Within this generic focus on producing useful information, the "Why?" of reintroduction monitoring involves several general purposes (Nichols and Armstrong 2012). First, monitoringprovidesinformationforstate-dependentdecision-

making. That is, the state of the population at each time step (e.g., abundance), as estimated via monitoring, may determine which, if any, management actions are implemented in response. Such management actions may

BOX 12.2 · Integrating Monitoring and Management to Enhance Black-Footed Ferret Reintroduction Success

Reintroduction of the black-footed ferret (*Mustela nigripes*) is one of the highest profile and longest running reintroduction attempts to date. It has successfully integrated monitoring within multiple facets of the long-term recovery process of this species in the Great Plains of North America. Initial attempts to breed the species in captivity during the 1970s failed due to effects of canine distemper and exposure to an untested vaccine (Carpenter et al. 1976). Loss of the last remaining extant wild population was linked to outbreaks of canine distemper and sylvatic plague (Jachowski 2014). Starting from an initial captive population of only seven genetically unrelated individuals, captive breeding has produced >8,700 ferrets over the past 27 years, with releases at >20 reintroduction sites in the western United States, Canada, and Mexico. The recovery goal for the species under the US Endangered Species Act is to establish 1,500 breeding adults in >10 populations of at least 30 adult animals each, with a more ambitious goal of at least 3,000 breeding adults in >30 populations to enable delisting of the species (USFWS 1988). Throughout this chapter, we discuss various aspects of monitoring focused on the ferret reintroductions. Here, we summarize core components of this monitoring.

Little was known about black-footed ferret ecology prior to reintroduction. Therefore, dedicated monitoring efforts at individual release sites have shed light on basic aspects of the biology of this species, such as dispersal, resource selection, recruitment, survivorship, and causes of mortality (Jachowski 2014). Across all reintroduction sites, annual or biannual post-release spotlight counts have provided important insight and have been invaluable for building hypotheses for future dedicated investigation. Using this protocol (Biggins et al. 2005), some sites have >20 years of annual monitoring data collected in a nearly identical manner. These long-term monitoring results have been used to guide and refine future releases, including the selection of release study sites and incorporation of dedicated studies on fine-scale ferret resource selection (Eads et al. 2011, Jachowski et al. 2011b, Chipault et al. 2012, Grassel et al. 2015) and disease ecology (Matchett et al. 2010).

During periods when spotlight monitoring effort is somewhat standardized across sites, resulting indices provide data for evaluation of large-scale patterns across reintroduction sites. For example, Jachowski et al. (2011a) used spotlight monitoring data from 11 reintroduction sites to evaluate relationships between ferret reintroduction success and disease outbreaks, release strategies, and prey availability. Prey abundance was the most effective predictor of reintroduction success; no reintroduction sites with <4,300 ha of habitat occupied by prairie dogs (*Cynomys* sp.) were successful. These findings drew into question the suitability for many current and planned reintroduction sites, which were almost all below the 4,300 ha size threshold. After revisions to the species recovery plan in 2013, the plan continues to set minimum ferret population size goals similar to the 1988 plan. In addition, the plan now explicitly calls for maintaining 100,000 ha of habitat occupied by prairie dogs to achieve down-listing of black-footed ferrets, and at least 200,000 ha for achieving species delisting (USFWS 2013).

Overall, based on the 30 year history of this reintroduction program, monitoring has shifted and adapted in priorities and intensity in response to threats facing the recovery of the species. The intensive spotlight monitoring discussed above was driven by an early focus on the establishment and persistence of ferrets following release. Participation in this monitoring was given high priority by the US Fish and Wildlife Service in allocating captive-reared ferrets to reintroduction sites (Jachowski and Lockhart 2009). Thus, site managers had guidance and motivation to conduct these relatively standardized surveys at least annually to ensure the allocation of animals to their site, and ensure sustained allocations over multiple years to supplement initial releases. As managers of many sites struggled to establish populations or withdrew from the program, competition among sites for captive-reared ferrets became less intense as more ferrets became available than could be released at suitable sites. This led to the allocation of captive-reared ferrets to sites that had minimal plans for monitoring or lacked resources for conducting monitoring. This resulted in considerable variation in monitoring frequency and intensity among sites, variation that persists to this day.

Monitoring over the past decade has emphasized surveillance and research on sylvatic plague. Similar to initial spotlighting efforts, plague-focused monitoring is currently highly prioritized and subsidized by state and federal government agencies. If the disease risk should be abated, it is likely that the funding for plague-based monitoring will slacken and present issues similar to those with spotlight monitoring, in terms of variation in effort and expertise among sites that limits potential cross-site comparisons.

These types of changes in monitoring priorities are likely to continue in coming years. Adaptation in the ongoing development of the black-footed ferret recovery program has been a relatively organic process, in large part due to the overall guidance for species recovery mandated under the US Endangered Species Act, which requires addressing challenges that face the overall, long-term recovery of the species. This adaptive synergy between monitoring and reintroduction activities is partly due to empowerment of an interdisciplinary reintroduction planning team linking scientists, managers, and policy-makers (Jachowski and Lockhart 2009). Keeping all partners up to speed on current research and monitoring priorities, and involving multiple sites in targeted monitoring programs, has helped participants to pool their resources and expertise in addressing daunting conservation challenges.

include release of more animals, predator control, habitat manipulations, cessation of a program deemed to be a failure, initiation of harvest, or simply continuation of the status quo. Second, monitoring seeks to determine whether the reintroduction is on track to meet management objectives, such as recovery criteria for the number of breeding adult black-footed ferrets (Box 12.2).

Third, monitoring allows scientific learning through the assessment of relative support for models based on alternative hypotheses about the system and through continued development of useful models. For example, if there are alternative hypotheses about relationships between current population density in an ongoing reintroduction versus expected site fidelity by subsequently released animals (as in an example from Hawaii briefly discussed in the next section), relative support for contrasting hypotheses could be updated iteratively with data from appropriately targeted monitoring. Such learning should produce more informed predictions about the likely effects of potential management actions considered in future decision-making, such as how many animals to release as the reintroduction effort continues. Monitoring information is used to update parameters in decision models (e.g., survival rates in demographic projection models), and learning occurs through quantitative comparison of the accuracy of predictions produced by alternative models of the system (e.g., decision models with or without parameters allowing predicted side fidelity to change as a function of population size). This is the foundation of adaptive management in the narrow sense of the term: structured decision-making with a focus on reducing uncertainty relevant to recurrent or other future decisions (key current references include McCarthy et al. 2012, Nichols and Armstrong 2012, Conroy and Peterson 2013, Runge 2013). Whether in the context of quantitative predictive models or without explicitly stated models, reintroduction monitoring should also help managers and scientists improve the success of future introductions (Nichols and Armstrong 2012).

What to Monitor

Because reintroduction monitoring should help reduce uncertainty affecting management decision-making, the precision with which management objectives have been defined dictates how clearly these objectives can guide development of the monitoring program. For example, the Missouri elk reintroduction program discussed later in this chapter demonstrates the value of tightly linking data-collection efforts to well-defined reintroduction objectives. State-dependent decision-making and assessment of reintroduction success is predicated on clear understanding of what "success" means in the program, and of what state parameters (e.g., population size, disease incidence) matter most in determining a management response. Useful decision-focused models are developed based on these parameters of highest management relevance and on hypothesized effects of management decisions and other factors on system dynamics.

Management objectives therefore provide the context for prioritizing among the numerous major attributes often considered for monitoring in reintroduction programs (Table 12.1). Quantitative reintroduction objectives also provide the context for determining the quality of information needed about the focal parameters (Box 12.1; Reynolds 2012). Chapter 6 in this volume provides guidance on setting useful objectives in the context of structured decision-making about reintroductions. Unfortunately, the literature review presented in Chapter 6 found that less than half of reintroduction efforts documented in the literature stated clearly quantifiable objectives. This problem is the most important obstacle to address for improving reintroduction monitoring.

Given that reintroductions seek to produce a viable population within some target area, usually the highest priority attributes for monitoring are parameters related to population size, population distribution (extent, patch occupancy), rates determining changes in these parameters (e.g., survival, fecundity, patch-level colonization and extinction), and factors affecting these rates. For example, the importance of information gained from demographic monitoring during reintroductions can be seen with the reintroduction of Pacific lamprey (*Entosphenus tridentatus*) to the Umatilla River in Oregon (Close et al. 2009). Although managers accounted for factors associated with the initial extirpation of the Pacific lamprey prior to

reintroduction, analyses of demographic monitoring data identified unforeseen factors limiting lamprey establishment and survival within different regions of the river.

Monitoring of movement, space use, habitat selection, and other aspects of individual behavior (Table 12.1) often illuminates factors affecting demographic success of the reintroduction, helps determine limitations to reintroduction success that may be mitigated (e.g., habitat improvements), and detects whether reintroduced animals are having undesired impacts. The Missouri elk, black-footed ferret (Box 12.2), and indigo snake (Box 12.3) examples illustrate several of these potential benefits of such finer-scale monitoring. In an example demonstrating that useful insights may be gained even with descriptive results from a small number of individuals, Tweed et al. (2003) used radiotelemetry to monitor short-term survival and movements of 14 reintroduced puaiohi (*Myadestes palmeri*), an endangered thrush endemic to the Hawaiian island of Kauai. Although survival was high, 8 of 14 birds had dispersed out of the reintroduction area within 10 weeks of release, suggesting that multiple releases may be needed in the area to develop a self-sustaining population. However, the authors hypothesized that post-release dispersal was related to availability of mates, leading to the prediction that fidelity would increase with future releases in the area.

Monitoring programs often include a focus on genetics (Biebach et al., Chapter 8, this volume), individual health and physiology (Jachowski et al., Chapter 9, this volume), and parasites and pathogens (Muths and McCallum, Chapter 10, this volume; Table 12.1). For example, genetic diversity of previously reintroduced insular populations of tuatara (*Sphenodon punctatus*) was investigated to determine influence of founder group size (Miller et al. 2009). A viability analysis was used to predict heterozygosity and allelic diversity following 10 generations. Because additional reintroductions to new areas are planned, these results for

TABLE 12.1
General Parameters of Frequent Emphasis in Reintroduction Monitoring, with Selected Published Examples

Category/Monitoring Parameters	Selected Examples
Distribution and Occurrence Dynamics	
Extent and proportion of area occupied	Nichols and Armstrong (2012)
Rate of spread	Haydon et al. (2008), Yott et al. (2011)
Patch-level colonization and extinction rates	Armstrong and Ewen (2002), Kramer-Schadt et al. (2005)
Demographic Attributes	
Population size or density	Castro et al. (2004), Armstrong et al. (2005), Taylor et al. (2005), Close et al. (2009)
Survival, causes of mortality	McKinstry and Anderson (2002), King et al. (2004), Slotta-Bachmayr et al. (2004), Deredec and Courchamp (2007), Frair et al. (2007), Devineau et al. (2010)
Presence of successful breeding	Sanz and Grajal (1998)
Reproductive rate	Richards and Short (2003), Mitchell et al. (2011)
Age, stage, and sex composition	Sarrazin and Legendre (2000), Towns and Ferreira (2001), Wronski et al. (2011)
Immigration and emigration rates	Mihoub et al. (2011)
Population growth rate	Dunham (1997), Armstrong et al. (2005), Somers et al. (2008)
Genetic Attributes and Processes	
Genetic diversity	Miller et al. (2009), De Barba et al. (2010), Jamieson (2011), Tracy et al. (2011), Andersen et al. (2014), Mowry et al. (2015)
Inbreeding and outbreeding depression	Jiménez et al. (1994), Marshall and Spalton (2000), Fredrickson et al. (2007), Brekke et al. (2010), Jamieson (2011)
Movements and Space Use	
Site fidelity	Tuberville et al. (2005), Frair et al. (2007), Armstrong et al. (2013), Berger-Tal and Saltz (2014)
Natal dispersal distance and location	Whitfield et al. (2009), Richardson et al. (2010)
Home range size	Dunham (1998), Perelberg et al. (2003), Benson and Chamberlain (2007), Hester et al. (2008)
Movement distances and rates	Berger-Tal and Saltz (2014)
Space-use overlap	Saltz et al. (2000), Dolev et al. (2002), Hirzel et al. (2004)
Habitat use and resource selection	Preatoni et al. (2005), Mihoub et al. (2009), Roe et al. (2010), Bennett et al. (2012)

(Continued)

TABLE 12.1
(Continued)

Category/Monitoring Parameters	Selected Examples
Individual Health	
Body condition	Mathews et al. (2006), Hardman and Moro (2006), Santos et al. (2009)
Stress levels	Teixeira et al. (2007), Franceschini et al. (2008), Dickens et al. (2009), Zidon et al. (2009), Gelling et al. (2010, 2012)
Pathogens and Parasites	
Disease presence	Goodman et al. (2012), Aiello et al. (2014)
Parasite incidence	Phillips and Scheck (1991), Moravec (2003), Mathews et al. (2006), Van Oosterhout et al. (2007), Wallach et al. (2008), Almberg et al. (2012), Goodman et al. (2012), Harris et al. (2014)
Community and Ecosystem Impacts	
Animal community composition	Gorman (2007), White and Garrott (2005), Smith and Tylers (2008)
Behavior and demography of resident species	Ripple et al. (2001), Fortin et al. (2005), Mao et al. (2005), Kauffman et al. (2010)
Habitat structure, composition, and function	Ripple et al. (2001), Ripple and Beschta (2003), Beyer et al. (2007), James and Eldridge (2007), Gibbs et al. (2008), Kemp et al. (2012)

the tuatara can inform decisions about future minimum founder group size.

Although this chapter focuses largely on post-release monitoring rather than prerelease assessments, we note the frequent importance of ongoing monitoring of genetics, individual health (including presence of infectious diseases), and demographic rates of captive populations and wild animals being translocated (e.g., Frankham 2008). For example, given that all captive-bred black-footed ferrets descend from an initial captive population of only seven genetically unrelated individuals (Box 12.2), captive ferrets have long been genetically monitored due to concerns about inbreeding depression (Wisely 2005). Although captive breeding has been viewed as largely successful for this species, decreased whelping rates by females and increased sperm abnormalities in males have been observed (Wisely et al. 2015).

To maximize genetic retention, pioneering efforts in sperm cryopreservation, artificial insemination, and husbandry have been developed for black-footed ferrets. Such efforts have served as a model for managing other critically endangered species (Howard et al. 2003, Santymire et al. 2006). More recently, black-footed ferrets have been proposed as an ideal case study for the use of reproductive cloning or interspecies somatic cell nuclear transfer (iSCNT) to reduce genetic depletion in a species subject to reintroduction efforts (Wisely et al. 2015).

As reintroductions are strong manipulations of ecosystems, studying their broader impacts may be of both management relevance and basic scientific value. Box 12.3 discusses ongoing implementation to look at ecosystem effects of indigo snake reintroduction. Monitoring the effect of reintroduced giant tortoises (*Geochelone*

BOX 12.3 · Monitoring the Reintroduction of a Top-Level Predator

12a Eastern indigo snake (*Drymarchon couperi*). Photo by Jim Godwin.

The eastern indigo snake (*Drymarchon couperi*; see figure 12a) likely was extirpated from the State of Alabama, United States, in the mid-twentieth century. Habitat loss and fire suppression in their open-canopy habitats were the primary drivers of the indigo snake's disappearance. Potentially contributing to their decline was over-collection for the pet trade as well as the application of gasoline to their winter refugia (gopher tortoise [*Gopherus polyphemus*] burrows) for rounding up rattlesnakes. However, Conecuh National Forest (CNF) in southern Alabama is undergoing a large-scale habitat restoration effort that includes the application of frequent prescribed fire and thinning of hardwood trees to recreate an open-canopy longleaf pine (*Pinus palustris*) ecosystem. Further, the threat of collection was alleviated with the federal listing of the indigo snake under the Endangered Species Act. When gassing of tortoise burrows became illegal in the state in 2009, the stage was set for a cooperative effort to reintroduce the indigo snake to Alabama.

As of 2015, 107 snakes have been released in the CNF. The program anticipates releasing approximately 200 additional snakes over the next few years. Initial post-release monitoring was conducted on the level of individual animals to evaluate release techniques, estimate survivorship, identify sources of mortality, quantify habitat selection, and determine whether released ani-

mals exhibit site fidelity. In addition, the reintroduction program has attempted to establish a long-term mark-recapture protocol to allow estimation of population demography and viability. To date, all released snakes contained passive integrated transponders (PIT tags) to allow identification of any recaptured snakes. In addition, most of the snakes released within the first few years of the project were equipped with radio transmitters allowing the animals to be monitored for up to approximately 36 months. However, monitoring efforts have faced challenges common to many reintroduction programs. For example, as the project expanded, logistical limitations prevented monitoring of all animals with radio transmitter, particularly over long time periods, leaving considerable uncertainty regarding the fate of individual animals.

To recapture snakes and monitor the population, the project now takes advantage of their winter habit of basking outside of gopher tortoise burrows. Searches target areas around where snakes have been released; recaptured animals are identified and released. Further monitoring is occurring via drift fence and trap arrays throughout the study area. Finally, local residents are encouraged to notify the project of any incidental indigo snake observations. Of the 107 individual snakes released so far, 71 have been located or otherwise observed at least once following release, with 29 mortalities documented. Recaptured animals grew considerably in the time since release, indicating that the animals obtained sufficient prey. Further, indigo snakes have been observed pursuing and consuming prey (Steen et al. 2016) and using suitable winter refugia, which provides some evidence that released animals display the requisite behavior for survival. Evidence of reproduction (i.e., adult females with viable eggs) has been gathered, but no indigo snakes without PIT tags (i.e., naturally recruited animals) have been captured. Given the importance of individual health to reintroduction success, recaptured snakes have been given cursory examinations in the field. Any snakes with obvious problems would be brought temporarily into captivity for a veterinary examination and treatment.

Monitoring of the release area will continue for the foreseeable future with the hope of obtaining sufficient recaptures to generate meaningful population estimates, although this is a difficult challenge for secretive animals such as snakes. Given a relatively low ability to recapture and monitor snakes with current methodologies and technology, this monitoring framework is adaptive (sensu Lindenmayer and Likens 2009) in that it may take advantage of relevant new tools and methods if they are developed. Similarly, increased natural history and population data about natural snake populations should help refine specific quantitative reintroduction objectives, which were relatively vague in the initial phases of the project.

Such objectives should guide decisions about future population monitoring as the program continues.

In addition to monitoring of individual growth and population demography, the program is investing significantly in monitoring the broader ecosystem effects of the reintroduction. As a top predator, the indigo snake likely exerts significant ecological impacts in the ecosystems it inhabits. Because indigo snakes extensively prey on other snake species, drift fence trap arrays are being used to monitor the reptile assemblages in the release area and elsewhere to evaluate whether reintroduced indigo snakes are reducing abundances of other snakes. This project will be expanded into a before-after control-impact study design to monitor reptile assemblages in an adjacent new release area. Because indigo snakes feed on other snake species that are the dominant predators of bird nests in the southeastern United States, the project seeks to monitor nest survival for ground and shrub-nesting birds and conduct point-count surveys to monitor occupancy of target bird species. Additional sampling focuses on rodents, which are likely important prey items for the snakes that indigo snakes are consuming. Overall, over the next few years continued monitoring will help to assess whether indigo snakes have formed a viable and self-sustaining population and to understand whether a trophic cascade was initiated in the longleaf pine ecosystem after their release.

---

nigra hoodensis) on vegetation in the Galapagos Islands has shown that tortoises may benefit ecosystem restoration efforts by aiding in the passive restoration of native cacti (Opuntia megasperma var. megasperma; Gibbs et al. 2008). Perhaps the flagship example of assessing ecosystem-wide impacts of a reintroduction is the general body of studies focused on effects of reestablished wolves in the Greater Yellowstone Ecosystem (e.g., see citations in Table 12.1 such as White and Garrott 2005, Ripple and Beschta 2012).

Besides contributing to understanding of ecosystems in which releases occur, post-release monitoring and research may provide the basis for broader conservation of such ecosystems.

One of the most profound results from monitoring associated with black-footed ferret reintroductions is the insight and guidance this monitoring provides related to broader prairie dog (Cynomys spp.) ecosystem conservation. Black-footed ferrets are specialists of prairie dog colonies, depending on prairie dogs as their primary prey base and on the burrows and habitat structure created by these rodents. As keystone species of the North American Great Plains that have been eliminated in much of their historic range, prairie dogs are of high conservation concern in their own right. However, prairie dog monitoring is most intensive (and in some US states only occurs) at ferret reintroduction sites. This intensive monitoring has allowed for the relatively rapid identification of the occurrence and spread of plague across the western landscape and for documenting its impacts on prairie dog populations. By rapidly identifying the location and time of exposure to these epizootic die-offs of prairie dogs, management actions can be directed to attempt to mitigate plague. Similarly, monitoring for ferrets assists in identifying human-induced declines in prairie dogs due to poisoning and land conversion. Further, due to the relatively large extent and abundance of prairie dogs required by ferrets, monitoring for ferrets and locating where ferrets persist help identify when conservation has reached an appropriately large scale for achieving a restored prairie dog ecosystem (Jachowski 2014). In this sense, monitoring for black-footed ferrets has cascading ecosystem-level benefits that have advanced conservation planning and decision-making for the Great Plains (Box 12.1).

Monitoring, Management, and Science

Again, although many things could be monitored (Table 12.1), the prioritization of what to monitor and the allocation of effort to different monitoring components should be driven by underlying management and scientific objectives. Animal reintroductions require decisions at all phases of the program, starting with the decision to initiate a reintroduction effort. Good decision-making revolves around the identification and consideration of clear fundamental objectives (Gregory et al. 2012). In wildlife and fisheries management, science and information partly are a means to the end of producing defensible decisions, and monitoring should serve as a purposeful tool for improving subsequent decisions in that system (Conroy and Peterson 2013). Readers developing a reintroduction program should take advantage of other literature on natural resource decision-making and the role of monitoring in reducing uncertainty (e.g., Conroy and Peterson 2013), particularly literature

focused specifically on decision-making in reintroductions (e.g., McCarthy et al. 2012, Converse et al. 2013, Runge 2013; Chapters 6 and 7, this volume).

Our focus on management-relevant learning is not intended to dismiss the value of structuring monitoring and reintroduction-focused research proactively to advance broader scientific learning, particularly learning that may improve the effectiveness of future reintroduction efforts (Sarrazin and Barbault 1996, Ewen and Armstrong 2007, Seddon et al. 2007, Armstrong and Seddon 2008). More often than not, structuring monitoring of multiple reintroductions so as to answer broader questions may bring in additional resources and expertise that will benefit each individual monitoring effort while advancing reintroduction science. For example, because sylvatic plague is likely the leading biological factor currently limiting black-footed ferret recovery (Biggins et al. 2011), disease monitoring and multisite experiments are testing specific hypotheses regarding disease dynamics (Matchett et al. 2010) and assessing vaccines in a field setting (Abbott et al. 2012, Tripp et al. 2015). This monitoring, using fixed methodological protocols, illustrates the potential for targeted monitoring to be conducted simultaneously with research to further advance reintroduction science and broader conservation efforts (Ewen and Armstrong 2007, Nichols and Armstrong 2012). This example comes from one of the highest profile reintroduction programs in existence, but even in a smaller-scale monitoring program tailored specifically to a localized reintroduction effort, we strongly recommend careful thought about how that monitoring might contribute to broader reintroduction science and improve future reintroductions.

Still, regardless of the broader scientific objectives for a reintroduction research and monitoring program, it is critical that scientific experts involved in, and sometimes driving, the development of a monitoring plan appreciate the intended practical role of monitoring in management decision-making. Without close collaboration by scientists and managers, the value and sustainability of a reintroduction monitoring program are at risk. Addressing the technical aspects of monitoring should involve experts with appropriate expertise in study design, taxa-specific methodology, data analysis, and decision modeling. This expertise in framing and effectively answering questions should complement the expertise and serve the needs of the natural resource managers and decision makers responsible for the overall reintroduction program (Lindenmayer and Likens 2010, Gitzen et al. 2012).

Monitoring programs traditionally have underestimated the amount of planning and operational resources needed to take monitoring field data and turn it into permanently usable and promptly used information (Lindenmayer and Likens 2010). Yet, "data and information are the primary products of an ecological monitoring program" (Fancy and Bennetts 2012, 491). A priori planning and sufficient investment are needed for aspects of information management such as creating and maintaining monitoring databases with suitable metadata, archiving data backups, ensuring quality control, summarizing monitoring data for managers in a timely manner, and reporting on the information produced (Fancy et al. 2009). Again, partnerships with managers are essential for planning on how, how often, and when monitoring information is needed for it to be useful in management decision-making.

We are focusing on monitoring as a means of gathering information relevant to decision-making and science. Certainly monitoring can contribute to other potential fundamental objectives of a reintroduction program, such as increasing public understanding of the associated ecosystem, or to broader objectives such as increasing public appreciation for fish and wildlife resources. Citizen-science approaches can be a valuable option for contributing to these other objectives, but require careful thought and planning to consider costs and benefits of citizen-science monitoring versus other approaches (Box 12.4).

SURVEY DESIGN FOR MONITORING REINTRODUCTIONS

"How to monitor" is an extremely broad topic covering all aspects of fish and wildlife survey design and analysis. In this section, we highlight a few important aspects of study design for monitoring reintroductions. Biologists involved in reintroduction monitoring should refer to more in-depth resources on design of ecological monitoring, such as Thompson et al. (1998), de Gruijter et al. (2006), McComb et al. (2010), and Gitzen et al. (2012). Reintroductions are population management actions, and reintroduction monitoring typically should provide information on population size, occurrence, and demographic rates (Sutherland et al. 2010). Therefore, we provide somewhat more detail on general approaches for monitoring demographic aspects of reintroduced populations.

Core steps in transparent, quantitative survey design (Box 12.1) apply to all monitoring situations. The general steps involve understanding what information is needed for the monitoring program to be useful; defining potential sources, and magnitudes, of bias and variability relevant to estimating the parameters of interest; assessing what approaches, effort, and allocation of effort (e.g., number of radio-transmitter locations per animal vs. number of animals radio-tagged) can produce the desired information, and assessing trade-offs in allocating resources. For example, such trade-offs may involve assessing whether the program can afford to adequately monitor both demographic dynamics of the reintroduced population and broader ecosystem responses to the introduction. This may result in hard decisions about how to focus available resources. We strongly believe that the effort involved in quantitative study planning will always pay off sufficiently to make the investment in planning worthwhile, particularly by ensuring that monitoring has a good chance of producing useful information.

Even simplified analyses of mock data can be highly useful as a starting point for evaluating

Budget and time constraints always limit the amount of monitoring that can be conducted by paid staff in reintroduction programs. The basic task of detecting whether reintroduced individuals remain alive in an area may be a major challenge for programs dealing with highly elusive species (e.g., Box 12.3), unless the animals can be tagged with radio transmitters that have long battery lives and broad signal ranges. Even well-funded monitoring programs typically can be sustained only for a limited time and generally must transition into less intensive operational monitoring (see Missouri elk example). Enlisting the aid of nonscientists in reintroduction monitoring can increase the scope and intensity of data collection. In addition, incorporating citizens into research and monitoring helps increase public awareness and support of science, and increases the public's scientific knowledge (Brossard et al. 2005; Bonney et al. 2009).

The use of citizen science is not a new concept. For example, the Christmas Bird Count (CBC) and North American Breeding Bird Survey (BBS) have been used to monitor North American bird populations for decades and have contributed to hundreds of published papers and reports (Silvertown 2009). Recently, heightened interest in phenology with respect to climate change has led to an increase in solicitation of observations from the general public, with phenological monitoring networks established in countries throughout the world. Similarly, the growing discipline of invasion ecology has recognized the value of utilizing volunteers to help detect introduced species before they become too abundant to be easily eradicated, and monitor the magnitude and speed of range expansion in already established invasive species (Delaney et al. 2008, Crall et al. 2010). This can take the simple form of soliciting information regarding nonindigenous species sightings.

However, citizen science can go well beyond passive calls to report wildlife sightings. For example, over the last two decades technological advances have revolutionized our ability to enlist the public in scientific endeavors (Dickinson et al. 2012; Roy et al. 2012). Smartphones allow the simultaneous capture of photos and associated GPS location data. Web applications can be created and used on smartphones and tablets by citizen scientists to report data in real time from the field. Many electronic devices can be equipped with sensors that measure environmental factors such as air- and water-quality parameters. Remote cameras and recording units (e.g., for recording bat, frog, and bird calls) are easily deployable by nonscientists and can quickly accumulate almost overwhelmingly large files of images or sounds. The combination of such powerful technology with motivated volunteers should receive increasing use in reintroduction monitoring.

A reintroduction program can use citizen science to maintain or expand existing monitoring studies or to generate entire datasets for quantifying the success of reintroductions. For example, Rich et al. (2013) used hunter surveys to gather data on sightings of gray wolves (*Canis lupus*) in Montana. Data generated through public assistance facilitated the creation of wolf occupancy and abundance models. In a large-scale application focused on extant native populations of koala (*Phascolarctos cinereus*), Sequeira et al. (2014) utilized presence-absence data gathered from citizen scientists to develop models of distribution, population size, and habitat suitability. The study solicited help of nonprofessionals through use of local radio, television, and newspaper outlets, social media, and education materials prepared for schools, and by creating communication networks of participants involved in the project.

In terms of information produced by reintroduction monitoring, the greatest contributions of a citizen-science approach might be to increase nonsystematic observations for assessing presence of reintroduced animals and their offspring within the general reintroduction area (e.g., as in the indigo snake example, Box 12.3), and to enhance the spatial or temporal coverage and intensity of systematic monitoring. The feasibility of incorporating citizen scientists into a reintroduction program depends on the species and questions of interest and the level of training and expertise needed to collect usable data or samples. Also important is the availability of citizen participants in the reintroduction area relative to the amount of help needed for the effort to be useful (Hochachka et al. 2012). From a study design standpoint, the same principles applicable to any other monitoring program (Box 12.1) should guide development of citizen-science approaches. A frequent reality of citizen-science monitoring is the need to sacrifice some control over where and when observations are collected (Hochachka et al. 2012). As with any other potentially important source of bias or noise, the effects of this reduced control on the utility of a citizen-science approach need to be evaluated carefully. A citizen-science approach may increase the cost-effectiveness or feasibility of reintroduction monitoring, but there are significant costs in staff time needed for recruiting, training, and working with volunteers, ensuring data quality control, and analyzing resulting data. The most important issue to consider is whether a citizen-science approach is an effective and efficient option for assessing progress toward meeting well-defined reintroduction objectives, and for meeting additional objectives such as increasing public support for the reintroduction program.

what quality of information may be obtainable with various potential levels of effort. For example, consider a hypothetical manager planning to release 100 animals, and wanting to monitor first-year survival and habitat selection. The manager, eager to use cutting-edge technology, can afford to deploy 10 satellite transmitters assigned randomly to individuals before release, with locations every 1 hour possible to ensure over 1 year of battery life. Defensible survival analyses will have to deal with uncertainty about whether some animals have died or left the study area and will use patterns of mortality events within the monitoring period to estimate survival functions. However, in our contrived scenario we will assume no failure of transmitters, perfect accounting of the number of animals alive, and a simple "quick-and-dirty" analysis focusing on estimation of the proportion of the 100 released animals alive after 1 year. Using mock data and standard statistical software with functions for analysis of proportions (and ignoring the finite-population correction), the manager assesses potential uncertainty in this 1-year population-level survival estimate, assuming all information about survival comes from the sample of 10 focal animals. If 5 of 10 study animals are alive after 1 year, the estimated proportion of the 100 released animals still alive would be 0.5 (corresponding to an estimate of 50 animals still alive), with a conservative 95% confidence interval of approximately (0.19, 0.81). Whether such high imprecision is a problem depends on the expected use of the information, but regardless, it serves as reality check. If 10 of 10 sample animals are alive after 1 year the estimated survival rate would be 1.0, but the confidence interval associated with this estimate will still be wide at (0.69, 1.0), due to the small sample and small proportion of the population being monitored. If the manager instead deploys 50 less expensive, very high frequency (VHF) radio transmitters and observes 25 out of 50 study animals alive after a year, a conservative confidence interval for population-level survival would be (0.36, 0.64), indicating a much more precise estimate. A more appropriate calculation incorporating the fact that half of the entire population has been sampled would shrink the width of this confidence interval further. Depending on the species our manager is reintroducing and on other factors affecting feasibility, various other survival monitoring approaches with or without transmitters, could be considered (Table 12.2) in place of, or along with, intensive radiotracking. Our point is not to criticize the original plan of deploying 10 satellite transmitters. It may be an appropriate choice, but this needs to be decided with a realistic view of what quality of information (in terms of precision) is obtainable for each focal parameter.

Comparative Surveys and Experiments

Frequently, managers need information not simply on the status and dynamics of the population, but also on the comparative impact of different management practices considered for use in the

TABLE 12.2

Selected General Study Approaches of Highest Utility for Monitoring Abundance, Density, and Demographic Rates of Reintroduced Populations

All Approaches Other Than Indices Incorporate Detection Probability in the Underlying Statistical Models to Produce Estimates Adjusted for Incomplete and Variable Detectability

General Approach	Brief Description	Selected Examples and References*
Mark-Recapture (Individually or Batch Marked Individuals)		
Radiotelemetry	If all released animals are marked, radiotracking may provide a census of the population and direct measures of survival, at least until reproduction starts occurring. If transmitter failure or movement outside the study area occurs, or radio-tagged animals are treated as a marked sample from a larger population including unmarked animals, then mark-recapture/mark-resight models will be appropriate.	Missouri elk and indigo snake (Box 12.1, *this chapter* White and Shenk (2001), *Buner and Schaub (2008)*
Active marking and recapture of individuals	Traditional mark-recapture sampling allows estimation of population size, survival, population growth rate, etc. Spatial mark-recapture methods incorporate location-of-capture data to support direct estimation of density without traditional ad hoc assumptions necessary for converting abundance to density estimates. Multistate models can estimate state-specific detection probabilities and demographic parameters, and transitions among states (e.g., diseased vs. not diseased states; Cooch et al. 2012)	Williams et al. (2001), Royle et al. (2014), *Armstrong and Ewen (2002), Kauffman et al. (2003)*
Mark-recapture based on identification of individuals from genetics or morphology	Follows the framework of traditional mark-recapture sampling but using natural occurring individual "marks" such as genetic identity or body coloration pattern. Typically, there is an increased risk of false identifications compared to traditional active-marking approaches, and available analytical models address this challenge.	Lukacs and Burnham (2005), Lampa et al. (2013), *Stenglein et al. (2010), Tsaparis et al. (2014)*
Mark-resight approaches	A sample of animals is marked initially, then subsequent detections are obtained during systematic surveys (e.g., via direct observation or camera surveys) without necessarily recapturing marked individuals or continuing to mark new individuals. Traditional mark-resight models estimate abundance, sometimes with conversion of abundance to density estimates based on ad hoc assumptions. However, Sollmann et al. (2013) combine mark-resight and telemetry data in an explicit density-estimation model. As with mark-recapture estimation, mark-resight monitoring can be structured to allow estimation of survival and multistate dynamics.	McClintock (2011), *Seddon et al. (2003)*

(Continued)

TABLE 12.2
(Continued)

General Approach	Brief Description	Selected Examples and References*
Sightability models	A detection probability (e.g., aerial sightability) model is developed from surveys of a sample of radio-tagged individuals. During subsequent operational surveys, this model is used to adjust survey counts for incomplete detectability as a function of covariates such as group size, habitat attributes, etc. Operational surveys typically will deal largely with an unmarked population.	Missouri elk example (*this chapter*); Steinhorst and Samuel (1989), Fieberg (2012)

Approaches Not Involving Individually Recognizable Marks

Distance sampling	Based on sighting distances at points or along transects, an estimated detectability vs. distance function becomes the basis for density estimation.	Buckland et al. (2001), *Seddon et al. (2003), Refoyo et al. (2015)*
Removal models	Removal models focus on a closely timed sequence of repeated surveys of a site during which individuals captured on each survey either are physically removed (e.g., stream electrofishing) or recaptures (or other subsequent observations) of an individual are ignored, such that the raw data analyzed are the counts of unique, newly detected individuals in each survey. For example, for avian point counts, the survey may be divided into several short survey intervals, with observers keeping track of individual birds detected and only the first observation of each individual bird included for analysis.	Williams et al. (2001), *Ward et al. (2008)*
Multiple-observer surveys	Essentially an extension of mark-recapture sampling, two or more observers simultaneously conduct independent or non-independent surveys at a plot and thereafter classify each individual animal observed during the survey as to which observer(s) saw the individual.	Williams et al. (2001)
N-mixture models	During spatially and temporally replicated surveys, counts of unique (usually unmarked) individuals are collected via direct observation, camera trapping, etc. Abundance estimation and open-population modeling of demographic rates are feasible.	Royle (2004), Dail and Madsen (2011)
Indices	Index-based methods include any approach that does not directly measure or estimate the parameter of interest, but instead focuses on a measurement assumed to be related closely enough to the underlying parameter to allow indirect monitoring of that parameter. Indices most frequently are intended for examining changes in abundance across space and time. Commonly used indices are counts of individuals observed not adjusted for incomplete detectability, and counts of tracks, scat, etc., without a formal model for relating the indirect counts to underlying density or abundance.	Skalski et al. (2005), *Grenier et al. (2009)*

* Italicized references are examples focusing on reintroduction monitoring.

current or subsequent reintroductions, such as relative survival of wild-caught versus captive-bred animals, or on the impact of the reintroduction on the broader ecosystem. As with any individual ecological study, the potential strength and scope of direct statistical inference about causal relationships (e.g., such as whether a soft-release strategy leads to higher survival post-release than a hard-release approach) is a function of the degree of experimental control, randomization, and replication obtained in the study (Skalski and Robson 1992). Logistical constraints often determine which study approaches are feasible in any given monitoring program. Randomized designed experiments with replication offer the most direct evaluation of cause-effect relationships, but in the reintroduction context their feasibility is mostly limited to situations such as randomly assigning animals, prior to release, to either a hard- or soft-release strategy to allow direct comparison of the strategies (e.g., Mitchell et al. 2011). Even when randomization is feasible, it may not be possible to replicate each treatment independently (e.g., Tuberville et al. 2005). Most frequently, comparative studies may lack any control or ability to randomize, such as when comparing survival of males versus females or of wild-caught versus captive-propagated animals in a single release area. However, these studies, like any observational or experimental study, still need to consider and manage potential sources of bias (e.g., incomplete detection probability, as discussed subsequently) and imprecision to the extent feasible and appropriate (Box 12.1).

With the before-after control-impact (BACI) design (e.g., Skalski and Robson 1992), no experimental control is feasible, but replicate sites (or other sampling units) in, for example, a reintroduction area and nearby spatial control areas are defined and monitored before and after the reintroduction or other manipulation (see Box 12.3 for an example). The BACI approach often may be a feasible way of getting information about impacts of a reintroduction on other species or components of the ecosystem, but the design must be established and implemented far enough ahead of the reintroduction so that adequate prerelease data collection can occur. Other impact-assessment designs may lack spatial controls in measuring some ecosystem attribute or process before and after reintroduction (e.g., Mao et al. 2005), or lack temporal controls, such as comparing a reintroduction area and "control" area after reintroduction but with no baseline prerelease measurements that would help increase confidence that observed effects are due to the reintroduction rather than other factors such as preexisting site differences.

Demographic Monitoring Design

In studies of animal populations (and in many cases, studies of plants and other organisms), it is rarely feasible to obtain a complete census of the population size or measure the presence of the species in survey units without any risk of false absences. For example, even with 100% initial tagging of released individuals, inference about survival may involve uncertainty in discriminating among transmitter failure, emigration, and mortality. Even when a census of animals can be obtained initially, establishment of a self-sustaining population comes with increased uncertainty about demographic status and rates, such as when the proportion of initially marked animals in the population declines as the reintroduced population grows (e.g., Missouri elk reintroduction, discussed later in this chapter).

Therefore, reintroduction monitoring sooner or later will face incomplete detectability (i.e., less than 1.0 probability of detecting each individual actually present during sampling of the population or of observing presence of the species in truly occupied survey units). In some types of studies, false detections also may be an issue, such as when species misidentification leads to a species being recorded wrongly as present at a site. In disease monitoring, a non-zero probability of false absence (failing to detect disease when present in an individual) or false presence from a diagnostic test is a similar issue, but uncertainty in assessing disease state and intensity may include broader overestimation or underestimation of the level of disease in an infected animal (Cooch et al. 2012, Miller et al. 2012). If detection probabilities are less than 1.0, then counts of animals observed during surveys or data on proportion of patches occupied by the species will underestimate actual abundance or occupancy, and may bias estimates of demographic rates or relationships between population status and covariates such as habitat characteristics (Tyre et al. 2003). If detection probabilities vary systematically over time or space—e.g., if probability of detecting species presence in occupied survey units increases as population density increases in occupied units with time since reintroduction— then inference about changes in the population based on the raw data uncorrected for variable detectability will be subject to bias. This has led to major concerns about use of population indices for assessing population status and dynamics (see, e.g., Anderson 2001, Kendall and Moore 2012, Nichols and Armstrong 2012). In the disease monitoring context, individual detection probabilities may vary with disease status, adding further potential for bias when seeking to estimate disease prevalence in the population based on data from a sample of captured animals. In this context, variation in individual detection probability and uncertainty in assessing disease status need to be addressed with an appropriate study design (Cooch et al. 2012, Miller et al. 2012, Bailey et al. 2014; Table 12.2).

One of the most exciting developments in quantitative ecology during the last 15 years has been the enormous expansion of statistical models and study approaches that explicitly seek to account for incomplete and variable individual or species-level detectability in estimating attributes of population status and dynamics. See Table 12.2 for an overview of demographic approaches (excluding occupancy monitoring); Nichols and Armstrong (2012) provide a broader overview of some relevant techniques in the context of reintroduction monitoring; MacKenzie (2012) provides a chapter-length introduction

to demographic and occurrence monitoring in general. Of special note are models such as occupancy and N-mixture models for estimating population/occurrence status and dynamics using types of data (raw counts of animals without individual identification of animals, observation of animal sign data such as tracks) that traditionally were useful only as indices of population parameters.

Large-scale monitoring programs often now focus on survey designs for estimating patch occupancy and local colonization/extinction probabilities from binary data on species detections in survey patches (observed/not observed; MacKenzie et al. 2006), including indirect sign data (Stanley and Royle 2005). Such designs are transferable to the context of quantifying habitat selection at finer scales, can integrate questions related to species interactions at the scale of habitat selection or patch-level species co-occurrence (MacKenzie et al. 2006), accommodate potential false-positive detections (Bailey et al. 2014), and can be extended to assessing occurrence of multiple states of occurrence (e.g., absent in a surveyed patch, present but not infected with some disease of concern, present and infected; absent vs. present without breeding vs. present with breeding; Nichols and Armstrong 2012). A focus on patch-level occupancy and occupancy dynamics may be most relevant to monitoring programs examining broader changes of occurrence as a reintroduced population becomes established and spreads in extent, and intensive study of marked individuals no longer is feasible or needed, or is limited to focal portions of the expanding population. The general occupancy design actually may allow monitoring of abundance and demographic rates, rather than simply patch-level presence/absence, when raw counts of individuals are collected during each visit, ideally in spatially or temporally replicated surveys (i.e., with N-mixture models, Table 12.2; Royle 2004, Chandler et al. 2011, Dail and Madsen 2011).

Increasingly, it is feasible to collect data in a way that allows quantification of detectability at little additional cost compared to relying on population indices, as long as the study design has been structured to meet the primary assumptions and requirements of relevant statistical estimation models. Therefore, index-based monitoring is not usually a justifiable option in reintroduction programs seeking defensible information. However, this does not mean that index-based monitoring is never a justifiable option (e.g., Skalski et al. 2005, 359–433, Johnson 2008). The potential for bias inherent in index-based monitoring must be considered in light of the quality of information needed to meet monitoring objectives, how the realities of the situation constrain the quality of information that can be collected, and the trade-offs associated with attempting to reduce this potential bias (Box 12.1). For example, if the objectives of a reintroduction are to achieve a population of at least 200 individuals and managers are concerned primarily with overestimation rather than underestimation of population size with respect to this threshold, then raw counts of animals observed provide a conservative measure of whether the threshold population size has been reached.

A large-scale example of using indices for population monitoring comes from the reintroduction programs for black-footed ferrets (Box 12.2). A standardized spotlight monitoring protocol (traveling through prairie dog colonies at night and detecting ferrets, especially by eye shine) has been used across sites to generate an index (minimum population size) of abundance (Biggins et al. 2005). For these annual standardized surveys, variation among years and sites in monitoring expertise and intensity should lead to concern as to the extent that resulting indices of minimum population size reflect true abundance. Regardless, they still provide data on minimum population size, which is vital for assessing down-listing recovery criteria for the species under the Endangered Species Act (Box 12.2). Thus, such indices have been critically important not only to site managers, but also to broader state, national, and international policy implementation and resource prioritization as mandated under the Endangered Species Act.

Programs considering index-based monitoring should proceed cautiously. Indices should not be presented as more information-rich than is warranted. Despite their apparent simplicity, indices entail at least as many assumptions as those needed to make reliable inference from more formal abundance-estimation methods. Use of indices should be based on proper study design, such as implementing consistent sampling effort across space and time, and measuring covariates hypothesized to explain variability in detectability (Skalski et al. 2005). Index-based monitoring should be chosen with full knowledge of the potential existence of traditional and modern alternatives, both in terms of potential data-collection designs and in terms of analytical approaches. For example, although mark-recapture data from any single year or site may be too sparse to allow meaningful population estimation that accounts for incomplete detectability, modern analytical methods allow integrated analysis of sparse data across multiple sites or times to produce defensible estimates of population status and changes that account for variation in detectability (e.g., Grenier et al. 2009, Royle and Converse 2014).

In all demographic monitoring—and most ecological monitoring—the existence of well-defined estimation models and the development of a field design that matches the structure assumed by the estimation models are simply starting points for quantitative monitoring design, not a guarantee of useful information. For example, the lower and more heterogeneous the per-survey detection probability, the more difficult it is to obtain useful information, i.e., to obtain estimates precise enough to meet the objectives of the monitoring effort. Examinations of precision and cost functions are essential during design of monitoring programs, to assess whether feasible survey effort is likely to produce anything useful and to allocate this survey effort to maximize the value of the resulting information (Box 12.1). Analytical variance-precision frameworks (e.g., Skalski and Robson 1992, MacKenzie et al. 2006) and simulation-based approaches (such as in program MARK or with user-written code) can be adapted to specific needs of each individual monitoring effort.

Adaptability in Monitoring Design

The longer a monitoring effort continues, the more likely it will have to deal with changes in core objectives and questions, improvements in available techniques, changes in budgets, and increased knowledge of sources of variability and potential bias faced by the study. For example, long-term monitoring programs invariably must wrestle with the trade-offs involved in switching to a new and improved data-collection method at the risk of disrupting the long-term continuity of the data set. Thus, all longer-term monitoring programs face pressure to adapt and evolve (Lindenmayer and Likens 2009). Reintroduction monitoring has a fundamental link to management decision-making, and management information needs are likely to change significantly over the life of the monitoring program (e.g., as demonstrated with monitoring related to black-footed ferrets, Box 12.2). Therefore, management-focused components of reintroduction monitoring programs generally will be more dynamically adaptive than traditional surveillance monitoring programs. If a management-focused study is no longer addressing core sources of decision-relevant uncertainty, the study likely needs to adapt or die.

As the monitoring program initially is developed, there should be extensive discussion about what sorts of changes are most likely to be faced by the program in the foreseeable future, and what steps can be taken to maximize the adaptability of the program in light of anticipated changes. For example, initially the entire released population may be marked, possibly with radio-tags. However as the population grows and tags fail or are lost, a monitoring program will face the challenge of ensuring information collected from a sample of animals can support conclusions about the broader population. This may necessitate a change in techniques, and typically involves transitioning to higher-uncertainty, less-intensive monitoring of an expanding population. With careful planning, data collected during the initial monitoring phase may support development of an updated monitoring design that can be implemented as longer-term, frequently less intensive, operational monitoring commences. This strategy is demonstrated by the Missouri elk reintroduction discussed later in the chapter.

A primary objective of most reintroductions is to produce a biological population that begins sustaining itself and growing in numbers and spatial extent. Thus, the basic nature of the reintroduction context calls for survey designs that can be expanded as the population grows, or that can be contracted as priorities and budgets change. For example, in planning a monitoring program, the statistical target population (e.g., area or collection of individuals) of interest should be defined relative to the expected life of the monitoring program. If a monitoring program uses spatial sampling (e.g., to select a large sample of stream reaches for assessing occupancy of a reintroduced fish), it could define the target population to encompass all areas expected to be of interest as the biological population expands. This may lead to explicit implementation of spatial stratification (but see, e.g., Johnson 2012, 65–66, for caveats) and probability sampling approaches that will allow the program to expand or contract the sampling effort as the extent of the target population changes. Such adjustments may also be needed as budget changes or, for example, when reevaluation of the design after initial data collection indicates more effort is needed to obtain sufficient precision (Olsen et al. 2012).

How Much Should Be Invested in Monitoring?

In conservation decision-making, the applied value of monitoring and research is a function of whether the resulting information leads to better decisions. That is, the question is whether information is likely to increase the expected utility of the decision outcome, where utility is a function of how well a decision choice satisfies management objectives (Conroy and Peterson 2013, 217–220). Thus, it is not taken as a foregone conclusion that "more information is better." Information has a price, in that resources devoted to monitoring reintroductions are resources that could be devoted to release of additional animals, habitat improvement in the reintroduction site, other agency priorities, etc. How much to invest in monitoring is one of many decisions across which to optimize allocation of resources in a reintroduction program, which in turn is simply one component of a broader organization such as a management agency.

As in general ecological monitoring, the expected value of monitoring information may change over time. Once some threshold population has been reached and the population appears to be growing or stable, or beyond a certain size, intensive demographic monitoring may be scaled back, for example, turning into routine assessment of population size without explicit monitoring of survival or birthing rates, as in the Missouri elk case study. The choice of whether to continue investing in monitoring should be made actively (Possingham et al. 2012), ideally based on careful pre-reintroduction planning regarding what possible outcomes would lead to a cessation or modification of monitoring.

CASE STUDY: MONITORING THE REINTRODUCTION OF ELK TO THE MISSOURI OZARKS

Background

The eastern subspecies of elk (*Cervus elaphus canadensis*) once ranged throughout most of eastern North America including Missouri, but was extinct by the late 1800s, primarily due to overharvest and habitat conversion (Murie 1951,

O'Gara and Dundas 2002). Almost immediately following extirpation and extinction, state and federal agencies began attempts to restore populations of elk in eastern North America (Witmer 1990). The subspecies of elk most often used in these restoration efforts has been Rocky Mountain elk (*C. e. nelsoni*), believed to be a suitable substitute for the extinct subspecies given the relatively low genetic variation among the various subspecies of North American elk (Cronin 1992). Many of these early attempts at restoration failed due to poaching, poor translocation practices, and low initial population sizes (Witmer 1990). However, interest in elk restoration in eastern North America has continued, and in general, reintroduction attempts made after 1980 have had more success (Witmer 1990). There are currently free-ranging reintroduced elk populations in nine US central and eastern states and in Ontario, Canada.

In 2000, the Missouri Department of Conservation (MDC) began exploring the possibility of reintroducing elk to the state, prompted by requests from citizens interested in a free-ranging elk population. A feasibility study conducted in 2000 by MDC concluded that elk restoration was biologically feasible in the Missouri Ozark region. However, at that time MDC decided to delay elk restoration due to concerns over chronic wasting disease (CWD) and habitat issues.

By 2010, the continued success of elk reintroductions in Arkansas, Kentucky, and Michigan, among others, prompted a renewed interest in elk restoration, and encouraged the Missouri Conservation Commission to move forward with elk reintroduction in October 2010. A 346 mi² Elk Restoration Zone (ERZ) was selected in the Missouri Ozark region due to the area's relatively high proportion of public land, low road density, low population density, and low row-crop acreage. These attributes were likely to minimize the potential for human-elk conflict. Prior to elk release, habitat management occurred in the ERZ to increase the amount of open land available to elk. Elk were captured from the successfully restored Kentucky

FIGURE 12.1. Reintroduced elk (*Cervus elaphus*) in the Missouri Ozarks. Photo courtesy of the Missouri Department of Conservation.

population during the winters of 2010–2011 (34 elk), 2011–2012 (34 elk), and 2012–2013 (39 elk). After an initial quarantine period, elk were moved each spring to the Missouri reintroduction area, where they were confined in holding pens for an additional quarantine period and to facilitate a soft release (figure 12.1).

Monitoring of the Elk Reintroduction Serves Agency Management Objectives

Throughout this chapter, we have emphasized that well-defined reintroduction objectives are the essential framework for deciding what attributes to monitor. This recommendation is well illustrated by the Missouri elk reintroduction effort. The general reintroduction objectives were restoring a free-ranging population of elk to the Missouri Ozarks, providing recreational opportunities associated with the elk population including wildlife viewing and hunting, and minimizing human-elk conflicts. To help assess progress and guide decisions aimed at meeting these restoration objectives, MDC and the University of Missouri developed a comprehensive research and monitoring plan for elk restoration. This work, primarily funded by the US Fish and Wildlife Service Wildlife Restoration

grant program, outlined eight research/monitoring objectives relevant to the restoration objectives (Table 12.3). University researchers, in collaboration with agency biologists, have been assessing demography (survival and reproduction), the utility of aerial surveys, movements, disease issues (particularly meningeal worm or brainworm [*Parelaphostrongylus tenuis*]), resource use (diet composition and quality, resource selection), and physiological stress response. The overall goal of this research and monitoring project has been to build the foundation necessary for effective management of the elk population.

The basis of the monitoring scheme was the use of GPS radio-collars, which were placed on all elk translocated from Kentucky to Missouri. In the initial stage of the reintroduction, the GPS-radiotelemetry approach essentially provided a continuous census of current population size and high-resolution data for measuring other parameters of interest. These GPS collars were used during the spring to identify females that had given birth so that elk calves could be captured, as female elk isolate themselves and make very limited movements during parturition (Vore and Schmidt 2001). Implanting vaginal-implant transmitters (VITs)

TABLE 12.3

Research and Monitoring Objectives for the Missouri Elk Restoration Project, Purpose Relating to Management Goals, and Methods Used in Monitoring

Research/Monitoring Objective	How the Objective Relates to Restoration Goals	Methods of Monitoring
Determine survival and reproductive rates	Will be used to determine if the restoration is/was successful (whether a population is likely to grow and become established).	Monitoring of elk fitted with GPS radio-collars.
Develop population model for projecting population growth	Uses survival and reproductive rates from demographic monitoring. Will be used to determine if and when it is appropriate to initiate harvest of the population to provide hunting opportunities while ensuring the population restoration is successful.	Stochastic stage-based (Lefkovitch) matrix model.
Build sightability model for estimating elk abundance	Accurate elk abundance estimates are a necessary check on abundance forecasts from the population model. Abundance projections and estimates serve to determine if the elk restoration is successful and the population can provide hunting opportunities.	Logistic regression model of sighting probability based on aerial sightability survey of GPS-radio-collared elk, incorporating habitat attributes at locations of sighted and unsighted elk.
Evaluate habitat selection of elk	Modeling habitat use will help Missouri Department of Conservation (MDC) proactively manage habitat for elk to prevent human-elk conflict.	Monitoring of elk fitted with GPS radio-collars.
Determine diet composition and quality	Understanding diet composition and preference will help MDC proactively manage habitat to prevent human-elk conflict.	Microhistological analysis of elk feces. Measurement of nitrogen and 2,6 diaminopimelic acid concentrations in elk feces.
Evaluate extent of physiological stress in the population	Understanding role of physiological stress in the population will help determine likelihood of successful establishment of population.	Laboratory analysis of cortisol content of elk feces.
Evaluate effects of human disturbance on elk movements	Understanding response of elk to human disturbance will help MDC effectively manage elk habitat and prevent human-elk conflict.	Monitoring of elk fitted with GPS radio-collars.
Determine prevalence of parasitic meningeal worms in the Ozarks and potential role in elk mortality	Understanding extent of meningeal worm prevalence in the Ozark region and role of the parasite in elk mortality will help determine likelihood of success of restoring a free-ranging elk population.	Field necropsies of dead elk, collection of harvested white-tailed deer heads to determine adult parasite prevalence in definitive host, collection of white-tailed deer feces to determine first-stage larval presence in definitive host, and collection of terrestrial gastropods to determine third-stage larval presence in intermediate host.

in pregnant females prior to release was not an option because elk were not sedated during routine testing and handling in the Kentucky facilities. Instead, elk calves were fitted with expandable VHF radio-collars that allowed monitoring of their survival and cause-specific mortality.

The goals of monitoring survival and reproductive rates were twofold. First, this monitoring sought to identify causes of mortality, which could guide management to alleviate causes potentially detrimental to herd growth. Second, demographic rates are being used in an interactive population model that will allow managers to simulate herd population growth and harvest strategies. The monitoring effort established a goal of collaring every elk in the Missouri population. Although this was an ambitious goal, especially with regard to calves born in the field, the initial sample of animals included nearly the entire population. This resulted in little uncertainty about population demographics, survival, cause-specific mortality rates, and movements during the initial years following elk translocation. High certainty about the herd's status was important to ensure the continued support of stakeholders in the Missouri elk restoration.

Assessing habitat selection in the herd was important to provide information for future habitat management, to make predictions about areas on the landscape where elk are anticipated to move as the population grows, and to identify any potential areas of conflict between elk and people. In the initial period after elk were released, information regarding elk locations and movements was distributed to agency personnel on a daily basis. This information transfer allowed agency personnel to respond proactively to potential human-elk conflicts and served to alert agency personnel to the presence of wandering elk. Prior to elk reintroduction, agency protocols regarding elk damage response were established, detailing which agency personnel were to be contacted to file the damage complaint and outlining steps to be initiated to prevent and alleviate elk damage.

Impact of Monitoring on Elk Management

The Missouri elk reintroduction uses monitoring as a component of adaptive management in the broad sense of the term. Preliminary monitoring results have already been used by MDC to guide population management. Survival and reproduction data have been used to build a population model to predict growth of the Missouri elk herd and provide estimates to MDC of when the population is likely to reach benchmark demographic and population growth rates. The agency plans to use hunting as the primary means to manage elk population size and range. However, before hunting can be initiated, the agency needs to be confident that the population is growing and that other demographic benchmarks are being met (e.g., bull-to-cow ratios, calf production rates). Based on population modeling using current vital rates and those observed in other elk restorations in eastern North America, the elk population is expected to reach these benchmarks several years after restoration. Such projects may enable MDC to provide the public with an estimate of the first hunting season and develop a hunting program before this time. The decision on when to proceed with a hunting season may involve dealing with uncertainty about the actual population size and demographic parameters as time progresses from the initial reintroduction.

Preliminary data on habitat selection by elk in the reintroduced population suggest that open land is a key resource for the Missouri elk herd, prompting the recommendation that habitat management for elk in Missouri should continue to focus on open-land management. Research on the effect of human-related disturbance on elk movements has demonstrated that the presence of white-tailed deer hunters did not cause significant or long-term dispersal by elk (Bleisch 2014), and thus this activity is not expected to be problematic to elk.

Adapting Monitoring to Future Needs

The second phase of elk research will be more of an operational monitoring approach. As operational monitoring is developed, specific questions of primary interest may change in response to what has been learned in the initial phase, and monitoring will be more tightly focused on reducing key sources of uncertainty relevant to management. For example, some initial components (e.g., stress hormone monitoring, diet composition and quality) will be de-emphasized, while demographic monitoring will continue to be a priority. Future estimates of survival and reproduction will be used as a check on current model projections of herd growth and to update predictions as vital rates change.

As discussed earlier, all released animals were radio-collared, essentially providing a census of herd size and initial demographic rates. However, as the herd grows and includes a mix of animals with and without collars, observation uncertainty about demographic status and rates will increase. As a step toward building effective operational demographic monitoring, a key initial research objective has been the development of an aerial sightability model that will be used by MDC to monitor the elk population on a longer-term basis. This sightability model will allow accounting for incomplete detectability in annual winter monitoring of herd size. This population estimate will be compared to the population estimate calculated using the population growth model. Both population estimates can be used in conjunction with one another to determine the confidence surrounding the elk population estimate. For example, if these two estimates are similar, managers can be confident in the population size estimate and in setting harvest rates based on that population estimate. However, if the two estimates vary significantly, a more conservative harvest rate based on the lowest population size estimate may be most appropriate to prevent population declines. Thus, the combined use of a population model and monitoring data is a core element of a general adaptive management approach.

As the elk population expands, resource use and elk distribution are likely to change. Elk will become less naïve to the location of resources and may develop movement patterns and habitat use different from the initial years following translocation. Monitoring movements and resource use during the operational phase will provide insights into the level at which landscapes become saturated with elk and the extent of habitat management required to support elk on public land and elk-friendly private property. Further, should funding for habitat management be reduced as the reintroduction transitions from the restoration to established phase, it will be imperative to optimize management to gain the greatest benefit from available funds.

Results from initial years of the elk restoration indicated that open lands (food plots) were key resources selected by elk. However, publicly owned open land is a limited resource in the ERZ, and the amount of biomass on these open lands decreases throughout the winter. As the elk population increases, elk may move farther to seek out forage, especially during winter. Some elk may seek out private pastureland to meet their energy needs. Understanding how an expanding elk population seeks out and uses resource patches is key to managers' ability to minimize and mitigate elk and human conflicts. Monitoring elk will also allow MDC to evaluate the effectiveness of any mitigation techniques implemented on private land.

The Missouri elk reintroduction and monitoring effort was the result of careful planning designed to address the major challenges faced in monitoring development. Every research proposal developed with MDC must be approved through an iterative proposal review process. The core stakeholders that would make appropriate and immediate use of the monitoring data were the management biologists in the area of elk restoration. The university research team established a close relationship with the management biologists early on, as they were involved on the research proposal committee. The management biologists were in close contact with the researchers during every aspect of the monitoring program, from helping with on-the-ground monitoring to application of the results to habitat management. The elk-reintroduction monitoring was planned to maximize the utility and longevity of the information produced, and its contribution to the overall reintroduction effort.

RECOMMENDATIONS FOR FUTURE RESEARCH AND DEVELOPMENT

In natural resources management and science, there are multiple valid purposes for ecological monitoring (e.g., monitoring focused on specific management impacts vs. broader-scale monitoring of status and trends beyond the context of specific management interventions). Given the continued development of large-scale ecological monitoring programs (e.g., Fancy and Bennetts 2012), there is need for a rapid expansion of approaches for integrating more focused, perhaps shorter-term monitoring and management experiments into the framework of such broader-scale monitoring programs. The point of such efforts would be to use background monitoring in conjunction with more focused monitoring efforts to increase the power of the latter to answer the management-focused questions of concern. Such approaches are served well by ever-more-accessible hierarchical modeling and joint-likelihood-type approaches that quantitatively integrate multiple sources of data to address more questions than would be possible from separate analyses of each individual data set. Strategies for integrating designs and data will be of high relevance to reintroduction monitoring programs seeking to assess impacts of the reintroduction on other species and the broader ecosystem.

On a similar note, we hope to see increasing integration of monitoring across individual reintroduction efforts as future efforts are planned. When a species or group of similar species (e.g., ferrets, mustelids, mammalian carnivores) is being reintroduced at multiple locations over some period of years, each attempt can be seen as a replicate for meta-analyses of factors affecting reintroduction success of that species group and for examination of other questions that cannot be examined meaningfully in any single reintroduction (e.g., Jachowski et al. 2011a, McCarthy et al. 2012, 262). Broadscale analyses will be most useful when monitoring has been designed to have some consistent elements across the replicate reintroductions. In line with the recommendations of Ewen and Armstrong (2007), the power of monitoring and research to advance reintroduction science should increase tremendously as conservation scientists work to design integrated studies directly addressing important a priori questions and hypotheses.

More generally, as in all aspects of natural resource management, the role of monitoring in reintroduction management decision-making should become more routinely formalized in structured decision-making/adaptive management frameworks (Conroy and Peterson 2013, Runge 2013). Part of this formalization will be a greater emphasis on quantifying the expected value of monitoring information, instead of simply assuming that any information obtained through monitoring will have value. As we have repeatedly emphasized, decisions about monitoring are part of the larger suite of decisions managers face in a reintroduction program. It remains to be seen how quickly such programs increase their use of formal decision analysis rather than implicit professional judgment to optimize the choices made, and whether formal optimization increases the effectiveness of reintroduction programs.

In any case, the first step for increasing the utility of monitoring is to increase the percentage of reintroduction programs that develop quantifiable reintroduction objectives (Chapter 6 in this volume). Without such clear management objectives, there is only an amorphous basis for designing reintroduction monitoring. Addressing this problem will allow programs to more effectively apply the incredible array of monitoring tools and approaches already available, and will significantly increase the conservation value of reintroduction monitoring efforts.

SUMMARY

An effective monitoring program will help maximize progress of a reintroduction program toward meeting its underlying conservation goals. Monitoring of a species reintroduction should be a highly targeted effort designed to assess progress toward meeting reintroduction objectives, determine the need for additional management actions to improve the probability of success, and reduce uncertainty about the effects of future management actions in the focal system and in other reintroduction efforts. Included in the last purpose are iterative learning through formal adaptive management and research aimed to advance reintroduction science. Carefully specified reintroduction objectives are the basis for prioritizing exactly what attributes should be monitored and how accurate the information produced by monitoring needs for it to be to be useful. Quantitative monitoring objectives are the basis for determining the amount and allocation of effort needed to achieve a high probability that the monitoring effort will produce useful information. The monitoring survey design partly determines the quality of information obtained through monitoring, within the constraints imposed by the system being monitored. The degree and rate of learning from monitoring partly is driven by how directly monitoring information is linked back into management decision-making and the strength of inference that can be obtained from the monitoring study. Development of an effective reintroduction monitoring program requires careful planning and difficult decisions, such as how much to invest in monitoring versus other aspects of the reintroduction program. Therefore, the time to start discussing potential monitoring strategies is early in the development of the overall reintroduction program. This is also the time to begin thinking about how the monitoring program may need to adapt over time to changes in management information needs and to changes in the extent and size of the reintroduced population.

MANAGEMENT RECOMMENDATIONS

- Reintroductions are active, costly interventions, usually with highly uncertain outcomes. Investment in monitoring is usually essential for assessing whether the desired outcome is being achieved and guiding management to increase likelihood of a successful reintroduction. The ability to increase the effectiveness of future reintroductions depends partly on having data for assessing factors affecting the success of past reintroductions.

- The first step in designing a monitoring program is to make sure that there are clearly defined objectives for the overall reintroduction program. In wildlife and fisheries management, science and information partly are a means to the end of producing defensible decisions, and monitoring should be seen as a purposeful tool for improving subsequent decisions in that system (Conroy and Peterson 2013). Good decision-making revolves around the identification and consideration of clear fundamental objectives (Gregory et al. 2012). Management objectives are the essential foundation for designing a reintroduction monitoring program.

- If a reintroduction monitoring program is intended to be relevant to management decisions, then managers should understand that they need to play a primary role in guiding the development of the program. Frequently this will involve close collaboration between scientific experts and managers as the program is developed and implemented.

- If data management, analysis, and reporting are treated as afterthoughts to worry about after data are collected, the monitoring program likely will fail. These aspects require careful planning and investment so that data collected by monitoring will promptly be turned into information useful to management and science.

- Monitoring is intended to produce information valuable enough to justify the programmatic resources spent getting that information. The decision to spend money collecting data does not guarantee that useful information will be obtained. The statistical design of a useful, efficient monitoring program requires careful thought not just about what to monitor, but also about the quality of information needed from the monitoring program, in terms of acceptable precision and potential for bias. Monitoring programs generally need to make the hard choice to monitor one or a few things well, rather than spreading monitoring resources too thin to get meaningful information on anything.

LITERATURE CITED

Abbott, R. C., J. E. Osorio, C. M. Bunck, and T. E. Rocke. 2012. Sylvatic plague vaccine: a new tool for conservation of threatened and endangered species. EcoHealth 9:243–250.

Aiello, C. M., K. E. Nussear, A. D. Walde, T. C. Esque, P. G. Emblidge, P. Sah, S. Bansal, and P. J. Hudson. 2014. Disease dynamics during wildlife translocations: disruptions to the host population and potential consequences for transmission in desert tortoise contact. Animal Conservation 17(Suppl. 1):27–39.

Almberg, E. S., P. C. Cross, A. P. Dobson, D. W. Smith, and P. J. Hudson. 2012. Parasite invasion following host reintroduction: a case study of Yellowstone's wolves. Philosophical Transactions of the Royal Society B: Biological Sciences 367:2840–2851.

Andersen, A., D. J. Simcox, J. A. Thomas, and D. R. Nash. 2014. Assessing reintroduction schemes by comparing genetic diversity of reintroduced and source populations: a case study of the globally threatened large blue butterfly (*Maculinea arion*). Biological Conservation 175:34–41.

Anderson, D. R. 2001. The need to get the basics right in wildlife field studies. Wildlife Society Bulletin 29:1294–1297.

Armstrong, D. P., R. S. Davidson, J. K. Perrott, J. Roygard, and Buchanan. 2005. Density-dependent population growth in a reintroduced population of North Island saddlebacks. Journal of Animal Ecology 74:160–170.

Armstrong, D. P., and J. G. Ewen. 2002. Dynamics and viability of a New Zealand robin population reintroduced to regenerating fragmented habitat. Conservation Biology 16:1074–1085.

Armstrong, D. P., N. McArthur, S. Govella, K. Morgan, R. Johnston, N. Gorman, R. Pike, and Y. Richard. 2013. Using radio-tracking data to predict post-release establishment in reintroductions to habitat fragments. Biological Conservation 168:152–160.

Armstrong, D. P., and P. J. Seddon. 2008. Directions in reintroduction biology. Trends in Ecology and Evolution 23:20–25.

Bailey, L. L., D. I. MacKenzie, and J. D. Nichols. 2014. Advances and applications of occupancy models. Methods in Ecology and Evolution 5:1269–1279.

Bennett, V. A., V. A. Doerr, E. D. Doerr, A. D. Manning, D. B. Lindenmayer, and H. J. Yoon. 2012. Habitat selection and post-release movement of reintroduced brown treecreeper individuals in restored temperate woodland. PLoS ONE 7:e50612.

Benson, J. F., and M. J. Chamberlain. 2007. Space use, survival, movements, and reproduction of reintroduced Louisiana black bears. Journal of Wildlife Management 71:2393–2403.

Berger-Tal, O., and D. Saltz. 2014. Using the movement patterns of reintroduced animals to improve reintroduction success. Current Zoology 60:515–526.

Beyer, H. L., E. H. Merrill, N. Varley, and M. S. Boyce. 2007. Willow on Yellowstone's northern range: evidence for a trophic cascade? Ecological Applications 17:1563–1571.

Biggins, D. E., J. L. Godbey, M. R. Matchett, L. R. Hanebury, T. M. Livieri, and P. Marinari. 2005. Monitoring black-footed ferrets during reestablishment of free-ranging populations: discussion of alternative methods and recommended minimum standards. Ppp.155–174. In J. E. Roelle, B. J. Miller, J. L. Godbey, and D. E. Biggins, eds., Recovery of the Black-Footed Ferret: Progress and Continuing Challenges. Proceedings of the Symposium on the Status of the Black-Dooted Ferret and Its Habitat. Scientific Investigations Report 2005–5293. US Geological Survey, Fort Collins, CO.

Biggins, D. E., T. M. Livieri, and S. W. Breck. 2011. Interface between black-footed ferret research and operational conservation. Journal of Mammalogy 92:699–704.

Bleisch, A. 2014. Initial movements and disturbance response of a newly reintroduced elk herd in the Missouri Ozarks. Master's thesis. University of Missouri, Columbia.

Bonney, R., C. B. Cooper, J. Dickinson, S. Kelling, T. Phillips, K. V. Rosenberg, and J. Shirk. 2009. Citizen science: a developing tool for expanding science knowledge and scientific literacy. BioScience 59:977–984.

Brekke, P., P. M. Bennett, J. Wang, N. Pettorelli, and J. G. Ewen. 2010. Sensitive males: inbreeding depression in an endangered bird. Proceedings of the Royal Society B: Biological Sciences 277:3677–3684.

Brossard, D., B. Lewenstein, and R. Bonney. 2005. Scientific knowledge and attitude change: The impact of a citizen science project. International Journal of Science Education 27:99–1121.

Buckland, S. T., D. R. Anderson, K. P. Burnham, J. L. Laake, D. L. Borchers, and L. Thomas. 2001. Introduction to Distance Sampling. Oxford University Press, Oxford, UK.

Buner, F., and M. Schaub. 2008. How do different releasing techniques affect the survival of reintroduced grey partridges *Perdix perdix*? Wildlife Biology 14:26–35.

Carpenter, J. W., M. J. G. Appel, R. C. Erickson, and M. N. Novilla. 1976. Fatal vaccine-induced canine distemper virus infection in black-footed ferrets. Journal of the American Veterinary Medical Association 169:961–964.

Castro, I., K. M. Mason, D. P. Armstrong, and D. M. Lambert. 2004. Effect of extra-pair paternity on effective population size in a reintroduced population of the endangered hihi, and potential for behavioural management. Conservation Genetics 5:381–393.

Chandler, R. B., J. A. Royle, and D. I. King. 2011. Inference about density and temporary emigration in unmarked populations. Ecology 92:1429–1435.

Chipault, J. G., D. E. Biggins, J. K. Detling, D. H. Long, and R. M. Reich. 2012. Fine-scale habitat use of reintroduced black-footed ferrets on prairie dog colonies in New Mexico. Western North American Naturalist 72:216–227.

Close, D. A., K. P. Currens, A. Jackson, A. J. Wildbill, J. Hansen, P. Bronson, and K. Aronsuu. 2009. Lessons from the reintroduction of a noncharismatic, migratory fish: Pacific lamprey in the upper Umatilla River, Oregon. Pp.233–253. In L. R. Brown, S. D. Chase, M. G. Mesa, R. J. Beamish, and P. B. Moyle, eds., Biology, Management, and Conservation of Lampreys in North America. American Fisheries Society Symposium 72, Bethesda, MD.

Conroy, M. J., and J. T. Peterson. 2013. Decision Making in Natural Resource Management: A Structured Adaptive Approach. Wiley-Blackwell, West Sussex, UK.

Converse, S. J., C. T. Moore, and D. P. Armstrong. 2013. Demographics of reintroduced populations: estimation, modeling, and decision analysis. Journal of Wildlife Management 77:1081–1093.

Cooch, E. G., P. B. Conn, S. P. Ellner, A. P. Dodson, and K. H. Pollock. 2012. Disease dynamics in wild populations: modeling and estimation: a review. Journal of Ornithology 152:S485–S509.

Crall, A. W., G. J. Newman, C. S. Jarnevich, T. J. Stohlgren, D. M. Waller, and J. Graham. 2010. Improving and integrating data on invasive species collected by citizen scientists. Biological Invasions 12:3419–3428.

Cronin, M. A. 1992. Intraspecific variation in mitochondrial DNA of North American cervids. Journal of Mammalogy 73:70–82.

Dail, D., and L. Madsen. 2011. Models for estimating abundance from repeated counts of an open metapopulation. Biometrics 67:577–587.

De Barba, M., L. P. Waits, E. O. Garton, P. Genovesi, E. Randi, A. Mustoni, and C. Groff. 2010. The power of genetic monitoring for studying demography, ecology and genetics of a reintroduced brown bear population. Molecular Ecology 19:3938–3951.

de Gruijter, J., D. Brus, M. F. P. Bierkens, and M. Knotters. 2006. Sampling for Natural Resource Monitoring. Springer, Dordrecht, the Netherlands.

Delaney, D. G., C. D. Sperling, C. S. Adams, and B. Leung. 2008. Marine invasive species: validation of citizen science and implications for national monitoring networks. Biological Invasions 10:117–128.

Deredec, A., and F. Courchamp. 2007. Importance of the Allee effect for reintroductions. Ecoscience 14:440–451.

Devineau, O., T. M. Shenk, G. C. White, P. F. Doherty, Jr., P. M. Lukacs, and R. H. Kahn. 2010. Evaluating the Canada lynx reintroduction programme in Colorado: patterns in mortality. Journal of Applied Ecology 47:524–531.

Dickens, M. J., D. J. Delehanty, and L. M. Romero. 2009. Stress and translocation: alterations in the stress physiology of translocated birds. Proceedings of the Royal Society B: Biological Sciences 276:2051–2056.

Dickinson, J. L., J. Shirk, D. Bonter, R. Bonney, R. L. Crain, J. Martin, T. Phillips, and K. Purcell. 2012.

The current state of citizen science as a tool for ecological research and public engagement. Frontiers in Ecology and the Environment 10:291–297.

Dolev, A., D. Saltz, S. Bar-David, and Y. Yom-Tov. 2002. Impact of repeated releases on space-use patterns of Persian fallow deer. Journal of Wildlife Management 66:737–746.

Dunham, K. M. 1997. Population growth of mountain gazelles *Gazella gazella* reintroduced to central Arabia. Biological Conservation 81:205–214.

Dunham, K. M. 1998. Spatial organization of mountain gazelles *Gazella gazella* reintroduced to central Arabia. Journal of Zoology 245:371–384.

Eads, D. A., J. J. Millspaugh, D. E. Biggins, T. M. Livieri, and D. S. Jachowski. 2011. Postbreeding resource selection by adult black-footed ferrets in the Conata Basin, South Dakota. Journal of Mammalogy 92:760–770.

Ewen, J. G., and D. P. Armstrong. 2007. Strategic monitoring of reintroductions in ecological restoration programmes. Ecoscience 14:401–409.

Fancy, S. G., and R. E. Bennetts. 2012. Institutionalizing an effective long-term monitoring program in the U. S. National Park Service. Pp. 481–497. In R. A. Gitzen, J. J. Millspaugh, A. B. Cooper, and D. S. Licht, eds., Design and Analysis of Long-term Ecological Monitoring Studies. Cambridge University Press, Cambridge, UK.

Fancy, S. G., J. E. Gross, and S. L. Carter. 2009. Monitoring the condition of natural resources in U. S. National Parks. Environmental Monitoring and Assessment 151:161–174.

Fieberg, J. R. 2012. Estimating population abundance using sightability models: R Sightability-Model package. Journal of Statistical Software 51:1–20.

Fischer, J., and D. B. Lindenmayer. 2000. An assessment of the published results of animal relocations. Biological Conservation 96:1–11.

Fortin, D., H. L. Beyer, M. S. Boyce, D. W. Smith, T. Duchesne, and J. S. Mao. 2005. Wolves influence elk movements: behavior shapes a trophic cascade in Yellowstone National Park. Ecology 86:1320–1330.

Frair, J. L., E. H. Merrill, J. R. Allen, and M. S. Boyce. 2007. Know thy enemy: experience affects elk translocation success in risky landscapes. Journal of Wildlife Management 71:541–554.

Franceschini, M. D., D. I. Rubenstein, B. Low, and L. M. Romero. 2008. Fecal glucocorticoid metabolite analysis as an indicator of stress during translocation and acclimation in an endangered large mammal, the Grevy's zebra. Animal Conservation 11:263–269.

Frankham, R. 2008. Genetic adaptation to captivity in species conservation programs. Molecular Ecology 17:325–333.

Fredrickson, R. J., P. Siminski, M. Woolf, and P. W. Hedrick. 2007. Genetic rescue and inbreeding depression in Mexican wolves. Proceedings of the Royal Society B: Biological Sciences 274:2365–2371.

Gelling, M., I. Montes, T. P. Moorhouse, and D. W. Macdonald. 2010. Captive housing during water vole (*Arvicola terrestris*) reintroduction: does short-term social stress impact on animal welfare? PLoS ONE 5:e9791.

Gelling, M., P. J. Johnson, T. P. Moorhouse, and D. W. Macdonald. 2012. Measuring animal welfare within a reintroduction: an assessment of different indices of stress in water voles *Arvicola amphibius*. PLoS ONE 7:e41081.

Gibbs, J. P., C. Marquez, and E. J. Sterling. 2008. The role of endangered species reintroduction in ecosystem restoration: tortoise–cactus interactions on Española Island, Galápagos. Restoration Ecology 16:88–93.

Gitzen, R. A., J. J. Millspaugh, A. B. Cooper, and D. S. Licht, eds. 2012. Design and Analysis of Long-Term Ecological Monitoring Studies. Cambridge University Press, Cambridge, UK.

Goodman, G., S. Girling, R. Pizzi, A. Meredith, F. Rosell, and R. Campbell-Palmer. 2012. Establishment of a health surveillance program for reintroduction of the Eurasian beaver (*Castor fiber*) into Scotland. Journal of Wildlife Diseases 48:971–978.

Gorman, M. L. 2007. Restoring ecological balance to the British mammal fauna. Mammal Review 37:316–325.

Grassel, S. M., J. L. Rachlow, and C. J. Williams. 2015. Spatial interactions between sympatric carnivores: asymmetric avoidance of an intraguild predator. Ecology and Evolution 5:2762–2773.

Gregory, R., L. Failing, M. Harstone, G. Long, T. McDaniels, and D. Ohlson. 2012. Structured Decision Making: A Practical Guide to Environmental Management Choices. Wiley-Blackwell, Chichester, West Sussex, UK.

Grenier, M. B., S. W. Buskirk, and R. Anderson-Sprecher. 2009. Population indices versus correlated density estimates of black-footed ferret abundance. Journal of Wildlife Management 73:669–676.

Hardman, B., and D. Moro. 2006. Optimising reintroduction success by delayed dispersal: Is the release protocol important for hare-wallabies? Biological Conservation 128:403–411.

Harris, N.C., T.M. Livieri, and R.R. Dunn. 2014. Ectoparasites in black-footed ferrets (*Mustela nigripes*) from the largest reintroduced population of the Conata Basin, South Dakota, USA. Journal of Wildlife Diseases 50:340–343.

Haydon, D.T., J.M. Morales, A. Yott, D.A. Jenkins, R. Rosatte, and J.M. Fryxell. 2008. Socially informed random walks: incorporating group dynamics into models of population spread and growth. Proceedings of the Royal Society B: Biological Sciences 275:1101–1109.

Hester, J.M., S.J. Price, and M.E. Dorcas. 2008. Effects of relocation on movements and home ranges of eastern box turtles. Journal of Wildlife Management 72:772–777.

Hirzel, A.H., B. Posse, P.A. Oggier, Y. Crettenand, C. Glenz, and R. Arlettaz. 2004. Ecological requirements of reintroduced species and the implications for release policy: the case of the bearded vulture. Journal of Applied Ecology 41:1103–1116.

Hochachka, W.M., D. Fink, and B. Zuckerberg. 2012. Use of citizen-science monitoring for pattern discovery and biological inference. Pp.460–477. In R.A. Gitzen, J.J. Millspaugh, A.B. Cooper, and D.S. Licht, eds., Design and Analysis of Long-Term Ecological Monitoring Studies. Cambridge University Press, Cambridge, UK.

Howard, J.G., Howard, P.E. Marinari, and D.E. Wildt, 2003. Black-footed ferret: model for assisted reproductive technologies contributing to in situ conservation. Pp.249–266. In W.V. Holt, A.R. Pickard, J.C. Rodger, and D.E. Wildt, eds., Reproductive Sciences and Integrated Conservation. Cambridge University Press, Cambridge, UK.

IUCN. 1998. Guidelines for Re-introductions. Prepared by the IUCN/SSC Reintroduction Specialist Group. IUCN, Gland, Switzerland.

Jachowski, D.S. 2014. Wild Again: The Struggle to Save the Black-footed Ferret. University of California Press, Berkeley.

Jachowski, D.S., R.A. Gitzen, S. Grassel, M.B. Grenier, B. Holmes, and J.J. Millspaugh. 2011a. Evaluating attempts to reintroduce black-footed ferrets to North America. Biological Conservation 144:1560–1566.

Jachowski, D.S., and J.M. Lockhart. 2009. Reintroducing the black-footed ferret *Mustela nigripes* to the Great Plains of North America. Small Carnivore Conservation 41:58–64.

Jachowski, D.S., J.J. Millspaugh, D.E. Biggins, T.M. Livieri, M.R. Matchett, and C.D. Rittenhouse. 2011b. Resource selection by black-footed ferrets in South Dakota and Montana. Natural Areas Journal 31:218–225.

James, A.I., and D.J. Eldridge. 2007. Reintroduction of fossorial native mammals and potential impacts on ecosystem processes in an Australian desert landscape. Biological Conservation 138:351–359.

Jamieson, I.G. 2011. Founder effects, inbreeding, and loss of genetic diversity in four avian reintroduction programs. Conservation Biology 25:115–123.

Jiménez, J.A., K.A. Hughes, G. Alaks, L. Graham, and R.C. Lacy. 1994. An experimental study of inbreeding depression in a natural habitat. Science 266:271–273.

Johnson, D.H. 2008. In defense of indices: the case of bird surveys. Journal of Wildlife Management 72:857–868.

Johnson, D.H. 2012. Monitoring that matters. Pp.54–73. In R.A. Gitzen, J.J. Millspaugh, A.B. Cooper, and D.S. Licht, eds., Design and Analysis of Long-Term Ecological Monitoring Studies. Cambridge University Press, Cambridge, UK.

Kauffman, M.J., J.F. Brodie, and E.S. Jules. 2010. Are wolves saving Yellowstone's aspen? A landscape-level test of a behaviorally mediated trophic cascade. Ecology 91:2742–2755.

Kauffman, M.J., W.F. Frick, and J. Linthicum. 2003. Estimation of habitat-specific demography and population growth for peregrine falcons in California. Ecological Applications 13:1802–1816.

Kemp, P.S., T.A. Worthington, T.E. Langford, A.R. Tree, and M.J. Gaywood. 2012. Qualitative and quantitative effects of reintroduced beavers on stream fish. Fish and Fisheries 13:158–181.

Kendall, W.L., and C.T. Moore. 2012. Maximizing the utility of monitoring to the adaptive management of natural resources. Pp.74–98. In R.A. Gitzen, J.J. Millspaugh, A.B. Cooper, and D.S. Licht, eds., Design and Analysis of Long-term Ecological Monitoring Studies. Cambridge University Press, Cambridge, UK.

King, R., C. Berg, and B. Hay. 2004. A repatriation study of the eastern massasauga (*Sistrurus catenatus catenatus*) in Wisconsin. Herpetologica 60:429–437.

Kramer-Schadt, S., E. Revilla, and T. Wiegand. 2005. Lynx reintroductions in fragmented landscapes of Germany: projects with a future or misunderstood wildlife conservation? Biological Conservation 125:169–182.

Lampa, S., K. Henle, R. Klenke, M. Hoehn, and B. Gruber. 2013. How to overcome genotyping errors in non-invasive genetic mark–recapture population size estimation: a review of available methods illustrated by a case study. Journal of Wildlife Management 77:1490–1511.

Lindenmayer, D. B., and G. E. Likens. 2009. Adaptive monitoring: a new paradigm for long-term research and monitoring. Trends in Ecology and Evolution 24:482–486.

Lindenmayer, D. B., and G. E. Likens. 2010. Effective Ecological Monitoring. CSIRO Publishing, Melbourne/Earthscan, London, UK.

Lukacs, P. M., and K. P. Burnham. 2005. Review of capture-recapture methods applicable to noninvasive genetic sampling. Molecular Ecology 14:3909–3919.

MacKenzie, D. I. 2012. Study design and analysis options for demographic and species occurrence dynamics. Pp.397–425. In R. A. Gitzen, J. J. Millspaugh, A. B. Cooper, and D. S. Licht, eds., Design and Analysis of Long-term Ecological Monitoring Studies. Cambridge University Press, Cambridge, UK.

MacKenzie, D. I., J. D. Nichols, J. A. Royle, K. H. Pollock, L. L. Bailey, and J. E. Hines. 2006. Occupancy Estimation and Modeling: Inferring Patterns and Dynamics of Species Occurrence. Academic Press, San Diego, CA.

Mao, J. S., M. S. Boyce, D. W. Smith, F. J. Singer, D. J. Vales, J. M. Vore, and E. H. Merrill. 2005. Habitat selection by elk before and after wolf reintroduction in Yellowstone National Park. Journal of Wildlife Management 69:1691–1707.

Marshall, T. C., and J. A. Spalton. 2000. Simultaneous inbreeding and outbreeding depression in reintroduced Arabian oryx. Animal Conservation 3:241–248.

Matchett, M. R., D. E. Biggins, V. Carlson, B. Powell, and T. Rocke. 2010. Enzootic plague reduces black-footed ferret (Mustela nigripes) survival in Montana. Vector-Borne and Zoonotic Diseases 10:27–35.

Mathews, F., D. Moro, R. Strachan, M. Gelling, and N. Buller. 2006. Health surveillance in wildlife reintroductions. Biological Conservation 131:338–347.

McCarthy, M. A., D. P. Armstrong, and M. C. Runge. 2012. Adaptive management of reintroduction. Pp.256–289. In J. G. Ewen, D. P. Armstrong, K. A. Parker, and P. J. Seddon, eds., Reintroduction Biology: Integrating Science and Management. Wiley-Blackwell, West Sussex, UK.

McClintock, B. 2011. Mark-resight models. Pp.1–34. In E. G. Cooch and G. C. White, eds., A Gentle Introduction to Program MARK. Ithaca, New York. Available online at: http://www.phidot.org/software/mark/docs/book.

McComb, B., B. Zuckerberg, D. Vesely, and C. Jordan. 2010. Monitoring Animal Populations and Their Habitats: A Practitioner's Guide. CRC Press, Boca Raton, FL.

McDonald, T. 2012. Spatial sampling designs for long-term ecological monitoring. Pp.101–125. In R. A. Gitzen, J. J. Millspaugh, A. B. Cooper, and D. S. Licht, eds., Design and Analysis of Long-Term Ecological Monitoring Studies. Cambridge University Press, Cambridge, UK.

McKinstry, M. C., and S. H. Anderson. 2002. Survival, fates, and success of transplanted beavers (Castor canadensis) in Wyoming. Journal of Wildlife Rehabilitation 116:60–68.

Mihoub, J. B., P. Le Gouar, and F. Sarrazin. 2009. Breeding habitat selection behaviors in heterogeneous environments: implications for modeling reintroduction. Oikos 118:663–674.

Mihoub, J. B., A. Robert, P. Le Gouar, and F. Sarrazin. 2011. Post-release dispersal in animal translocations: Social attraction and the "vacuum effect." PLoS ONE 6:e27453.

Miller, D. A., B. L. Talley, K. R. Lips, and E. H. Campbell Grant. 2012. Estimating patterns and drivers of infection prevalence and intensity when detection is imperfect and sampling error occurs. Methods in Ecology and Evolution 3:850–859.

Miller, K. A., N. J. Nelson, H. G. Smith, and J. A. Moore. 2009. How do reproductive skew and founder group size affect genetic diversity in reintroduced populations? Molecular Ecology 18:3792–3802.

Mitchell, A. M., T. I. Wellicome, D. Brodie, and K. M. Cheng. 2011. Captive-reared burrowing owls show higher site-affinity, survival, and reproductive performance when reintroduced using a soft-release. Biological Conservation 144:1382–1391.

Moravec, F. 2003. Observations on the metazoan parasites of the Atlantic salmon (Salmo salar) after its reintroduction into the Elbe River basin in the Czech Republic. Folia Parasitologica 50:298–304.

Mowry, R. A., T. M. Schneider, E. K. Latch, M. E. Gompper, J. Beringer, and L. S. Eggert. 2015. Genetics and the successful reintroduction of the Missouri river otter. Animal Conservation 18:196–206.

Murie, O. L. 1951. The Elk of North America. Stackpole Co., Harrisburg, PA.

Muths, E., and V. Dreitz. 2008. Monitoring programs to assess reintroduction efforts: a critical component in recovery. Animal Biodiversity and Conservation 31.1:47–56.

Nichols, J. D., and D. P. Armstrong. 2012. Monitoring for reintroductions. Pp.223–254. In J. G. Ewen, D. P. Armstrong, K. A. Parker, and P. J. Seddon, eds., Reintroduction Biology: Integrating Science and Management. Wiley-Blackwell, West Sussex, UK.

Nichols, J. D., and B. K. Williams. 2006. Monitoring for conservation. Trends in Ecology and Evolution 21:668–673.

O'Gara, B. W., and R. G. Dundas. 2002. Distribution: past and present. Pp.67–119. In D. E. Toweill and J. W. Thomas, eds., North American Elk: Ecology and Management. Smithsonian Institute Press, Washington, DC.

Olsen, A. R., T. M. Kincaid, and Q. Payton. 2012. Spatially balanced survey designs for natural resources. Pp.126–150. In R. A. Gitzen, J. J. Millspaugh, A. B. Cooper, and D. S. Licht, eds., Design and Analysis of Long-term Ecological Monitoring Studies. Cambridge University Press, Cambridge, UK.

Perelberg, A., D. Saltz, S. Bar-David, A. Dolev, and Y. Yom-Tov. 2003. Seasonal and circadian changes in the home ranges of reintroduced Persian fallow deer. Journal of Wildlife Management 67:485–492.

Phillips, M. K., and J. Scheck. 1991. Parasitism in captive and reintroduced red wolves. Journal of Wildlife Diseases 27:498–501.

Possingham, H. P., R. A. Fuller, and L. N. Joseph. 2012. Choosing among long-term ecological monitoring programs and knowing when to stop. Pp.498–508. In R. A. Gitzen, J. J. Millspaugh, A. B. Cooper, and D. S. Licht, eds., Design and Analysis of Long-term Ecological Monitoring Studies. Cambridge University Press, Cambridge, UK.

Preatoni, D., A. Mustoni, A. Martinoli, E. Carlini, B. Chiarenzi, S. Chiozzini, S. Van Dongen, L. A. Wauters, and G. Tosi. 2005. Conservation of brown bear in the Alps: space use and settlement behavior of reintroduced bears. Acta Oecologica 28:189–197.

Refoyo, P., C. Olmedo, I. Polo, P. Fandos, and B. Muñoz. 2015. Demographic trends of a re-introduced Iberian ibex Capra pyrenaica victoriae population in central Spain. Mammalia 79:139–146.

Reynolds, J. H. 2012. An overview of statistical considerations in long-term monitoring. Pp.23–53. In R. A. Gitzen, J. J. Millspaugh, A. B. Cooper, and D. S. Licht, eds., Design and Analysis of Long-Term Ecological Monitoring

Studies. Cambridge University Press, Cambridge, UK.

Rich, L. N., R. E. Russell, E. M. Glenn, M. S. Mitchell, J. A. Gude, K. M. Podruzny, C. A. Sime, K. Laudon, D. E. Ausband, and J. D. Nichols. 2013. Estimating occupancy and predicting numbers of gray wolf packs in Montana using hunter surveys. Journal of Wildlife Management 77:1280–1289.

Richards, J. D., and J. Short. 2003. Reintroduction and establishment of the western barred bandicoot Perameles bougainville (Marsupialia: Peramelidae) at Shark Bay, Western Australia. Biological Conservation 109:181–195.

Richardson, K., J. G. Ewen, D. P. Armstrong, and M. E. Hauber. 2010. Sex-specific shifts in natal dispersal dynamics in a reintroduced hihi population. Behaviour 147:1517–1532.

Ripple, W. J., and R. L. Beschta. 2003. Wolf reintroduction, predation risk, and cottonwood recovery in Yellowstone National Park. Forest Ecology and Management 184: 299–313.

Ripple, W. J., and R. L. Beschta. 2012. Trophic cascades in Yellowstone: the first 15 years after wolf reintroduction. Biological Conservation 145:205–213.

Ripple, W. J., E. J. Larsen, R. A. Renkin, and D. W. Smith. 2001. Trophic cascades among wolves, elk and aspen on Yellowstone National Park's northern range. Biological Conservation 102:227–234.

Roe, J. H., M. R. Frank, S. E. Gibson, A. O. Attum, and B. A. Kingsbury. 2010. No place like home: an experimental comparison of reintroduction strategies using snakes. Journal of Applied Ecology 47:1253–1261.

Roy, H. E., M. J. O. Pocock, C. D. Preston, D. B. Roy, J. Savage, J. C. Tweddle, and L. D. Robinson. 2012. Understanding Citizen Science and Environmental Monitoring: Final Report on Behalf of UK Environmental Observation Framework. NERC Centre for Ecology and Hydrology and National History Museum, London, UK.

Royle, J. A. 2004. N-mixture models for estimating population size from spatially replicated counts. Biometrics 60:108–115.

Royle, J. A., R. B. Chandler, R. Sollmann, and B. Gardner. 2014. Spatial Capture-Recapture. Academic Press, Waltham, MA.

Royle, J. A., and S. J. Converse. 2014. Hierarchical spatial capture-recapture models: modelling population density in stratified populations. Methods in Ecology and Evolution 5:37–43.

Runge, M. C. 2013. Active adaptive management for reintroduction of an animal population. Journal of Wildlife Management 77:1135–1144.

Saltz, D., M. Rowen, and D. I. Rubenstein. 2000. The effect of space-use patterns of reintroduced Asiatic wild ass on effective population size. Conservation Biology 14:1852–1861.

Santos, T., J. Pérez-Tris, R. Carbonell, J. L. Tellería, and J. A. Díaz. 2009. Monitoring the performance of wild-born and introduced lizards in a fragmented landscape: implications for ex situ conservation programmes. Biological Conservation 142:2923–2930.

Santymire, R. M., P. E. Marinari, J. S. Kreeger, D. E. Wildt, and J. G. Howard. 2006. Sperm viability in the black-footed ferret (*Mustela nigripes*) is influenced by seminal and medium osmolality. Cryobiology 53:37–50.

Sanz, V., and A. Grajal. 1998. Successful reintroduction of captive-raised yellow-shouldered Amazon parrots on Margarita Island, Venezuela. Conservation Biology 12:430–441.

Sarrazin, F., and R. Barbault. 1996. Reintroduction: challenges and lessons for basic ecology. Trends in Ecology and Evolution 11:474–478.

Sarrazin, F., and S. Legendre. 2000. Demographic approach to releasing adults versus young in reintroductions. Conservation Biology 14:488–500.

Seddon, P. J., D. P. Armstrong, and R. F. Maloney. 2007. Developing the science of reintroduction biology. Conservation Biology 21:303–312.

Seddon, P. J., K. Ismail, M. Shobrak, S. Ostrowski, and C. Magin. 2003. A comparison of derived population estimate, mark-resighting and distance sampling methods to determine the population size of a desert ungulate, the Arabian oryx. Oryx 37:286–294.

Sequeira, A. M., P. E. Roetman, C. B. Daniels, A. K. Baker, and C. J. Bradshaw. 2014. Distribution models for koalas in South Australia using citizen science-collected data. Ecology and Evolution 4:2103–2114.

Silvertown, J. 2009. A new dawn for citizen science. Trends in Ecology and Evolution 24:467–471.

Skalski, J. R., and D. S. Robson. 1992. Techniques for Wildlife Investigations: Design and Analysis of Capture Data. Academic Press, San Diego, CA.

Skalski, J. R., K. E. Ryding, and J. J. Millspaugh. 2005. Wildlife Demography: Analysis of Sex, Age, and Count Data. Elsevier Science, San Diego, CA.

Slotta-Bachmayr, L., R. Boegel, P. Kaczensky, C. Stauffer, and C. Walzer. 2004. Use of population viability analysis to identify management priorities and success in reintroducing Przewalski's horses to southwestern Mongolia. Journal of Wildlife Management 68:790–798.

Smith, D. W., and D. B. Tyers. 2008. The beavers of Yellowstone. Yellowstone Science 16:4–14.

Sollmann, R., B. Gardner, A. W. Parsons, J. J. Stocking, B. T. McClintock, T. R. Simons, K. H. Pollock, and A. F. O'Connell. 2013. A spatial mark–resight model augmented with telemetry data. Ecology 94:553–559.

Somers, M. J., J. A. Graf, M. Szykman, R. Slotow, and M. Gusset. 2008. Dynamics of a small re-introduced population of wild dogs over 25 years: Allee effects and the implications of sociality for endangered species' recovery. Oecologia 158:239–247.

Stanley, T. R., and J. A. Royle. 2005. Estimating site occupancy and abundance using indirect detection indices. Journal of Wildlife Management 69:874–883.

Steen, D. A., J. A. Stiles, S. H. Stiles, J. C. Godwin, and C. Guyer. 2016. Observations of feeding behavior by reintroduced Indigo Snakes in southern Alabama. Herpetological Review 47:11–13.

Steinhorst, K. R., and M. D. Samuel. 1989. Sightability adjustment methods for aerial surveys of wildlife populations. Biometrics 45:415–425.

Stenglein, J. L., L. P. Waits, D. E. Ausband, P. Zager, and C. M. Mack. 2010. Efficient, noninvasive genetic sampling for monitoring reintroduced wolves. Journal of Wildlife Management 74:1050–1058.

Sutherland, W. J., D. Armstrong, S. H. Butchart, J. M. Earnhardt, J. Ewen, I. Jamieson, C. G. Jones et al. 2010. Standards for documenting and monitoring bird reintroduction projects. Conservation Letters 3:229–235.

Taylor, S. S., I. G. Jamieson, and D. P. Armstrong. 2005. Successful island reintroductions of New Zealand robins and saddlebacks with small numbers of founders. Animal Conservation 8:415–420.

Teixeira, C. P., C. S. De Azevedo, M. Mendl, C. F. Cipreste, and R. J. Young. 2007. Revisiting translocation and reintroduction programmes: the importance of considering stress. Animal Behaviour 73:1–13.

Thompson, W. L., G. C. White, and C. Gowan. 1998. Monitoring Vertebrate Populations. Academic Press, San Diego, CA.

Towns, D. R., and S. M. Ferreira. 2001. Conservation of New Zealand lizards (Lacertilia: Scincidae) by translocation of small populations. Biological Conservation 98:211–222.

Tracy, L. N., G. P. Wallis, M. G. Efford, and I. G. Jamieson. 2011. Preserving genetic diversity in threatened species reintroductions: how many individuals should be released? Animal Conservation 14:439–446.

Tripp, D. W., T. E. Rocke, S. P. Streich, R. C. Abbott, J. E. Osorio, and M. W. Miller. 2015. Apparent field safety of a raccoon poxvirus-vectored plague vaccine in free-ranging prairie dogs (*Cynomys* spp.), Colorado, USA. Journal of Wildlife Diseases 51:401–410.

Tsaparis, D., N. Karaiskou, Y. Mertzanis, and A. Triantafyllidis. 2014. Non-invasive genetic study and population monitoring of the brown bear (*Ursus arctos*) (Mammalia: Ursidae) in Kastoria region: Greece. Journal of Natural History 49:393–410.

Tuberville, T. D., E. E. Clark, K. A. Buhlmann, and J. W. Gibbons. 2005. Translocation as a conservation tool: site fidelity and movement of repatriated gopher tortoises (*Gopherus polyphemus*). Animal Conservation 8:349–358.

Tweed, E. J., J. T. Foster, B. L. Woodworth, P. Oesterle, C. Kuehler, A. A. Lieberman, A. T. Powers et al. 2003. Survival, dispersal, and home-range establishment of reintroduced captive-bred puaiohi, *Myadestes palmeri*. Biological Conservation 111:1–9.

Tyre, A. J., B. Tenhumberg, S. A. Field, D. Niejalke, K. Parris, and H. P. Possingham. 2003. Improving precision and reducing bias in biological surveys: estimating false-negative error rates. Ecological Applications 13:1790–1801.

USFWS (US Fish and Wildlife Service). 1988. Black-footed Ferret Recovery Plan. US Fish and Wildlife Service, Denver, CO.

USFWS. 2013. Recovery Plan for the Black-footed Ferret (*Mustela nigripes*). US Fish and Wildlife Service, Denver, CO.

Van Oosterhout, C., A. M. Smith, B. Haenfling, I. W. Ramnarine, R. S. Mohammed, and J. Cable. 2007. The guppy as a conservation model: implications of parasitism and inbreeding for reintroduction success. Conservation Biology 21:1573–1583.

Vore, J. M., and E. M. Schmidt. 2001. Movements of female elk during calving season in northwest Montana. Wildlife Society Bulletin 29:720–725.

Wallach, A. D., U. Shanas, K. Y. Mumcuoglu, and M. Inbar. 2008. Ectoparasites on reintroduced roe deer *Capreolus capreolus* in Israel. Journal of Wildlife Diseases 44:693–696.

Ward, D. M., K. H. Nislow, and L. K. Holt. 2008. Do native species limit survival of reintroduced Atlantic salmon to historic rearing streams? Biological Conservation 141:146–152.

White, G. C., and T. M. Shenk. 2001. Population estimation with radio-marked animals. Pp.329–350. In J. J. Millspaugh and J. M. Marzluff, eds., Radio Tracking and Animal Populations. Academic Press, San Diego, CA.

White, P. J., and R. A. Garrott. 2005. Yellowstone's ungulates after wolves: expectations, realizations, and predictions. Biological Conservation 125:141–152.

Whitfield, D. P., A. Douse, R. J. Evans, J. Grant, J. Love, D. R. McLeod, R. Reid, and J. D. Wilson. 2009. Natal and breeding dispersal in a reintroduced population of white-tailed eagles *Haliaeetus albicilla*. Bird Study 56:177–186.

Williams, B. K., J. D. Nichols, and M. J. Conroy. 2001. Analysis and Management of Animal Populations: Modeling, Estimation and Decision Making. Academic Press, San Diego, CA.

Wisely, S. M. 2005. The genetic legacy of the black-footed ferret: past, present, and future. Pp.37–43. In J. E. Roelle, B. J. Miller, J. L. Godbey, and D. E. Biggins, eds., Recovery of the Black-Footed Ferret: Progress and Continuing Challenges. Proceedings of the Symposium on the Status of the Black-Footed Ferret and Its Habitat. Scientific Investigations Report 2005-5293. US Geological Survey, Fort Collins, CO.

Wisely, S. M., O. A. Ryder, R. M. Santymire, J. F. Engelhardt, and B. J. Novak. 2015. A road map for 21st century genetic restoration: gene pool enrichment of the black-footed ferret. Journal of Heredity 106:581–592.

Witmer, G. W. 1990. Re-introduction of elk in the United States. Journal of the Pennsylvania Academy of Science 64:131–135.

Wronski, T., H. Lerp, and K. Ismail. 2011. Reproductive biology and life history traits of Arabian oryx (*Oryx leucoryx*) founder females reintroduced to Mahazat as-Sayd, Saudi Arabia. Mammalian Biology-Zeitschrift für Säugetierkunde 76:506–511.

Yott, A., R. Rosatte, J. A. Schaefer, J. Hamr, J. Fryxell. 2011. Movement and spread of a founding population of reintroduced elk (*Cervus elaphus*) in Ontario, Canada. Restoration Ecology 19:70–77.

Zidon, R., D. Saltz, L. S. Shore, and U. Motro. 2009. Behavioral changes, stress, and survival following reintroduction of Persian fallow deer from two breeding facilities. Conservation Biology 23:1026–1035.

THIRTEEN

Management of Reintroduced Wildlife Populations

Matt W. Hayward and Rob Slotow

ALTHOUGH GLOBAL REVIEWS historically have tended to suggest a poor success rate (Griffith et al. 1989, Wolf et al. 1998, Fischer and Lindenmayer 2000), reintroductions have become a fundamental tool in the conservation managers' toolbox (Seddon and Armstrong, Chapter 2, this volume). This implies disagreement between conservation managers and academics—a mismatch in the quest for evidence-based, best-practice implementation of specific management actions. With the poor success rates shown in the published literature, there seems no justification for managers to conduct reintroductions. Yet conservation managers continue to use reintroduction to improve the status of the species they manage—at increasing rates (Soorae 2008, 2011, 2013). The most likely explanation for this discrepancy is that conservation managers have been busy undertaking reintroductions rather than writing about them, as scientific, synthetic reviews of reintroduction success invariably rely on the primary literature.

Scientists play an important role in conservation reintroductions, but this is not necessarily formalized through academic institutions. Organizations that are heavily involved in reintroductions invariably employ trained scientists, including those with PhDs (e.g., Australian Wildlife Conservancy [AWC; Box 13.1], South African National Parks; New Zealand Department of Conservation, and US Fish and Wildlife Service). While these scientists may have affiliations with academic institutions, their priority is not always to publish the results of their work. Instead, much of their time is spent monitoring the outcomes of management interventions, ideally via an adaptive management framework (Roux and Foxcroft 2011; McCarthy et al. 2012; Chauvenet et al., Chapter 6, this volume). Hence, while using prior experience based on robust, peer-reviewed research is important and will lead to improved efficiencies (e.g., Delsink et al. 2013a), we argue that relying exclusively on the primary literature will not enhance the field, as conservation managers have learnt lessons that are not recorded there.

In this chapter, we use our experiences working within teams of conservation managers on large-scale reintroduction projects to identify some of the management challenges that scientists and managers face in their efforts to ensure the long-term success of wildlife reintroductions. We have found that designing management strategies must begin with a detailed knowledge of what originally caused the extirpation in the first place, and how success for a given reintroduction attempt is defined. Following release, a wide suite of management intervention tools can be implemented to help wildlife reintroductions succeed, and to limit the risk of overpopulation. We conclude by reviewing the lessons learnt during our experiences in wildlife reintroduction, and identify information deficiencies that could be better addressed through greater engagement of scientists with conservation managers to increase the peer-reviewed reporting on conservation actions that has hitherto not reached the primary literature.

HOW THE DEFINITION OF SUCCESS AFFECTS MANAGEMENT OF REINTRODUCED WILDLIFE

There are a variety of ways that people have defined the success of reintroductions, including breeding by wild-born individuals; recruitment exceeding mortality over 3 years; an unsupported population exceeding 500 individuals; or a self-sustaining population (Chauvenet et al., Chapter 6, this volume; Seddon 1999). Definitions of success of reintroduction programs are challenging because of their duration, the life history of the reintroduced species, and the limitation of the release site in the number of individuals that it can sustain (Hayward et al. 2007b). For example, the reintroduced population of African wild dogs (*Lycaon pictus*) in Shamwari Private Game Reserve (South Africa) was considered successful based on the definition of a 3-year breeding program exceeding mortality (Hayward et al.

2007b); however, since then, wild dogs have been removed from the reserve because of concerns around their impact on the prey populations, rendering the reintroduction a failure (J. O'Brien, Shamwari Private Game Reserve, South Africa, pers. comm.). Furthermore, the estimated carrying capacity for this site and others in the Eastern Cape, South Africa, where predators have been reintroduced, is below that which would signify success based on any of the other demographic criteria (Hayward et al. 2007c), illustrating that, irrespective of duration, these sites individually will never support large enough populations to be considered reintroduction successes. Indeed, given the focus of reintroduction programs on threatened species, it is not assured that "successful" reintroductions will ever exceed the effective population size necessary to theoretically avoid genetic problems, especially if the latest estimates of this are valid (Traill et al. 2007).

Irrespective of the definition of success, the key to successful reintroduction is removing the key threatening process (Box 13.1). Caughley (1994) highlighted that the key to conservation success is separating the species of conservation concern from the agent(s) of its decline. The same is true for reintroduction success, with great levels of success occurring when we have a better understanding of these key threatening processes, and where there are fewer threats, or where these have been identified and mitigated. For example, although studies in Australia often attribute the decline of mammals in the critical weight range (i.e., those weighing between 35 g and 5.5 kg) to numerous causes (Burbidge and McKenzie 1989, Lunney 2001), reintroduction programs that treated just one cause (introduced predators: red foxes, *Vulpes vulpes*, and feral cats, *Felis catus*) invariably succeeded (Short 2009, Frank et al. 2014, Hayward et al. 2014). The same is largely true for New Zealand (Innes et al. 2012, 2015). In South Africa, overhunting/persecution appears to be the primary agent of the decline of large mammals and, in the absence of this threat,

BOX 13.1 · Reintroduction Lessons from the Australian Wildlife Conservancy: The Importance of Addressing the Agent of the Initial Decline

Separating the species of conservation concern from the agent(s) of its decline is critical to achieving reintroduction success. For example, the Australian Wildlife Conservancy (AWC) is a private, not-for-profit conservation agency that purchases land and undertakes intensive conservation actions on its properties (Innes et al. 2015). AWC has conducted more than 60 mammal translocations (AWC 2014) involving over 20 species and over 2,000 individuals, but few of these are reported in the primary literature. The success rate (based on established populations) of these translocations has been high (~73% as of 2010) as they have generally been on mammal-free, "mainland islands" (Hayward et al. 2010). Below we use two examples that have yielded important lessons for the practical side of reintroduction involving numbats (*Myrmecobius fasciatus*) and bridled nailtail wallabies (*Onychogalea fraenata*).

Numbats are small (600–900 g), diurnal, myrmecophagous, dasyurid marsupials that once occurred across the southern half of Australia. The arrival of the fox, *Vulpes vulpes*, and cat, *Felis catus*, decimated populations until the species was only found in tiny remnant forest and woodland patches in Western Australia (Friend et al. 2013). The species has been reintroduced to several areas in Western Australia with limited success, but AWC and Earth Sanctuaries Limited reintroduced numbats to Yookamurra Sanctuary in South Australia in the early 1990s (Hayward et al. 2015b). This population became established, and was used as the founder stock, for a reintroduction to Scotia Sanctuary in far western New South Wales, which now numbers over 400 individuals. The reason for these two successful projects is primarily due to the agents of the numbat's decline no longer being present—that is, foxes and cats were excluded by conservation fences. In the absence of these introduced predators, numbats have survived resource scarcity through the Millennium Drought, and then the drought-breaking floods.

Bridled nailtail wallabies have also thrived inside the conservation fences at Scotia; however, this area is limited at 8,000 ha and the population is at carrying capacity. Bridled nailtails are listed as Endangered on the IUCN Red List and only survive in four localized sites, with Scotia supporting approximately 75% of the global population. To enable them to increase in abundance to satisfy more secure status listings, the species needs new areas to inhabit. Consequently, AWC decided to translocate bridled nailtails beyond the conservation fences into an area where foxes and cats were intensively controlled through the use of broad-scale poison baiting. Four even-sex groups of 40 bridled nailtails were released over successive months, with 30 fitted with VHF radio-collars. We tested several factors hypothesized as affecting reintroduction success (soft vs. hard release; wild-caught vs. captive-origin stock; supplementary feeding vs. not; predator trained vs. predator naïve). While predator training yielded marginally longer survivorship, the animals were rapidly killed by red foxes, rendering the other factors moot.

In total, seven previously extirpated species have been reintroduced to Scotia, and all have successfully established self-sustaining populations. The key reason for this is that the agents of their initial decline (foxes and cats) were completely eradicated from within the conservation fences. Beyond the conservation fences—even in areas where enormous effort has been expended in controlling introduced species and where the habitat is exactly the same as inside the fences (although potentially with lower browsing pressure given the high density of reintroduced animals inside the fences)—the agents of the decline of critical weight range fauna remain and very few native species are able to persist.

reintroductions appear highly successful, even in highly disturbed habitats (Penzhorn 1971, Rowe-Rowe 1992, Hofmeyr et al. 2003, Hayward et al. 2007b). The success of the gray wolf (*Canis lupus*) reintroduction to Yellowstone National Park, United States, is also largely due to the removal of a single threat: human persecution (Smith and Bangs 2009). However, the ability to effectively mitigate the key threatening processes is likely to vary between countries and taxa. It may also be that this effect is context-dependent, whereby persecution may be easier to treat in countries with good governance, as appears to have occurred with the carnivores of Europe (Chapron et al. 2014).

MANAGEMENT INTERVENTIONS FOR REINTRODUCED WILDLIFE

Once animals have been reintroduced, the program should not be considered completed. As discussed in Chapter 12, a post-release monitoring program is an essential part of a reintroduction, and should be designed to ensure that the actions to meet the program's objectives are measureable, have time schedules attached, are sufficiently resourced, and have responsibilities allocated (IUCN/SSC 2013). The monitoring program must also utilize robust methods to avoid conservation controversies, like that around the use of dingoes (*Canis lupus dingo*) to limit the impacts of introduced red foxes and feral cats (Hayward et al. 2015a).

Continued management may also be a necessity once a reintroduction has occurred. This management can be aimed at maintaining or expanding the founder population, at limiting overpopulation, or at other issues that may arise, such as human-wildlife conflict. It is only with adequate, robust monitoring that the necessity and/or success of the common management activities that we describe below can be confidently assessed (i.e., intensive management, fencing, introduced species management, apex predators, habitat manipulation, conflict mitigation, reinforcement, metapopulation management, and overpopulation). Fur-

thermore, ongoing management provides us with the opportunity to learn how species that are reintroduced, and that are often threatened, respond to such intervention, thereby providing a source of evidence for approaching best practices in the future (Kettles and Slotow 2009, Miller et al. 2013). In general, both monitoring and management should be designed to assess specific objectives, determine management needs to improve success rates, and reduce uncertainty surrounding future management actions (Gitzen et al., Chapter 12, this volume).

Management to Maintain Reintroduced Populations

In an ideal reintroduction program, the animals would be released and rapidly breed to reach a stable carrying capacity of a satisfactory effective population size (N_e), without causing any problems to other species or the local ecosystem, or resulting in reducing ecosystem goods and services, or increasing human-wildlife conflict. However, this ideal achievement of objectives rarely occurs, and so, below, we describe some common strategies that have been implemented to facilitate and mitigate achievement of objectives for reintroduced wildlife populations.

Fencing

Fencing is a widely applied tool for mammal reintroduction and conservation, but it is not without controversy within conservation circles (Creel et al. 2013, Pfeifer et al. 2014, Woodroffe et al. 2014). Although there are undoubted impacts of fencing on wildlife conservation, particularly those fences not designed specifically for conservation purposes (Hayward and Kerley 2009), the data that have been presented suggest conservation fencing is important to the conservation of difficult-to-conserve large carnivores like lions (*Panthera leo*; Packer et al. 2013) and African wild dogs (Gusset et al. 2008). Similarly, cost-benefit analysis suggests it is also a valuable wildlife conservation strategy in Nepal (Sapkota et al. 2014). In Australasia, given that the high failure rate of reintro-ductions is attributed to introduced predators (Short et al. 1992), "marooning" reintroductions on off-shore islands (Short and Smith 1994, Rankmore et al. 2008) and the creation of fenced mainland "islands" have become popular to separate native fauna from their major threatening process: predation (Saunders and Norton 2001, Dickman 2011, Frank et al. 2014). Finally, it is legislated in South Africa that areas conserving dangerous animals (large predators/megaherbivores) must be fenced, and reserves require certification of adequate enclosure under Section 35 of the *Nature and Environmental Conservation Ordinance* (No. 19 of 1974), *National Environmental Management and Biodiversity Act* (10 of 2004) (Threatened or Protected Species Regulations) and Operational Guideline *Certificate of Adequate Enclosure and Dangerous Game Specifications* 25/2/2008. Such boundary fences are required to prevent human-wildlife conflict, but conflict can also be reduced, even within reserves, by using "virtual" fences such as alert polygons from radio-collared animals (Jachowski et al. 2014).

The key rationale behind implementing fencing for conservation purposes is the separation of biodiversity from the factors threatening it (Hayward and Kerley 2009). Clearly, there are some threatening processes that are more readily treated with hard-boundary separation (e.g., invasive mammalian predators or human persecution; Hayward 2012). Alternatively, innovative approaches may reduce the hard edge effect of fencing. For example, African elephants (*Loxodonta africana*) can be excluded from sensitive habitats by a permeable fence that prevents their movement, but allows movement of other species (Slotow 2011; see also Jachowski et al. 2014, for broader discussion). However, there are other threats to reintroduced populations that clearly cannot be solved by fencing (e.g., chytridiomycosis of frogs [Sapsford et al. 2015], climate change or pollution impacts [e.g., Urbanek et al. 2014], etc.). Thus, it is necessary to determine whether conservation fencing will reduce key threats for

reintroduction programs before they are implemented, and to identify potential impacts of the creation of conservation fencing. There also need to be sufficient resources allocated for regular fence checking, maintenance, and, ultimately, replacement (Grant et al. 2007, Slotow 2011). The issue of conservation fences also has implications for IUCN Red List assessments, where some assessors include populations protected by fences, while others do not, due to individual definitions of "self-sufficiency" in the reintroduced population.

Introduced Species Control

In Australia, fencing is one way of separating threatened wildlife from introduced predators, but introduced predators are controlled over a far greater area by way of poisoning (Possingham et al. 2004, Mahon 2010). This broadscale predator control has led to delisting of previously threatened species, such as the southern brown bandicoot (*Isoodon obesulus*), and western quoll (*Dasyurus geoffroii*) in Western Australia (Possingham et al. 2004). However, not all species have responded positively in the long term. The woylie (*Bettongia penicillata*) is a small wallaby from Western Australia that was listed as Endangered on the IUCN Red List in 1996, before broadscale fox control was implemented and numbers rebounded (Wayne et al. 2013). Woylies had again declined precipitously by 2002 when they were relisted as Endangered (Wayne et al. 2013), but this seems to be because fox control was so successful that it allowed mesopredators, such as feral cats and varanid lizards, to increase and take an equally heavy toll on the species (de Tores and Marlow 2011, Sutherland et al. 2011, Marlow et al. 2015). Witmer and Fuller (2011) provide a useful review of control of introduced species in the United States, and the challenges in dealing with invasive species. We describe additional examples of limiting introduced species within the discussion below.

Restoring Apex Predators

An alternative to broadscale poisoning is the restoration of apex predators to suppress mesopredators (Crooks and Soule 1999). The return of gray wolves to parts of the United States has reduced the abundance, and hence impacts, of coyotes (*Canis latrans*) on several small mammal species of conservation concern (Ripple et al. 2013). Attempts at finding similar beneficial effects in Australia between dingoes, red foxes, and feral cats have yielded contradictory results (Letnic et al. 2009, Allen et al. 2014). Nonetheless, calls have been made to reintroduce dingoes to conserve threatened native wildlife (Ritchie et al. 2012).

Apex predators may provide some protection for prey species shared with mesopredators; however, they are also predators in their own right and are also likely to have some impact on prey species themselves (Allen and Fleming 2012). For apex predators to yield benefits to prey species, they need to provide net reductions in predation levels (Hayward et al. 2015a). Further, it is clearly important to account for behavioral and physiological responses of individuals following reintroduction if we are to understand the mechanisms behind the demographic success or failure of the project (Jachowski et al., Chapter 9, this volume). Thus, an understanding of trophic interactions among reintroduced apex predators and prey is critical to forming appropriate management strategies.

Habitat Manipulation

There are many ways habitat can be manipulated to maximize its benefits to reintroduced populations. Here, we discuss examples of providing artificial water points, shelter, and supplementary food, and controlling fire. The provision of water can allow species to survive in areas that are otherwise too dry for them or where fences block access to water (Slotow 2011, Gadd 2012), although this can cause substantial ecological impacts through piosphere effects (i.e., impacts of grazing on vegetation and soils) or altering the outcome of competition for rare species (Landsberg et al. 1997, Smit et al. 2007).

Fire management is another method of manipulating habitat to the benefit of reintroduced species. Recent fires open up the vegeta-

tion to allow predators easy access to prey that lack sufficient refuge, as occurred with feral cats in the Western Australian Kimberly region (McGregor et al. 2014). Conversely, long periods without fire can drive habitat into successional states that also lack suitable predation refuges, as occurs in long-unburnt swamp shrublands for quokkas (*Setonix brachyurus*; Hayward et al. 2005). Indeed, the reintroduction of native marsupials in Australia may reduce the impacts of fire and reduce the management costs associated with fire risk minimisation (Hayward et al. 2016). Clearly, knowledge of the ecology of the reintroduced species is of paramount importance before implementing such conservation management actions.

Supplementary food provisioning is another method of manipulating habitat that can benefit reintroduced populations. European bison (*Bison bonasus*) in Poland's Białowieża Primeval Forest have been provisioned with hay, since their reintroduction, to improve their survival over winter (and then sickly looking individuals are culled over summer; Hayward et al. 2011). Red kites (*Milvus milvus*) in Wales were fed upon their release, and this practice continues today despite the species being secure in the United Kingdom (Orros and Fellowes 2015). Feeding stations for vultures in Europe and Africa have provided a reliable food source that is safe from poisons and pesticides, and have led to population increases (Kane et al. 2014, Moreno-Opo et al. 2015). As with the provisioning of water, there are community-level impacts for supplementary feeding that should be considered, whereby species interactions are affected by human management decisions (Selva et al. 2014, Yarnell et al. 2014).

Artificial shelter is another strategy frequently used to improve survivorship and philopatry of founder populations. At the AWC's Scotia Sanctuary (Box 13.1), managers constructed artificial nests for the greater stick-nest rat, *Leporillus conditor*, but these were soon ignored in favor of a nest they constructed themselves (Matt W. Hayward, pers. obser.). Artificial shelters have also been used for reintroduction of arboreal primates

(Konstant and Mittermeier 1982) and fossorial European ground squirrels (*Spermophilus citellus*; Gedeon et al. 2011). While shelter types tend to be species-specific, this management action again leads to the potential creation of ecological imbalances favoring individual species at the expenses of others.

Conflict Mitigation

Resident individuals often have superiority over new immigrants, and hence, in territorial species, tend to win fights through better knowledge of the local environment and refuge areas (i.e., the resident advantage; Oyegbile and Marler 2005). This can lead to high rates of mortality for newly reintroduced individuals, but mitigation measures are available. For example, black rhinoceros (*Diceros bicornis*) are territorial and the high mortality rate of newly reintroduced individuals can be reduced using the territorial marks (odors) of other individuals to direct them away from areas of conflict (Linklater and Hutcheson 2010, Linklater et al. 2011). Similarly, African wild dogs can also be managed to avoid conflict with conspecifics through the use of scent (Jackson et al. 2012). An extreme example of active conflict mitigation comes from Shamwari Private Game Reserve, where cheetahs (*Acinonyx jubatus*) were habituated by blowing a whistle when they were fed so they associated the whistle with food (Hayward et al. 2007a). Subsequently, when managers identified cheetah that were cornered in a section of the reserve where lions could target them, reserve managers moved them away from the zone of conflict using the whistle (Hayward et al. 2007a).

Conservation managers have also intervened when reintroduced elephants killed over 50 rhinoceros in both Pilanesberg National Park and Hluhluwe-iMfolozi Park, South Africa (Slotow and van Dyk 2001, Slotow et al. 2001). In this case, young elephants that were orphaned during culling operations in Kruger National Park were reintroduced to Pilanesberg, and as they became sexually mature, started killing black and white rhinoceros (Slotow et al. 2001). The introduction of older male elephants to these populations moderated the musth of the young males and halted the killing of the rhinoceros (Slotow et al. 2000).

Reinforcement

While one-off reintroductions sometimes work, there are times when repeated reinforcement is needed to supplement/augment/restock the founder population (Seddon et al. 2012). The standard example of reinforcement is the New Zealand black stilt (*Himantopus novaezelandiae*) that is sustained in the wild via repeated releases of captive-bred birds (Keedwell et al. 2002). Similarly, the annual release of captive-bred orange-bellied parrots (*Neophema chrysogaster*) from zoos in Australia was assumed to be reinforcing the wild population, but this was not the case (Smales et al. 2000). Continual reinforcement raises the question of exit strategy, that is, at what point is investing further resources no longer justifiable, despite management adjustments to improve the situation (Seddon et al. 2012).

There is an increasing body of evidence that emphasizes the need to take into account the social structure of the population when conducting initial and subsequent releases. As seen in the example above, killing of rhinoceroses by reintroduced young male elephants required the reinforcement of the elephant population with older males that dominated the existing males to resolve the problem (Slotow et al. 2001). This example underscores a broader need to understand sociality in planning initial release attempts and subsequent reinforcement in restoring highly social species. An additional example comes from attempts to establish a population of African wild dogs in Hluhluwe-iMfolozi Game Reserve. These were not successful until reinforcements allowed a critical mass of groups to develop, whereupon the population became self-sustaining (Gusset et al. 2008, 2009, Somers et al. 2008). This highlights two important considerations for managing highly social species. Firstly, the number of groups may be more important than the number of individuals (Ferreira and Hofmeyr 2014), and, secondly, social groups are fundamental to the appropriate behavior of species and reintroductions need to account for this (Shannon et al. 2013). This divergence in effective population size (groups vs. individuals) is not something routinely considered by managers, but insights arose due to the involvement of researchers in reintroduction programs.

Ecological Restoration

The ultimate goal of many reintroduction programs is to allow wildlife to interact and restore ecosystem function without the need for human intervention. As such, the restoration of the complete fauna is often a desired goal, and the inclusion of predators (especially apex predators) can restore lost trophic relationships and often limit the need for unpalatable culling or hunting of herbivores. This was one of the drivers of the Yellowstone wolf reintroduction (Smith and Bangs 2009), as well as the reintroduction of top-order predators to many sites in South Africa (Hayward et al. 2007b). Indeed, the management of reintroduced predators in South Africa has now moved beyond simply determining when the abundance of a species exceeds the amount of food available to it, to a situation whereby natural dispersal processes can now be mimicked in management activities by translocating surplus animals out of confined populations (Ferreira and Hofmeyr 2014).

Metapopulation Management

Often, reintroduction sites are not large enough to support a viable and sustainable population. In such cases, individual populations could still be important components if they are managed as part of a metapopulation, provided this is formally constructed and managed as such (Davies-Mostert et al. 2009, Lindsey et al. 2011). Metapopulation management means each individual population does not have to be sustainable alone; collectively they can constitute a sustainable population; however, this does carry a management burden of ongoing interventions to move animals among populations where those populations are isolated by

humanity (see Ferreira and Hofmeyr 2014, for a conceptual model of lion metapopulation management in South Africa). Restoring naturally occurring metapopulations is somewhat easier in that animals do not need to be moved provided that corridors of connectivity are maintained. However, it is increasingly difficult to restore species where movement corridors are intact, and the presence of introduced species can limit such unassisted movement (Hayward et al. 2003).

The value of managing small, isolated populations as a metapopulation is perhaps best illustrated in African wild dog restoration in South Africa (Davies-Mostert et al. 2009). The Wild Dog Action Group (WAG) established new wild dog populations and reinforced existing ones using a series of isolated (fenced) reserves where each population acted independently (with regard to colonization and extinction), but contributed to the regional population through movements driven by conservation managers (Davies-Mostert et al. 2009).

Conversely, the absence of metapopulation management, and the often inappropriate, selection of founder stock in the lion population of South Africa's smaller game reserves, has raised concerns over the value of these individuals to the conservation of the species (Hunter et al. 2007, Slotow and Hunter 2009). Similar concerns exist about the lack of metapopulation management for fenced, reintroduced populations of marsupials in Australia (e.g., for boodies, *Bettongia lesueur*; bridled nailtail wallabies, *Onychogalea fraenata*; bilbies, *Macrotis lagotis*; and mala, *Lagorchestes hirsutus*; Hayward et al. 2014).

Metapopulation management requires proper planning; however, metapopulation management tends to start after the reintroduction process is well initiated, and thereafter involves "catching up." For example, rhinoceros were reintroduced throughout South Africa with no metapopulation management plan. This was not a problem where the founder population rapidly increased to a sustainable size and thereby retained most of its genetic diversity, but problems arose where the recipient site was too small to support a sustainable population. Subsequently, a metapopulation approach is evolving (Emslie et al. 2009), although this still mostly entails moving animals from a donor population to a recipient population, rather than shifting animals back and forth among populations to promote genetic diversity.

Similarly, with lions there was no metapopulation management plan at the start of reestablishing the species throughout South Africa (Slotow and Hunter 2009), but this is now in place and has improved the conservation status of the small lion populations (Miller et al. 2013). A metapopulation management plan was developed for African wild dogs in South Africa by the WAG, but this included only the reserves initially affiliated with WAG. Additional small populations were established and managed independently of WAG (and at one stage contributed more dogs to founding new populations than did the metapopulation [Rob Slotow, pers. obser.]). Planning around wild dogs in South Africa became more inclusive, and has since matured, such that there is a real metapopulation. This is a success story (even if some reserves have removed dogs, see above). However, it was only when a more academic slant was incorporated into the metapopulation management (Gusset et al. 2008) that it developed the desired momentum and the program was fully supported and implemented. Management of reintroduced cheetah, on the other hand, despite similarly being reintroduced into small, isolated reserves, has only made preliminary attempts to apply a metapopulation approach (Marnewick et al. 2009, Lindsey et al. 2011). We suspect the lack of implementation for cheetahs is due to the lack of an organization driving and coordinating the process to date.

We described above how metapopulation management can be used to increase reintroduced populations; however, the technique can also act to limit overpopulation. African wild dog packs tend to split when they become too large with either sex dispersing (McNutt 1996).

WAG used the splitting of packs to avert wild dog overpopulation at individual sites (Davies-Mostert et al. 2009).

Overall, metapopulation management should be initiated when several small populations are established that, individually, would be below an effective population size and therefore risk a loss of genetic diversity and ultimate extinction. This is advantageous, as smaller sites can then be used to support viable populations. The disadvantage is that it requires costly, intensive, and often invasive ongoing management. Thus, entering into metapopulation management requires persistent management intervention and collaboration. This approach is most likely to succeed if it is based on metapopulation modeling that incorporates decision analysis. For example, Di Minin et al. (2013) use metapopulation modeling in conjunction with multicriteria decision analyses to provide a more integrated approach to planning for conservation of reintroduced, fragmented small populations of a range of species within the Maputaland–Pondoland–Albany Biodiversity Hotspot.

Management to Limit Overpopulation of Reintroduced Wildlife

Overpopulation of reintroduced wildlife is often considered a problem (Burbidge et al. 2011), particularly in written reintroduction plans in some countries (Matt W. Hayward, pers. obser.). Clearly, monitoring is critical to confidently conclude overpopulation is occurring, and an appropriate decision structure must be in place to ensure a timely decision occurs about what action to take (Martin et al. 2012). Overpopulation of reintroduced fauna is clear evidence of the success of reintroduction, and where threatened species are involved, it could, and arguably should, be considered a positive outcome because of the clear improvement in conservation status this is likely to yield. At the same time, overpopulation undoubtedly raises some challenges to managers, governments, and society of how best to manage it.

One way to avoid this perception problem is through management of population growth rather than population size. We encourage managers to focus on what the population will do in the future rather than what it is at present. In terms of best practice, interventions should be about dealing with limiting growth before population size becomes a problem (i.e., dealing with growth and reducing it in such a manner that it does not become a problem). We describe such methods above, including relocating individuals to found new populations to reinforce existing ones, or to implement meta-population management. One aspect that is rarely considered in reintroduced populations is the issue of irruptive populations, where, because all reintroduced populations are newly established, as soon as the adults reproduce, and again when their young reproduce, the average age of the population reduces relative to stable populations. This age structure means that future reproductive potential is continually increasing and average mortality is continually decreasing (it takes longer for senescence to kick in, and mortality from old age to occur as all the animals are young). These factors are exacerbated by abundant resources for a relatively small population, which can result in earlier age of first breeding and shorter interbirth intervals (e.g., elephants [Slotow et al. 2005] and lions [Trinkel et al. 2010]). This results in an irruptive population growth (Caughley 1970), and this has been highlighted in multiple elephant population management studies (Mackey et al. 2006, Slotow et al. 2008). It should also be noted that there is some evidence for growth of reintroduced populations to slow down due to density dependence, either from ecological (e.g., lions [Trinkel et al. 2010]) or social (e.g., black rhinoceros [Greaver et al. 2014]) reasons.

Below we discuss some methods that have been used to manage overpopulation.

Park Expansion

One strategy conservation managers can use to avoid overpopulation is to expand the reintro-duction site. While not technically a reintroduction, the expansion of South Africa's Addo Elephant National Park (Addo) was essential to avoid the overpopulation of elephants (Kerley and Boshoff 1997). The park was proclaimed in 1931 after Major P. J. Pretorius had shot 114 elephants in the area in 2 years; however, the remaining 11 elephants required coercion to remain within the park boundary until 1954 when it was fenced and the elephant population was 22 (Bradfield, n.d.). The elephant population expanded in the absence of hunting, but several phases of park expansion reduced overpopulation (Gough and Kerley 2006) and the population now exceeds 600 individuals. Similar expansion has been adopted in smaller reserves (e.g., Makalali Private Game Reserve [Delsink et al. 2013b] and Phinda Private Game Reserve [Druce et al. 2008]), and it has been recommended for other species such as lion (Slotow and Hunter 2009) and wild dog (Lindsey et al. 2004). Indeed, fence dropping to increase size and connectedness is key to managing isolated populations of charismatic large mammals in an integrated conservation land-use plan for northern Kwazulu-Natal, South Africa (Di Minin et al. 2013). This should optimally be implemented via a conservation planning framework, rather than leaving it to ad hoc processes on individual reserves (Di Minin et al. 2013). Furthermore, expanding protected areas can be linked to species metapopulation management plans, which often depend on reserve expansion prior to translocations (Miller et al. 2013, Ferreira and Hofmeyr 2014).

We acknowledge that this option is limited because there are limited sites remaining available for protected area expansion. This is particularly challenging in open, unfenced landscapes where the constraints to park expansion are land ownership issues in addition to fencing.

Restocking New Reserves

Once it is carefully determined that there are "surplus" animals available, those individuals can be used to establish new or reinforce existing populations at other locations. For example, lions are highly fecund and can rapidly overpopulate reintroduction sites (Hayward et al. 2007b). One strategy Addo managers have used to avoid overpopulation is to use those lions bred from the six original founders to create three additional, isolated populations in recent expansions of the park. Similarly, excess lions are used to found new populations across the broader regional metapopulation (Miller et al. 2013). Also, elephant and buffalo (*Syncerus caffer*) from Addo that are free from corridor disease and bovine tuberculosis have also been used to found new populations within other nearby game reserves. Using individuals from populations approaching overpopulation to found new and reinforce existing populations was instrumental in the successful conservation of black and white rhinoceros in South Africa (Hall-Martin and Knight 1994). Similarly, the AWC's Karakamia Wildlife Sanctuary (286 ha) in Western Australia is one of the few sites where woylies have not declined, and, to avoid overpopulation of this continually increasing population, it has been used as a source for numerous translocations across Australia (e.g., AWC 2004). The general lesson here is that by perceiving overpopulation at a site positively, other avenues of ecological restoration are available.

Contraception

Contraception is an important tool in management of population growth, as it has the potential to rapidly slow growth of small populations. Implementation of contraception has been studied in a range of reintroduced species, including elephants (Delsink et al. 2007), cheetahs (Bertschinger et al. 2002), and lions (Orford and Perrin 1988, Kettles and Slotow 2009, Miller et al. 2013). Especially in elephants, contraception has been demonstrated to successfully control growth (Delsink et al. 2006, 2013b, Druce et al. 2011), and research has demonstrated that concerns over negative consequences on behavior of elephants from contraception are unfounded (Delsink et al. 2013b, Druce et al. 2013). Besides hormonal or immunocontraception of females, additional approaches to reproductive control are being tested. For example, the possibility of sterilization and surgical intervention on ovaries is being considered as a mechanism to reduce litter sizes, for example, in lions (Miller et al. 2013). Interventions in male reproduction have also been tested; for example, vasectomies in lions (Kettles and Slotow 2009, Miller et al. 2013) and elephants (Rubio-Martínez et al. 2014), and chemical interventions in males of several African carnivores (Bertschinger et al. 2002).

Sport Hunting and Culling

The reintroduction of gray wolves to Yellowstone National Park, United States, has been highly successful, such that sport hunting of wolves is now occurring in surrounding areas. Although the hunting of wolves is highly controversial (Lute and Gore 2014) and not currently based on sound science (Creel et al. 2015), hunting is likely to be effective in reducing wolf abundance (Creel and Rotella 2010). Furthermore, managers must consider that sport hunting has negative social perceptions that limit its use (Garrott et al. 1993) and that hunting wide-ranging species has impacts well beyond the kill site, with 7 of 10 wolf packs inhabiting Yellowstone National Park (the site of the original reintroduction and where hunting is not allowed) affected by hunting adjacent to the park (National Park Service, n.d.). In Africa, Kettles and Slotow (2009) identified hunting as a potential management intervention in lions. However, hunting can alter dominance hierarchies and social structure of highly social species, and have ramifications on population dynamics within protected areas. Selective harvesting of male lions in the buffer zone of Zimbabwe's Hwange National Park, for example, led to infanticidal pride takeovers that can threaten populations (Loveridge et al. 2007). Similar impacts are likely to occur in leopards, *Panthera pardus* (Balme et al. 2010) and elephants (Selier et al. 2014).

A potential benefit of sport hunting is that it increases the land available for conservation. In South Africa, 10,000 farms now operate as hunting lodges (Carruthers 2008) and the "protected" area estate there increased to 23% of the country, from just 7%, when private areas that include hunting are included (Snijders 2012).

Culling can be an effective method of population reduction, but, similar to sport hunting, it can be controversial and have unintended negative consequences for social species such as elephants (Slotow et al. 2008). The elephants of South Africa's Kruger National Park were limited by culling until the mid-1990s (van Aarde et al. 1999); however, international pressure drove the government to ban this until recently (Slotow et al. 2008). In South Africa, culling of elephant is now viewed as a last resort, which can only be implemented after a range of other interventions have already been ruled out (DEAT 2008).

LESSONS LEARNED

Monitoring

Monitoring is key to determine the success or failure of reintroduction programs and any management interventions that are implemented (Gitzen et al., Chapter 12, this volume). Programs are frequently criticized for inadequate monitoring (Nichols and Armstrong 2012), so interaction between managers and scientists during the development of a monitoring plan and prior to the reintroduction is more likely to yield a monitoring program that satisfies all stated objectives. Furthermore, the long planning timeframes offer excellent opportunities for before-after control-impact experiments, yet these rarely occur, despite the benefits in robust knowledge they afford. Monitoring must be strategic to ensure value for money given the limited funds available for reintroduction programs (Ewen and Armstrong 2007). Further, monitoring at the intensity necessary for useful results is likely to be very expensive for conservation managers to undertake alone. A cheaper option would be to partner with academic/research institutions; however, these

institutions are also limited by funds and so there will not be a complete removal of necessary costs. Regardless, there needs to be regular dialogue between researchers and managers to ensure that sufficient lead-in time is available to collect baseline data before management actions are implemented, such as is in a structured decision-making framework (Chauvenet et al., Chapter 6, this volume).

Importance of Evidence-Based Best Practices for Decision-Making

Reintroductions are complex interventions, which often have unintended consequences. In this chapter, we have highlighted a number of techniques that can be used to manage populations of reintroduced species, a domain that often happens well after the initial introductions. As such, the planning team that may be assembled to plan and initiate introductions may have dispersed, leaving managers on the ground to deal with future decisions. Too often, interventions are based on insufficient evidence (i.e., gut feelings of managers). For example, approximately 340 black-backed jackals (*Canis mesomelas*) were culled in South Africa's Addo and Karoo National Parks on the assumption that they were responsible for declines in populations of small ungulates; however, this was not tested (Bega 2010). Removal of large predators from African reserves was also based on gut feelings until we had methods to determine whether they were exceeding carrying capacity (Hayward et al. 2007c), where now we can use this knowledge to implement more natural interventions (Ferreira and Hofmeyr 2014).

Because monitoring programs are often designed to determine and enhance initial population growth, they may be inappropriate to inform later management challenges. Managers are often compelled to act in response to increasing populations, with limited resources that have to be drawn from their routine budgets. This makes planning interventions, implementing them, and learning from them secondary to immediate action limited by local capacity and resources. Reintroductions often involve high-profile species, such that any interventions to manage them once populations are increasing can be contentious, reducing transparency and consultation in the processes. In addition, because of the expense of maintaining long-term monitoring or of conducting long-term research projects, these tend to have dissipated at exactly the time when learning from doing would be most valuable. Further, long-term monitoring data often are needed to establish the benchmarks required to obtain a mechanistic understanding of ecological processes (Jachowski et al., Chapter 9, this volume). Therefore, there is often a strong disconnect between management, and knowledge and understanding, both in terms of using broad knowledge and understanding to crystalize the best possible approaches in the specific circumstances, as well as the ability to learn from any intervention to improve practice in the future.

Our personal experience in working on reintroduced species has highlighted this conundrum. Much of what we have learned has been from studying reintroduced populations in partnership with managers responsible for dealing with them on the ground. This has been focused around key, substantial interventions that have often been innovative solutions, and by studying them as they unfolded we have had the opportunity to feed into the design of the intervention, into the monitoring programs associated with them, as well as in assessing and evaluating the outcome with a mechanistic understanding of factors influencing success. The peer review process then makes the conclusions more focused, more robust, and more accessible for shaping future thinking and policy. For example, based on the understanding that older elephant males are required as part of a functional population, South African Norms and Standards for the management of elephant now require reserves to introduce older males (DEAT 2008). By working with managers through a number of different inter-

ventions on species, in reserves and/or outside of them, that provide ideal case studies for learning, we, and other authors who have been working with a similar approach, have generated a body of knowledge, which could and should inform best practice in managing reintroduced populations.

We have been fortunate in that as researchers, we have often been able to live and work on-site with managers, which both foster these relationships and allow data to be collected almost as soon as new interventions are planned. We recognize that this is not always the case. We believe that it is critical that, as reintroduced populations grow, managers draw on scientists to inform best-practice options, and provide further opportunity for adaptive learning. Such partnerships are essential given the complexity of natural systems and unpredictable responses that are often unintended consequences of interventions. Given the replication that often occurs in reintroduction programs and the disparate research teams that study them, there is scope for more collaboration among researchers to replicate their studies and thereby increase the robustness of the results.

Requirement for Management Interventions

Even in the largest conservation areas, ongoing conservation management may be required. Kruger National Park in South Africa is 21,000 km² (as big as Wales), but may require management of some species (e.g., elephants [Biggs et al. 2008]). Yellowstone National Park in the United States is 8983 km², but the wolf reintroduction there has been so successful that the government has now implemented management activities to limit their abundance outside the park (National Park Service, n.d.). At the same time, the assumption that conservation managers can afford a "hands-off" approach to allow reintroduced populations in relatively small areas to increase is not valid. Management decisions, including the decision to do nothing and let the system run, should be grounded in evidence-based best practice aimed not only at achieving the objectives of the reintroduction program for that species, but also at achieving the broader objectives of that conservation area in general. Due consideration needs to be given as to whether we need to act as god, as well as ensuring that we take responsibility to intervene when necessary to achieve success—a balanced approach framed in conservation objectives, and based on best practice, is needed.

Constraints from Government

Government agencies are inherently risk averse, yet the escalating biodiversity catastrophe means we need to act quickly, effectively, and innovatively (Steffen et al. 2011). Governments need to become part of the solution rather than inhibiting conservation actions with red tape and bureaucracy. For example, plans to reintroduce beavers (*Castor castor*) to Scotland (Scottish Beaver Trial) have been ongoing since 1995, and by May 2016 a decision still had not been made. The Welsh Beaver Trial has been in planning for 13 years, and beavers still are not officially allowed to live in Wales. Populations have turned up throughout the United Kingdom through unofficial channels, making the official government decision-making process redundant, and illustrating the waste of time and resources it has caused. In this context, we are past the decision to reintroduce and should be planning effectively for management of the population to achieve longer-term objectives. This problem similarly exists in North America, where returning wood bison (*Bison bison*) to Alaska has taken even longer—23 years (Joling 2015).

Private land owners and organizations can be a driver of reintroductions and facilitate novel reintroduction initiatives. The example cited above on African wild dogs is illustrative. While planning driven by large conservation agencies was delaying action, private reserves proceeded with dog reintroductions, which were successful (including using captive-bred animals) and which provided more dogs for additional introductions to more reserves. The endpoint was a successful metapopulation that did not compromise the large Kruger Park population through removals.

Disparity between Countries

The risk aversion mentioned above varies between countries. Some countries are highly risk averse, whereas others have led the way with intensive and innovative conservation management actions. For example, compare the plans of making New Zealand feral-free (Russell et al. 2015), with the acceptance of and hesitancy to control introduced gray squirrels (*Sciurus carolinensis*) in the United Kingdom or foxes and cats in Australia. This disparity may be driven by differences in the key threatening processes affecting different countries, where some are easier to mitigate than others. However, minimizing the bureaucratic differences between countries is likely to assist in improving the status of threatened species—such as has been seen in the recovery of the endangered black-footed ferret (*Mustela nigripes*) in North America (Jachowski and Lockhart 2009), and been proposed for elephant metapopulation management in Southern Africa (van Aarde and Jackson 2007).

The private sector traditionally leads innovation where competition is high, but conservation tends to lack competition so ideas stagnate and action is slow (Innes et al. 2015). The presence of private conservation organizations that own and manage land for conservation (e.g., ecotourism reserves in South Africa, nongovernment organizations like AWC in Australia, and The Nature Conservancy in the United States), alongside government-run conservation agencies, is likely to stimulate the competition necessary to improve conservation outcomes.

SUMMARY

This chapter describes a range of conservation management interventions that have been used to improve the success of reintroduced populations, and to limit problems arising. We discuss the role of fencing for conservation, introduced predator control, restoring apex predators, habitat manipulation with water and fire, conflict mitigation to reduce mortality rates of territorial species, continued reinforcement of small populations, and metapopulation management as strategies to sustain reintroduced populations. We also discuss the need to alter negative perceptions of overpopulation following reintroduction, and the use of park expansion and creating new reserves to house overabundant species, metapopulation management, and culling and/or sport hunting to reduce populations. We highlight some lessons learnt (including the importance of longer-term monitoring of reintroductions); the ongoing management requirements of reintroductions to recognize that overabundance is not a failure of a reintroduction program, but rather a great success that provides broader opportunities; and the need to reduce disparity among countries in using reintroduction as a conservation strategy, which may be driven by excessive bureaucracy. We conclude that reintroduction is a successful conservation intervention and that more should be done to minimize the bureaucratic red tape limiting such actions. Furthermore, to reduce costs and maximize monitoring value, we suggest tighter links between conservation managers and academic/research institutions over long time frames to ensure adequate data are collected for adaptive management, as well as ongoing communication to conservation stakeholders.

MANAGEMENT RECOMMENDATIONS

- Reintroduction is a successful conservation action that has improved the status of numerous threatened species (Hayward 2011, Luther et al. 2016). Yet implementing a reintroduction program requires an acknowledgment that ongoing conservation management will probably be necessary and that this needs to be adequately funded.

- Fundamental to reintroduction success are the separation of reintroduced individuals from the key threatening process(es) and adequate monitoring of the program. We need to structure reintroduction programs better in terms of clarity of objectives, interventions to achieve these objectives, monitoring of their success, and robust reporting on what has been achieved and learned. We need to move from a siloed approach to a more open, transparent, and accountable one, where we draw on broad expertise in planning reintroductions to stipulate what we hope to achieve from a population perspective, and how we will implement and monitor the program to ensure adaptive learning.

- Metapopulation approaches offer strong opportunity, especially for large species that require extensive areas for sustainable population sizes, which are clearly not available in many parts of the world within the distributions of many threatened species that could benefit from reintroduction.

- Manipulative interventions, such as contraception and culling, are likely to be increasingly important as we manage dwindling biodiversity in smaller and more isolated habitat fragments, where the risk of overpopulation looms large and therefore requires management.

- Clearly, with close interaction between managers and researchers there are huge opportunities for advances in our understanding of ecological systems when such manipulations are incorporated into robust experimental designs.

LITERATURE CITED

Allen, B. L., L. R. Allen, R. M. Engeman, and L. K. P. Leung. 2014. Sympatric prey responses to lethal top-predator control: predator manipulation experiments. Frontiers in Zoology 11:56.

Allen, B. L., and P. J. S. Fleming. 2012. Reintroducing the dingo: the risk of dingo predation to threatened vertebrates of western New South Wales. Wildlife Research 39:35–50.

AWC (Australian Wildlife Conservancy). 2004. Wildlife Matters October 2004. Australian Wildlife Conservancy, Subiaco East, WA, 8pp.

AWC. 2014. Wildlife Matters Summer 2013/14. Australian Wildlife Conservancy, Subiaco East, WA, 12pp.

Balme, G., R. Slotow, and L. Hunter. 2010. Edge effects and the impact of non-protected areas in carnivore conservation: leopards in the Phinda–Mkhuze Complex, South Africa. Animal Conservation 13:315–323.

Bega, S. 2010. SANParks caught in jackal cull row. Independent Online. Available online at: http://www.iol.co.za/news/science/sanparks-caught-in-jackal-cull-row-1.722678#.VQLezmP-uSo South Africa.

Bertschinger, H., T. Trigg, W. Jochle, and A. Human. 2002. Induction of contraception in some African wild carnivores by downregulation of LH and FSH secretion using the GnRH analogue deslorelin. Reproduction: Cambridge Supplement 60:41–52.

Biggs, H. C., R. Slotow, B. Scholes, J. Carruthers, R. Van Aarde, G. Kerley, W. Twine, D. Grobler, H. Bertschinger, and C. Grant. 2008. Towards integrated decision-making for elephant management. Pp.530–578. In R. J. Sholes and K. G. Mennell, eds., Assessment of South African Elephant Management. Witwatersrand University Press, Johannesburg, South Africa.

Bradfield, M. n.d. Information for Guides Operating in the Addo Elephant National Park. South African National Parks, Addo, South Africa.

Burbidge, A. A., M. Byrne, D. Coates, S. T. Garnett, S. Harris, M. W. Hayward, T. G. Martin et al. 2011. Is Australia ready for assisted colonisation? Policy changes required to facilitate translocation under climate change. Pacific Conservation Biology 17:259–269.

Burbidge, A. A., and N. L. McKenzie. 1989. Patterns in the modern decline of Western Australia's vertebrate fauna: causes and conservation implications. Biological Conservation 50:143–198.

Carruthers, J. 2008. "Wilding the farm or farming the wild"? The evolution of scientific game ranching in South Africa from the 1960s to the present. Transactions of the Royal Society of South Africa 63:160–181.

Caughley, G. 1970. Eruption of ungulate populations, with emphasis on Himalayan tahr in New Zealand. Ecology 51:53–72.

Caughley, G. 1994. Directions in conservation biology. Journal of Animal Ecology 63:215–244.

Chapron, G., P. Kaczensky, J. D. Linnell, M. Von Arx, D. Huber, H. Andrén, J. V. López-Bao, M. Adamec, F. Álvares, and O. Anders. 2014. Recovery of large carnivores in Europe's modern human-dominated landscapes. Science 346:1517–1519.

Creel, S., M. Becker, S. Durant, J. M'soka, W. Matandiko, A. Dickman, D. Christianson, E. et al. 2013. Conserving large populations of lions: the argument for fences has holes. Ecology Letters 16:1413–e3.

Creel, S., and J. J. Rotella. 2010. Meta-analysis of relationships between human offtake, total mortality and population dynamics of gray wolves (Canis lupus). PLoS ONE 5(9):e12918.

Creel, S., M. Becker, D. Christianson, E. Droge, N. Hammerschlag, M. W. Hayward, U. Karanth et al. 2015. Questionable policy for large carnivore hunting. Science 350:1473–1475.

Crooks, K. R., and M. E. Soule. 1999. Mesopredator release and avifaunal extinctions in a fragmented system. Nature 400:563–566.

Davies-Mostert, H. T., M. G. L. Mills, and D. W. Macdonald. 2009. South Africa's wild dog Lycaon pictus meta-population management programme. Pp.10–42. In M. W. Hayward and M. J. Somers, eds., The Reintroduction of Top-Order Predators. Wiley-Blackwell, Oxford, UK.

de Tores, P. J., and N. J. Marlow. 2011. The relative merits of predator exclusion fencing and repeated 1080 fox baiting for protection of native fauna: five case studies from Western Australia. Pp.21–42. In M. J. Somers and M. W. Hayward, eds., Fencing for Conservation. Springer, New York.

DEAT. 2008. National Norms and Standards for the Management of Elephants in South Africa. Government Gazette, Department of Environment and Tourism, Pretoria, South Africa.

Delsink, A. K., J. Kirkpatrick, J. Van Altena, H. J. Bertschinger, S. M. Ferreira, and R. Slotow. 2013b. Lack of spatial and behavioral responses to immunocontraception application in African elephants (Loxodonta Africana). Journal of Zoo and Wildlife Medicine 44:S52–S74.

Delsink, A. K, J. Van Altena, D. Grobler, H. Bertschinger, J. Kirkpatrick, and R. Slotow. 2006. Regulation of a small, discrete African elephant population through immunocontraception in the Makalali Conservancy, Limpopo, South Africa. South African Journal of Science 102:403–405.

Delsink, A. K., J. Van Altena, D. Grobler, H. J. Bertschinger, J. F. Kirkpatrick, and R. Slotow.

2007. Implementing immunocontraception in free-ranging African elephants at Makalali Conservancy. Journal of the South African Veterinary Association 78:25–30.

Delsink, A., A. T. Vanak, S. Ferreira, and R. Slotow. 2013a. Biologically relevant scales in large mammal management policies. Biological Conservation 167:116–126.

Di Minin, E., L. T. Hunter, G. A. Balme, R. J. Smith, P. S. Goodman, and R. Slotow. 2013. Creating larger and better connected protected areas enhances the persistence of big game species in the Maputaland-Pondoland-Albany biodiversity hotspot. PLoS ONE 8:e71788.

Dickman, C. R. 2011. Fences or ferals? Benefits and costs of conservation fencing in Australia. Pp.43–64. In M. J. Somers and M. W. Hayward, eds., Fencing for Conservation. Springer, New York.

Druce, H. C., R. Mackey, K. Pretorius, and R. Slotow. 2013. The intermediate-term effects of PZP immunocontraception: behavioural monitoring of the treated elephant females and associated family groups. Animal Conservation 16:180–187.

Druce, H. C., R. L. Mackey, and R. Slotow. 2011. How immunocontraception can contribute to elephant management in small, enclosed reserves: Munyawana population as a case study. PLoS ONE 6:e27952.

Druce, H. C., K. Pretorius, and R. Slotow. 2008. The response of an elephant population to conservation area expansion: Phinda Private Game Reserve, South Africa. Biological Conservation 141:3127–3138.

Emslie, R. H., R. Amin, and R. Kock, eds. 2009. Guidelines for the in situ re-introduction and translocation of African and Asian rhinoceros. IUCN, Gland, Switzerland. Available online at: https://portals.iucn.org/library/efiles/html /SSC-OP-039/cover.html (accessed on November 18, 2015).

Ewen, J. G., and D. P. Armstrong. 2007. Strategic monitoring of reintroductions in ecological restoration programmes. Ecoscience 14:401–409.

Ferreira, S. M., and M. Hofmeyr. 2014. Managing charismatic carnivores in small areas: large felids in South Africa. South African Journal of Wildlife Research 44:32–42.

Fischer, J., and D. B. Lindenmayer. 2000. An assessment of the published results of animal relocations. Biological Conservation 96:1–11.

Frank, A. S. K., C. N. Johnson, J. M. Potts, A. Fisher, M. J. Lawes, J. C. Z. Woinarski, K. Tuft et al. 2014. Experimental evidence that feral cats cause local extirpation of small mammals in Australia's tropical savannas. Journal of Applied Ecology 51:1486–1493.

Friend, J. A., M. Page, and S. Gilfillan. 2013. National Recovery Plan for the Numbat. Department of Environment, Water, Heritage and the Arts, Canberra, Australia.

Gadd, M. E. 2012. Barriers, the beef industry and unnatural selection: a review of the impacts of veterinary fencing on mammals in southern Africa. Pp.153–186. In M. W. Hayward and M. J. Somers, eds., Fencing for Conservation: Restriction of Evolutionary Potential or a Riposte to Threatening Processes? Springer, New York.

Garrott, R. A., P. White, and C. A. Vanderbilt White. 1993. Overabundance: an issue for conservation biologists? Conservation Biology 7:946–949.

Gedeon, C., O. Váczi, B. Koósz, and V. Altbäcker. 2011. Morning release into artificial burrows with retention caps facilitates success of European ground squirrel (Spermophilus citellus) transloca-tions. European Journal of Wildlife Research 57:1101–1105.

Gough, K., and G. I. H. Kerley. 2006. Lack of density dependent regulation in Addo's elephant population. Oryx 40:464–441.

Grant, C. C., R. Bengis, D. Balfour, M. J. S. Peel, W. Mostert, H. Killian, R. Little, I. P. J. Smit, M. Garai, and M. Henley. 2007. Controlling the distribution of elephant. Pp.1–51. In R. J. Scholes and K. G. Mennell, eds., Assessment of South African Elephant Management. South African Government, Pretoria, South Africa.

Greaver, C., S. M. Ferreira, and R. Slotow. 2014. Density-dependent regulation of the critically endangered black rhinoceros population in Ithala Game Reserve, South Africa. Austral Ecology 39:437–447.

Griffith, B., J. M. Scott, J. W. Carpenter, and C. Reed. 1989. Translocation as a species conservation tool: status and strategy. Science 245:477–480.

Gusset, M., O. Jakoby, M. Muller, M. J. Somers, R. Slotow, and V. Grimm. 2009. Dogs on the catwalk: modeling the re-introduction of endangered wild dogs in South Africa. Biological Conservation 142:2774–2781.

Gusset, M., S. J. Ryan, M. Hofmeyr, G. Van Dyk, H. T. Davies-Mostert, J. A. Graf, C. Owen et al. 2008. Efforts going to the dogs? Evaluating attempts to re-introduce endangered wild dogs in South Africa. Journal of Applied Ecology 45:100–108.

Hall-Martin, A., and M. Knight. 1994. Conserva-tion and management of black rhinoceros in South African national parks. Proceedings of a

Symposium on Rhinos as Game Ranch Animals, September 9–10, Onderstepaart, South Africa, pp.11–19.

Hayward, M. W. 2011. Using the IUCN Red List to determine effective conservation strategies. Biodiversity and Conservation 20:2563–2573.

Hayward, M. W. 2012. Perspectives on fencing for conservation based on four case studies: marsupial conservation in Australian forests; bushmeat hunting in South Africa; large predator reintroduction in South Africa; and large mammal conservation in Poland. Pp.7–21. In M. J. Somers and M. W. Hayward, eds., Fencing for Conservation. Springer, New York.

Hayward, M. W., J. Adendorff, J. O'Brien, A. Sholto-Douglas, C. Bissett, L. C. Moolman, P. Bean et al. 2007a. Practical considerations for the reintroduction of large, terrestrial, mammalian predators based on reintroductions to South Africa's Eastern Cape Province. The Open Conservation Biology Journal 1:1–11.

Hayward, M. W., J. Adendorff, J. O'Brien, A. Sholto-Douglas, C. Bissett, L. C. Moolman, P. Bean et al. 2007b. The reintroduction of large carnivores to the Eastern Cape Province, South Africa: an assessment. Oryx 41:205–214.

Hayward, M. W., L. Boitani, N. D. Burrows, P. Funston, K. U. Karanth, D. MacKenzie, K. Pollock, and R. W. Yarnell. 2015a. Ecologists need to use robust survey design, sampling and analysis methods. Journal of Applied Ecology 52:286–290.

Hayward, M. W., P. J. de Tores, and P. B. Banks. 2005. Habitat use of the quokka Setonix brachyurus (Macropodidae: Marsupialia) in the Northern Jarrah Forest of Australia. Journal of Mammalogy 86:683–688.

Hayward, M. W., P. J. de Tores, M. J. Dillon, and B. J. Fox. 2003. Local population structure of a naturally occurring metapopulation of the quokka (Setonix brachyurus Macropodidae: Marsupialia). Biological Conservation 110:343–355.

Hayward, M. W., K. Herman, and E. Mulder. 2010. Update of Australian Wildlife Conservancy Re-introductions. Reintroduction Specialist Group e-Newsletter 1:11–12.

Hayward, M. W., and G. I. H. Kerley. 2009. Fencing for conservation: restriction of evolutionary potential or a riposte to threatening processes? Biological Conservation 142:1–13.

Hayward, M. W., R. Kowalczyk, Z. A. Krasiński, M. Krasińska, J. Dackiewicz, and T. Cornulier. 2011. Restoration and intensive management have no effect on evolutionary strategies. Endangered Species Research 15:53–61.

Hayward, M. W., K. E. Moseby, and J. L. Read. 2014. The role of predator exclosures in the conservation of Australian fauna. Pp.363–379. In A. S. Glen and C. R. Dickman, eds., Carnivores of Australia. CSIRO Publishing, Heidelberg/Melbourne, Australia.

Hayward, M. W., J. O'Brien, and G. I. H. Kerley. 2007c. Carrying capacity of large African predators: predictions and tests. Biological Conservation 139:219–229.

Hayward, M. W., A. S. L. Poh, J. Cathcart, C. Churcher, J. Bentley, K. Herman, L. Kemp et al. 2015b. Numbat nirvana: the conservation ecology of the endangered numbat (Myrmecobius fasciatus) (Marsupialia: Myrmecobiidae) reintroduced to Scotia and Yookamurra Sanctuaries, Australia. Australian Journal of Zoology 63(4):258–269.

Hayward, M. W., G. Ward-Fear, F. L'Hotellier, K. Herman, A. P. Kabat, & J. P. Gibbons. (2016) Could biodiversity loss have increased Australia's bushfire threat? Animal Conservation Online early DOI: 10.1111/acv.12269.

Hofmeyr, M., R. Davies, P. Nel, and S. Dell. 2003. Operation Phoenix: the introduction of larger mammals to Madikwe Game Reserve. Pp.8–20. In M. Brett, ed., Madikwe Game Reserve: A Decade of Progress. North West Parks and Tourism Board, Rustenberg, South Africa.

Hunter, L. T. B., K. Pretorius, L. C. Carlisle, M. Rickelton, C. Walker, R. Slotow, and J. D. Skinner. 2007. Restoring lions Panthera leo to northern KwaZulu-Natal, South Africa: short-term biological and technical success but equivocal long-term conservation. Oryx 41:196–204.

Innes, J., B. Burns, A. Sanders, and M. W. Haywar. 2015. The impact of private sanctuary networks on reintroduction programmes in Australia and New Zealand. Pp.185–199. In D. P. Armstrong, M. W. Hayward, D. Moro, and P. J. Seddon, eds., Reintroduction Biology in Australia and New Zealand. CSIRO Publishing, Melbourne, Australia.

Innes, J., G. Lee, B. Burns, C. Campbell-Hunt, C. Watts, H. Phipps, and T. Stephens. 2012. Role of predator-proof fences in restoring New Zealand's biodiversity: a response to Scofield et al. (2011). New Zealand Journal of Ecology 36:232–238.

IUCN/SSC. 2013. Guidelines for Reintroductions and other Conservation Translocations. Reintroduction Specialist Group of the IUCN Species Survival Commission, Gland, Switzerland.

Jachowski, D. S., and J. M. Lockhart. 2009. Reintroducing the black-footed ferret (Mustela nigripes) to the Great Plains of North America. Small Carnivore Conservation 41:58–64.

Jachowski, D. S., R. Sloto, and J. J. Millspaugh. 2014. Good virtual fences make good neighbors: opportunities for conservation. Animal Conservation 17:187–196.

Jackson, C. R., J. W. McNutt, and P. J. Apps. 2012. Managing the ranging behaviour of African wild dogs (*Lycaon pictus*) using translocated scent marks. Wildlife Research 39:31–34.

Joling, D. 2015. Alaska prepares for wood bison return after a century. KLS.com. Available online at: http://www.ksl.com/?nid=1012&sid=33902569.

Kane, A., A. Jackson, A. Monadjem, M. Colomer, and A. Margalida. 2014. Carrion ecology modelling for vulture conservation: are vulture restaurants needed to sustain the densest breeding population of the African white-backed vulture? Animal Conservation 18:279–286.

Keedwell, R. J., R. F. Maloney, and D. P. Murray. 2002. Predator control for protecting kaki (*Himantopus novaezelandiae*): lessons from 20 years of management. Biological Conservation 105:369–374.

Kerley, G. I. H., and A. F. Boshoff. 1997. A proposal for a Greater Addo National Park: a regional and national conservation and development opportunity. Terrestrial Ecology Research Unit, Department of Zoology, University of Port Elizabeth, Port Elizabeth, South Africa.

Kettles, R., and R. Slotow. 2009. Management of free-ranging lions on an enclosed game reserve. South African Journal of Wildlife Research 39:23–33.

Konstant, W. R., and R. A. Mittermeier. 1982. Introduction, reintroduction and translocation of Neotropical primates: past experiences and future possibilities. International Zoo Yearbook 22:69–77.

Landsberg, J., C. James, S. R. Morton, T. J. Hobbs, J. Stol, A. Drew, and H. Tongway. 1997. The Effects of Artificial Sources of Water on Rangeland Biodiversity. CSIRO Division of Wildlife and Ecology; and Biodiversity Convention and Strategy Section of the Biodiversity Group. Environment Australia, Canberra.

Letnic, M., M. S. Crowther, and F. Koch. 2009. Does a top-predator provide an endangered rodent with refuge from an invasive mesopredator? Animal Conservation 12:302–312.

Lindsey, P. A., J. T. du Toit, and M. G. L. Mills. 2004. Area and prey requirements of African wild dogs under varying habitat conditions: implications for reintroductions. South African Journal of Wildlife Research 34:77–86.

Lindsey, P. A., C. J. Tambling, R. Brummer, H. T. Davies-Mostert, M. W. Hayward, K. A. Marnewick, and D. M. Parker. 2011. Minimum prey and area requirements of cheetahs: implications for reintroductions and management of the species as a managed metapopulation. Oryx 45:587–599.

Linklater, W. L., K. Adcock, P. du Preez, R. R. Swaisgood, P. R. Law, M. H. Knight, J. V. Gedir, and G. I. H. Kerley. 2011. Guidelines for large herbivore translocation simplified: black rhinoceros case study. Journal of Applied Ecology 48:493–502.

Linklater, W. L., and I. R. Hutcheson. 2010. Black rhinoceros are slow to colonize a harvested neighbour's range. South African Journal of Wildlife Research 40:58–63.

Loveridge, A. J., A. W. Searle, F. Murindagomo, and D. Macdonald. 2007. The impact of sport-hunting on the population dynamics of an African lion population in a protected area. Biological Conservation 134:548–558.

Lunney, D. 2001. Causes of the extinction of native mammals of the western division of New South Wales: an ecological interpretation of the nineteenth century historical record. Rangeland Journal 23:44–70.

Lute, M. L., and M. L. Gore. 2014. Knowledge and power in wildlife management. The Journal of Wildlife Management 78:1060–1068.

Luther, D. A., T. M. Brooks, S. H. M. Butchart, M. W. Hayward, M. E. Kester, J. Lamoreux, and A. Upgren. 2016. Determinants of bird conservation action implementation and associated population trends of threatened species. Conservation Biology. doi:10.1111/cobi.12757.

Mackey, R. L., B. R. Page, K. J. Duffy, and R. Slotow. 2006. Modelling elephant population growth in small, fenced, South African reserves. South African Journal of Wildlife Research 36:33–43.

Mahon, P. S. 2010. Targeted control of widespread exotic species for biodiversity conservation: the red fox (*Vulpes vulpes*) in New South Wales, Australia. Ecological Management and Restoration 10:S59–S69.

Marlow, N. J., N. D. Thomas, A. A. E. Williams, B. Macmahon, J. Lawson, Y. Hitchen, J. Angus, and O. Berry. 2015. Cats (*Felis catus*) are more abundant and are the dominant predator of woylies (*Bettongia penicillata*) after sustained fox (*Vulpes vulpes*) control. Australian Journal of Zoology 63:18–27.

Marnewick, K. A., M. W. Hayward, D. Cilliers, and M. J. Somers. 2009. Survival of cheetahs relocated from ranchland to fenced protected areas in South Africa. Pp. 282–306. In M. W. Hayward and M. J. Somers, eds., The Reintroduction of Top-Order Predators. Wiley-Blackwell, Oxford, UK.

Martin, T. G., S. Nally, A. A. Burbidge, S. Arnall, S. Garnett, S. Harris, M. W. Hayward et al. 2012. Acting fast avoids extinction: plight of the Christmas Island pipistrelle and Orange-bellied Parrot. Conservation Letters 5:274–280.

McCarthy, M. A., D. P. Armstrong, and M. C. Runge. 2012. Adaptive management of reintroduction. Pp.156–289. In J. G. Ewen, D. P. Armstrong, K. A. Parker, and P. J. Seddon, eds., Reintroduction Biology: Integrating Science and Management. Wiley-Blackwell, London, UK.

McGregor, H. W., S. Legge, M. E. Jones, and C. N. Johnson. 2014. Landscape management of fire and grazing regimes alters the fine-scale habitat utilisation by feral cats. PLoS ONE 9:e109097.

McNutt, J. W. 1996. Sex-biased dispersal in African wild dogs, *Lycaon pictus*. Animal Behaviour 52:1067–1077.

Miller, S., C. Bissett, A. Burger, B. Courtenay, T. Dickerson, D. Druce, S. Ferreira, P. Funston, D. Hofmeyr, and P. Kilian. 2013. Management of reintroduced lions in small, fenced reserves in South Africa: an assessment and guidelines. South African Journal of Wildlife Research 43:138–154.

Moreno-Opo, R., A. Trujillano, Á. Arredondo, L. M. González, and A. Margalida. 2015. Manipulating size, amount and appearance of food inputs to optimize supplementary feeding programs for European vultures. Biological Conservation 181:27–35.

National Park Service. n.d. Information on the 2012-13 wolf hunt near Yellowstone National Park. Available online at: http://www.nps.gov /yell/learn/nature/wolfhunt.htm.

Nichols, J. D., and D. P. Armstrong. 2012. Monitoring for reintroductions. Pp.223–255. In J. G. Ewen, D. P. Armstrong, K. A. Parker, and P. J. Seddon, eds., Reintroduction Biology: Integrating Science and Management. Wiley-Blackwell, London, UK.

Orford, H. J. L., and M. R. Perrin. 1988. Contraception, reproduction and demography of free-ranging Etosha lions (*Panthera leo*). Journal of Zoology 216:717–733.

Orros, M. E., and M. D. Fellowes. 2015. Widespread supplementary feeding in domestic gardens explains the return of reintroduced Red Kites *Milvus milvus* to an urban area. Ibis 157:230–238.

Oyegbile, T. O., and C. A. Marler. 2005. Winning fights elevates testosterone levels in California mice and enhances future ability to win fights. Hormones and Behavior 48:259–267.

Packer, C., A. J. Loveridge, S. Canney, T. M. Caro, S. T. Garnett, M. Pfeifer, K. K. Zander et al. 2013.

Conserving large carnivores: dollars and fence. Ecology Letters 16:635–641.

Penzhorn, B. L. 1971. A summary of the re-introduction of ungulates into South African National Parks (to 31 December 1970). Koedoe 14:145–159.

Pfeifer, M., C. Packer, A. C. Burton, S. T. Garnett, A. J. Loveridge, D. MacNulty, and P. J. Platts. 2014. In defense of fences. Science 345:389.

Possingham, H. P., P. Jarman, and A. Kearns. 2004. Independent review of Western Shield – February 2003. Conservation Science Western Australia 5:2–11.

Rankmore, B., A. Griffiths, J. Woinarski, B. Ganambarr, R. Taylor, K. Brennan, K. Firestone, and M. Cardoso. 2008. Island Translocation of the Northern Quoll *Dasyurus hallucatus* as a Conservation Response to the Spread of the Cane Toad *Chaunus (Bufo) marinus* in the Northern Territory, Australia. Report to the Australian Government's Natural Heritage Trust. Department of Natural Resources, Environment and the Arts, Darwin, Australia.

Ripple, W. J., A. J. Wirsing, C. C. Wilmers, and M. Letnic. 2013. Widespread mesopredator effects after wolf extirpation. Biological Conservation 160:70–79.

Ritchie, E. G., B. Elmhagen, A. S. Glen, M. Letnic, G. Ludwig, and R. A. McDonald. 2012. Ecosystem restoration with teeth: what role for predators? Trends in Ecology and Evolution 27:265–271.

Roux, D. J. and L. C. Foxcroft. 2011. The development and application of strategic adaptive management within South African National Parks. Koedoe 53:1–5.

Rowe-Rowe, D. T. 1992. The Carnivores of Natal. Natal Parks Board, Pietermaritzburg, South Africa.

Rubio-Martínez, L. M., D. A. Hendrickson, M. Stetter, J. R. Zuba, and H. J. Marais. 2014. Laparoscopic vasectomy in African Elephants (*Loxodonta africana*). Veterinary Surgery 43:507–514.

Russell, J. C., J. G. Innes, P. H. Brown, and A. E. Byrom. 2015. Predator-free New Zealand: conservation country. BioScience 65:520–525.

Sapkota, S., A. Aryal, S. R. Baral, M. W. Hayward, and D. Raubenheimer. 2014. Economic analysis of electric fencing for mitigating human-wildlife conflict in Nepal. Journal of Resources and Ecology 5:237–243.

Sapsford, S. J., M. J. Voordouw, R. A. Alford, and L. Schwarzkopf. 2015. Infection dynamics in frog populations with different histories of decline caused by a deadly disease. Oecologia 179:1099–1110.

Saunders, A., and D. Norton. 2001. Ecological restoration at mainland islands in New Zealand. Biological Conservation 99:109–119.

Seddon, P. J. 1999. Persistence without intervention: assessing success in wildlife reintroductions. Trends in Ecology and Evolution 14:503.

Seddon, P. J., W. M. Strauss, and J. Innes. 2012. Animal translocations: what are they and why do we do them. Pp.1–32. In J. Ewen, D. P. Armstrong, K. Parke, and P. J. Seddon, eds., Reintroduction Biology: Integrating Science and Management. Wiley-Blackwell, Oxford, UK.

Selier, S. A. J., B. R. Page, A. T. Vanak, and R. Slotow. 2014. Sustainability of elephant hunting across international borders in southern Africa: a case study of the greater Mapungubwe Transfrontier Conservation Area. The Journal of Wildlife Management 78:122–132.

Selva, N., T. Berezowska-Cnota, and I. Elguero-Claramunt. 2014. Unforeseen effects of supplementary feeding: ungulate baiting sites as hotspots for ground-nest predation. PLoS ONE 9:e90740.

Shannon, G., R. Slotow, S. M. Durant, K. N. Sayialel, J. Poole, C. Moss, and K. McComb. 2013. Effects of social disruption in elephants persist decades after culling. Frontiers in Zoology 10:62.

Short, J. 2009. The characteristics and success of vertebrate translocations within Australia. Department of Agriculture, Fisheries and Forestry, Canberra, Australia.

Short, J., S. D. Bradshaw, J. Giles, R. I. T. Prince, and G. R. Wilson. 1992. Reintroduction of macropods (Marsupialia: Macropodoidea) in Australia: a review. Biological Conservation 62:189–204.

Short, J., and A. P. Smith. 1994. Mammal decline and recovery in Australia. Journal of Mammalogy 75:288–297.

Slotow, R. 2011. Fencing for purpose: a case study of elephants in South Africa. Pp.91–104. In M. J. Somers and M. W. Hayward, eds., Fencing for Conservation. Springer, New York.

Slotow, R., D. Balfour, and O. Howison. 2001. Killing of black and white rhinoceroses by African elephants in Hluhluwe-Umfolozi Park, South Africa. Pachyderm 31:14–20.

Slotow, R, M. E Garaï, B. Reilly, B. Page, and R. D. Carr. 2005. Population ecology of elephants re-introduced to small fenced reserves in South Africa. South African Journal of Wildlife Research 35:23–32.

Slotow, R., and L. T. B. Hunter. 2009. Reintroduction decisions taken at the incorrect social scale devalue their conservation contribution: African lion in South Africa. Pp.43–71. In M. W. Hayward and M. J. Somers, eds., The Reintroduction of Top-order Predators. Wiley-Blackwell, Oxford, UK.

Slotow, R., and G. van Dyk. 2001. Role of delinquent young "orphan" male elephants in high mortality of white rhinoceros in Pilanesberg National Park, South Africa. Koedoe 44:85–94.

Slotow, R., G. van Dyk, J. Poole, B. Page, and A. Klocke. 2000. Older bull elephants control young males. Nature 408:425–426.

Slotow, R., I. J. Whyte, M. Hofmeyr, G. I. H. Kerley, and T. Conway. 2008. Lethal management of elephant. Pp.370–405. In R. J. Scholes and K. G. Mennell, eds., Assessment of South African Elephant Management. South African Government, Pretoria, South Africa.

Smales, I., P. Brown, P. Menkhorst, M. Holdsworth, and P. Holz. 2000. Contribution of captive management of orange-bellied parrots to the recovery programme for the species in Australia. International Zoo Yearbook 37:171–178.

Smit, I. P. J., C. C. Grant, and B. J. Devereux. 2007. Do artificial waterholes influence the way herbivores use the landscape? Herbivore distribution patterns around rivers and artificial surface water sources in a large African savanna park. Biological Conservation 136:85–99.

Smith, D. W., and E. E. Bangs. 2009. Reintroduction of wolves to Yellowstone National Park: history, values and ecosystem restoration. Pp.92–125. In M. W. Hayward and M. J. Somers, eds., The Reintroduction of Top-Order Predators. Wiley-Blackwell, Oxford, UK.

Snijders, D. 2012. Wild property and its boundaries—on wildlife policy and rural consequences in South Africa. Journal of Peasant Studies 39:503–520.

Somers, M. J., J. A. Graf, M. Szykman Gunther, R. Slotow, and M. Gusset. 2008. Dynamics of a small re-introduced population of wild dogs over 25 years: Allee effects and the implications of sociality for endangered species; recovery. Oecologia 158:239–247.

Soorae, P. S. 2008. Global Re-introduction Perspectives: Re-introduction Case-Studies from around the Globe. IUCN/SSC Re-introduction Specialist Group, Abu Dhabi, UAE.

Soorae, P. S. 2011. Global Re-introduction Perspectives: 2011. More case studies from around the globe. IUCN/SSC Re-introduction Specialist Group, Gland, Switzerland/ Environment Agency, Abu Dhabi, UAE.

Soorae, P. S. 2013. Global Re-introduction Perspectives: 2013. Further Case Studies From Around the Globe. IUCN/SSC Re-introduction Specialist Group, Gland, Switzerland/Environment Agency, Abu Dhabi, UAE.

Steffen, W., J. Grinevald, P. Crutzen, and J. McNeill. 2011. The Anthropocene: conceptual and historical perspectives. Philosophical Transactions of the Royal Society A: Mathematical, Physical and Engineering Sciences 369:842–867.

Sutherland, D. R., A. S. Glen, and P. J. De Tores. 2011. Could controlling mammalian carnivores lead to mesopredator release of carnivorous reptiles? Proceedings of the Royal Society of London (Series B) 278:641–648.

Traill, L. W., C. J. A. Bradshaw, and B. W. Brook. 2007. Minimum viable population size: a meta-analysis of 30 years of published estimates. Biological Conservation 139:159–166.

Trinkel, M., P. Funston, M. Hofmeyr, D. Hofmeyr, S. Dell, C. Packer, and R. Slotow. 2010. Inbreeding and density-dependent population growth in small, isolated lion populations. Animal Conservation 13:374–382.

Urbanek, R. P., E. K. Szyszkoski, and S. E. Zimorski. 2014. Winter distribution dynamics and implications to a reintroduced population of migratory whooping cranes. Journal of Fish and Wildlife Management 5:340–362.

van Aarde, R. J., and T. P. Jackson. 2007. Megaparks and metapopulations: addressing the causes of locally high elephant numbers in southern Africa. Biological Conservation 134:289–297.

van Aarde, R. J., I. J. Whyte, and S. L. Pimm. 1999. Culling and the dynamics of the Kruger National Park African elephant population. Animal Conservation 2:287–294.

Wayne, A. F., M. A. Maxwell, C. G. Ward, C. V. Vellios, B. G. Ward, G. L. Liddelow, I. Wilson, J. C. Wayne, and M. R. Williams. 2013. Importance of getting the numbers right: quantifying the rapid and substantial decline of an abundant marsupial, *Bettongia penicillata*. Wildlife Research 40:169–183.

Witmer, G. W., and P. L. Fuller. 2011. Vertebrate species introductions in the United States and its territories. Current Zoology 57:559–567.

Wolf, C. M., T. Garland, Jr., and B. Griffith. 1998. Predictors of avian and mammalian translocation success: reanalysis with phylogenetically independent contrasts. Biological Conservation 86:243–255.

Woodroffe, R., S. Hedges, and S. M. Durant. 2014. To fence or not to fence. Science 344:46–48.

Yarnell, R. W., W. L. Phipps, S. Dell, L. M. MacTavish, and D. M. Scott. 2014. Evidence that vulture restaurants increase the local abundance of mammalian carnivores in South Africa. African Journal of Ecology 53:287–294.

FOURTEEN

Outreach and Environmental Education for Reintroduction Programs

Anna L. George and Estelle A. Sandhaus

I F A REINTRODUCED CONDOR FLIES over the chaparral and no one is around to see it, does it still make a difference? Of course, the return of the California condor (*Gymnogyps californianus*) matters to both the species and the ecosystem, but the impact of a reintroduction program can be greatly amplified through targeted education and outreach programs. In some cases, the reintroduction may only be considered successful because of the associated education or public awareness campaign (Sanz and Grajal 1998). Whether the purpose is general environmental education or influencing policymakers for long-term support, outreach is a critical step in the success of most reintroduction programs.

However, outreach and environmental education programs have not been fully integrated into reintroduction guidelines, strategies, and monitoring standards that exist today (e.g., Armstrong and Seddon 2008, Sutherland et al. 2010, IUCN/SSC 2013). Community education has long been recognized as an important component of reintroduction. In the early 1990s, 70% of reintroductions considered to be successful contained a community education compo-

nent (Beck et al. 1994). Perhaps for some biologists, it enters an uncomfortable gray area between science and advocacy or requires a skill set that is not always taught or valued in academic programs (Van Heezik and Seddon 2005, Muir and Schwartz 2009). Others may mistakenly believe that conservation work should solely be an environmental science endeavor without consideration of the social sciences (Mascia et al. 2003, Heberlein 2012). Public education and outreach activities also are often not rewarded through the primary metrics that traditional science institutions use to evaluate scholarship: publications and grants (O'Meara 2005, Uriarte et al. 2007). Finally, money, time, and other resources continue to be large barriers to education as much as to conservation (Ardoin and Heimlich 2013). Despite these obstacles, this step can be exceptionally important in achieving a successful reintroduction. Most conservation practitioners value the ability to communicate to nonscientists more highly than do conservation academics (Muir and Schwartz 2009), and nowhere is the importance of this communication as clear as in reintroduction programs.

At the most basic level, outreach can be used to raise awareness of the target species for reintroduction. Thus, the first goal of an outreach program may simply be to introduce the broader public to that species. Very commonly, the reintroduced animal is used as a flagship or focal species to raise more concern in the general public about a problem or critical habitat that is threatened by human actions (Dietz et al. 1994, Lambeck 1997). This approach is a common tactic used by zoos and aquariums, particularly to guide exhibit messaging for charismatic animals. For instance, outreach was used with reintroduction of the yellow-shouldered Amazon parrot (*Amazona barbadensis*) to highlight the illegal pet trade (Sanz and Grajal 1998), while the Tennessee Aquarium used a small fish, the Barrens topminnow (*Fundulus julisia*), to educate on the role of indicator species (George et al. 2013). Public displays also can highlight some of the steps required to restore an endangered species, such as a program developed in a number of zoos for golden lion tamarins (*Leontopithecus rosalia*) that provided them with a complex rearing environment in which to learn appropriate behaviors while simultaneously allowing the visiting public to learn from an integral part of the reintroduction efforts (Stoinski et al. 1997).

A second goal for education and communications programs is to move from awareness of the program to influencing behavior of the general public. There are a number of human behaviors that may need to be addressed in order to protect reintroduced animals and their habitat. Since humans are the cause of death in 50% of cases involving reintroduced mammals, addressing negative attitudes toward the target species, such as with gray wolves (*Canis lupus*), can be critical to ensuring their survival (Jule et al. 2008). If animals are familiarized with humans during captive breeding, they may be at even more risk for negative interactions with the public (Fa et al. 2011). Sometimes, outreach needs to be targeted to a specific group of stakeholders, such as messages to funders and legislators in order to promote long-term support for the reintroduction effort or conservation policies, or citizen scientists, who may help monitor the reintroduced animal. Private breeders may merit specialized outreach, particularly where their breeding efforts might either advance propagation knowledge or yield additional individuals for reintroduction (Cade 1980, Borneman and Lowrie 2001, Rinkevich 2005).

Outreach and environmental education should be key outputs in reintroduction programs, much like publishing is a key output for scientific research. However, these have been overlooked components of reintroduction to date. Of the 239 projects detailed in 4 years of IUCN reintroduction case studies, less than 20% set education as one of the goals of the project (IUCN 2008, 2010, 2011, 2013). While there are few trends in which programs set education goals by taxonomic group, the IUCN-defined region with the highest percentage of projects with education goals was Meso and South America (Table 14.1). Meanwhile, 26 projects (11%) self-reported educational difficulties even though they did not set educational goals. Clearly, those undertaking a reintroduction should start early in their planning process to set outreach as a goal to help ensure success. While few reintroduction programs currently have staff devoted solely to education (Sutton and Lopez 2014), it is our hope that this topic becomes a more established part of program design and funding in the future.

Our purpose in this chapter is to provide a framework and case studies for incorporating education and outreach programs into reintroductions to help achieve success. While Chapter 4 draws a broad overview as to why human dimensions must be incorporated into every stage of reintroduction, this chapter more closely addresses specific education and outreach efforts that impact reintroduction success immediately prior to and during the program. Our proposed framework for reintroduction outreach is four-part, including (1) assessing current knowledge and attitudes, (2) identifying the goals and target audience(s), (3) crafting and executing programs, and (4) evaluating

TABLE 14.1

A Summary of 239 Self-reported IUCN Reintroduction Case Studies (IUCN 2008, 2010, 2011, 2013) Separated by Geographic Region (above) and Taxonomic Group (below) Showing the Number and Percentage for Each Group That Set Education Goals at the Onset of the Program

Geographic Region	Total	With Education Goals	Percent
North America and Caribbean	48	9	19
West Europe	43	9	21
South and East Asia	31	7	23
Oceania	53	6	11
Africa	27	5	19
West Asia	18	4	22
East Europe, North and Central Asia	10	1	10
Meso and South America	9	5	56
Taxonomic Group	Total	With Education Goals	Percent
Amphibians	14	1	7
Birds	46	6	13
Fishes	28	6	21
Invertebrates	18	4	22
Mammals	67	18	27
Plants	40	4	10
Reptiles	26	7	27

and adapting. This outreach framework assumes that the reintroduction program has passed the conceptualization stage discussed in Chapter 4. Thus, we expand upon designing and implementing education, outreach, and communications that aid in the success of the reintroduction during releases and subsequent monitoring efforts.

While many of the partners engaged in a reintroduction effort are already likely to be undertaking environmental education and outreach as part of their mission to protect natural resources, we believe that the reintroduction team is ultimately responsible for creating and implementing outreach and education specific to the reintroduction program as part of the steps necessary for reintroduction best practices (George et al. 2009, Maschinski et al. 2012, IUCN/SSC 2013). Deciding which part-

ners in the team are best suited for implementing different aspects of outreach is an important part of the overall plan, which we recommend be addressed by the reintroduction team once specific audiences have been identified. Therefore, this chapter is intended for conservation practitioners and managers who are designing and implementing a reintroduction program with little experience in education and communication, as well as for environmental educators who are partnering with a reintroduction team specifically to assist in designing effective outreach.

DEFINING OUTREACH AND EDUCATION

There are many different options available to those who want to share information about their reintroduction program, ranging from targeted education programs to social marketing campaigns that aim to shift behaviors. Reintroduction outreach should be flexible and diverse, targeting all stakeholders for a reintroduction program, from schoolchildren to funders to policymakers to resource users to citizen scientists. Reaching a diverse audience must also be a key consideration, in order to build a broad base of support for the program (Bonta and Jordan 2007).

We discuss principles from education, outreach, and social marketing that may be useful tools for those wanting to start or expand knowledge of reintroduction programs and support for them. For the purposes of this chapter, we distinguish *education* as programs with the goals of imparting awareness, knowledge, values, skills, and actions in participants (UNESCO 1978, Monroe et al. 2007). These efforts tend to take place in a more formal learning setting, such as classroom programs and teacher training, are often longer term, and encourage the development of critical thinking to prepare students to make informed decisions as citizens (Jacobson et al. 2006). Education programs tend to involve a two-way transmission of knowledge between teacher and student (Scott and Gough 2003, Jiménez et al. 2015). In contrast, we define *outreach* as communication or educational approaches that seek to increase awareness and inform an audience. Outreach may include a broad set of techniques that often rely on one-way transmission of information, from short programs in informal settings, such as field trips, interpretation, blogs, or science cafes, to marketing and communication pieces including media stories, advertisements, and public meetings (Jacobson et al. 2006, Jiménez et al. 2015).

We also draw a difference between both education and outreach as defined above and *social marketing* (Zaltman et al. 1972), with the end goal for the target audience to change a behavior or take action, even if the participant does not understand why that change matters. In contrast, the end goal of education is for the student to gain knowledge and skills that allow

informed decision-making about behaviors and actions, though the individual is left to determine what the best actions are (Monroe et al. 2007). Social marketing uses marketing principles to promote a targeted action that is thought to be beneficial for society, such as seat-belt use or antismoking campaigns across the United States. To date, environmental social marketing has largely been targeted at promoting sustainable behaviors in the public for energy or water conservation, and has not often been used at a species conservation level (McKenzie-Mohr 2000). Though social marketing has not been heavily used for reintroduction to date, knowledge of marketing principles can help reintroduction teams create more effective outreach programs when considering audiences and behaviors.

STAGE 1: ASSESS CURRENT ATTITUDES AND KNOWLEDGE

Before developing an educational or outreach program, program leaders should understand current attitudes and knowledge concerning the animal and its habitat. As extensively covered in Chapter 4, this is ideally undertaken prior to implementation of reintroduction for planning purposes as this can affect the success of the entire reintroduction. If an attitude assessment was not considered in the early conceptualization of the program, however, it should be undertaken prior to outreach as the resulting data will help guide future outreach plans, as well as be used for evaluation and adaptive management. For example, surveys about attitudes prior to gray wolf reintroduction programs indicated that farmers and ranchers held the most negative attitudes, indicating a key outreach need (Williams et al. 2002; Riley and Sandström, Chapter 4, this volume). In addition, attitude surveys can be used during and after the reintroduction to assess whether education and outreach are meeting established goals, and if not, what needs to be changed. Interestingly, as the population becomes more urban, attitudes toward nature may improve, as direct interaction with wild animals generally causes a decrease in positive attitudes (Williams et al. 2002, Louv 2008). However, this relationship can be complicated by those with no experience or strong feelings toward wildlife, demonstrating that attitudes may vary widely in different audiences, and even in the same audience at different time periods, depending on experiences those individuals have with the animal or its habitat (Heberlein and Ericsson 2005, Heberlein 2012). Because attitude and knowledge assessments were thoroughly covered in Chapter 4 and should be conducted prior to releases, we move on to a discussion of how to use the results in subsequent programming.

STAGE 2: IDENTIFY PURPOSE AND AUDIENCE

Attitudes toward the reintroduced animal and knowledge of it can impact behaviors undertaken by the general public and specific stakeholders toward the reintroduction program, ultimately determining success. Shaping the public's knowledge and behaviors is thus one of the major drivers for outreach. In some cases, the goal may be creating behaviors that help those managing the program, such as reporting a species sighting, or deterring behaviors that impact the reintroduced species, such as poaching. Some outreach may simply be to help residents recognize an animal that has been missing from the landscape for quite some time or understand that new regulations exist (Traweek 2012). Thus, one of the first steps in planning reintroduction outreach is combining results from the attitude surveys with knowledge of biological and management needs for the species to determine what human behaviors will be helpful or harmful to the species. In addition, these surveys will help distinguish specific audiences that are most likely to exhibit those specific behaviors so that outreach can be finely targeted.

As specific behaviors are identified, consideration should be given to what barriers exist to these behaviors in the population you are trying to change (McKenzie-Mohr 2000). A classic marketing model to examine where these barriers occur is abbreviated as AIDA: Awareness, Interest, Desire, and Action (Strong 1925). Is the goal of your education and outreach to raise awareness of a problem, create further interest in addressing the problem, increase the desire of the person to see the problem changed, or cause the person to act and change the problem on their own? Once the barriers to action in different audiences have been identified, you can design a program that works to remove them, whether it is creating awareness of an issue or making conservation actions easier. For example, one goal may be to raise awareness in the general public about the program, where the barrier may simply be a lack of knowledge that the species exists. Another goal may be to encourage specific user groups to report sightings of the reintroduced species, such as fishermen with game species where barriers may be lack of knowledge about how to report a sighting or fear of legal ramifications for handling a sensitive species (Box 14.1). Surveys and focus groups can be helpful in determining what the actual barrier is for a specific audience.

"In the end we will conserve only what we love; we will love only what we understand; and we will understand only what we are taught." Many conservation groups and environmental educators have been inspired by this beautiful quote delivered by Baba Dioum at the 1968 meeting of the IUCN (in Norse and Crowder 2005) to create programs that address awareness and interest in their audiences. However, conservation psychology research indicates that environmental knowledge is necessary, but not sufficient, for producing more positive environmental attitudes and pro-environmental behavior (Hines et al. 1986, Kollmuss and Agyeman 2002, Bamberg and Möser 2007, Levine and Strube 2012). Hines et al. (1986) identified six factors that influence pro-environmental behavior: a greater knowledge of environmental issues, greater awareness of impact of personal actions on environmental issues, a stronger belief that personal actions can make a difference, pro-environmental attitudes, verbal pledges of an intention to behave in an environmentally friendly way, and a greater sense of personal responsibility. While Bamberg and Möser (2007) indicate that awareness of the problem is an indirect step toward pro-environmental action, other factors are more directly linked, including social norms, guilt, and personal responsibility for the problem. These results suggest that education and outreach programs must be designed so that they move beyond simply increasing awareness in the audience. Instead, programs must be effective at establishing emotional connections, instilling personal responsibility, and engendering the belief that personal actions will have an impact. Participants may be encouraged to write a pledge and read it out loud to the rest of the class about how they will change their behavior to help a specific animal, such as promising to lower the thermostat in winter to help save polar bears from carbon emissions (Dillahunt et al. 2008). Outreach programs may need to show participants that they have both the ability and a responsibility to protect the reintroduced animal or its habitat.

Furthermore, the needs of local communities must be considered as part of the barrier assessment, particularly in regions where communities rely heavily on natural resources for sustenance. Basic human needs must be met if local communities are to engage in activities that benefit wildlife, and local communities must be included in developing an integrated conservation plan for a species (Kaimowitz and Sheil 2007, Savage et al. 2010). As part of an integrated cotton-top tamarin (*Saguinus oedipus*) conservation program in Colombia, Savage et al. (2010) describe working with local communities to develop an alternative cookstove program that would reduce the amount of wood harvested from the forest while providing other benefits to its users such as less smoke produced. In Papua New Guinea, the Tree Kangaroo Conservation Program has worked to form long-term partnerships with local communities, providing both immediate and long-term investments in community education and human health care (Ancrenaz et al. 2007). In turn, local landowners have pledged portions of intact habitat on their land to establish the country's first conservation area, which has improved the wild population of Matschie's tree kangaroo (*Dendrolagus matschiei*) (Ancrenaz et al. 2007). In both of these examples, the basic needs of the audience were an integral part of addressing barriers that could inhibit behavior change.

A clear understanding of what behaviors exist in which audience that may be harmful to the reintroduced species, as well as the best way to remove those barriers, will help the team design the most effective program. As stated earlier, different audiences will exhibit different behaviors and respond to different programs. Identifying those distinct audiences is necessary before starting outreach so that programs are targeted and coherent.

Working with Children

In contrast to reintroduction outreach programs geared toward adults, there are easy entry points for programs with children and students, as well as a wealth of research identifying many different strategies that will help reintroduction professionals design programs that reach children most effectively. While formal settings in schools present the most obvious opportunity to connect with children through programs, there are countless opportunities in informal environments, from scout and guide troops and summer camps to after-school programs, field trips, and festivals. An important step in working with children is creating and delivering messages that are developmentally appropriate for different ages rather than relying more closely on marketing models.

Sobel (1996) describes three stages of a child's relationship with nature largely based upon his own observations and attitude surveys toward animals (Kellert 1984). In early childhood (ages 4–7), educators should focus on helping children develop empathy for nature. Programs should use common animals and familiar surroundings, as the child's home is incredibly central to his or her life. In addition, connecting children to "baby" animals or neotenic animals (those that exhibit juvenile traits and behaviors as adults) may facilitate a connection between young children and nature (Sobel 1996, Bexell et al. 2007). Interestingly, these surveys (Kellert 1984) suggest this age group has the most exploitative and least caring attitudes toward animals; if this trend still persists today, the focus on empathy is particularly important.

During the elementary years (ages 8–11), Sobel (1996) recommends that educators should focus on exploration. At this stage, children are beginning to discover more of their world, and their innate curiosity can be harnessed during unstructured play. Chawla (1988) notes that the two most common responses by many environmentalists to early influences are free time outdoors and the influence of a parent or mentor to explore nature. Education programs targeting this age group should also consider incorporating any of seven design principles developed by Sobel (2008) that embody types of play children already employ in their lives. These include adventure, fantasy and imagination, animal allies, maps and paths, special places, small worlds, and hunting and gathering. Sobel's intention is for educators to use child's play to inspire and enrich environmental education. For example, as elementary schoolchildren explore their expanding worlds, they might be encouraged to draw a map of their neighborhood. This could be adapted for a reintroduction outreach program to include maps of animal habitats, or mapping where reintroduced animals have been sighted by biologists or the public.

Prior to this developmental stage, many persons recommend avoiding teaching about "tragedies" such as endangered species, habitat destruction, or climate change (Sobel 1996,

Armstrong 2006). However, the saturation of contemporary culture by media makes it exceptionally unlikely that children are not hearing of environmental problems at young ages. Therefore, environmental problems may be brought up in classes, even if not the goal of a program. Educators should be ready to help children process the information, think creatively about solutions, and envision a different future as part of processing the information (Armstrong 2006). In particular, educators need to help children become part of a larger community so that they do not feel the entire burden of the problem. Because reintroduction programs are solving a problem in conservation, there may be many opportunities for getting this age group involved in joining the solution, such as adopting an animal through a reintroduction program (National Zoological Park 2015). Throughout all of this, educators must be ready to help children process emotions and feelings that may even be similar to grief from death or divorce in the family— particularly in children who are attached to the environment. Many scientists are accustomed to remaining detached and impartial to maintain objectivity. However, if scientists are working directly with children as part of education and outreach, it may be important to discuss their emotions about the environment, especially when discussing problems, in order to help younger students develop a stronger emotional connection to nature.

Finally, tweens and teens (ages 12–17) are more inspired by activities that allow them to begin to influence and shape their world. Sobel (1996) recommends that environmental education with this age group focus on social action, or teaching teens and young adults how to influence their local environment. This corresponds with public opinion research from The Ocean Project (2009, 2011) that suggests that teens and tweens have the highest level of concern about the state of the ocean, compared to other age groups, and also feel the most strongly about the importance of individual actions. These youth are also increasingly the environmental decision-makers in their families, providing adults in the household and community with advice on environmental matters. At a practical level, it can be hard to engage high school students using off-site field trips because of conflicts with other courses, sports, jobs, and the many other extracurricular activities dominating teen life. While reintroduction outreach fits well with educational goals for this group, programs may need to be developed that can easily be incorporated by teachers into the classroom (e.g., Colorado Division of Wildlife 2005, Wildlife Conservation Society 2015) or that involve teens as volunteers (see Box 14.1). Young adults should not be overlooked for creative outreach programs, particularly given their ability and willingness to influence their families.

Working with children and young adults can be daunting for any professional that has not received training in educational methods for the appropriate age group. For this reason, partnerships with teachers and other experienced educators are especially helpful in designing successful programs (Brewer 2002). Scientists and other reintroduction practitioners must also decide how much time to devote to the outreach program. Many scientists consider outreach a form of volunteering that is secondary to their job responsibilities, often because they are not professionally encouraged to conduct these activities (Andrews et al. 2005). However, there are resources to help scientists become better educators, in terms of both program creation and content delivery (e.g., Jacobson et al. 2006, Sobel 2008). As with a new research initiative, the process may need to be adapted throughout the program based on direct experiences, feedback from teachers and students, and more formal evaluations as appropriate.

STAGE 3: CRAFT AND EXECUTE

Once the goal(s) and audience(s) of the reintroduction outreach have been determined, the next step is to craft specific programs and begin implementing them in the community. If it has not already become apparent based upon the work to date, it is time to determine who will lead the delivery of the programs. Many individual member organizations within a reintroduction team likely have some outreach mandates, such as government agencies charged with protecting wildlife resources or zoos or aquariums that are rearing animals for reintroduction. If multiple partners are able to help, we recommend that team members having the highest credibility with the target audience(s), as well as those with the most expertise in design and delivery, conduct the program. More commonly, however, outreach and education are heavily reliant on partnerships with all team members contributing what they are able. Roughly 65% of surveyed wildlife reintroduction programs had no staff devoted solely to education and outreach (Sutton and Lopez 2014). Partners commonly used print media (72%), internet media (59%), audiovisual media (58%), primary schools (57%), and secondary schools (55%) in achieving their goals, though the survey did not ask about partnerships with informal educational organizations (Sutton and Lopez 2014). There are many creative ways for all reintroduction team members to become involved.

The timing and duration of the outreach effort also needs to be discussed. Leaders in an Amazon yellow-shouldered parrot reintroduction program stated that "a significant portion of the success of this program rests on 5 years of previous work on environmental education, public awareness, and ecological studies" (Sanz and Grajal 1998). In some cases, outreach and education must begin before any reintroduction activities, such as to help prevent harming of reintroduced animals. In other cases, outreach may wait until the new population is more established with signs of biological success. Regardless of when outreach begins, the reintroduction team should expect that these efforts will continue as long as, or longer than, active reintroduction of the species.

Below, we provide examples of different education and outreach programs for reintroduction,

· Incorporating Outreach into the Lake Sturgeon
Reintroduction Program in Tennessee

14a Visitors to the Tennessee Aquarium get the chance to touch a lake sturgeon while
hearing about the reintroduction program from docents. Interpretive signage around the
exhibit provides additional information on lake sturgeon biology and the reintroduction
program. Photo by Todd Stailey, Tennessee Aquarium.

Once wide-ranging across eastern North America, lake sturgeons (*Acipenser fulvescens*) were extirpated from the southeastern United States in the 1960s and 1970s due to degraded water quality, overfishing, and the construction of dams. During the 1980s and 1990s, a number of changes to the water quality in the Tennessee River, attributed to regulations under the Clean Water Act and implementation of Tennessee Valley Authority's Reservoir Release Improvement Program to provide steady, oxygenated tailwater releases from dams, led biologists to believe that reestablishment of lake sturgeon in the southeast was again possible (Scott et al. 1996). The Tennessee Lake Sturgeon Reintroduction Working Group (TLSRWG) formed in 1995 to guide the effort, as similar programs in other southeastern US states were also beginning to restore the southern range of this species.

Attitude surveys were not conducted prior to reintroduction. Regardless, the TLSRWG quickly agreed that one of the most important audiences for outreach on the lake sturgeon program was the fishing community in the southeast, both recreational and commercial. This group was selected due to the higher frequency of encounters this group would

have with the animals and the need for education on releasing the animals unharmed and reporting their location to the reintroduction team. Because the Tennessee Aquarium was one of the members in the TLSRWG, Aquarium visitors were also identified as a key audience for outreach, as lake sturgeon could easily be displayed on-site, helping to reach recreational anglers. Finally, the environmental educators at the aquarium also selected an elementary school near a lake sturgeon release site as an additional audience for outreach programs.

During the early stages of the reintroduction program, work with the commercial fishing community demonstrated why outreach can be such a critical component to successful reintroductions, as they were more likely to recapture lake sturgeon than the TLSRWG. During the mid-2000s, one pair of commercial fishermen worked with the team by sending reports of all caught lake sturgeon, including size, age estimate (noted by scute removal from the fish), location of capture, and even scanning for or inserting passive integrated transponder (PIT) tags for further monitoring (George et al. 2009). The TLSRWG also refined methods for capture of lake sturgeon based on recommendations from this pair, leading to more successful and standardized monitoring efforts.

Outreach to the recreational fishing community for the Tennessee Lake Sturgeon Reintroduction Program emphasizes releasing the fish, safe and careful handling of a lake sturgeon without harming it, and reporting the catch to the state agency to help monitor the population (George et al. 2008). Local anglers are asked to report catches to the state management agency by text in the fishing regulations, signage at boat ramps, wallet-sized cards handed out with fishing license sales, and through media stories (George et al. 2013). Recently, the TLSRWG began creation of an online reporting platform, iCaughtOne.org, to standardize catch reports across the southeastern region and make them widely available to all program partners. These reports are now linked to iNaturalist.org to increase the citizen science component of monitoring, allowing reporters to view their contributions to the monitoring effort on a social platform (Dickinson et al. 2012).

At the Tennessee Aquarium, an interactive exhibit is used to engage visitors with ongoing reintroduction programs (George et al. 2013). A lake sturgeon touch tank offers visitors the chance to feel the fish while a docent explains the program and ensures the safe handling of the animals (figure 14a). This exhibit relies on verbal interpretation, though there is also signage around the exhibits that explain the purpose and goals of the program while also offering suggestions on ways the visitors can get involved. Some of the docents at the tank are teen volunteers at the aquarium, and they have prompted the development of a high school outreach program on lake sturgeon movement because they have invited the reintroduction biologists to speak at their schools. In addition to the new high school program, staff work most closely with an elementary school located three miles from the release site of reintroduced lake sturgeon (George et al. 2013). Though this is 120 mi from the aquarium, the students get to raise a sturgeon in their classroom and attend the release, helping to protect the animals that live in their backyards. When releases are held in different sites, nearby schoolchildren and the general public are invited to help release animals, providing education not only to those in attendance, but also to the broader public through the subsequent media stories (George et al. 2013). Finally, several assistantships are offered each year for undergraduate or graduate students to gain experience in conservation aquaculture by rearing young sturgeon at the Tennessee Aquarium.

Because funding for the program has primarily been devoted to reintroduction, evaluation of outreach results is lagging behind other aspects of the project. However, the number of recreational fishing catch reports is increasing (Collier et al. 2011), suggesting that either the population is increasing, the public is more knowledgeable about reporting captures, or a combination of both. Private marketing surveys conducted for the Tennessee Aquarium indicate that "sturgeon" was one of the top three conservation programs associated with the Aquarium for regional residents, indicating general awareness of the long-missing animal is increasing (S. Corwon, IMPACTS Research & Development, pers. comm., 2015). While further refinement is always needed, the lake sturgeon program has set some standards for working with reintroduction of game fishes.

---

grouped by common types of partnerships, adapted from Sutton and Lopez (2014).

Traditional and Social Media

Reintroduction outreach using both traditional and internet media to spread conservation messages is typically geared toward the general public, with the expectation of reaching teens, young adults, and adults. Though interviews may seem daunting at first, journalists are often excellent partners for sharing reintroduction stories with a broad audience (Sutton and Lopez 2014). Media can be invited to cover key events, such as actual reintroduction with release of the animals, as well as to share stories about other reintroduction outreach, such as when a school group is involved with monitoring for the animals. Work with traditional media requires preparing interviewees for the greatest impact, as well as helping scientists and educators deliver key messages about the program (see Corbett 2006, Jacobson 2009, Baron 2010). Media stories can have a powerful effect on the public's mood toward reintroduction projects. Though media outlets in Ireland initially supported the concerns of the farming community about reintroduction of white-tailed sea eagles (*Haliaeetus albicilla*), widespread poisoning of the animals led the media to share graphic images of dead eagles that changed national sentiment toward the project (O'Rourke 2014). In some cases, negative stories such as these can be helpful. While regional and national media have a broad reach, specialized forms of traditional media, such as hunting or fishing magazines, can help target specific audiences for outreach goals. Developing contacts in the journalism industry will help ensure that stories are well placed and thoughtfully covered, and most partners in reintroduction programs, such as universities, agencies, and nongovernmental organizations (NGOs), have public relations staff to help with that process.

With the rise of social media, reintroduction partners have a variety of options for directly reaching both specific and general audiences. Simple options like blogs or Facebook pages can be useful in broadly sharing information about the program and the animal (Caro and Sherman 2011, Bik and Goldstein 2013). A Facebook page can even be "managed" by the reintroduced animal, such as Luna the Lamprey (US Fish and Wildlife Service 2014). Originally started as a way to document the spawning migration of Pacific lamprey (*Entosphenus tridentatus*) in the Columbia River watershed in the northwestern United States, Luna now shares other conservation stories from the region through her Facebook feed. Hunting and fishing forums or other online communities often manage pages where reintroduction staff can share information or solicit input from target audiences (Box 14.1).

The *Year of the Frog* campaign led by the Amphibian Ark sought to raise awareness of the need for global amphibian breeding programs in order to increase conservation funding (Griffiths and Pavajeau 2008, Pavajeau et al. 2008). They began an aggressive campaign in 2008 to highlight the needs of amphibians, using their many partners of zoos, aquariums, government agencies, universities, and even pet dealers to share a common message about amphibian breeding programs (Pavajeau et al. 2008, Zippel et al. 2011). Celebrity endorsements (Sir David Attenborough, Jeff Corwin, Jean-Michel Cousteau, Jane Goodall, and Kermit the Frog) were used for online and network public service announcements, earning an estimated US$1 million in advertising value from the news stories alone (Zippel et al. 2011). Amphibian Ark helped set up funding partnerships between zoos and aquariums in developed and developing nations to augment and stabilize captive breeding programs (Zippel et al. 2011), and they continue to use social media to promote fundraising and global reintroduction of amphibians (Amphibian Ark 2015).

Schools

Many outreach and education programs targeting children for an audience begin with a simple visit to a classroom or a field trip by the students to the reintroduction site. Though short classroom programs may not be sufficient to change attitudes and behaviors, they are a useful step for getting started with the program (Jacobson et al. 2006, Goodwin et al. 2010). There are several existing environmental education curriculum guides that may be a useful starting place for short outreach programs, including Project WILD (2010), focusing on North American wildlife, Windows on the Wild (World Wildlife Fund 2001), focusing on global biodiversity, and Project WET (2011), focusing on water quality education. Each of these guides contains numerous activities that can be completed in the classroom or the field, some of which are either already suitable for, or easily adaptable to, different reintroduction targets. Project WET and Project WILD, available in all 50 US states, require program leaders to attend a daylong training workshop, which can help create confidence in program delivery for those new to environmental education. In addition, the programs list discrete learning objectives for each activity that often correlate closely with state standards. Interestingly, one of the common requests from teachers to Project WILD is to provide more assistance in delivering action steps for students to take after they have learned about conservation problems for wildlife (Pitman 2004). We recommend starting here for inspiration and training in creating and delivering education programs about the reintroduction effort.

For the purposes of reintroduction outreach, it is typically preferable to reach children who live where the project occurs. In situ education efforts can incorporate place-based learning by teaching students about their backyard wildlife, with the goal of educating those who have the ability to make a conservation behavior change that matters for the program (Box 14.1). In addition, this helps avoid overwhelming young children with global tragedies, instead focusing on local problems that have conservation solutions in which they can participate (Armstrong 2006, The Ocean Project 2011). In Montana, the Adopt-A-Trout outreach program allowed students in K-12 classrooms to follow westslope cutthroat trout (*Oncorhynchus clarki lewisi*) and bull trout (*Salvelinus confluentus*) that were assisted past barriers in their spawning migrations (Schmetterling and Bernd-Cohen 2002). In addition to lessons in the classrooms using interactive websites, students were taken on field trips to the habitat. Interactive field trips were, not surprisingly, considered some of the most essential opportunities for students to understand the program, as students were engaged in real-world application of their classroom learning.

Reintroduction biologists also should look for opportunities to use school groups as citizen scientists to help with monitoring efforts. As an educational tool, citizen science can result in the development of science literacy; one study of 700 participants in a Cornell Lab of Ornithology program found almost 80%

used thought processes similar to those used in scientific investigations (Trumbull et al. 2000). The student-collected data also can be helpful to the program; student groups assisting in a leatherback sea turtle (*Dermochelys coriacea*) tracking program in Costa Rica were only 1–2% off from shell morphometric measurements taken by professionals while gathering 60% more data than could be completed by the staff alone (Brewer 2002). If concerns persist over the quality of data collected, the best option is to use dedicated training with clear instructions to reduce those concerns (Mayer 2010).

Finally, it is important to note that reintroduction outreach for children does not have to relate only to science learning; interdisciplinary activities, particularly when used for place-based learning, can help learning occur in a broader context (Ardoin 2006). Youth art contests have been used for designing postcards, calendars, or stamps to share information about endangered species (Maschinski et al. 2012, Endangered Species Coalition 2015). Students in the Adopt-A-Trout program in Montana created songs, skits, and art projects (paper maché trout mobiles) to tell stories about where their fish traveled (Schmetterling and Bernd-Cohen 2002). On Margarita Island, Venezuela, the annual carnival parade features a parrot float to celebrate the reintroduction of yellow-shouldered Amazon parrots (Sanz and Grajal 1998). While all of these examples involve the visual and performing arts, there are many opportunities to create programs that tap into creative writing, geography, history, or social studies.

Informal Educational Organizations

Another common outlet for reintroduction outreach is using informal educational organizations to reach a broad audience of all ages. Because zoos and aquariums were early pioneers in captive husbandry techniques to maintain collections, and because many have conservation and education missions, many zoos, aquariums, and museums (ZAMs), as well as botanical gardens and nature centers, are actively engaged in reintroduction partnerships. In fact, one of the earliest documented reintroduction projects, the return of American bison (*Bison bison*) to western North America, began when a 1908 census revealed 1,116 bison in zoos and 25 in the wild (Kisling 2001, Seddon et al. 2007; Seddon and Armstrong, Chapter 2, this volume). The New York Zoological Park started the reintroduction effort by donating 15 bison to the Wichita Forest Reserve in Oklahoma; within 25 years, there were more than 21,000 bison living in managed or protected herds (Kisling 2001). Currently, up to one in seven threatened species is displayed at a zoo or aquarium, providing a global captive population that can be used for reintroduction breeding (Conde et al. 2011). Many ZAMs have developed their capacity to manage many aspects of reintroduction programs, especially educational goals (Wilson and Price 1994, Mazur and Clark 2001, Seddon et al. 2007, Conde et al. 2011).

More than 700 million people visit zoos and aquariums worldwide each year, a group that is often looking for conservation education (Falk et al. 2007, Gusset and Dick 2011). Though recreation and entertainment are major factors driving visitor attendance at ZAMs, educational opportunities are an important visitor motivation as well (Kisling 2001, Falk et al. 2008). In addition, accreditation standards set by the Association of Zoos and Aquariums (AZA) require member institutions to meet both education and conservation goals (AZA 2015). For example, members must have a written conservation plan, which may include in situ efforts, participation in AZA-run species survival or taxon advisory plans, or interpretive educational programs (AZA 2015). In the United Kingdom, mid- to large-sized zoos also have a mandate to incorporate educational elements into in situ conservation projects (Fa et al. 2011). In many cases, ZAMs are excellent places to start outreach programs because they correspond so closely to their missions and their visitor expectations (Miller et al. 2004, Patrick et al. 2007).

A first step for reintroduction outreach at ZAMs is developing exhibits that display the reintroduced species (Boxes 14.1 and 14.2). While educational displays and programs at zoos and aquariums tend to use biodiversity as a prevailing topic for conservation education (Heimlich et al. 2013), the animal may also be used an ambassador for the habitat. In addition, content of zoo education programs and displays has shifted from biological facts about animals to instilling a desire for caring or concern about animals and their habitats (Ballantyne et al. 2007, Fa et al. 2011). At the Tennessee Aquarium, the entire breeding program for the Barrens topminnow is on display for visitors behind a half-wall in a recreated lab. Husbandry staff are in the room and able to answer questions for a few hours each day, while at other times a lifelike video of a "scientist" explains the program in a 6-minute loop (George et al. 2013). Other ZAMs in the United States work to support breeding facilities overseas, such as the relationship between the Houston Zoo, Atlanta Botanical Gardens, and El Valle Amphibian Conservation Center in Panama (Gagliardo et al. 2008). However, one note of caution is that displays for education may conflict with and preclude conservation breeding if standards for population management cannot be maintained due to mixed-individual or mixed-species displays (Griffiths and Pavajeau 2008).

Signage accompanying the exhibit is still the most common path by which visitors receive key conservation messages (Mony and Heimlich 2008). Signs that are simple with positive conservation messaging are most likely to lead to behavioral change in visitors (Fa et al. 2011). Conservation messages must suggest realistic actions that visitors could take to help the animal or habitat, especially when the reintroduction effort is located far from the captive breeding site with the education program. However, as most visitors learn more from other people, via personal experience or videos, than they do from signage, investment in both staff and volunteers is vital for on-site education (Fa et al. 2011, Perdue et al. 2012).

In addition to signs or staff at the exhibits, ZAMs often have an education staff that can help create programs for children or adults about the reintroduction program, using established formats such as formal programs in the zoo or classroom, overnight stays or summer camps, professional development and teacher training, or seminars and lectures for the public (Fa et al. 2011). Animal adoption programs, often used by ZAMs for fundraising, may be rebranded to help provide support for the needs of the reintroduction team while also delivering educational messaging about the target species (Frank et al. 2009, Fa et al. 2011).

Naturalist or Wildlife Enthusiast Organizations

On a different note, reintroduction biologists may face less of an uphill battle and even find voluntary support for reintroduction by working with resource groups that are positively affected by the reintroduction. Similar to programs developed for youth, adults, particularly those who already have an inclination for the outdoors, can be enlisted to help with survey and monitoring work as citizen scientists (Cornwell and Campbell 2012, Dickinson et al. 2012, Maschinski et al. 2012, Sequeira et al. 2014). Survey work may be as unstructured as simply encouraging user groups to report sightings of the reintroduced animal (Box 14.1). Volunteers have also been used to help plant propagated trees for reintroduction (Maschinski et al. 2012). Volunteers can even be tapped for structured, long-term scientific data collection, such as recording ethogram-based behavioral data on nesting California condors (Sandhaus 2013). As seen with children, citizen science helps improve scientific thinking and reasoning in adults (Trumbull et al. 2000, Cornwell and Campbell 2012). Finally, presentations at local chapters of conservation organizations, such as the Audubon Society, Trout Unlimited,

or Ducks Unlimited, can be good ways to share key messages and recruit additional volunteers if needed (Caro and Sherman 2011).

Though not specific to reintroduction, an Adopt-A-Herring program in Massachusetts (United States) combined several education principles into an adult outreach program that reached key stakeholders with information about watershed restoration (Frank et al. 2009). The group specifically targeted an audience that was already interacting with the riverine ecosystem in some way, including anglers, local businesses, and watershed groups. Members of these organizations were asked to adopt a herring in order to follow the progress of "their" fish in the river as part of a radio-telemetry program. The scientists were also able to raise over US$2,500 from the program, as the herring adoption fee was $10 per person (Smith 2010). In return, participants could name their adopted herring, receive weekly updates on where their herring was, and participate in releases (Frank et al. 2009). Because of the fish adoption program, different community members were brought into discussions about fisheries restoration and the role of human actions on watershed conservation.

Government Agencies

When more focused audiences for program goals are necessary, methods of outreach need to become more refined. Working with private landowners such as farmers or ranchers is often an important step of reintroduction outreach, particularly in the western United States. In dealing with private landowners, rather than trying to change hearts and minds in a community alone, reintroduction biologists should consider partnerships with an already established agency or organization working in that area. For example, the Natural Resources Conservation Service (NRCS) in the United States already has many local relationships in place to provide advice or assistance to landowners. While most NRCS outreach is rural, the agency does run some advertising campaigns that reach a more national audience, including suburban and urban landowners (Newton 2001). However, most NRCS works tend to be targeted at a very local level, including "kitchen table talks" where a farmer working on a conservation project is asked to invite neighbors to meet with the NRCS agent to discuss options for their land (Newton 2001). Because it can be difficult to develop trust without support from inside the community, making use of existing relationships on the ground is critical to helping new ideas gain traction (Newton 2001). Ideally, these efforts all combine to provide small gains in community support for the reintroduction program over the long term.

Private landowners may be enlisted to help protect critical resources for reintroduced animals or to protect the animals themselves. Reintroduction programs for predators may need to focus on human behaviors that impact their prey, such as preventing hunting of prairie dogs (*Cynomys* spp.) in order to ensure prey availability for reintroduced black-footed ferrets (*Mustela nigripes*) in the western United States or reducing lead exposure from bullets in prey (Jachowski and Lockhart 2009; see Box 14.2). In some cases, the focus needs to be on preventing intentional poisoning of the reintroduced animal, such as the white-tailed sea eagle (*Haliaeetus albicilla*) in Ireland (O'Rourke 2014). One common form of outreach to landowners involves payment to compensate for loss or damage due to reintroduction efforts, such as for livestock taken as prey (Wagner et al. 1997). However, receiving compensatory payment does not appear to make recipients more tolerant of gray wolf reintroduction programs; prior attitudes and cultural background appear to be more important in shaping behaviors toward the species (Naughton-Treves et al. 2003, Gangaas et al. 2014). Often private landowners have strong concerns about potential legal ramifications from having an endangered species on their property (Maschinski et al. 2012); providing resources or indemnity to assuage their fears (whether grounded or not)

can have major effects on the success of a program. In the United States, habitat conservation plans have been used with private landowners to allow for incidental take of endangered species under the Endangered Species Act if certain conservation measures are established on the property to protect the species or its habitat (Wilhere 2002). While using trusted agencies for outreach delivery may not ensure the success of the program in changing behavior, it is often still the best option for reaching some audiences.

Influential Stakeholders

Finally, a critical audience for outreach includes local politicians and funding partners, who are often extremely influential stakeholders. Politicians can be very influential in promoting policies necessary to protect the wildlife (Chapters 4 and 5). If they are opposed to the project, they may be able to use their platforms to harm public opinion, as seen in the western United States when politicians opposed to gray wolf reintroduction spoke out negatively to feed the fear generated by the program (Williams 2013). When working with politicians, it is critical to demonstrate why the reintroduction will improve the livelihood of their constituents, or simply to be prepared to counter negative attitudes that may exist (Caro and Sherman 2011). Though politicians may not have control over funding allocations, their support for a program could lead to changes in policies or political opinions that can impact success over the long term.

Funding partners, both current and potential, are clearly important to ensure the long-term success of the program. They should be sent regular updates on progress, invited to milestone events, or provided updates through the media, in addition to regular grant reports as required. Unless requested otherwise, consider funding partners as members of the reintroduction team and acknowledge them in outreach publications and presentations. Inviting funding partners to observe or participate in

The California condor, extinct in the wild just over two decades ago, is often cited as a species saved from the brink of extinction through captive breeding and reintroduction efforts. However, ongoing anthropogenic threats to survival and recruitment currently require intensive hands-on management to avert significant mortality in the reintroduced populations (Walters et al. 2010, Finkelstein et al. 2012). Lead toxicosis, which was of major concern in the rapid decline in the remnant condor population in the 1980s (Snyder and Snyder 2000, Rideout et al. 2012, Johnson et al. 2013), remains the most important source of mortality for the combined geographic populations of juvenile and adult free-flying California condors (Rideout et al. 2012). Studies indicate that the main source of lead toxicosis in these condors is incidental ingestion by condors of spent lead ammunition fragments from animal carcasses that were shot with lead (Church et al. 2006, Cade 2007, Green et al. 2008, Finkelstein et al. 2012). Condor program partners employ a variety of environmental education and outreach approaches to teach multiple audiences about the California condor and the recovery efforts. This includes natural history education at condor displays in zoos and traveling exhibits in museums, traditional and social media outreach, and curricula developed in partnership with schools. The audience identified as the most urgent to involve in outreach efforts, however, are those individuals who engage in taking of wildlife with firearms in the range of the condor.

The taking of wildlife with firearms in and of itself is not a barrier to condor recovery. Regulated hunting, for instance, is the foundation of the North American model of conservation (Heffelfinger et al. 2013), and regulated hunting with non-lead ammunition, such as copper ammunition, in the range of the condor is not only compatible with condor recovery, but also, as argued by Walters et al. (2010), may be critical to recovery by providing an ongoing clean source of food in an altered landscape in which humans are now the dominant predators.

From a biological and ecological framework, compliance with use of non-lead ammunition, whether through strictly voluntary participation or through mandate, must be high. Because the condor is an obligate scavenger (Snyder and Snyder 2005) that feeds communally (Koford and National Audubon Society 1966), contamination of just a single carcass can result in a poisoning event that affects multiple birds. Taking the condor's long life span into consideration, Finkelstein et al. (2012) predicted that very low carcass contamination rates will be necessary to avoid high probabilities of lead poisoning within the condor population. However, the human dimensions surrounding this desired behavioral change are complex. Multiple stakeholders, including federal and state wildlife agencies, sportsmen's groups, and wildlife NGOs are invested in the outcome, making a simple solution agreeable to all stakeholders highly unlikely (see discussion in Leong et al. 2012).

Broadly, two different approaches have been used in engaging sportsmen in removing lead in two general US regions where condors have been reintroduced: (1) California and (2) northern Arizona/southern Utah ("southwest"). Lead reduction efforts in the southwest condor range, where the release population is designated nonessential experimental population under the Endangered Species Act, have been voluntary (Southwest Condor Review Team 2012), while voluntary, and subsequently, mandated, approaches have been used in the California range (Hill 2009, Walters et al. 2010). It has been argued that, given the workload and funding challenges faced by today's game wardens, voluntary compliance with fish and wildlife laws is increasingly important (Filteau 2012). In regard to the need for very high hunter participation rates in use of non-lead ammunition, it is clear that widespread hunter acceptance is crucial whether the framework is mandatory, as in California currently, or voluntary, as in the southwest. Though the frameworks differ, both the California and southwest regions have used a number of similar outreach strategies, including direct hunter outreach through attendance at sportsman's events, hosting of non-lead shooting events that give hunters the opportunity to view demonstrations and try non-lead ammunition, and free non-lead ammunition giveaways.

The Arizona Game and Fish Department and The Peregrine Fund, along with program cooperators, implemented a voluntary lead reduction program within the condor range in northern Arizona in 2003. They initially employed surveys to assess baseline hunter awareness and focus groups to guide effective outreach and communications plans, and check station and post-hunt surveys for program assessment (Sullivan et al. 2007, Sieg et al. 2009, Southwest Condor Review Team 2012). The Southwest Condor Review Team (2012) reported significant change in familiarity with non-lead ammunition and the voluntary lead reduction program over time and increased hunter participation rates in voluntary lead reduction during 2007–2011, underscoring the importance of building assessment into outreach programs from the outset.

A number of organizations have implemented or have been involved with lead outreach efforts in California, including government agencies and NGOs. While a lack of reports in the peer-reviewed literature may reflect less of an emphasis on formal assessment of outreach efforts in California, there are findings of interest in summary and management reports. For instance, in a survey of 2012 non-lead ammunition giveaway participants, Ventana Wildlife Society received feedback that

although 62% of the hunters completing the survey had not previously tried the products they ordered through the program, 97% of those surveyed would consider purchasing the same products in the future. Ventana Wildlife Society (2012) also gained insight into the top three factors identified by hunters as most likely to prevent them from switching to non-lead ammunition: cost, performance, and availability. By identifying these and other barriers, condor managers have gained valuable insights into where to focus their evolving outreach efforts.

Though lead reduction efforts for condor recovery are ongoing, progress has been made toward hunter consideration of and participation in lead reduction efforts in the range of the condor. The contrasting approaches in California and the southwest provide fertile material for study of two different frameworks—voluntary and mandatory—to reducing spent lead ammunition in the condor's environment, and the process and outcomes are relevant to a broad range of reintroduction programs facing anthropogenic threats to species recovery.

outreach activities, particularly with young students, can help increase support for the outreach efforts in addition to the reintroduction efforts. While this audience is often one of the smallest in terms of group size, it can easily be one of the most influential on long-term success for the specific program, as well as for future reintroduction efforts.

STAGE 4: EVALUATE AND ADAPT

One of the biggest challenges faced by outreach programs may be evaluation of the results, particularly tracking long-term changes (Frank et al. 2009, Heimlich 2010). However, this is typically one of the most critical parts, as it will determine if goals are met, learning targets are achieved, behaviors are changed, time is well spent, and investments by donors are tracked (Heimlich 2010, Khalil and Ardoin 2011). If these objectives are not being reached, then aspects of the program need to change. Simply

put, evaluation of outreach should be viewed as part of the adaptive management process (Walters 1986).

Because the education and outreach programs described in this chapter, as well as those used more broadly in the environmental education field, have a wide variety of audiences and goals, it is hard to provide one clear set of evaluation standards for reintroduction outreach (Heimlich 2010). For classroom programs, consider evaluations from the perspective of the teacher rather than that of individual students as a first step in gathering feedback (Schmetterling and Bernd-Cohen 2002). Resources such as Project WILD offer evaluation options for use with their shorter classroom programs, such as discussion questions, writing assignments, or follow-up activities (Project WILD 2010). ZAMs may have staff trained in exhibit or program evaluation who can adapt these procedures for purposes of the reintroduction outreach (Khalil and Ardoin

2011). Both the reintroduction practitioners and target audiences will guide the changes and additions to education and outreach programs through the realization of new management needs, unpredicted opportunities for work, discovery of barriers to participation, and even direct requests from the audience itself (Box 14.1).

To become more adept at education evaluation, seek opportunities for longer workshops such as through the online Applied Environmental Education Program Evaluation (AEEPE) or the National Conservation Training Center (NCTC) run by the US Fish and Wildlife Service (Monroe 2010). Many resources specific to environmental education evaluation can also aid in the design of a more rigorous evaluation program that may be required for larger grants (e.g., Jacobson et al. 2006, Ernst et al. 2009, National Audubon Society 2011).

Evaluating the impact of other types of outreach, particularly for the general public, is often best folded into continuing social science studies on public attitudes (Serfass et al. 2014). In addition, reintroduction teams may wish to track traditional media stories through project mentions and audience sizes for different outlets, and share the results with project funders. While metrics for tracking social media are evolving as quickly as the platforms themselves, there are opportunities to evaluate whether the strategy is effective and leading to a growing, engaged audience (Peters et al. 2013, Effing and Spill 2016). Regardless of how the evaluation is designed, it is a critical, but often overlooked, part of education and outreach.

FUTURE RESEARCH AND DEVELOPMENT

Though outreach and environmental education have been recognized as critical to the success of a number of reintroduction programs, they have yet to be fully integrated into existing guidelines, strategies, and monitoring standards. Because many conservation challenges are anthropogenic in nature, it is critical that outreach and environmental education pro-grams, and their evaluation, be fully integrated into those guidelines and standards. We encourage integration across disciplines in the development of these frameworks, including but not limited to formal and informal education, outreach, social marketing, and other aspects of human dimensions. Because there is already some recognition in the literature that these elements are important to reintroduction success, and identification of some of the barriers to their implementation in relation to biologists' training and backgrounds (see Clark 2009, for a discussion), we encourage creative solutions toward overcoming those barriers, and, above all, quantitative assessments and publication of those solutions and assessments.

We see much potential for further study of promotion of conservation-related behaviors in audiences of different types in different venues. We encourage exploration of interdisciplinary approaches and further assessment of efficacy of various modalities of outreach and environmental education. There is still much to be learned about how to move the audiences beyond basic knowledge acquisition and how to promote lasting change in conservation-related behaviors. Particularly in ZAMs and school programming, we encourage further investigation into the relationship between immediate conservation actions taken on-site (at the ZAM or school) and sustained behavioral change off-site (at home or in the range of the reintroduced animal). Given space limitations and regulatory barriers to displaying some protected species in ZAMs, one important avenue of investigation may be the relative efficacy of surrogate species, rather than the target species, for display and outreach. We also encourage further research into the efficacy of voluntary versus mandatory approaches toward behavioral change on various issues involving reintroduced species (see Box 14.2).

Rapidly evolving technologies, including interactive digital platforms and use of social media, present many opportunities for innovation and assessment. What are the effects on young children of wide availability of conservation messaging through digital programming? Which types of technology are most effective in delivering our messages surrounding reintroduction programs, or perhaps better, encouraging a dialogue that allows reintroduction biologists and the wider community to work together on solutions to reintroduction challenges? For all lines of inquiry, we encourage attention to differences in technology use across generations.

SUMMARY

Outreach is a critical step in the success of reintroduction programs. It can serve purposes ranging from influencing behavior in the general public that help protect reintroduced animals to encouraging resource users to report sightings of the animal. The reintroduced animal can also be used as a flagship species to raise more concern in the general public about a critical habitat that is threatened by human actions.

While outreach has traditionally been targeted toward children as an offshoot of environmental education, there are many other audiences that deserve consideration. Goals for reintroduction outreach may include increasing the knowledge of the audience about the animal or habitat, or may be targeted toward behavior change. Most importantly, outreach should be flexible and diverse, targeting all stakeholders for a reintroduction program, from schoolchildren to funders to policymakers to resource users to citizen scientists. Serfass et al. (2014) suggested that a thorough and coordinated approach to stakeholder outreach about river otter (*Lontra canadensis*) reintroduction in Pennsylvania contributed to positive attitudes toward otters, a result not seen in more limited programs elsewhere in the United States. Social science, education, and marketing principles can all be adapted and blended to create a thoughtful outreach approach for reintroduction programs.

Before beginning outreach, program leaders should understand current attitudes and knowledge toward the animal and habitat, including what existing barriers may affect attitudes or actions taken by the audience. Ideally, this is undertaken for planning purposes prior to implementation of reintroduction, as the resulting data will help in identifying the audiences, selecting key messages, and developing the outreach program. In addition, these attitude surveys can be used during and after the reintroduction to assess whether outreach is working and what needs to be changed in order to meet identified goals.

Outreach programs can take a variety of forms depending on key audiences and messages, as well as what partnerships are available to develop effective programs. Partners for reintroduction outreach often include traditional and social media, school groups, informal educational organizations, government agencies, naturalist and citizen science groups, and influential stakeholders. If the goal is to reach the general public, then media, schools, and informal educational organizations are excellent first steps. As more specific audiences are identified, more specialized programs may be created, such as stories in hunting or fishing magazines, landowner compensation, or direct work with politicians. Ideally, these efforts all combine to provide broad and targeted support for the reintroduction program over the long term. A well-planned and executed outreach program can result in a more successful reintroduction, addressing and mitigating the anthropogenic factors that are the root cause of the conservation problem at hand.

MANAGEMENT RECOMMENDATIONS

We recommend the following steps as a starting place for incorporating outreach and education into reintroduction programs:

- Set public awareness as one measure of success for the program. Without this accountability, outreach will not be clearly listed as an action step and given the necessary attention (Fraser 2008, Goring et al. 2014).

- Determine the public's current knowledge, considering the broader public as well as key stakeholders. Do they already have positive attitudes or negative concerns about the project that should be addressed (Decker et al. 2010)? If they are already educated about the animal or habitat, create a message that builds upon their current attitudes and awareness, though be careful that all assumptions of knowledge are accurate.

- Determine your specific audience(s). Who needs to learn about the program? Are you trying to reach a specific resource user group, such as policymakers, schoolchildren, or funders? In some cases, reaching one group (often schoolchildren) can lead to media stories with a much broader audience, though often a more diluted message. If multiple audiences are necessary, multiple programs likely are needed to reach your goals.

- Determine the desired action each audience will take. Are you sharing information to raise awareness of the animal or habitat? Or do you need your audience to take a more active role in the project, such as reporting sightings of reintroduced animals? Be clear about what outcome you would like from the outreach or education program.

- Determine barriers that your audience faces. If you desire for your audience to take action following the education program, what factors are currently preventing that action? For example, those who feared bison were more than 20 times more likely to oppose reintroduction than those who did not (Decker et al. 2010). With that knowledge, the outreach program may be designed to reduce fear of bison rather than specifically inform about reintroduction.

- If the reintroduction program does not have educational staff available to lead outreach, find partners to design and implement the educational strategy. Zoos, aquariums, museums, and nature centers are potential partners to turn to for their environmental education and public relations staff, but

many government agencies can offer similar assistance. Graduate students are another option for program delivery, as their development of these communication skills is an important part of workforce preparation that is often not taught in a traditional curriculum (Muir and Schwartz 2009).

- Develop a specific message for outreach or education based upon your audience, its knowledge, and its barriers. Identify how best to deliver this message, using as many methods as possible given your resources. Repeat this task as necessary for each audience you are trying to reach.

- Evaluate your program and adapt as needed. Adaptive management is required for all aspects of population management for reintroduced species (Chapters 6 and 7), including the education and outreach program(s). If your audience is not gaining in knowledge or changing behaviors, it is time to determine what you can try to be more successful. Most importantly, please share your program with other environmental education and reintroduction practitioners so that the field can be advanced.

LITERATURE CITED

Amphibian Ark. 2015. Amphibian Ark Facebook page. Available online at: https://www.facebook.com/AmphibianArk?fref=ts (accessed September 21, 2015).

Ancrenaz, M., L. Dabek, and S. O'Neil. 2007. The costs of exclusion: recognizing a role for local communities in biodiversity conservation. PLoS Biology 5:e289.

Andrews, E., A. Weaver, D. Hanley, J. H. Shamatha, and G. Melton. 2005. Scientists and public outreach: participation, motivations, and impediments. Journal of Geoscience Education 53:281–293.

Ardoin, N. M. 2006. Toward an interdisciplinary understanding of place: Lessons for environmental education. Canadian Journal of Environmental Education 11:112–126.

Ardoin, N. M., and J. E. Heimlich. 2013. Views from the field: conservation educators' and practitioners' perceptions of education as a strategy for achieving conservation outcomes. Journal of Environmental Education 44:97–115.

Armstrong, C. L. 2006. No tragedies before grade four? Expert opinion on teaching climate change to children. Master's thesis, Royal Roads University, Victoria, BC.

Armstrong, D. P., and P. J. Seddon. 2008. Directions in reintroduction biology. Trends in Ecology and Evolution 23:20–25.

AZA (Association of Zoos and Aquariums). 2015. The Accreditation Standards and Related Policies. Available online at: https://www.aza.org/uploadedFiles/Accreditation/AZA-Accreditation-Standards.pdf (accessed February 2, 2015).

Ballantyne, R., J. Packer, K. Hughes, and L. Dierking. 2007. Conservation learning in wildlife tourism settings: lessons from research in zoos and aquariums. Environmental Education Research 13:367–383.

Bamberg, S., and G. Möser. 2007. Twenty years after Hines, Hungerford, and Tomera: a new meta-analysis of psycho-social determinants of pro-environmental behaviour. Journal of Environmental Psychology 27:14–25.

Baron, N. 2010. Escape from the Ivory Tower: A Guide to Making Your Science Matter. Island Press, Washington, DC.

Beck, B. B., L. G. Rapaport, M. S. Stanley Price, and A. C. Wilson. 1994. Reintroduction of captive born animals. Pp.256–286. In P. J. S. Olney, G. M. Mace, and A. T. C. Feistner, eds., Creative Conservation: Interactive Management of Wild and Captive Animals. Chapman & Hall, London, UK.

Bexell, S. M., O. S. Jarrett, L. Lan, H. Yan, E. A. Sandhaus, Z. Zhihe, and T. L. Maple. 2007. Observing panda play: implications for zoo programming and conservation efforts. Curator 50:287–297.

Bik, H. M., and M. C. Goldstein. 2013. An introduction to social media for scientists. PLoS Biology 11:e1001535.

Bonta, M., and C. Jordan. 2007. Diversifying the American environmental movement. Pp.13–33. In E. Enderle, ed., Diversity and the Future of the U. S. Environmental Movement. Yale University School of Forestry and Environmental Studies, New Haven, CT.

Borneman, E. H., and J. Lowrie. 2001. Advances in captive husbandry and propagation: an easily utilized reef replenishment means from the private sector? Bulletin of Marine Science 69:897–913.

Brewer, C. 2002. Outreach and partnership programs for conservation education where endangered species conservation and research occur. Conservation Biology 16:4–6.

Cade, T. J. 1980. The husbandry of falcons for return to the wild. International Zoo Yearbook 20:23–35.

Cade, T. J. 2007. Exposure of California condors to lead from spent ammunition. Journal of Wildlife Management 71:2125–2133.

Caro, T., and P. W. Sherman. 2011. Endangered species and a threatened discipline: behavioural ecology. Trends in Ecology and Evolution 26:111–118.

Chawla, L. 1988. Children's concern for the environment. Children's Environments Quarterly 5:13–20.

Church, M. E., R. Gwiazda, R. W. Risebrough, K. Sorenson, C. P. Chamberlain, S. Farry, W. Heinrich, B. A. Rideout, and D. R. Smith. 2006. Ammunition is the principal source of lead accumulated by California condors re-introduced to the wild. Environmental Science & Technology 40:6143–6150.

Clark, J. D. 2009. Aspects and implications of bear reintroduction. Pp.126–146. In M. W. Hayward and M. J. Somers, eds., Reintroduction of Top-Order Predators. Wiley-Blackwell, Oxford, UK.

Collier, W. R., P. W. Bettoli, and G. D. Scholten. 2011. Dispersal and dam passage of sonic-tagged juvenile lake sturgeon in the upper Tennessee River. Proceedings of the Annual Conference of the Southeastern Association of Fish and Wildlife Agencies 65:143–147.

Colorado Division of Wildlife. 2005. Return of the Snow Cat: Reintroduction of the Lynx to Colorado. Colorado Division of Wildlife, Denver.

Conde, D. A., N. Flesness, F. Colchero, O. R. Jones, and A. Scheuerlein. 2011. An emerging role of zoos to conserve biodiversity. Science 331:1390–1391.

Corbett, J. B. 2006. Communicating Nature: How We Create and Understand Environmental Messages. Island Press, Washington, DC.

Cornwell, M. L., and L. M. Campbell. 2012. Co-producing conservation and knowledge: citizen-based sea turtle monitoring in North Carolina, USA. Social Studies of Science 42:101–120.

Decker, S. E., A. J. Bath, A. Simms, U. Lindner, and E. Reisinger. 2010. The return of the king or bringing snails to the garden? The human dimensions of a proposed restoration of European bison (*Bison bonasus*) in Germany. Restoration Ecology 18:41–51.

Dickinson, J. L., J. Shirk, D. Bonter, R. Bonney, R. L. Crain, J. Martin, T. Phillips, and K. Purcell. 2012.

The current state of citizen science as a tool for ecological research and public engagement. Frontiers in Ecology and the Environment 10:291–297.

Dietz, J. M., L. A. Dietz, and E. Y. Nagagata. 1994. The effective use of flagship species for conservation of biodiversity: the example of lion tamarins in Brazil. Pp.32–49. In P. J. S. Olney, G. M. Mace, and A. T. C. Feistner, eds., Creative Conservation: Interactive Management of Wild and Captive Animals. Chapman & Hall, London, UK.

Dillahunt, T., G. Becker, J. Mankoff, and R. Kraut. 2008. Motivating environmentally sustainable behavior changes with a virtual polar bear. Pervasive 2008 Workshop Proceedings 8: 58–62.

Effing, R., and T. A. Spil. 2016. The social strategy cone: Towards a framework for evaluating social media strategies. International Journal of Information Management 36:1–8.

Endangered Species Coalition. 2015. 2015 Saving endangered species youth art contest. Available online at: http://www.endangered.org/campaigns/endangered-species-day/2015-saving-endangered-species-youth-art-contest/ (accessed February 2, 2015).

Ernst, J. A., M. C. Monroe, and B. Simmons. 2009. Evaluating Your Environmental Education Programs: A Workbook for Practitioners. North American Association of Environmental Educators, Washington, DC.

Fa, J. E., S. M. Funk, and D. O'Connell. 2011. Zoo Conservation Biology. Cambridge University Press, Cambridge, UK.

Falk, J. H., J. E. Heimlich, and K. Bronnenkant. 2008. Zoo and aquarium visitors' meaning-making. Curator 51:55–79.

Falk, J. H., E. M. Reinhard, C. L. Vernon, K. Bronnenkant, N. L. Deans, and J. E. Heimlich. 2007. Why Zoos and Aquariums Matter: Assessing the Impact of a Visit. Association of Zoos and Aquariums, Silver Spring, MD.

Filteau, M. R. 2012. Deterring defiance: "don't give a poacher a reason to poach." International Journal of Rural Criminology 1:236–255.

Finkelstein, M. E., D. F. Doak, D. George, J. Burnett, J. Brandt, M. Church, J. Grantham, and D. R. Smith. 2012. Lead poisoning and the deceptive recovery of the critically endangered California condor. Proceedings of the National Academy of Sciences 109:11449–11454.

Frank, H. J., M. E. Mather, R. M. Muth, S. M. Pautzke, J. M. Smith, and J. T. Finn. 2009. The adopt-a-Herring program as a fisheries conservation tool. Fisheries 34:496–507.

Fraser, D. J. 2008. How well can captive breeding programs conserve biodiversity? A review of salmonids. Evolutionary Applications 1:535–586.

Gagliardo, R., P. Crump, E. Griffith, J. Mendelson, H. Ross, and K. Zippel. 2008. The principles of rapid response for amphibian conservation, using the programmes in Panama as an example. International Zoo Yearbook 42:125–135.

Gangaas, K. E., B. P. Kaltenborn, and H. P. Andreassen. 2014. Environmental attitudes associated with large-scale cultural differences, not local environmental conflicts. Environmental Conservation 42:41–50.

George, A. L., C. Echevarria, L. S. Friedlander, G. Scholten, E. M. Scott, Jr., and T. B. Sinclair, Jr. 2008. Saving the sturgeon: re-introduction of lake sturgeon to the Tennessee River, USA. Pp.30–33. In P. S. Soorae, ed., Global Re-Introduction Perspectives: Re-Introduction Case-Studies from around the Globe. IUCN/SSC Re-introduction Specialist Group, Abu Dhabi, UAE.

George, A. L., M. T. Hamilton, and K. F. Alford. 2013. We all live downstream: engaging partners and visitors in freshwater fish reintroduction programmes. International Zoo Yearbook 47:140–150.

George, A. L., B. R. Kuhajda, J. D. Williams, M. Cantrell, P. L. Rakes, and J. R. Shute. 2009. Guidelines for using propagation and translocation for reintroduction or augmentation (PTRA) for the conservation of southeastern fishes. Fisheries 34:529–545.

Goodwin, M. J., S. Greasley, P. John, and L. Richardson. 2010. Can we make environmental citizens? A randomized control trial of the effects of a school-based intervention on the attitudes and knowledge of young people. Environmental Politics 19:392–412.

Goring, S. J., K. C. Weathers, W. K. Dodds, P. A. Soranno, L. C. Sweet, K. S. Cheruvelil, J. S. Kominoski, J. Rüegg, A. M. Thorn, and R. M. Utz. 2014. Improving the culture of interdisciplinary collaboration in ecology by expanding measures of success. Frontiers in Ecology and the Environment 12:39–47.

Green, R. E., W. G. Hunt, C. N. Parish, and I. Newton. 2008. Effectiveness of action to reduce exposure of free-ranging California condors in Arizona and Utah to lead from spent ammunition. PLoS ONE 3:e4022.

Griffiths, R. A., and L. Pavajeau. 2008. Captive breeding, reintroduction, and the conservation of amphibians. Conservation Biology 22:852–861.

Gusset, M., and G. Dick. 2011. The global reach of zoos and aquariums in visitor numbers and conservation expenditures. Zoo Biology 30:566–569.

Heberlein, T. A. 2012. Navigating Environmental Attitudes. Oxford University Press, New York.

Heberlein, T. A., and G. Ericsson. 2005. Ties to the countryside: accounting for urbanites attitudes toward hunting, wolves, and wildlife. Human Dimensions of Wildlife 10:213–227.

Heffelfinger, J. R., V. Geist, and W. Wishart. 2013. The role of hunting in North American wildlife conservation. International Journal of Environmental Studies 70:399–413.

Heimlich, J. E. 2010. Environmental education evaluation: reinterpreting education as a strategy for meeting mission. Evaluation and Program Planning 33:180–185.

Heimlich, J. E., V. C. Searles, and A. Atkins. 2013. Zoos and aquariums and their role in education for sustainability in schools. Pp.199–210. In R. McKeown and V. Nolet, eds., Schooling for Sustainable Development in Canada and the United States. Springer, Dordrecht.

Hill, H. J. 2009. Taking the lead on lead: Tejon Ranch's experience switching to non-lead ammunition. P.350. In R. T. Watson, M. Fuller, M. Pokras, and W. G. Hunt, eds., Ingestion of Lead from Spent Ammunition: Implications for Wildlife and Humans. The Peregrine Fund, Boise, ID.

Hines, J. M., Hungerford, H. R., and A. N. Tomera. 1986. Analysis and synthesis of research on responsible pro-environmental behavior: a meta-analysis. Journal of Environmental Education 18(2):1–8.

IUCN. 2008. Global Re-Introduction Perspectives: Re-Introduction Case-Studies from around the Globe. IUCN/SSC Re-introduction Specialist Group, Abu Dhabi, UAE.

IUCN. 2010. Global Re-Introduction Perspectives: 2010. Additional Case-Studies from around the Globe. IUCN/SSC Re-introduction Specialist Group, Abu Dhabi, UAE.

IUCN. 2011. Global Re-Introduction Perspectives: 2011. More Case-Studies from around the Globe. IUCN/SSC Re-introduction Specialist Group, Abu Dhabi, UAE.

IUCN. 2013. Global Re-Introduction Perspectives: 2013. Further Case-Studies from around the Globe. IUCN/SSC Re-introduction Specialist Group, Abu Dhabi, UAE.

IUCN/SSC. 2013. Guidelines for Reintroductions and Other Conservation Translocations. Version 1.0. IUCN Species Survival Commission, Gland, Switzerland.

Jachowski, D. S., and J. M. Lockhart. 2009. Reintroducing the black-footed ferret Mustela nigripes to the Great Plains of North America. Small Carnivore Conservation 41:58–64.

Jacobson, S. K. 2009. Communication Skills for Conservation Professionals. Island Press, Washington, DC.

Jacobson, S. K., M. D. McDuff, and M. C. Monroe. 2006. Conservation Education and Outreach Techniques. Oxford University Press, Oxford, UK.

Jiménez, A., M. J. Díaz, M. C. Monroe, and J. Benayas. 2015. Analysis of the variety of education and outreach interventions in biodiversity conservation projects in Spain. Journal for Nature Conservation 23:61–72.

Johnson, C. K., T. R. Kelly, and B. A. Rideout. 2013. Lead in ammunition: a persistent threat to health and conservation. EcoHealth 10:455–464.

Jule, K. R., L. A. Leaver, and S. E. G. Lea. 2008. The effects of captive experience in reintroduction survival in carnivores: a review and analysis. Biological Conservation 141:355–363.

Kaimowitz, D., and D. Sheil. 2007. Conserving what and for whom? Why conservation should help meet basic human needs in the tropics. Biotropica 39:567–574.

Kellert, S. R. 1984. Attitudes toward animals: age-related development among children. Pp.43–60. In M. W. Fox and L. D. Minckley, eds., Advances in Animal Welfare Science 1984/85. The Humane Society of the United States, Washington, DC.

Khalil, K., and N. Ardoin. 2011. Programmatic evaluation in Association of Zoos and Aquariums-accredited zoos and aquariums: a literature review. Applied Environmental Education and Communication 10:168–177.

Kisling, V. N., Jr. 2001. Zoological gardens of the United States. Pp.147–180. In V. N. Kisling, Jr., ed., Zoo and Aquarium History: Ancient Animal Collections to Zoological Gardens. CRC Press, Boca Raton, FL.

Koford, C. B., and National Audubon Society. 1966. The California Condor. Dover, New York.

Kollmuss, A., and J. Agyeman. 2002. Mind the gap: why do people act environmentally and what are the barriers to pro-environmental behavior? Environmental Education Research 8:239–260.

Lambeck, R. J. 1997. Focal species: a multi-species umbrella for nature conservation. Conservation Biology 11:849–856.

Leong, K. M., D. J. Decker, and B. T. Lauber. 2012. Stakeholders as beneficiaries of wildlife management. Pp.26–40. In D. J. Decker, S. J. Riley, and W. F. Siemer, eds., Human Dimensions of

Wildlife Management. 2nd ed. Johns Hopkins University Press, Baltimore, MD.

Levine, D. S., and M. J. Strube. 2012. Environmental attitudes, knowledge, intentions and behaviors among college students. Journal of Social Psychology 152:308–326.

Louv, R. 2008. Last Child in the Woods: Saving Our Children from Nature-Deficit Disorder. Algonquin Books, Chapel Hill, NC.

Maschinski, J., S. J. Wright, and C. Lewis. 2012. The critical role of the public: Plant conservation through volunteer and community outreach projects. Pp.53–69. In J. Maschinski and K. E. Haskins, eds., Plant Reintroduction in a Changing Climate: Promises and Perils. Island Press, Washington, DC.

Mascia, M. B., J. P. Brosius, T. A. Dobson, B. C. Forbes, L. Horowitz, M. A. McKean, and N. J. Turner. 2003. Conservation and the social sciences. Conservation Biology 17:649–650.

Mayer, A. 2010. Phenology and citizen science. BioScience 60:172–175.

Mazur, N., and T. Clark. 2001. Zoos and conservation: policy making and organizational challenges. Bulletin Series Yale School of Forestry and Environmental Studies 105:185–201.

McKenzie-Mohr, D. 2000. Promoting sustainable behavior: An introduction to community-based social marketing. Journal of Social Issues 56:543–554.

Miller, B., W. Conway, R. P. Reading, C. Wemmer, D. Wildt, D. Kleiman, S. Monfort, A. Rabinowitz, B. Armstrong, and M. Hutchins. 2004. Evaluating the conservation mission of zoos, aquariums, botanical gardens, and natural history museums. Conservation Biology 18:86–93.

Monroe, M. C. 2010. Challenges for environmental education evaluation. Evaluation and Program Planning 33:194–196.

Monroe, M. C., E. Andrews, and K. Biedenweg. 2007. A framework for environmental education strategies. Applied Environmental Education and Communication 6:205–216.

Mony, P. R. S., and J. E. Heimlich. 2008. Talking to visitors about conservation: Exploring message communication through docent–visitor interactions at zoos. Visitor Studies 11:151–162.

Muir, M. J., and M. W. Schwartz. 2009. Academic research training for a nonacademic workplace: a case study of graduate student alumni who work in conservation. Conservation Biology 23:1357–1368.

National Audubon Society. 2011. Tools of Engagement: A Toolkit for Engaging People in Conservation. National Audubon Society, New York.

National Zoological Park. 2015. Adopt-a-Species. How adopting helps animals. Available online at: http://nationalzoo.si.edu/Support/AdoptSpecies/HowADOPTHelps/default.cfm (accessed September 13, 2015).

Naughton-Treves, L., R. Grossberg, and A. Treves. 2003. Paying for tolerance: rural citizens' attitudes toward wolf depredation and compensation. Conservation Biology 17:1500–1511.

Newton, B. J. 2001. Environmental education and outreach: experiences of a federal agency. BioScience 51:297–299.

Norse, E. A. and L. B. Crowder. 2005. Preface. Pp.xvii–xx. In E. A. Norse and L. B. Crowder, eds., Marine Conservation Biology. Island Press, Washington, DC.

The Ocean Project. 2009. America, the ocean, and climate change: New research insights for conservation, awareness, and action. Available online at: http://theoceanproject.org/communication-resources/market-research/ (accessed January 15, 2015).

The Ocean Project. 2011. America and the ocean: annual update 2011. Available online at: http://theoceanproject.org/communication-resources/market-research/ (accessed January 15, 2015).

O'Meara, K. A. 2005. Encouraging multiple forms of scholarship in faculty reward systems. Research in Higher Education 46:479–509.

O'Rourke, E. 2014. The reintroduction of the white-tailed sea eagle to Ireland: people and wildlife. Land Use Policy 38:129–137.

Patrick, P. C., C. E. Matthews, D. F. Ayers, and S. D. Tunnicliffe. 2007. Conservation and education: prominent themes in zoo mission statements. The Journal of Environmental Education 38:53–60.

Pavajeau, L., K. C. Zippel, R. Gibson, and K. Johnson. 2008. Amphibian ark and the 2008 *Year of the Frog* campaign. International Zoo Yearbook 42:24–29.

Perdue, B. M., T. S. Stoinski, and T. L. Maple. 2012. Using technology to educate zoo visitors about conservation. Visitor Studies 15:16–27.

Peters, K., Y. Chen, A. M. Kaplan, B. Ognibeni, and K. Pauwels. 2013. Social media metrics—a framework and guidelines for managing social media. Journal of Interactive Marketing 27:281–298.

Pitman, B. J. 2004. Project WILD: A summary of research findings 1983–1995 and 1996–2003. Available online at: http://www.projectwild.org/%5C/documents/ProjectWILDSummaryofResearchFindings1983-1995and1996-2003_000.pdf (accessed February 2, 2015).

Project WET. 2011. Project WET: Curriculum and Activity Guide 2.0. Project WET Foundation, Bozeman, MT.

Project WILD. 2010. Project WILD: K-12 Curriculum and Activity Guide. Council for Environmental Education, Houston, TX.

Rideout, B. A., I. Stalis, R. Papendick, A. Pessier, B. Puschner, M. E. Finkelstein, D. R. Smith et al. 2012. Patterns of mortality in free-ranging California condors (*Gymnogyps californianus*). Journal of Wildlife Diseases 48:95–112.

Rinkevich, B. 2005. Conservation of coral reefs through active restoration measures: recent approaches and last decade progress. Environmental Science & Technology 39:4333–4342.

Sandhaus, E. A. 2013. Nesting Behavior in a Reintroduced Population of California Condors. Dissertation, Georgia Tech, Atlanta, GA.

Sanz, V., and A. Grajal. 1998. Successful reintroduction of captive-raised yellow-shouldered Amazon parrots on Margarita Island, Venezuela. Conservation Biology 12:430–441.

Savage, A., Guillen, R., Lamilla, I., and L. Soto. 2010. Developing an effective community conservation program for cotton-top tamarins (*Saguinus oedipus*) in Colombia. American Journal of Primatology 72:379–390.

Schmetterling, D. A., and T. Bernd-Cohen. 2002. Native species conservation through education: the Adopt-A-Trout program in Montana. Fisheries 27(9):10–15.

Scott, E. M., K. D. Gardner, D. S. Baxter, and B. L. Yeager. 1996. Biological and Water Quality Responses to Tributary Tailwaters to Dissolved Oxygen and Minimum Flow Improvements. Resource Group, Water Management, Tennessee Valley Authority, Norris, TN.

Scott, W., and S. Gough. 2003. Rethinking relationships between education and capacity-building: remodelling the learning process. Applied Environmental Education and Communication 2:213–219.

Seddon, P. J., D. P. Armstrong, and R. F. Maloney. 2007. Developing the science of reintroduction biology. Conservation Biology 21:303–312.

Sequeira, A. M. M., P. E. J. Roetman, C. B. Daniels, A. K. Baker, and C. J. A. Bradshaw. 2014. Distribution models for koalas in South Australia using citizen science-collected data. Ecology and Evolution 4:2103–2114.

Serfass, T. L., J. A. Bohrman, S. S. Stevens, and J. T. Bruskotter. 2014. Otters and anglers can share the stream! The role of social science in dissuading negative messaging about reintroduced predators. Human Dimensions of Wildlife 19:532–544.

Sieg, R., K. A. Sullivan, and C. N. Parish. 2009. Voluntary lead reduction efforts within the northern Arizona range of the California condor. Pp.341–349. In R. T. Watson, M. Fuller, M. Pokras, and G. Hunt, eds., Ingestion of Lead from Spent Ammunition: Implications for Wildlife and Humans. The Peregrine Fund, Boise, ID.

Smith, J. M. 2010. I want to adopt a herring! Available online at: https://sites.google.com/site/ipswichriverherring/i-want-to-adopt-a-herring (accessed February 3, 2015).

Snyder, N., and H. Snyder. 2000. The California Condor: A Saga of Natural History and Conservation. Academic Press, San Diego, CA.

Snyder, N., and H. Snyder. 2005. Introduction to the California Condor. University of California Press, Berkeley.

Sobel, D. 1996. Beyond Ecophobia: Reclaiming the Heart in Nature Education. Orion Society, Great Barrington, MA.

Sobel, D. 2008. Childhood and Nature: Design Principles for Educators. Stenhouse Publishers, Portland, ME.

Southwest Condor Review Team. 2012. A Review of the Third Five Years of the California Condor Reintroduction Program in the Southwest (2007–2011). Prepared for the United States Fish and Wildlife Service, Pacific Southwest Office (Region 8), Sacramento, CA.

Stoinski, T. S., B. B. Beck, M. D. Bowman, and J. Lenhardt. 1997. The gateway zoo program: a recent initiative in golden lion tamarin reintroductions. Pp.113–130. In J. Wallis, ed., Primate Conservation: The Role of Zoological Parks. American Society of Primatology, Chicago, IL.

Strong, E. K. 1925. Theories of selling. Journal of Applied Psychology 9:75–86.

Sullivan, K., R. Sieg, and C. Parish. 2007. Arizona's efforts to reduce lead exposure in California condors. Pp.109–120. In A. Mee, L. A. Hall, and J. Grantham, eds., California Condors in the 21st Century. Series in Ornithology No. 2. American Ornithologists' Union, Washington, DC/Nuttall Ornithological Club, Cambridge, MA.

Sutherland, W. J., D. Armstrong, S. H. M. Butchart, J. M. Earnhardt, J. Ewen, I. Jamieson, C. G. Jones et al. 2010. Standards for documenting and monitoring bird reintroduction projects. Conservation Letters 3:229–235.

Sutton, A. E., and R. Lopez. 2014. Findings from a survey of wildlife reintroduction practitioners. F1000 Research 3:29.

Traweek, K. 2012. Oregon grey wolf reintroduction, conservation and management evaluation. Senior thesis, Western Oregon University, Monmouth.

Trumbull, D. J., R. Bonney, D. Bascom, and A. Cabral. 2000. Thinking scientifically during participation in a citizen-science project. Science Education 84:265–275.

Uriarte, M., H. A. Ewing, V. T. Eviner, and K. C. Weathers. 2007. Scientific culture, diversity and society: suggestions for the development and adoption of a broader value system in science. BioScience 57:71–78.

UNESCO. 1978. Final Report. Intergovernmental Conference on Environmental Education, United Nations Educational, Scientific, and Cultural Organization with United Nations Environment Program in Tbilisi, USSR, 14–26 October 1977. ED/MD/49. UNESCO, Paris, France.

US Fish and Wildlife Service. 2014. Luna the lamprey. Available online at: http://www.fws.gov/pacific/Fisheries/sphabcon/lamprey/Luna.html (accessed February 3, 2015).

Van Heezik, Y., and P. J. Seddon. 2005. Structure and content of graduate wildlife management and conservation biology programs: An international perspective. Conservation Biology 19:7–14.

Ventana Wildlife Society. 2012. First-year results of a free non-lead ammunition program to assist California condor recovery in central California. Available online at: http://www.ventanaws.org/images/species/species_condor_lead/Free_Non-Lead_Program_2012.pdf (accessed February 16, 2015).

Wagner, K. K., R. H. Schmidt, and M. R. Conover. 1997. Compensation programs for wildlife damage in North America. Wildlife Society Bulletin 25:312–319.

Walters, C. 1986. Adaptive Management of Renewable Resources. Macmillan, New York.

Walters, J. R., S. R. Derrickson, D. M. Fry, S. M. Haig, J. M. Marzluff, and J. M. Wunderle, Jr. 2010. Status of the California condor (*Gymnogyps californianus*) and efforts to achieve its recovery. The Auk 127:969–1001.

Wildlife Conservation Society. 2015. Wild explorations. Available online at: http://www.wildexplorations.com (accessed September 13, 2015).

Wilhere, G. F. 2002. Adaptive management in habitat conservation plans. Conservation Biology 16:20–29.

Williams, C. K., G. Ericsson, and T. A. Heberlein. 2002. A quantitative summary of attitudes toward wolves and their reintroduction (1972–2000). Wildlife Society Bulletin 30:575–584.

Williams, T. 2013. Should wolves stay protected under Endangered Species Act? Available online at: http://e360.yale.edu/feature/should_wolves_stay_protected_under_endangered_species_act/2674/ (accessed February 3, 2015).

Wilson, A. C., and M. R. S. Price. 1994. Reintroduction as a reason for captive breeding. Pp.243–264. In P. J. S. Olney, G. M. Mace, and A. T. C. Feistner, eds., Creative Conservation: Interactive Management of Wild and Captive Animals. Chapman & Hall, London, UK.

World Wildlife Fund. 2001. Windows on the Wild: Biodiversity Basics. An Educator's Guide to Exploring the Web of Life. Acorn Naturalists, Tustin, CA.

Zaltman, G., P. Kotler, P, and I. Kaufman, eds. 1972. Creating Social Change. Holt, Rinehart, and Winston, New York.

Zippel, K., K. Johnson, R. Gagliardo, R. Gibson, M. McFadden, R. Browne, C. Martinez, and E. Townsend. 2011. The amphibian ark: a global community for ex situ conservation of amphibians. Herpetological Conservation and Biology 6:340–352.

FIFTEEN

The Future of Animal Reintroduction

David S. Jachowski, Rob Slotow, Paul L. Angermeier,
and Joshua J. Millspaugh

THE NUMBER AND DIVERSITY OF species subject to reintroduction has dramatically increased (Chapter 2). At the same time, our knowledge of the science and practice of reintroducing animals, too, has increased rapidly in many dimensions. These advancements have likely played an important role in higher reported success rates over the past couple of decades, contradicting low historical estimates of reintroduction success rates published in the 1980s that gave rise to the dogma that "most reintroduction attempts fail" (Chapter 2).

In this chapter, we attempt to build on this strong foundation of knowledge by highlighting key recent advancements and providing a new synthetic conceptual model for animal reintroduction planning and implementation. First, we believe that two recent innovations in reintroduction biology highlighted in this volume are worth considering further—namely, integration of the "Reintroduction Landscape" (Chapter 5) and use of "Structured Decision-Making" in reintroduction decisions (Chapters 6 and 7). Second, to address current and future

challenges, multiple chapters in this volume have offered conceptual frameworks for addressing issues related to human (Chapter 4) and institutional/political dimensions (Chapter 5), disease (Chapter 10), release decisions (Chapter 11), and management (Chapter 13) in reintroduction decisions. Here we attempt to integrate and synthesize these various frameworks, along with introducing the topic of reintroduction ethics, by presenting a new conceptual model that that can be used to enhance the long-term sustainability of reintroductions.

As with any evolving field, it is important to review past lessons, and also to look forward to the future, and how it can be guided by the past. To that end, in this chapter we also offer our perspective on how embracing a process-based approach to species and system restoration could help direct this dynamic field of conservation into the future. Finally, reintroduction has expanded beyond just a species or ecosystem recovery tool, to being the underlying framework for a variety of pioneering conservation initiatives (Chapter 2). We conclude by discussing the various ways in which experiences and frameworks developed for reintroduction practitioners can be used to craft solutions to current and emerging issues in conservation biology.

KEY RECENT INNOVATIONS AND ADVANCEMENTS

Reintroduction Landscapes: Enhancing Sustainability by Placing Reintroductions in a Broader Socioeconomic Context

A consistent theme of this volume is the need for a broad, interdisciplinary approach when reintroducing a species. In part, this is reflected in the diversity of expertise represented by chapter authors, who include not only ecologists, but also historians (Chapter 3), social scientists (Chapters 4 and 14), and policymakers (Chapter 5). Additional chapters could have been added to this volume by economists, anthropologists, philosophers, and experts in other disciplines. Given the strong influence of sociopolitical factors on reintroduction implementation and success (Chapters 4 and 5), we expect these fields will have a growing role in reintroduction planning and implementation.

However, multidisciplinary thinking is only a mechanism or process that gets at a broader need to place reintroductions in a sociopolitical context. Traditionally, reintroductions have been driven by biologists with a conservation agenda (figure 15.1). As addressed in multiple chapters, reintroductions often need long-term support beyond the initial release of individuals. Such sustained efforts require sustained resources to conduct monitoring and management (Chapters 12 and 13). To address this challenge, in Chapter 5, Dunham et al. coin the term "reintroduction landscape" to represent a comprehensive approach that puts species reintroduction planning within a broader sociopolitical context. They astutely suggest that reintroduction practitioners can gain much from the field of landscape ecology, which necessarily considers the "inextricable nature of social, institutional, and ecological contexts, as well as interactions among these processes in land-

scapes." We agree with Dunham et al., and further encourage the reframing of reintroduction benefits from a strictly biological perspective to one that encompasses a broad range of societal benefits, as represented in the notion of natural capital (www.naturalcapitalcommittee.org). This will require a new model for reintroduction in practice that is based on the concept that only by integrating reintroductions with broader societal outcomes will it be possible to achieve long-term commitments needed for reintroductions to succeed (figure 15.1).

Structured Decision-Making: Linking Modeling, Monitoring, and Management

As addressed in Part 2 of this volume, setting goals and assessing progress in attaining them are vital not only to a reintroduction program, but also to informing future reintroduction attempts. Reported successes, along with failures, illustrate how outcomes are dependent on a wide suite of complex ecological, social, and institutional factors (Chapter 5).

Despite these challenges, a number of frameworks are emerging that have the potential to advance our ability not only to define success, but also to manage target populations to achieve success in the future. The integration of research, monitoring, and management within a structured decision-making (SDM) or adaptive management framework shows great potential in advancing the field, and is already being applied in several pioneering reintroduction projects (Chapters 6 and 7). An especially compelling aspect of adaptive management is that management objectives and goals can be modified to match shifting social values, which we cannot always predict. Thus, progress in carrying out a reintroduction can be reassessed and adjustments can be made continually based on relevant, current social values. In the future, it is likely that modelers will build on these frameworks by integrating information from many sources, and even multiple reintroduction programs (Chapter 7).

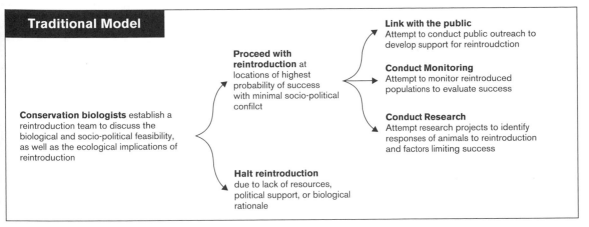

Traditional Model

Conservation biologists establish a reintroduction team to discuss the biological and socio-political feasibility, as well as the ecological implications of reintroduction

Proceed with reintroduction at locations of highest probability of success with minimal socio-political confilct

Halt reintroduction due to lack of resources, political support, or biological rationale

Link with the public
Attempt to conduct public outreach to develop support for reintroudction

Conduct Monitoring
Attempt to monitor reintroduced populations to evaluate success

Conduct Research
Attempt research projects to identify responses of animals to reintroduction and factors limiting success

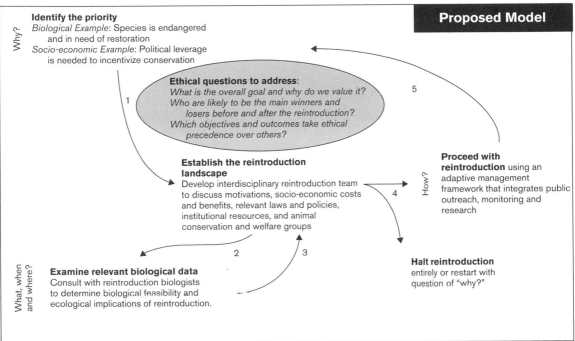

Proposed Model

Why?

Identify the priority
Biological Example: Species is endangered and in need of restoration
Socio-economic Example: Political leverage is needed to incentivize conservation

Ethical questions to address:
What is the overall goal and why do we value it?
Who are likely to be the main winners and losers before and after the reintroduction?
Which objectives and outcomes take ethical precedence over others?

1

5

Establish the reintroduction landscape
Develop interdisciplinary reintroduction team to discuss motivations, socio-economic costs and benefits, relevant laws and policies, institutional resources, and animal conservation and welfare groups

4

Proceed with reintroduction using an adaptive management framework that integrates public outreach, monitoring and research

How?

What, when and where?

Examine relevant biological data
Consult with reintroduction biologists to determine biological feasibility and ecological implications of reintroduction.

2

3

Halt reintroduction entirely or restart with question of "why?"

FIGURE 15.1. A comparison of the generalized traditional model for justifying and carrying out species reintroduction (top panel) and the proposed future model (lower panel). In the proposed conceptual model we provide an ethical framework for approaching animal reintroduction that begins with explicitly stating the rationale for undertaking a reintroduction. Once a rationale is identified, an early step (step 1) is to develop an interdisciplinary reintroduction team that consists of representative stakeholders such as policy makers, land owners, community members, economists, biologists, and animal conservation and welfare groups. The composition of this team and initial discussions should focus on addressing three key ethical questions that inform the broader socioeconomic reintroduction landscape. Once a decision to further investigate the feasibility of a reintroduction is made (step 2), reintroduction biologists provide information on the relative feasibility of a release and long-term management needs. This information can then be related to the interdisciplinary reintroduction team (step 3) to support a decision on how to proceed with a reintroduction, or whether to halt it and potentially reassess the rationale for why such an action is a priority (step 4). If a reintroduction is attempted, we (and other authors in this volume) recommend implementing an adaptive management framework. Further, that process should be linked to the fundamental ethical questions that need to be continuously reassessed by stakeholders within the established reintroduction landscape (step 5).

Additionally, as outlined in Chapter 5, there is great potential in adopting basic principles of landscape ecology to link sociopolitical processes with conventional ecological and evolutionary approaches to improve reintroduction decision-making and outcomes. Through adoption of novel decision-making frameworks and integration with fields other than animal ecology (e.g., sociology, ecological economics, landscape ecology, conservation genetics), we are beginning to build more appropriate and sophisticated approaches to species reintroduction that should enhance the long-term probability of success (Chapter 7). Notably, as mentioned above, this integrative approach to reintroduction also requires development of multidimensional measures of success that link ecological, social, economic, and institutional metrics. How we meet this new challenge remains to be seen, but we propose that a starting place might rest in education by ensuring newly minted conservation biologists and wildlife managers are conversant in these disparate sciences during their undergraduate and graduate training (e.g., see Williams and Brown 2012, Conroy and Peterson 2013). However, this will require the growth of interdisciplinary instruction in most current university programs for fish and wildlife professionals.

A SYNTHETIC FRAMEWORK FOR ANIMAL REINTRODUCTIONS

Reintroduction biology is fundamentally an ode to the past sung in a present that is increasingly pulled in different directions and reinvented as social values change over time (Chapter 2). The goals for reintroduction are similarly shifting. For example, Goble et al. (2012) recently estimated that 84% of endangered or threatened species are unlikely to be restored to the point of no longer requiring management action, thus remaining "conservation reliant." This means that the reintroduction of a species back to some historical state should not be seen as a short-term event, but rather a long-term endeavor, in many cases requiring continuous management to reach a restoration goal (Chapter 13).

Accordingly, as addressed throughout this volume, thorny questions remain about what the goals of species reintroduction should be and how to ensure that reintroductions are successful and sustainable over the long term. In practice, we suggest that the answers to why, what, and where reintroductions should occur lie at the intersection of societal values and scientific feasibility, both of which continue to evolve. As discussed in Chapter 5, answers to these largely normative questions, as well as the management actions that follow from them, must be informed broadly by diverse and representative portions of society as well as by biologists. Thus, a key challenge for the future of reintroduction biology is to ensure that scientific information is accessible to decision-makers, practitioners, and the public as they develop reintroduction programs and address the fundamental questions of why, what, and where.

Why?

Similar to other fields of fish and wildlife management, human values drive which animal species are reintroduced (Chapter 4) and often how success should be defined (Chapter 6). Motivation for reintroduction comes from a range of intrinsic and utilitarian human values. Proponents of species reintroduction for conservation purposes often assume there is something intrinsically valuable about the evolutionary past (Soulé 1985). They view human disturbance as a disruptive process that needs to be reversed or mitigated. Curiously, within this movement there are differing perspectives on when that disturbance occurred or was greatest and on what time period should be selected as a goal for restoration, whether it might be the industrial revolution, European expansion and colonization, or Pleistocene extinction (Angermeier 2000, Truett and Phillips 2009). Utilitarian proponents for reintroduction focus on what is useful about reintroduction, such as providing natural capital, ecosystem services, or direct financial benefits for people, such as nature-based tourism (Hayward et al. 2007).

Regardless of the underlying ethic, it is evident that reintroduction goals are limited by the socioeconomic and political bounds we erect (Chapter 5). In this sense, the question of why to reintroduce is increasingly being redefined over time, to include multiple goals ranging from species and population recovery (IUCN 1998) to the restoration of lost processes and social benefits. We suggest that future decisions as to why a reintroduction should proceed be based on meeting one or more of the following criteria. First, reintroductions, as generally currently practiced, are justified when they exhibit the capacity to sustainably restore species or ecological processes back to some predetermined historical state. This has historically been the goal of most reintroductions and the broader field of restoration ecology in general, and typically driven by conservation biologists with conservation agendas (figure 15.1). Going forward, we see the role of scientists as providing not only ecological rationales, but also sociological and economic rationales including the relative costs and benefits of reintroduction. Thus, rather than having conservation biologists push forward reintroductions in relative isolation, as has often been done in the past, we see sustainable reintroductions succeeding over the long term only when they are driven and supported by policy-makers and the general public (figure 15.1).

Second, beyond restoring a population of a species, in some instances reintroductions can be justified based on their broader ecological and conservation benefits. As the preeminent twentieth-century wildlife ecologist George Schaller wrote, "conservation problems are social and economic, not scientific" (2007, 27). Sometimes, a decision to reintroduce puts into motion several collateral conservation benefits that can represent significant advancements for conservation. For example, in South Africa, southern white rhinoceros (*Ceratotherium simum simum*) and south-eastern black rhinoc-

eros (*Diceros bicornis minor*) are shared among parks and with other African countries for reintroduction or to augment existing populations, with key strategic goals of building broader political support for conservation and relieving poaching pressure within the country (Ferreira et al. 2015). Other examples exist in North America where black-footed ferret (*Mustela nigripes*) reintroductions have been initiated on marginally suitable sites to incentivize broader grassland conservation initiatives (Jachowski 2014). These types of leverage-based decisions often focus on sociopolitical relationships that are sought even in cases where sustainability of the reintroduced population is in doubt. While these motivations for reintroduction are not as well documented or publicly discussed, such decisions are an important part of the reintroduction prioritization process.

What and Where?

Many times, in practice, the questions of "what" and "where" to reintroduce are inextricably tied to normative discussions surrounding "why." In our personal experience, consensus building around the question of "why" is the first critical step to advancing or halting further discussion about reintroduction. In the parlance of SDM, answers to these questions reflect stakeholders' "fundamental objectives" (Conroy and Peterson 2013). Once the question of why is addressed and a decision to reintroduce is made, the questions of "what" and "where" can be used as filters to choose the species and localities that offer the best chances of success.

Compared to answering questions of "why," which are often sociopolitical and can require substantial input from economists, social scientists, and nonscientist stakeholders—especially for species lacking formal regulatory requirements—deciding "what" species and "where" to reintroduce it offers a much larger role for the participation of natural scientists, as their expert knowledge is crucial to cost-effective outcomes (figure 15.1). We view this as a critical distinction from previous attempts or frameworks for reintroduction, particularly in impoverished countries where conservationists (sometimes with narrow agendas) from developed countries have driven the answers to why, what, and where reintroductions have occurred in the past. In these instances, it has been difficult to sustain reintroduction efforts in developing countries, and even in developed countries. We suggest that, in general, the development of fundamental objectives for reintroduction is outside the purview of scientists, whereas the means of achieving those objectives may be informed by scientists (Conroy and Peterson 2013).

Within this defined role for natural scientists, the questions that can act as filters for deciding "what" to reintroduce are often context specific. At a minimum, they all generally include consideration of (1) availability and cost of animals that provide a source for reintroduction; (2) probability of success based on species biology and previous experience; and (3) probability of undesirable side effects. Thus, the primary role of natural scientists in reintroduction is to synthesize this knowledge and develop decision-support tools that enable practitioners to objectively weigh and score potential criteria so they can make good choices (e.g., Jachowski and Lockhart 2009). Fortunately, when approached as a "reintroduction landscape" as discussed above and in detail in Chapter 5, these questions can be iteratively addressed in an explicit, multidisciplinary framework (figure 15.1). Furthermore, because answers to many of these questions will not be readily apparent, they may require experimentation, which might best be conducted within the frameworks of SDM (Conroy and Peterson 2013) or adaptive management (Williams and Brown 2012). In particular, SDM facilitates reaching decisions by carefully analyzing problems associated with meeting objectives, dealing explicitly with uncertainty, and responding transparently to legal mandates and public preferences or values. If a decision is made to implement some management action, that action can be treated as an experiment to be conducted and assessed scientifically so that new useful knowledge about the action's outcome is gained. To the extent that subsequent management actions are informed by the experiment, the entire process, including feedbacks, is adaptive.

Incorporating Reintroduction Values and Ethics

The modern practice of reintroduction derives its strength from the science of conservation biology, particularly from efforts to develop the science of restoring extirpated species. The motivation for reintroducing species, as is so for all conservation, is founded on human values (which reflect beliefs) and ethics (value-driven rules or modes of behavior). Although there has been substantial growth in scientific knowledge, there has been relatively little advancement in developing the ethical foundation for how, why, what, when, and where reintroductions should be undertaken. For example, a question originating from our discussion above is whether it is ethically responsible to reintroduce animals at sites with low probability of success simply as political leverage for the greater conservation good. The "correct" answer is likely to be case specific, but deriving it would require careful consideration of relevant values (i.e., conservation benefits) as well as ethics (i.e., appropriate rules for action). A broader, persistent ethical question for those considering reintroductions in underdeveloped countries is if resources directed toward expensive reintroductions should instead be spent directly on basic human needs for health, food, and education. In this latter scenario, even if a reintroduction is pushed forward by conservation organizations, there is a need for those reintroduction advocates to adapt their ethical framework and reintroduction priorities to local sociopolitical conditions if conservation is to be sustainable. This need is illustrated in the trade-offs resulting from reintroductions of large, fierce predators such as tigers (*Panthera tigris*) and the subsequent negative impacts on local human welfare versus the more diffuse and global positive benefits that accrue to people via existence and aesthetic values attached to such predators (Johnsingh and Madhusudan 2009).

Furthermore, multiple current or emerging topics related to reintroduction biology, such as rewilding and de-extinction (see Chapter 2) to date have largely also avoided important ethical questions. With the advancement of genetic engineering, it seems inevitable that de-extinction of fish and wildlife species will become increasingly feasible. Scientific momentum is underway and some biological guidance is in place (see Seddon et al. 2014), but important ethical questions remain unanswered (but see Diehm 2015). Do we expect genetically reengineered thylacines (*Thylacinus cynocephalus*) to be treated as if they were true reincarnations of extinct species? While it seems logical that such curiosities would populate zoos and private game farms, are they to be released into the wild, and if so, who should be involved in making and influencing those decisions? How important is it that the species fulfill its former ecological niche? Although de-extinction has great popular appeal because of its technological sophistication, at what point does it become too expensive to play a major role in publicly backed conservation of biodiversity?

To put these species reintroduction, and broader restoration, questions into context, it is important to review the current relevant ethical frameworks that have been developed. The ethical foundation for animal reintroduction can be broadly categorized into one or more of three possible ethical frameworks. First, one of the longest established ethical frameworks is that of restitutive justice or to make amends for a past wrong (Taylor 1986). While this rationale has primarily been applied to individual organisms, it is not hard to expand it to include close genetic relatives (Taylor 1986), such that reintroductions can be viewed as restitution for past human error. Second, rather than focusing on the individual, ethical frameworks for reintroduction can be, and often are, focused on preservation of species (Rolston 1988). Such species- or biodiversity-based thinking has long been encouraged by conservation biologists

(Soulé 1985) and is further emphasized in current policy such as the US Endangered Species Act. Third, many conservationists value ecological systems as the unit that should be preserved or restored. This ethical framework dates back to Aldo Leopold's (1970) "Land Ethic," and has been similarly adopted repeatedly by current conservation biologists (Soulé 1985). While each of these frameworks is somewhat distinctive, in practice, reintroductions can overlap or touch on multiple ethical frameworks.

We suggest that to strengthen the socioeconomic foundation of reintroduction, a collective and easily integrated ethical framework is needed that builds on and complements the SDM and adaptive management frameworks outlined above. Specifically, we encourage each proposed reintroduction to simultaneously and repeatedly consider the questions of why, what, when, and where (figure 15.1). The rationale for why a reintroduction is important needs to be explicitly developed, followed by consultation with representative stakeholders who could be impacted by the reintroduction and who are vital to its long-term sustainability. At a minimum, we suggest that such stakeholders include policy-makers, landowners, and relevant management agencies, as well as community members, economists, biologists, and animal welfare groups that can help reach a decision on whether to proceed with the proposed reintroduction (figure 15.1). During this process, we further encourage that three key ethical questions be explicitly asked and be revisited repeatedly by stakeholders:

1. What is the overall goal and why do we value it?

2. Who are likely to be the main winners and losers before and after the reintroduction?

3. Which objectives and outcomes take ethical precedence over others?

Concurrently, or once a decision to reintroduce is made, we encourage biologists to synthesize existing data and offer guidance on the best reintroduction practices for addressing questions of what individuals to release, when, and where. For example, answers to question 2 above ultimately derive from human values and how they are weighted; these factors are commonly explicated in SDM (Conroy and Peterson 2013). Addressing questions of what, when, and where is a typical outcome of adaptive management (Williams and Brown 2012). This information, including informed estimates of the relative feasibility of a release and long-term management needs, may be used by an interdisciplinary reintroduction team in deciding how to proceed with reintroduction under an adaptive management framework, or whether to halt the reintroduction and reassess the rationale for why such an action is needed (figure 15.1).

EMBRACING PROCESS-BASED APPROACHES TO REINTRODUCTION

One of the key benefits from reintroducing species is that we gain knowledge of their biology. Many of the chapters in the book detail increased knowledge and understanding of organisms and systems. For example, only through simultaneous attempts to reintroduce the desert pupfish (*Cyprinodon macularius*) and Gila topminnow (*Poeciliopsis occidentalis*), two species that historically occurred together, did it become evident that the Gila topminnow outcompetes and predates on endangered desert pupfish—thus limiting success of desert pupfish restoration (Robinson and Ward 2011). Furthermore, reintroductions can serve as quasi-experiments, allowing us to assess ecosystem-level response to what happens when we lose a species, and what happens when we put it back. This has been most dramatically observed in Yellowstone National Park (United States), where reintroduction of gray wolves (*Canis lupus*) has provided some of the strongest evidence to date of the existence of trophic cascades within terrestrial ecosystems (Ripple and Beschta 2003, 2012). Similarly, predator reintroduction into South Africa has provided

critical insight into the nuances of the currently popular "landscape of fear" model in predator-prey ecology (Moll et al. 2016). Collectively, it is evident that biological information gained through reintroductions adds greatly to our knowledge base of species- and ecosystem-level processes, particularly for threatened or endangered species.

Building on this foundation of an initial emphasis on the recovery of endangered species, going forward we anticipate species reintroduction to be increasingly focused on the processes involved in ecological functions. In addition to being more ecologically justifiable to scientists, reintroduction rationales based on ecological functionality may have broader societal appeal than those based narrowly on either aesthetic or utilitarian value of individual species. Below, we highlight several instances where the primary rationale for conservation introduction or reintroduction is to restore impaired or lost ecological processes. Furthermore, we discuss how knowledge of these ecological processes can improve and guide future reintroductions.

Ecological Replacements

While reintroduction of native, extirpated species has been a general objective of conservation to date (Seddon et al. 2012), in some instances managers might need to use surrogate species to fill gaps in assemblage structure or ecological processes. As addressed in Chapter 2, the introduction of exotic Aldabra giant tortoises (*Aldabrachelys gigantea*) to the Mauritius islands to replace extinct native tortoise species illustrates the potential for such practices to restore lost ecological functions. Similarly, in Europe, the process of "rewilding" has proceeded in some areas where large herbivores such as Heck cattle (*Bos taurus*) and Konik horses (*Equus ferus caballus*) have been introduced to replace lost aurochs (*Bos primigenius*), tarpan (*E. f. ferus*), and wisent (*Bison bonasus*) (Navarro and Periera 2012, Smit et al. 2015).

Conservationists have the capacity to place species or surrogate species back into highly managed settings, but a key uncertainty is whether they will perform the same way as in past ecological states. An important challenge for ecological replacement is the continual evaluation of success in restoring identified ecological processes (e.g., predator-prey interactions). Reintroduction of wolves into Yellowstone National Park, for example, is widely viewed as a rewilding success by virtue of their cascading ecological effects (Ripple and Beschta 2012). However, trophic cascades are not known to occur in all systems, and the cascading ecological effects of species like wolves are likely highly context specific (Borer et al. 2005, Kauffman et al. 2010). In the future, we encourage practitioners not only to monitor the recovery of target populations, but also to monitor for the hypothesized ecological effects of those recovered populations. In this sense, we expect reintroduction biology, and conservation translocations in general, to increasingly focus less on individual populations and species, as in endangered species recovery efforts, and broaden their perspective to consider ecosystem and landscape outcomes of reintroduction. We expect this perspective to draw significantly from the field of ecological restoration, perhaps culminating in the integration of restoration ecology and reintroduction biology to produce practical knowledge at multiple ecological levels.

Re-introducing Nonnative Species

The complex interaction between restoration of ecological processes and species reintroduction is further complicated when nonnative species with desirable conservation impacts are considered for re-introduction following extermination. We use the term "re-introduction" here to differentiate from true reintroductions, which we define in Chapter 1 as "the process of releasing a species back to where it historically occurred but had been extirpated by humans." This issue has most recently arisen concerning the role of the dingo (*Canis lupus dingo*) in por-

tions of Australia. Dingos are a relatively recent arrival to Australia, introduced approximately 3,500 years ago, that have been linked to the decline of several native vertebrates from the mainland around that time (Johnson and Wroe 2003). Similar to large canids elsewhere in the world, dingos have been persecuted for their impact on livestock and broader human disdain to the point of being extirpated from portions of their former range (Letnic et al. 2012). Where dingos are abundant, nonnative mesopredators such as red fox (*Vulpes vulpes*) and cats (*Felis catus*) are rare (Glen and Dickman 2005). Thus, as top predators, dingos can have beneficial, cascading effects on other nonnative wildlife that otherwise limit populations of native fauna (Johnson et al. 2007, Letnic et al. 2012). This has led to calls for re-introduction of what can be considered a nonnative carnivore where it has been extirpated (Dickman et al. 2009, Newsome et al. 2015).

This example of the dingo brings up a number of broader conservation questions that will need to be addressed by reintroduction practitioners in the future. First, similar to previous discussions regarding the importance of "naturalness" in conservation decision-making (Angermeier 2000), how do we determine what is natural and what functions make some nonnatives or perhaps invasive species deserving of reintroduction? A subsequent question is at what point has a nonnative species been present in a system long enough to be considered native? Here again, reintroduction practitioners face fundamental questions about why to reintroduce (or in this case re-introduce) and what the restoration goal is, and we foresee the need for multidisciplinary ecological and socioeconomic discussions to guide future conservation translocation efforts.

Matching Reintroduction Attempts with Current Ecological Contexts

Process-based approaches to restoration are not limited to highly charismatic and controversial large mammals and the equally controversial topics of trophic cascades or keystone species. Understanding and accounting for current species interactions and community-level processes can directly improve reintroduction success for a wide variety of species. For example, as addressed in Chapter 10, disease is a major obstacle to establishing reintroduced populations. It is increasingly apparent that we need not only to address the original cause of decline prior to attempting reintroduction, but also to anticipate potential negative effects of reintroduction on the surrounding biotic community in ecological contexts that may be very different from those in which the species of concern initially declined. For example, the Panamanian golden frog (*Atelopus zeteki*) is not only highly vulnerable to the amphibian-killing chytrid fungus (*Batrachochytrium dendrobatidis*), but also likely acts as a "superspreader" for the disease (DiRenzo et al. 2014). Thus, successful reintroduction of this species not only is unlikely due to the persistence of chytrid fungus at the reintroduction area, but it could also potentially harm other amphibian species by amplifying the number of zoospores on the landscape. Anticipating such unintended, negative impacts from reintroduction requires a nuanced, mechanistic understanding of a wide variety of ecological fields, ranging from animal physiology (Chapter 9) to disease ecology (Chapter 10).

Individual- and Population-Level Processes

Factors affecting reintroduction success are not limited to species interactions, but include individual- and population-level behaviors such as site-specific habitat use and migration, which evolved to be adaptive under specific environmental regimes. Indeed, conservation of these highly adapted processes is widely seen as one of the most urgent and complicated problems facing animal conservation, the loss of which often serves as a warning that a population or species is facing extinction (Berger 2004, Harris et al. 2009). Experience from restoration of whooping cranes *(Grus americana)*, and other

nearly extirpated species, illustrates how such behavioral processes are dynamic, often being difficult to define and recreate in increasingly human-altered landscapes with novel selective pressures (Soulé et al. 2003, Ellis et al. 2013, Jachowski et al. 2015). However, as is evident with the whooping crane, restoration of migratory behavior can be done, albeit at great cost (Chapter 9). Going forward, we foresee the need for in-depth behavioral studies of extant species as a key priority so that we can document these often highly context-specific processes, as well as their links to individual fitness, before they are lost. Such knowledge can help managers formulate tactics that ensure the continued adaptive evolution of reintroduced populations despite rapidly changing environments.

Traditionally, reintroduction and biodiversity conservation in general have focused on the demographic recovery of a population or species (Soulé 1985). However, if more than 84% of endangered species are likely to be conservation reliant into the future (Scott et al. 2010), it seems unrealistic to expect full recovery of a reintroduced species to a self-sustaining state. This quandary dictates that conservationists often need to reexamine the goals and objectives of reintroduction. While simply avoiding extinction is an obvious goal, we suggest that, in some cases, conserving and restoring highly adapted behavioral processes is prerequisite to species recovery itself. For species like the whooping crane, behavioral processes such as migration are inextricably linked to species persistence. For others, global change and intensive management have altered behavioral processes, resulting in persistent populations but lost behaviors. For example, in South Africa where African elephants (*Loxodonta africana*) and other large mammals have been extensively reintroduced into fenced reserves, fences not only restrict potential migratory behavior, but in the case of elephants can also lead in some cases to extremely reclusive behavior away from grasslands, in forested environments (Jachowski et al. 2013). Here again, scientists can provide evidence on when and if these processes existed, but the desirability or value of restoring or preserving these processes is fundamentally a social problem. A dialogue is needed on the relative importance of mere population persistence versus persistence in a state that allows the reintroduced animals to behave as they had in the past.

Prioritizing Processes in Practice

The restoration of ecological processes is important in justifying long-term support for animal reintroductions, and conservation translocations in general. As discussed above, socioeconomic support is a key element of successful reintroductions. Use of process-based justifications for reintroduction such as the potential future harvest of reintroduced migratory fishes, or carbon storage potential for reintroduced elephants, has clear ties to natural capital that can incentivize support for reintroduction.

Using restoration of ecological processes as a focal justification for species reintroduction presents several important scientific challenges. Reintroduction biology will need to develop a fuller understanding of (a) functions of individual species in ecological processes, (b) functional similarities among species, (c) consequences for ecosystem processes when species are lost, (d) capacities of the functions of species to evolve, (e) human social values assigned to individual species, and (f) human social values assigned to ecological processes, as distinct from those assigned to individual species. Moreover, this knowledge and relevant decision-support tools will need to be made accessible so that stakeholders can use them to decide if a proposed restoration is desirable and appropriate. In this context, we expect lost species that were highly interactive (e.g., keystone species) to be both strong and controversial candidates.

If social values shift in ways that make ecological processes a top priority for species reintroduction, there would need to be changes in conceptual and institutional frameworks. For example, conventional objectives for conserving biodiversity, which emphasize genetic, species, and ecosystem diversity (e.g., Noss 1990, Angermeier and Karr 1994), would need to be revised to address diversity of ecological processes. Moreover, for institutional conservation priorities to shift away from traditional recovery of endangered species (such as through the US Endangered Species Act) to the recovery of ecological processes, a new legal and regulatory framework would need to be developed.

BEYOND RESTORATION

One of the first and most logical extensions of reintroduction biology has been to use established translocation practices to prevent extinction or extirpation of a species through assisted colonization. Variously termed assisted colonization, assisted migration, or conservation introduction, these practices all differ from conventional conservation in that they involve translocation of species to environments outside their previously known range. While calls for assisted colonization have increased due to rapid global change and recent advances in predictive modeling (e.g., Chauvenet et al. 2013), there is ongoing debate among scientists and managers about how to balance the potential benefits of preventing a species' extinction against the risk of a translocated species causing detrimental ecological effects similar to those caused by invasive species (Ricciardi and Simberloff 2009, Lunt et al. 2013, Hancock and Gallagher 2014). While some advances in ways to simulate natural processes have been made (e.g., Venter et al. 2014), the scientific debate seems to be well ahead of the social debate about whether assisted colonization should be a management option, and whether political and managerial support exists for such practices (Neff and Larson 2014). Therefore, although reintroduction biology can provide technical tools and expertise to carry out species translocations, a healthy public debate, informed by science, is needed to determine the contexts in which assisted colonization is desirable and appropriate. There is a clear role for experts in outreach and environmental education to help shape this debate (Chapter 13). Finally, we see a need for these challenges to be addressed using an ethical framework similar to the one we outline above and in figure 15.1.

SUMMARY

It is evident that reintroduction biology has given us a framework and tools to evaluate a whole suite of new approaches to conservation through translocation. One of the primary benefits of species reintroduction has been the tremendous gain in knowledge of species, ecosystem, and evolutionary biology. Experiences in reintroducing fish and wildlife have also revealed multiple obstacles to the long-term sustainability of reintroductions. To overcome these challenges, we see the need for a broader socioeconomic foundation of support, as opposed to past reliance on political will, intrinsic value of species, or "top-down" regulations to achieve conservation goals. This can be accomplished, in part, by approaching reintroduction with multidisciplinary teams that include nonscientist stakeholders and address the complex, and often interactive, ecological and socioeconomic questions that undergird the reintroduction landscape. Once reintroductions are undertaken, to further enhance their success, there is a need for rapid growth in the use of SDM to link modeling, monitoring, and management of reintroduced populations.

To address thorny questions about the goals of reintroduction and enhance long-term success and sustainability of reintroductions, we propose that there is a need to develop a novel ethical reintroduction framework (figure 15.1). This new model for undertaking reintroductions encourages the formation of interdisciplinary reintroduction teams that first assesses the question of why to reintroduce based around addressing three ethical questions:

1. What is the overall goal and why do we value it?
2. Who are likely to be the main winners and losers before and after the reintroduction?
3. Which objectives and outcomes take precedence over others?

Interaction among a diverse range of stakeholders about these fundamental questions, and associated reintroduction-related biological and social issues, can help establish the broader reintroduction landscape. Furthermore, our proposed model provides guidance on the iterative process of deciding what, when, where, and how to proceed with reintroduction under an adaptive management framework or whether to halt the reintroduction and reassess the rationale for why such an action is needed.

Finally, growth of the field of reintroduction biology provides a window into the cutting edge of conservation practices and promises to enhance conservation's effectiveness. From an initial focus on restoring endangered species, reintroduction biology has given rise to a wide spectrum of conservation translocations ranging from traditional reintroductions to assisted colonization in novel environments (Chapter 2). We anticipate a major future emphasis on the conservation of ecological processes to guide reintroduction attempts. These approaches continue to expand and redefine what is possible to achieve while at the same time presenting several scientific and ethical challenges.

LITERATURE CITED

Angermeier, P. L. 2000. The natural imperative for biological conservation. Conservation Biology 14:373–381.

Angermeier, P. L. and J. R. Karr. 1994. Biological integrity versus biological diversity as policy directives: protecting biotic resources. BioScience 44:690–697.

Berger, J. 2004. The last mile: how to sustain long-distance migration in mammals. Conservation Biology 18:320–331.

Borer, E. T., E. W. Seabloom, J. B. Shurin, K. E. Anderson, C. A. Blanchette, B. Broitman, S. D.

Cooper, and B. S. Halpern. 2005. What determines the strength of a trophic cascade? Ecology 86:528–537.

Chauvenet, A. L., J. G. Ewen, D. Armstrong, and N. Pettorelli. 2013. Saving the hihi under climate change: a case for assisted colonization. Journal of Applied Ecology 50:1330–1340.

Conroy, M. J., and J. T. Peterson. 2013. Decision Making in Natural Resource Management: A Structured, Adaptive Approach. John Wiley and Sons, Hoboken, NJ.

Dickman, C. R., A. S. Glen, and M. Letnic. 2009. Reintroducing the dingo: can Australia's conservation wastelands be restored. Pp.238–269. In M. W. Hayward and M. J. Somers, eds., Reintroduction of Top-Order Predators, Wiley-Blackwell, London, UK.

Diehm, C. 2015. Should extinction be forever? Restitution, restoration, and reviving extinct species. Environmental Ethics 37:131–143.

DiRenzo, G. V., P. F. Langhammer, K. R. Zamudio, and K. R. Lips. 2014. Fungal infection intensity and zoospore output of Atelopus zeteki, a potential acute chytrid supershedder. PLoS ONE 9:e93356.

Ellis, D. H., W. J. Sladen, W. A. Lishman, K. R. Clegg, J. W. Duff, G. F. Gee, and J. C. Lewis. 2013. Motorized migrations: the future or mere fantasy? BioScience 53:260–264.

Ferreira, S. M., C. Greaver, G. A. Knight, M. H. Knight, I. P. Smit, and D. Pienaar. 2015. Disruption of rhino demography by poachers may lead to population declines in Kruger National Park, South Africa. PLoS ONE 10:e0127783.

Glen, A. S., and C. R. Dickman. 2005. Complex interactions among mammalian carnivores in Australia, and their implications for wildlife management. Biology Reviews 80:387–401.

Goble, D. D., J. A. Wiens, J. M. Scott, T. D. Male, and J. A. Hall. 2012. Conservation-reliant species. BioScience 62:869–873.

Hancock, N., and R. Gallagher. 2014. How ready are we to move species threatened from climate change? Insights into the assisted colonization debate from Australia. Austral Ecology 39:830–838.

Harris, G. S., J. G. C. Thirgood, J. P. Hopcraft, G. M. Cromsigt, and J. Berger. 2009. Global decline in aggregated migrations of large terrestrial mammals. Endangered Species Research 7:55–76.

Hayward, M. W., G. I. Kerley, J. Adendorff, L. C. Moolman, J. O'brien, A. Shoto-Douglas, C. Bissett et al. 2007. The reintroduction of large carnivores to the Eastern Cape, South Africa: an assessment. Oryx 41:205–214.

IUCN (World Conservation Union). 1998. Guidelines for Re-introductions. IUCN/SSC Re-introduction Specialist Group, Gland, Switzerland.

Jachowski, D. S. 2014. Wild Again: The Struggle to Save the Black-footed Ferret. University of California Press, Berkeley.

Jachowski, D. S., D. C. Kesler, D. A. Steen, and J. R. Walters. 2015. Rethinking baselines in endangered species recovery. Journal of Wildlife Management 79:3–9.

Jachowski, D. S., and J. M. Lockhart. 2009. Reintroducing the black-footed ferret Mustela nigripes to the Great Plains of North America. Small Carnivore Conservation 41:58–64.

Jachowski, D., R. Slotow, and J. J. Millspaugh. 2013. Delayed physiological acclimatization by African elephants following reintroduction. Animal Conservation 16:575–583.

Johnsingh, A. J. T., and M. D. Madhusudan. 2009. Tiger reintroduction in India: Conservation tool or costly dream? Pp.146–163. In M. W. Hayward and M. J. Somers, eds., Reintroduction of Top-Order Predators, Wiley-Blackwell, London, UK.

Johnson, C. N., J. L. Isaac, and D. O. Fisher. 2007. Rarity of a top predator triggers continent-wide collapse of mammal prey: dingoes and marsupials in Australia. Proceedings of the Royal Society B: Biological Sciences 274:341–346.

Johnson, C. N., and S. Wroe. 2003. Causes of extinction of vertebrates during the Holocene of mainland Australia: arrival of the dingo, or human impact? The Holocene 13:941–948.

Kauffman, M. J., J. F. Brodie, and E. S. Jules. 2010. Are wolves saving Yellowstone's aspen? A landscape-level test of a behaviorally mediated trophic cascade. Ecology 91:2742–2755.

Leopold, A. 1970. Sand County Almanac. Ballantine Books, New York.

Letnic, M., E. G. Ritchie, and C. R. Dickman. 2012. Top predators as biodiversity regulators: the dingo Canis lupus dingo as a case study. Biological Reviews 87:390–413.

Lunt, I. D., M. Byrne, J. J. Hellmann, N. J. Mitchell, S. T. Garnett, M. W. Hayward, and K. K. Zander. 2013. Using assisted colonization to conserve biodiversity and restore ecosystem function under climate change. Biological Conservation 157:172–177.

Moll, R. J., A. K. Killion, R. A. Montgomery, C. J. Tambling, and M. W. Hayward. 2016. Spatial patterns of African ungulate aggregation reveal complex but limited risk effects from reintroduced carnivores. Ecology 97:1123–1134.

Navarro, L. M., and H. M. Pereira. 2012. Rewilding abandoned landscapes in Europe. Ecosystems 15:900–912.

Neff, M. W., and B. M. Larson. 2014. Scientists, managers, and assisted colonization: Four contrasting perspectives entangle science and policy. Biological Conservation 172:1–7.

Newsome, T. M., G. A. Ballard, M. S. Crowther, J. A. Dellinger, P. J. Fleming, A. S. Glen, A. C. Greenville et al. 2015. Resolving the value of the dingo in ecological restoration. Restoration Ecology 23:201–208.

Noss, R. F. 1990. Indicators for monitoring biodiversity: a hierarchical approach. Conservation Biology 4:355–364.

Ricciardi, A., and D. Simberloff. 2009. Assisted colonization is not a viable conservation strategy. Trends in Ecology and Evolution 24:248–253.

Ripple, W. J., and R. L. Beschta. 2003. Wolf reintroduction, predation risk, and cottonwood recovery in Yellowstone National Park. Forest Ecology and Management 184:299–313.

Ripple, W. J., and R. L. Beschta. 2012. Trophic cascades in Yellowstone: the first 15 years after wolf reintroduction. Biological Conservation 145:205–213.

Robinson, A. T., and D. L. Ward. 2011. Interactions between desert pupfish and Gila topminnow can affect reintroduction success. North American Journal of Fisheries Management 31:1093–1099.

Rolston, H. III. 1988. Environmental Ethics: Duties to and Values in the Natural World. Temple University Press, Philadelphia, PA.

Schaller, G. B. 2007. A Naturalist and Other Beasts: Tales from a Life in the Field. Sierra Club Books, San Francisco, CA.

Scott, J. M., D. D. Goble, A. M. Haines, J. A. Wiens, and M. C. Neel. 2010. Conservation-reliant species and the future of conservation. Conservation Letters 3:91–97.

Seddon, P. J., A. Moehrenschlager, and J. Ewen. 2014. Reintroducing resurrected species: selecting deextinction candidates. Trends in Ecology and Evolution 29:140–147.

Seddon, P. J., W. M. Strauss, and J. Innes. 2012. Animal translocations: what are they and why do we do them? Pp.1–32. In J. G. Ewen, D. P. Armstrong, K. A. Parker, and P. J. Seddon, eds., Reintroduction Biology: Integrating Science and Management. Wiley-Blackwell, London, UK.

Smit, C., J. L. Ruifrok, R. van Klink, and H. Olff. 2015. Rewilding with large herbivores: the importance of grazing refuges for sapling establishment and wood-pasture formation. Biological Conservation 182:134–142.

Soulé, M. E. 1985. What is conservation biology? BioScience 35:727–734.

Soulé, M. E., J. A. Estes, J. Berger, and C. M Del Rio. 2003. Ecological effectiveness: Conservation goals for interacting species. Conservation Biology 17:1238–1250.

Taylor, P. W. 1986. Respect for Nature: A Theory of Environmental Ethics. Princeton University Press, Princeton, NJ.

Truett, J., and M. Phillips. 2009. Beyond historic baselines: restoring Bolson tortoises to Pleistocene range. Ecological Restoration 27:144–151.

Venter, J. A., H. H. Prins, D. A. Balfour, and R. Slotow. 2014. Reconstructing grazer assemblages for protected area restoration. PLoS ONE 9:e90900.

Williams, B. K., and E. D. Brown. 2012. Adaptive Management: The U. S. Department of Interior Applications Guide. US Department of Interior, Washington, DC.

INDEX